GLOBAL CHANGE AND PROTECTED AREAS

ADVANCES IN GLOBAL CHANGE RESEARCH

VOLUME 9

The titles in this series are listed at the end of this volume.

GLOBAL CHANGE AND PROTECTED AREAS

Edited by

Guido Visconti
Department of Physics,
University of L'Aquila, L'Aquila, Italy

Martin Beniston
University of Fribourg,
Fribourg, Switzerland

Emilio D. Iannorelli
Regione Abruzzo, L'Aquila, Italy

and

Diego Barba
Parco Scientifico e Tecnologico d'Abruzzo,
L'Aquila, Italy

KLUWER ACADEMIC PUBLISHERS
DORDRECHT / BOSTON / LONDON

A C.I.P. Catalogue record for this book is available from the Library of Congress.

ISBN 0-7923-6918-1

Published by Kluwer Academic Publishers,
P.O. Box 17, 3300 AA Dordrecht, The Netherlands.

Sold and distributed in North, Central and South America
by Kluwer Academic Publishers,
101 Philip Drive, Norwell, MA 02061, U.S.A.

In all other countries, sold and distributed
by Kluwer Academic Publishers,
P.O. Box 322, 3300 AH Dordrecht, The Netherlands.

Printed on acid-free paper

Printed in the Netherlands.

Contents

Contributors

ANTONELLA BALERNA, INFN, Laboratori Nazionali di Frascati, via E. Fermi 40, 00044 Frascati, Italy

ALFRED BECKER, Potsdam-Institut fuer Klimafolgenforschung, Postfach 60 12 03, D-14412 Potsdam, Germany

MARTIN BENISTON, Department of Geography, University of Fribourg, Perolles, CH 1700 Fribourg, Switzerland

ENRICO BERNIERI , INFN - Laboratori Nazionali di Frascati, Via E. Fermi 40, 00044 Frascati Italy;

MICHELE BRUNETTI, CNR - Istituto ISAO, Via P. Gobetti, 101, 40129 Bologna, ITALY

GILBERTO CALDERONI, Dipartimento di Scienze della Terra, Università degli Studi di Roma "La Sapienza", P.le Aldo Moro 5, 00185 Roma, Italy

GIACOMO CAVUTA, Università G. D'Annunzio, Dipartimento di Economia e Storia del territorio, Viale Pindaro, 42, 65127 Pescara, Italy

LEANDRO D'ALESSANDRO, Dipartimento di Scienze della Terra, Università di Cheti, Madonna delle Piane - via Dei Vestini, 66013 Chieti Scalo, Italy

VLADIMIR DAUVALTER, Institute of North Industrial Ecology Problems, Russian Academy of Sciences, Kola Science Centre-Russian Academy of Sciences, 14 Fersman Street, 184200 Apatity, Murmansk Region, Russia

BRIGITTA ERSCHBAMER, Institute of Botany, University of Innsbruck, Sternwartestrasse, 15, A 6020 Innsbruck, Austria

MARINA FUSCHI, Università G. D'Annunzio, Dipartimento di Economia e Storia del territorio, Viale Pindaro, 42, 65127 Pescara, Italy

FILIPPO GIORGI, International Center of Theoretical Physics, Physics of Weather and Climate Group, POB 586, 34100 Trieste, Italy

CARLO GIRAUDI, ENEA, CR., Casaccia, P.O. Box 2400, 00100 Roma A.D., Italy

GEORGE GRABHERR, Dept. Of Vegetation Ecology and Conservation Biology, Institute of Plant Physiology -University of Vienna, Althanstrasse 14, 1090 Wien, Austria

LORETTA GRATANI, Dipartimento di Biologia Vegetale, Università degli Studi di Roma "La Sapienza", P.le A. Moro 5, 00185 Roma, Italy

PHILIPPE GROSVERNIER, Natura, Applied biology and ecological engineering, CH-2722 Les Reussilles, Switzerland

LEE HANNAH, Climate Change Group, Ecology and Conservation, P/ bag x7, Claremont, South Africa

CHARLES HARRIS, Department of Earth Sciences, Cardiff University, P.O. Box 914, Park Place, Cardiff CF1 3YE, United Kingdom

ERIK R. HAUGE, ETEAM, 30378 Appaloosa Drive, Evergreen, CO 80439-8635, USA

MARION JANIGA, Tatra National Park Research Center, 059 60 Tatransk Lomnica, Slovakia

RAMASAMY KRISHNAMOORTY, Institute for Ocean Management, POB 5327, College of Engineering, Anna University, Madras (Chennai) 600025, India

FAUSTO. MANES, Univeristà di Roma "La Sapienza", Department of Plant Biology, P.le A. Moro, 5, 00185 Rome, Italy

THOMAS MATHIS, Institute of Geobotany, Section Paleoecology, Universty of Bern, Altenbergrain 21, CH 3013 Bern, Switzerland

BARRY MEATYARD, Science and plants for schools-University of Warwick, Environmental Sciences Research and Education Unit, Warwick Institute of Education, University of Warwick, Coventry CV4 7AL, United Kingdom

PEDRO MENDEZ-GUARDADO, Departamento de Geografia y Ordenacion Territorial, C.U.C.S.H., Universidad de Guadalajara, Av. De los Maestros y Av. Mariano Barena, C.P. 44260 Guadalajara, Jalisco, Mexico

GUY MIDGLEY, Climate Change Group, Ecology and Conservation-National Botany Institute, P/bag x7, Claremont, South Africa

EDWARD A.D. MITCHELL, Department of Plant Ecology, Institute of Botany, University of Neuchâtel, Rue Emile-Argand 11, 2007 Neuchâtel, Switzerland

SANDRA NOTARO, Istituto Agrario di S. Michele a/Adige, Via Mach, 1, 38010 S. Michele a/Adige, Italy

HERALD PAULI, Department of Vegetation Ecology and Conservation Biology, Institute of Plant Physiology, University of Vienna, Althanstrasse, 14, A-1091 Wien, Austria

LORETTA PACE, Dipartimento di Scienze Ambientali, Università degli Studi di L'Aquila, Via Vetoio Loc. Coppito, L'Aquila, Italy

MASSIMO PECCI, SPEL-DIPIA, Dept. of Production Plants, Interaction with the Environment, Via Urbana, 167, 00184 Roma, Italy

ROGER A. PIELKE, Department of Atmospheric Science, Colorado State University, Fort Collins, CO 80523, USA

EMANUELA PIERVITALI, Istituto di Fisica dell'Atmosfera, CNR, Via del Fosso del Cavaliere, 100, 00133 Roma, Italy

MARTIN PRICE, Centre for Mountain Studies, Perth College, University of the Highlands and Islands, Crieff Road, Perth PH1 2NX, United Kingdom

KARL REITER, University of Vienna, Institute of Plant Physiology, Dept. of Vegetation Ecology and Conservation Biology, Althanstrasse, 14, A-1090 Vienna, Austria

SCOTT SLOCOMBE, Geography and Environmental Studies, Wilfrid Laurier University, 75 University Avenue, W. Waterloo, ON, Canada, N2L 3C5, USA

A. SIEGENTHALER, Dept. of Plant Ecology, Institute of Botany, University of Neuchâtel, Neuchâtel, Switzerland

LENKA SOUKUPOVA, Institute of Botany, Czech Academy of Science, 252 43 Pruhonice, Czech Republic

BARBARA TOMASSETTI, Dipartimento di Fisica, Università degli Studi di L'Aquila, via Vetoio Coppito, L'Aquila, Italy

DANIEL VONDER MUHLL, Lab. of Hydraulics, Hydrology and Glaciology (VAW), Swiss Federal Institute of Technology (ETH), Gloriastrasse 37/39, CH-8092 Zurich, Switzerland

DAVID WELCH, Physical Science Advisor, Parks Canada, 25 Eddy Street, 4th floor, Hull, Quebec, K1A 0M5, Canada

BARBEL ZIERL, Swiss Federal Institute for Forest, Snow and Landscape Research, Zuercherstrasse, 111, 8903 Birmensdorf, Switzerland

Acknowledgements

The idea to have a meeting on Global Change and Protected Areas originated from the peculiarity of the Abruzzo Region, in central Italy. This region has devoted more than 30% of its territory to natural parks that include the National Park of Abruzzi, Gran Sasso and Monti della Laga, Maiella and the Regional Park of Sirente – Velino. The first acknowledgment goes to Stefania Pezzopane responsible for the Regional Department on Natural Parks who enthusiastically supported the idea and who has worked hard at the regional and national level. The support of the Parks President and Directors has been important, especially in connection with the idea that within their structure the parks may include a stable and continuing research on Global Change. Miriam Balaban of International Science Services has made available all her professionalism for the very complex organization of the meeting. In particular we would like to thank Alessia Copersini, Ortensia Ferella and Pierlugi Strinella of the International Science Service for their dedicated work to each detail of the organization. A very dedicated and warm acknowledgment goes to Simona Marinangeli of G. Visconti's research group. She has been responsible for most of the initial contact but mainly for the final format of the manuscripts. Simona made a still timely and very important publication possible.

Section 1

Climatic and Environmental Changes

Global Change and Mountain Regions – an IGBP Initiative for Collaborative Research

ALFRED BECKER* AND HARALD BUGMANN**,
**Potsdam Institute for Climate Impact Research, P.O. Box 601203, D-14412 Potsdam, Germany, phone: +49-331-288-2541; fax: +49-331-288-2600; email: becker@pik-potsdam.de*
***Mountain Forest Ecology, Swiss Federal Institute of Technology Zürich, ETH-Zentrum, CH-8092 Zürich, Switzerland, phone: +41-1-632-3239; fax: +41-1-632-1146; email: bugmann@fowi.ethz.ch*

Key words: Mountain regions, interdisciplinary global change research, altitudinal gradients, indicators of change, comparative regional studies.

Abstract: Mountain regions are of special importance for global change research. Due to the strong altitudinal gradients many mountain regions provide unique opportunities to detect and analyse global change processes and phenomena. Therefore, integrated interdisciplinary collaborative research activities are suggested to be implemented globally in a well coordinated way to understand, model and predict environmental change processes in mountain regions and, where needed, make proposals towards sustainable land, water and resources management. The required research is suggested to be structured around four activities and a number of specific tasks to be briefly described in the following. Moreover, suggestions will be made for the implementation and international coordination of the research.

1. INTRODUCTION

Recognizing the significance of mountain regions for global change research, the IGBP core projects BAHC and GCTE, together with START/SASCOM, organized in March/April 1996 a workshop in Kathmandu, Nepal, which resulted in IGBP Report #43: Predicting Global

G. Visconti et al. (eds.), Global Change and Protected Areas, 3–9.

Change Impacts on Mountain Hydrology and Ecology (Becker and Bugmann (Eds.), 1997, IGBP, Stockholm/Sweden).

Immediately after the workshop, the results were discussed in a special session at the first IGBP Congress (Bad Münstereifel, Germany, 18 – 22 April 1996), which was attended by representatives of the IGBP core projects BAHC, GCTE, LUCC and PAGES, and by the IGBP Secretariat, in particular the IGBP Executive Director. The session participants welcomed the results of the Kathmandu Workshop and representatives of LUCC and PAGES enthusiastically expressed an interest to participate in the further development of a joint proposal for Global Change research in mountain regions. Accordingly a follow-up workshop was held in Pontresina, Switzerland (16-18 April 1998) with the support of the Swiss Academy of Natural Sciences (SANW). It was attended by representatives from BAHC, GCTE, LUCC and PAGES. As a result of that workshop a proposal was prepared for an "Initiative for Collaborative Research on Global Change and Mountain Regions". This initiative was formally endorsed by the four IGBP Core Projects BAHC, GCTE, PAGES and LUCC at the Second IGBP Congress in Shonan Village (Japan) in May 1999. A publication has been prepared and will be issued soon.

In parallel, first steps are taken to implement the initiative in close collaboration with other programs, interested in mountain research, in particular the IHDP, UNESCO/IHP and MAP, FAO and WCRP/GEWEX. A Scientific Advisory Group (SAG) has been established for this purpose, in which for the first phase BAHC is represented by Alfred Becker, GCTE by Harald Bugmann, LUCC by Lisa Graumlich and PAGES by Bruno Messerli. For specific information and advise the CPO's of BAHC, Sabine Lütkemeier, and PAGES, Frank Oldfield, may be contacted.

2. RATIONALE AND OBJECTIVES OF THE MOUNTAIN RESEARCH INITIATIVE

Mountain regions occupy about one-fifth of the Earth's surface and provide goods and services to about half of humanity. Accordingly, they receive particular attention in the United Nations system, lastly by the UN Declaration for the year 2002 to become the International Year of Mountains.

The strong altitudinal gradients in mountain regions provide unique and sometimes the best opportunities to detect and analyze global change processes and phenomena because

- meteorological, hydrological, cryospheric and ecological conditions change strongly over relatively short distances; thus biodiversity tends to be high, and characteristic sequences of ecosystems and cryospheric systems are found along mountain slopes. The boundaries between these systems experience shifts due to environmental change and thus may be used as indicators of such changes.

- the higher parts of many mountain ranges are not affected by direct human activities. These areas include many national parks and other protected environments. They may serve as locations where the environmental impacts of climate change alone, including changes in atmospheric chemistry, can be studied directly.

- mountain regions are distributed all over the globe, from the Equator almost to the poles and from oceanic to highly continental climates. This global distribution allows us to perform comparative regional studies and to analyze the regional differentiation of environmental change processes as characterized above.

Therefore, the IGBP Initiative for Collaborative Research on Global Change and Mountain Regions strives to achieve an integrated approach for observing, modelling and investigating global change phenomena and processes in mountain regions, including their impacts on ecosystems and socio-economic systems.

The ultimate objectives of the Initiative are

- to develop a strategy for detecting signals of global environmental change in mountain environments;

- to define the consequences of global environmental change for mountain regions as well as lowland systems dependent on mountain resources (highland-lowland interactions); and

- to make proposals towards sustainable land, water, and resource management for mountain regions at local to regional scales.

To achieve the above objectives, the research under the Mountain Initiative will be structured around four Activities, each of which is divided into a small number of specific tasks:

3. ACTIVITY 1: LONG-TERM MONITORING AND ANALYSIS OF INDICATORS OF ENVIRONMENTAL CHANGE IN MOUNTAIN REGIONS

This Activity will be accomplished through the coordination of ongoing research and, where required, the initiation of new projects in mountain regions around the world. A set of four mountain-specific indicator groups of environmental change is considered:

- Cryospheric indicators related to snow conditions, glaciers, permafrost and solifluction processes (Task 1.1);
- Terrestrial ecosystems, particularly mountain plant communities and soils (Task 1.2);
- Freshwater ecosystems, in particular high mountain streams and lakes (Task 1.3);
- Watershed hydrology, i.e. water balance components of high mountain watersheds/headwater basins (Task 1.4).

Contemporary monitoring will be arranged within the context of reconstructions of longer-term past trends and variability, provided through close collaboration with relevant aspects of the IGBP core project PAGES.

4. ACTIVITY 2: INTEGRATED MODEL-BASED STUDIES OF ENVIRONMENTAL CHANGE IN DIFFERENT MOUNTAIN REGIONS

To achieve the overall goals of the Initiative, it is necessary to develop a framework that permits to analyze and predict hydrological, cryological and ecological characteristics and their linkages with land use and climate at various spatial and temporal scales. Accordingly, this Activity is organized in the following four research themes:

- Development of coupled ecological, hydrological, cryological and land use models for the simulation of land cover and land surface processes in complex mountain landscapes and river basins under current and changing atmospheric and socio-economic conditions (Task 2.1);

- Development of regional scale atmospheric models for mountain regions capable of providing high resolution areal distribution patterns of atmospheric driving forces, in particular precipitation, for the study of land surface processes (Task 2.2);

- Integrated analysis of environmental change in mountain regions by means of fully coupled land surface-atmosphere models, where feasible and appropriate, or by qualitative assessments (Task 2.3);

- Regional scale mountain land surface experiment to support the development, application and validation of the above models (Task 2.4).

5. ACTIVITY 3: PROCESS STUDIES ALONG ALTITUDINAL GRADIENTS AND IN ASSOCIATED HEADWATER BASINS

Ecological and hydrological, including glaciological field studies and experiments, including manipulative ones, along altitudinal gradients and at sensitive sites can provide invaluable data on potential responses of mountain ecosystems to anthropogenically induced environmental change as well as increasing understanding of the associated biotic feedbacks. They are also required to support modelling (Activity 2) and for the identification of indicators of global change. Research themes to be addressed within this Activity include:

- Development of indicators of mountain ecosystem response to environmental forcing factors, based on an improved process understanding of these unique systems insofar as they are sensitive to global change forcings and for a process-related interpretation of historical and paleorecords (Task 3.1);
- Assessment of runoff generation and flowpath dynamics in and on hill slopes and in headwater catchments, including the examination of the role of biogeochemical 'hot spots', for instance for N transformation in mountain areas (Task 3.2);
- The relationship between diversity and ecosystem function, taking advantage of the strong changes of diversity along altitudinal gradients and an assessment of the associated changes in ecosystem functions (Task 3.3).

Paleo-archives will be used to explore system responses to both natural variability and anthropogenic impacts.

6. ACTIVITY 4: SUSTAINABLE LAND USE AND NATURAL RESOURCES MANAGEMENT

The overall objective of this Initiative is to evaluate and enhance sustainable land, water, and resource management strategies for mountain regions. Three priority areas are suggested for assessment:

- Changes in forest resources, with potential implications for agriculture, rates of erosion, slope stability and magnitude of floods, and biodiversity (Task 4.1);
- Intensification and/or extensification of agriculture (including grazing), with potential implications for food security, rates of erosion, slope stability and magnitude of floods, and biodiversity (Task 4.2);
- Changes in water resources due to factors such as changing agricultural practices, increasing temporary or permanent population, and/or increasing energy generation, with implications for downstream water supply, energy availability, flooding, and sediment transfer (Task 4.3).

Work on these linked themes will include paleo-research, local knowledge and scientific investigation, e.g. with respect to evaluating optimal combinations of traditional and innovative land use and resource management systems.

7. STEPS TOWARDS THE IMPLEMENTATION OF THE INITIATIVE

The following first steps are envisaged to be taken soon towards the implementation of the initiative:

1. to actualize, combine and complete existing data bases with addresses etc. of research institutions, organizations and scientists active in mountain research, especially in global change research in mountain regions,
2. to provide a certain administrative support for work under the initiative, most probably through the IPO's of PAGES and BAHC,
3. to make an inventory of existing research sites, stations, river basins, regional studies etc., which may serve as a basis or component for future research,
4. to develop plans for interdisciplinary, integrating mesoscale regional research projects,
5. to prepare an international workshop in 2000/2001.

It is clear that the participation of the international mountain research community is crucial for the implementation of the initiative. Therefore, one of the first main tasks is to develop the required contacts based on 1. and 3. above.

8. ACRONYMS (INCLUDING THOSE OF POTENTIAL COOPERATING INSTITUTIONS AND ORGANIZATIONS)

AMA - African Mountain Association (Univ. of Bern, Switzerland)

AMA - Andean Mountain Association (Univ. of Athens, USA)

BAHC - Biospheric Aspects of the Hydrological Cycle (IGBP Core Project)

CIP - Centro Internacional de la Papa, Lima/Peru

CONDESAN - Consortium for the Sustainable Development of the Andean Ecoregion, coordinated by CIP, Lima/Peru

CPO - IGBP Core Project Office

DIVERSITAS - International Programme of Biodiversity Science, co-sponsored by IUBS, SCOPE, UNESCO, ICSU, IGBP and IUMS

FAO - Food and Agriculture Organization of the United Nations: Forestry Department is Task Manager for Chapter 13 of AGENDA 2000 (Mountain Agenda); other departments/divisions are also active in mountain areas

GEWEX - Global Energy and Water Cycle Experiment (component of WCRP)

GCTE - Global Change and Terrestrial Ecosystems (IGBP Core Project)

GTOS - Global Terrestrial Observing System

IAHS - International Association of Hydrological Sciences

ICIMOD - International Centre for Integrated Mountain Development, Kathmandu, Nepal

ICMH - International Committee on Mountain Hydrology (of WMO and IAHC)

ICRAF - International Centre for Research on Agroforestry, Nairobi, Kenia

ICSU - International Council for Scientific Unions

IGBP - The International Geosphere-Biosphere Programme

IGU - International Geographical Union

IHDP - International Human Dimensions Programme on Global Environmental Change

IHP - International Hydrological Programme of UNESCO

IUBS - International Union of Biological Sciences

IUCN - The International Union of the Conservation of Nature

IUFRO - International Union of Forestry Research Organizations

IUMS - The International Union of Microbiological Societies

LUCC - Land Use/Land Cover Change (joint IGBP and IHDP project)

MAB - Man and the Biosphere Programme of UNESCO

PAGES - Past Global Changes (IGBP Core Project)

SASCOM - South Asian START Committee

SCHC - Standing Committee on Headwater Control

SCOPE - Scientific Committee on Problems of the Environment

START - Global Change System for Analysis, Research and Training (IGBP component)

UN - United Nations

UNESCO - United Nations Educational, Scientific and Cultural Organization

UNU - United Nations University

WCRP - World Climate Research Programme

WMO - World Meteorological Organization

Climate Variations in Italy in the Last 130 Years

MICHELE BRUNETTI (1), LETIZIA BUFFONI (2), FRANCA MANGIANTI (3), MAURIZIO MAUGERI (4), TERESA NANNI (1)
(1) ISAO-CNR - via Gobetti, 101 - I40129, Bologna, ITALY
(2) Osservatorio Astronomico di Brera - via Brera, 28 - I20121, Milan, ITALY
(3) Ufficio Centrale di Ecologia Agraria - via del Caravita, 7A - I00186 Roma, ITALY Ph.: +39 06 6793880
(4) Istituto di Fisica Generale Applicata - via Brera, 28 - I20121, Milan, ITALY

Key words: Climate Variation, Temperature, Daily Temperature Range, Precipitation, Italy, rend.

Abstract: Series of annual and seasonal temperature and precipitation representing, respectively, northern and southern Italy are compared for trend in the period 1865-1996. Temperature and precipitation trends are almost always apposite except for the northern winter where they have a correlated behaviour till about 1980. The result is that the Italian climate is becoming warmer and drier. Monthly mean values of daily minimum and maximum temperature and of daily temperature range (DTR) have a positive trend. In particular the DTR reaches its maximum values in about 1945 then it decreases increasing suddenly in the last ten years.

1. INTRODUCTION

In 1995, the authors began a research program on the reconstruction of the past climate in the Mediterranean area with the main purpose of setting up an Italian climatological database to reconstruct monthly mean, maximum and minimum temperature (1865 - today) and precipitation (1833 - today) average series for two Italian sub-regions: Northern Italy and Central-Southern Italy. A detailed discussion of the results of the research is reported in Maugeri et al., Buffoni et al. Brunetti et al. and Brunetti et al. [1],

G. Visconti et al. (eds.), Global Change and Protected Areas, 11–17.

[2], [3] and [4]. The aim of this paper is to synthesise the results of the project.

2. DATA

Excluding some works on single series, the most relevant project of reconstruction and digitisation of Italian meteorological series was carried out in the '70s by the Italian National Research Council (CNR). The project supported the digitization of daily and monthly minima, maxima and mean temperature and precipitation data of 27 secular series [5]. Daily series usually began from 1870and lacked a lot of data, while monthly ones were often longer and more complete.

In this context the first step in our work on temperature and precipitation in Italy was to create a new monthly series data-base with the aims of adding more series, updating existing series to 1996, filling some gaps in the existing series by means of new data sources, cheeking the data and correcting any errors.

The series included in the new data-base are listed in Table 1 and their locations are shown in Figure 1. The series include monthly mean values of daily minimum, mean and maximum temperatures (Tmin, Tmin, Tmax) and monthly total precipitation (P). They can be divided in two climatic areas [6] - Northern Italy (N) and Central-Southern Italy (S) - that correspond, respectively, to the continental and the peninsular zones of the country.

After establishing the new database, the Craddock homogeneity test was applied to the temperature (Tmin, T and Tmax) and precipitation series [7]. Some series were then homogenized [8], [9] both on the basis of the test results and of the stations history.

After homogenization, the temperature (precipitation) series were completed over the period 1865- 1996 (1833-1996) by means of a procedure described in Maugeri et al. [1] and in Buffoni et al. [2]. With the completed data, we calculated monthly mean values of the daily temperature range (DTR) from Tmin and Tmax series. Following the procedure of [1] the series (T, Tmin, Tmax, DTR and P) were then averaged over N and S and seasonal and yearly anomalies and their 5-y running means were calculated.

Seasonal and yearly N and S average anomalies were analysed with the Mann-Kendall non parametric test, as described in Sneyers [10], to look for a trend. The slopes of the trends were calculated by least square linear fitting. The Mann-Kendall test was also used for a progressive analysis as already done in Maugeri et al. [1].

The correlation between seasonal and yearly DTR and seasonal and yearly precipitation and mean temperature was also performed.

Figure 1. Map of the stations distributed over the Italian territory, numbered as in Tab. 1.

3. TRENDS

As far as mean temperature is concerned, on a yearly basis there is a positive trend with 0.99 significance level (sl) both for N and S; on a seasonal basis, considering a 0.99 sl, T has a positive trend in all four seasons in S, while in N it has a positive trend in autumn, winter and, using a 0.95 sl, in spring.

Table 1. List of the stations available.

	Station Location	Climatic Area	Temperature	Precipitation
			Series length	Series length
1	Alessandria	N	1867-1984	1857-1985
2	Arezzo	S	1877-1992	1876-1992
3	Belluno	N	1873-1993	1875-1993
4	Benevento	S	-	1870-1996
5	Bologna	N	1866-1996	1813-1988
6	Cagliari	S	1879-1996	1853-1996
7	Cuneo	N	1877-1993	1877-1992
8	Ferrara	N	1879-1996	1865-1983
9	Genova	N	1865-1981	1833-1980
10	L'Aquila	S	1874-1994	1874-1994
11	Livorno	S	1866-1994	1857-1994
12	Mantova	N	1866-1993	1840-1995
13	Messina	S	1881-1992	1866-1992
14	Milano	N	1763-1995	1764-1996
15	Napoli	S	1865-1989	1821-1993
16	Padova	N	1866-1979	1725-1987
17	Palermo	S	1866-1992	1806-1992
18	Parma	N	1866-1993	1833-1987
19	Pavia	N	1866-1980	1812-1979
20	Perugia	S	1866-1993	1811-1993
21	Pesaro	N	1871-1991	1866-1991
22	Piacenza	N	1872-1995	1872-1992
23	Reggio C.	S	1879-1996	1868-1996
24	Reggio E.	N	1876-1996	1867-1996
25	Roma	S	1831-1994	1782-1996
26	Rovigo	N	1879-1993	1878-1993
27	Sassari	S	1876-1997	1876-1996
28	Siracusa	S	1878-1988	1869-1989
29	Torino	N	1866-1995	1803-1996
30	Trieste	N	-	1841-1996
31	Vallombrosa	S	-	1872-1989
32	Venezia	N	-	1836-1980

The T trends in the annual temperature series, calculated by least squares linear fitting, range from 0.4° C/100 for N to 0.7° C/100y for S. For the winter season the slopes are greater, ranging from 0.7° C/100y (N) to 0.9°

C/100y (S), while for the summer season they are lower and in some cases not significant.

The progressive application of the Mann-Kendall test allows a more detailed analysis of the series. A complete discussion of this analysis is reported in Maugeri et al. [1]; the synthesis of the results is that both for N and S the positive temperature trend seems to stare around 1920. After 1920 the temperature rises rapidly till 1950, then it is more or less constant from 1950 to 1985, with only a slight drop in the period 1970-1980. In the last 5-10 years, it begins to rise again in all seasons.

As concerns DTR, the results of the analysis are discussed in detail in Brunetti et al. [4]. The results of Mann-Kendall test indicate that DTR has a positive trend (sl >95%) with the only exception of winter in N (negative) and of spring and summer in S (not significant). The increase in the DTR in the period 1865-1996 is weak but significant (0.22°C for N and 0.12°C for S).

The comparison of these results with literature shows that the Italian situation is anomalous, because generally the DTR is characterised by a negative trend [11], [12]. A detailed comparison of the results for the period 1865-1996 is however hampered by the lack of data.

A more detailed analysis can be obtained with the progressive application of the Mann Kendall test to the DTR series. The results indicate that in the last decades of the 19'h century the DTR trend has generally been negative. After the initial decrease, in all the season the DTR trend (except the winter one) begins to increase from a date included in the period 1920-1940 for N and in the period 1900-1920 for S. Then it continues to increase till around 1970 in N and around 1950 in S. In the last decades the trends are generally constant in N, whereas in S they are constant in autumn and decrease in spring and in summer. In winter both there is a tendency to a negative trend in N and a positive one in S.

As far as precipitation is concerned, on a yearly basis a negative trend (sl 0.99) is evident both for N and S; on a seasonal basis, there is negative trend in spring, summer and autumn, whereas in winter the trend is not significant (S) or positive (N). The slopes of the P yearly series, calculated by least squares linear fitting, range between -104 mm/100 y for S and -47 mm/100 y for N giving estimated decreases in the period 1866-1995 of 135 and 61 mm. These values correspond, respectively, to 18% and 7% of S and N yearly mean values. Both for N and S, spring and autumn have the steepest trends. As temperature series, also precipitations ones were studied by means of a progressive application of the Mann-Kendall test. A complete discussion of this analysis is reported in Buffoni et al. [2]; the most interesting result regards P in S whose high significant negative trend seems to be mainly caused by a strong precipitation decrease in the last 50 years.

4. RELATIONSHIP BETWEEN DTR TEMPERATURE AND DTR PRECIPITATION

The comparison of the behaviour of T, P and DTR for the period 1865-1996 has been deeply analysed by Brunetti et al. [4]. The correlation between yearly and seasonal DTR and P is always negative and highly significant (> 99%) whereas the one between DTR and T is positive (significance > 95%) in spring and summer for N and in spring, summer and autumn for S. The negative P - DTR correlation is more significant in N than in S whereas the positive T - DTR one is comparable in the two different geographical areas. The correlation among T, P and DTR is mainly due to high frequent variability but the same behaviour that is present for the seasonal and yearly data is evident also on longer time scales. Both the yearly and the secular correlated behaviours are probably caused by the same changes in atmospheric circulation, with warm and dry conditions (high T and DTR, low P) being related to an increase of the frequency of subtropical anticyclones over the western Mediterranean basin.

5. CONCLUSIONS

The main results of the research are:

Temperature has a significant (>95%) positive trend with more pronounced slopes in S. The temperature generally rises rapidly from 1920 to 1950, is more or less constant from 1950 to 1985 and begins to rise again in the last 1 0 years. DTR has a significant (> 95%) positive trend, higher for N; it is negative for northern winter and not significant for southern spring and summer. For N the trend reaches positive values between 1950 and 1970, then remains constant while for S it reaches positive values between 1930 and 1950, then it decreases in spring and summer and remains constant in autumn. Precipitation has a significant (> 95 %) negative trend. As for temperature, the slopes are more pronounced for S. Especially for S the high significant negative trend seems to be mainly caused by a strong precipitation decrease in the last 50 years. DTR series have a good anticorrelation (significance > 99%) with P series and a significant correlation (significance > 95% in spring and in summer both for N and S and in autumn for S) with T series.

6. ACKNOWLEDGEMENTS:

Many thanks to Uffici Idrografici of Bologna, Cagliari, Catanzaro,

Palermo, Parma, Pescara, Pisa, Reggio Calabria, Roma, Torino, Venezia, for their kind collaboration in making data not yet published in year books available.

7. REFERENCES

Anzaldi C., L. Mirri, V. Trevisan, (Eds.), Archivio Storico delle osservazioni meteorologiche, Pubblicazione CNR AQ/5/27, Rome, 1980.

Auer I., (Ed.), Experience with the completion and homogenization of long term precipitation series in Austria, Central European Research Initiative - Project group Meteorology - working paper 1, Vienna, 1992.

Bohm, R., (Ed.), Description of the Procedure of Homogenizing Temperature Time Series in Austria, Central European Research Initiative - Project group Meteorology - working paper 2, Vienna, 1992.

Brunetti M., L. Buffoni, M. Maugeri, T. Nanni, Theor. Appl. Climatol submitted (1 999b).

Brunetti M., M. Maugeri, T. Nanni, Theor. Appl. Climatol. in press (1999a).

Buffoni L., M. Maugeri, T. Nanni, Theor. Appl. Climatol. in press (1999).

Craddoock J. M., Weather 34 (1979) 332-346.

Easterling D. R., B. Horton, P. D. Jones, T. C. Peterson, T.R. Karl, D. E. Parker, M. J. Salinger, V. Razuvayev, N. Plummer, P. Jamason, C.K. Folland, Science 277 (1997) 364-367.

Karl T.R., P. D. Jones, R. W. Knight, G. Kukla, N. Plummer, V. Razuvayev, K. P. Gallo, J. Lindseay, R. J. Charlson, T. C. Peterson, Bull. Am. Meteorol. Soc. 74 (1993) 1007-1023.

Lo Vecchio G., T. Nanni, Theor. Appl. Climatol. 51 (1995) 159-165.

Maugeri M., T. Nanni, Theor. Appl. Climatol. 61 (1998) 191-196.

Sneyers R., (Ed.), On the statistical analysis of series of observation, WMO, Technical Note N. 143, Geneve,1990.

Dendroclimatic Information on Silver Fir (Abies Alba Mill.) in the Northern Apennines

MICHELE BRUNETTI, DANIELE GAMBETTI, GUIDO LO VECCHIO, TERESA NANNI
ISAO-CNR - via Gobetti, 101 - 140129, Bologna, ITALY

Key words: Dendroclimatology, Apennines, silver fir.

Abstract: The main studies on silver fir were aimed at investigating the relationship between annual ring growth rhythms and climate factors. The aim of the present work is to know how the silver fir responds to both the environmental factors and meteorological parameters as a function of altitude and soil characteristics. The results are that the ring growth depends mainly on winter mean temperature of the same year considered in the chronologies, from summer precipitation and from mean summer temperature of the preceding year. The influence of winter mean temperature appears more pronounced in shallow soil at high altitude. The effect of mean summer temperature and of summer precipitation is greater at low altitude, and at high altitude only in shallow soil.

1. INTRODUCTION

A great deal of dendroclimatic studies on silver fir of Apennines have been performed during the last fifty years [1], [2], [3], [4], [5], [6], [7], [8], [9]. In particular some researches [1], [3], [5] regard Northern Apennines woodlands.

The quoted researches, chiefly devoted to detecting silver fir radial growth trends and their link with climate forcing, were developed by periodical observations of living tissues using the microscope and by more and more sophisticated statistical analysis methods.

G. Visconti et al. (eds.), Global Change and Protected Areas, 19–27.

Table I. Sampling sites and meteo-stations.

PLACE	STAND	TOPONYM	COMPART MENT	*ALTITUDE (m.s.l)*	*EXP.*	SLOPE
ABETONE		*Meteorological station*: altit.: 1340 m; - lat.: 44° 07' N; long.10° 09' E (Green.)				
	I	Sestaione V. - *Fortezza*	380-372	1530	E	0-5°
Geogr.Co-ord.	II	" - "	380	1620	E-SE	40°
- 44° 07' 00" N	III	" - *Fosso del Pianaccio*	421-422	1280	E	15°
- 10° 09' 00" E	IV	" - *Fosso della Rena*	416	1240	E	45°
CAMPIGNA		*Meteorological station*: - altit.: 1068 m; - lat.: 43° 53' N; - long. 11° 42' E (Green.)				
	I	Bidente delle Celle V. - *Poggio Martino*	9	1380	E-NE	18°
Geogr.Co-ord	II	" - *Costa Poggio Aggio Grosso*	6	1320	SE	43°
- 43° 53' 17" N	III	" - *Fosso delle Capanne Vecchie*	22	1020	NE	0-5°
- 11° 42' 00" E	IV	" " " "	21	950	W-NW	44°
VALLOMBROSA		*Meteorological station*: altit.: 955 m; lat.: 43° 43' N; long. E (Green.) 11° 35'				

At the present the most significant results of these studies give a background understanding of the factors working on silver fir growth cambial rhythms in relation to the climatic signal, but many problems are open. In particular: the problem of the provenance, examined in some works [3], [4], [5], that could explain some disagreement in the results of the correlation analysis between climate and radial growth; the influence of the sample size (number of examined trees); the role of environmental factors in comparable climatic scenarios [10].

Nevertheless, it is commonly accepted that climate has a determining role on these phenomena.

In a previous paper [9] we analysed a ring width chronology obtained [6] in order to detect a ring growth climatologically limiting factor. The more general purpose of the present work is to study the response of silver fir to environmental (altitude, soil deepness) and climatological (temperature, precipitation) factors in two areas of Northern Apennines: Abetone and Campigna.

2. DATA AND METHODS

2.1 Description and identification of sites

The samplings were made in the Campigna and Abetone woodlands. In Campigna the sites from which the samplings area is located on the Northern

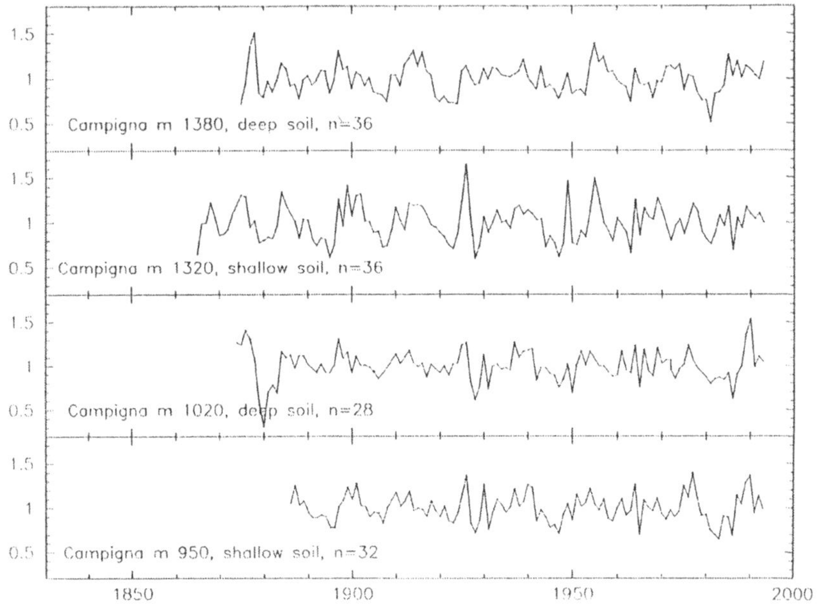

Figure 1. Tree ring width index chronologies of Campigna (n = number of samples).

Apennines, near the watershed. The highest mountains are M. Falco (1658 m a. s. l.) and M. Falterona (1654 m a. s. l.). Geologically this side of the Apennines have sandstone banks (Oligocene) intercalated with marl ones (Miocene). Ground is characterised by steeply sloping land with several parallel streams.

Soils are principally acid-brown and podzol-brown types *(District cambisols/Ubric leptosols* - FAO classification). Vegetation is represented by a mixed wood with silver fir *(Abies alba* Mill.), beech *(Fagus sylvatica* L.) and other tree species *(Acer pseudoplatanus L., Ulmus glabra Hudson, Fraxinus excelsior* L., etc.). Other typical species are: *Carpinus betulus L., Dryopterixfilix-mas* (L.) Schott., *Sanicula europea L., Prenanthes purpurea* L.. In the widely spaced areas, where the soil is shallow there are *Vaccinus mirtillus L., Luzula nivea* (L.) DC., *Deschampsiaflexuosa* (L.) Trin., *Festuca heterophylla (Lam.)* etc.

The climate is temperated-axeric-cold, with a small annual thermal range, cold winter and fresh and rainy summer. Prevalent winds come from Southwest and Northeast. The Campigna meteorological station is located at 1068 m a. s. I. (about 2.500 m in straight line from sampling site, on the same principal slope). Characteristics mean values (referring to the period 1953-1982) from Campigna are: mean annual temperature:8.4°C; yearly precipitation: 1870mm.

Table 2. Correlation coefficients between Campigna chronologies and Vallombrosa seasonal meteorological parameters. "t"= growth year, "t- 1 "= preceding year, WT=mean winter temperature, ST= mean summer temperature, SP= summer precipitation. The confidence limit is: >99% for boldface data, between 99% and 95% otherwise.

CAMPIGNA	WT	ST	SP
I 1380 m s.l.m. Deeper soil	**Year t: r = 0,34**	Year t: - - Year t-1: r = - 0,21	Year t: - - **Year t-1: r = 0,36**
II 1320 m s.l.m. Shallow soil on steep slope	**Year t: r = 0,28**	Year t: r = - 0,18 **Year t-1: r = - 0,29**	**Year t: r = 0,34** **Year t-1: r = 0,49**
III 1020 m s.l.m. Deeper soil	**Year t: r = 0,29**	Year t: r = - 0,23 **Year t-1: r = - 0,23**	**Year t: r = 0,38** **Year t-1: r = 0,50**
IV 950 m s.l.m. Shallow soil on steep slope	**Year t: r = 0,30**	**Year t: r = - 0,32** Year t-1: r = - 0,20	**Year t: r = 0,36** **Year t-1: r = 0,31**

The Abetone area is about 90 km far (Northwest direction) from Campigna and similar in morphology, geology, pedology and hydrography. Samples were collected from the Apennines Southern slope, near the watershed. The highest mountains are: Alpe delle Tre Potenze (1890 m a. s. I.), Libro Aperto (1936 m a. s. I.) and M.te Cimone (2165 m a. s. I.). Vegetation is similar to that of Campigna. In a climatic sense, Abetone differs from Campigna in its lower mean temperature, more annual rain and frequent local thermal-inversion phenomenon. The Abetone meteorological station is located in Boscolungo, at 1340 m a. s. I.. Characteristic mean values from Abetone (referring to the period 1953-1982) are: mean annual temperature: 6.9°C; yearly precipitation: 2597 mm

2.2 Samplings and characterisation of samples

Samplings were carried out in Abetone (Sestaione valley (PT)) and Campigna (Bidente valley (FO)) sites. We selected different tree groups with particular structures in order to dissociate and differentiate climatic parameters from environmental ones. Both in Campigna and Abetone we took samples from four groups of trees: two in a higher site (I and II) and two in a lower one (III and IV). At the same altitude we considered trees living both in deep and in shallow soil. For each group we sampled 17 trees extracting two cores from each tree; 147 trees and 294 cores all together.

Table 1 indicates the characteristics of Campigna and Abetone sampling

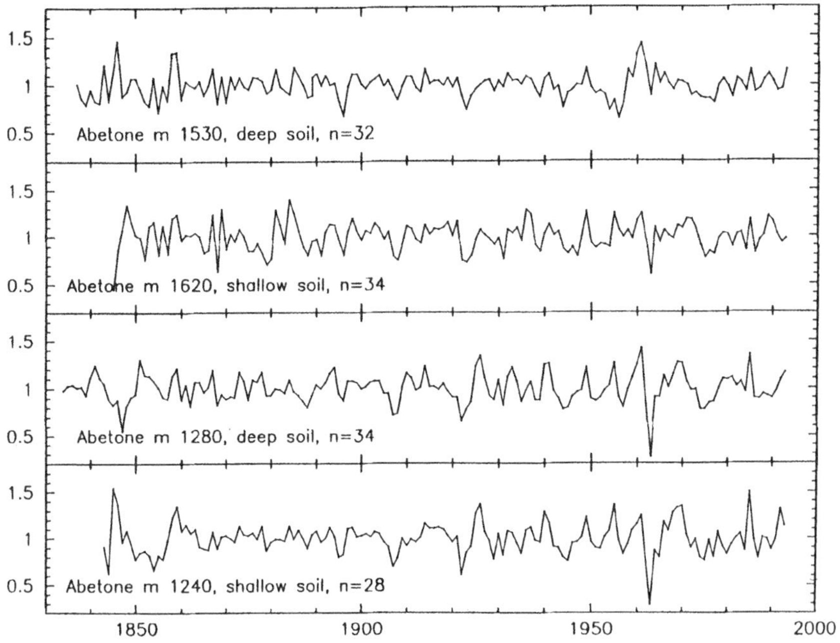

Figure 2. Tree ring width index chronologies of Abetone (n = number of samples).

sites, where the samplings were carried out.

2.3 Climatic data

For climatic series (temperature and precipitation) we have data from the Campigna and Abetone stations of the National Hydrographic Service, and from Vallombrosa (these series were reconstructed from the period 1872-1989, tested and published by Gandolfo and Sulli [11]), but:

- the Campigna series only cover a short time span (1953-1984);

1. The Abetone series, statistically tested, are not reliable enough. Besides they cover a short time period (1927-1987) and have missing data that can produce false effects in estimating the shifted correlation. So we did not utilise them. Therefore we tested the correlation between the Vallombrosa data and those from Campigna ($r > 0.7$, significance level > 99.9%) and from Abetone ($r > 0.5$, significance level >97.7%). Taking into account the good correlation values we decided to use the Vallombrosa series without missing data, which are reliable cover a period comparable with the chronologies.

We computed the annual and seasonal mean temperatures and annual and seasonal precipitation. Annual temperature and precipitation were conventionally made to correspond to the period from December 1st to November 30th while the date of the solar year in which January occurred was assigned to it; winter to the month of December-January-February, spring to March-April-May, summer to June-July-August, and autumn to September-October-November.

2.4 Methods

Tree ring width was measured using a microscope (mag. 16x) 11100 mm resolution. Tests were carried out on single curves, synchronisation, cross-dating, construction and dating of mean curves (eight chronologies, one for every site). For each of the eight chronologies the 13-y running mean was calculated and standardisation was obtained by dividing the original value by the smoothed one. In figure 1 and 2 the chronologies obtained in this way are plotted both for Campigna and for Abetone. Correlation tests between each chronology and the climatic data were performed.

3. RESULTS AND DISCUSSION

The results of correlation tests between chronologies and seasonal meteorological parameters are shown in table 2 and 3. Mean winter temperature (WT) presents a fairly good correlation with the chronologies of Campigna and Abetone. For both locations the best correlation value is relative to the highest site. Mean summer temperature (ST) presents a slight anticorrelation with the chronologies of Campigna and Abetone shifted one year forward, a slight correlation with the non- shifted chronologies for the two lowest sites (Campigna deep and shallow soil), no correlation with the chronology at the highest site (Abetone, deep soil). Summer precipitation (SP) present a good correlation with the chronologies of Campigna and Abetone shifted one year forward. For Campigna 11, III, IV and Abetone IV sites there is also a correlation with non-shifted chronologies. With regards to the Abetone chronologies, for ST and SP, the correlation with the meteorological parameters increases with decreasing altitude, on the other hand, concerning WT, it increases with increasing altitude in shallow soil. As concerns Campigna chronologies, for ST and SP, the correlation increases with decreasing altitude. This behaviour can be explained bearing in mind that the Abetone sites are about 250 m c. a. higher than those of Table 4 (period 1890-1989) shows the similitude matrixes of our eight chronologies with regards to the correlation and sign test.

A-I and C-IV appear not significantly correlated with Campigna and

Table 3. Like table 2 for Abetone.

ABETONE	WT	ST	SP
I 1530 m s.l.m. Deeper soil	Year t: r = 0,25	Year t: r = - -	Year t: r = - - Year t-1: r = 0,20
II 1620 m s.l.m. Shallow soil on steep slope	Year t: r = 0,36	Year t: r = - - Year t-1: r = - 0.22	Year t: r = - - Year t-1: r = 0,23
III 1280 m s.l.m. Deeper soil	Year t: r = 0,21	Year t: r = - - Year t-1: r = - 0.27	Year t: r = - - Year t-1: r = 0.3
IV 1240 m s.l.m. Shallow soil on steep slope	Year t: r = 0,27	Year t: r = - Year t-1: r = - 0.2	Year t: r = 0,24 Year t-1: r = 0.28

Abetone areas respectively.

This fact can be explained by the different environmental characteristics of the two sites. A-I stands grow in deep soil at 1580 m a. s. l., probably the best environmental situation for the silver fir in that area, so firs are slightly affected by climatic conditions. On the contrary the Campigna sites are lower (from 950 ma. S. l. to 1380 m a. s. l.) than A-I and they are more affected by climatic factors: this difference can justify the low correlation values. In a similar way we can explain the low correlation between C-IV and the Abetone area.

In fact C-IV is strongly conditioned by climatic parameters and has a low correlation with Abetone sites less influenced by climate. These are typical situations where the ring growth depends mainly on environmental conditions (altitude and soil depth) than on climate parameters. We reach very similar conclusions considering separately the periods 1890-1939 and 1940-1989.

4. CONCLUSIONS

On the basis of the results from both the chronology-chronology and the chronology- meteoparameters analyses, confirmed by the single year analysis, we can confirm that:

Table 4. Matrix of similarity of eight tree ring width chronologies for the period 1890-1989: sign test (left) and correlation (right). The confidence limit is: >99% for boldface data, between 99% and 95% underlined data, < 95% italic data.

	C-I	C-II	C-III	C-IV	A-I	A-II	A-III	A-IV
C-I		**0.71/0.60**	**0.70/0.41**	**0.65/0.37**	0.55/0.12	**0.64/0.42**	0.54/**0.33**	0.56/**0.38**
C-II			**0.78/0.61**	**0.70/0.62**	0.60/0.22	**0.65/0.48**	**0.65/0.50**	**0.63/0.57**
C-III				**0.75/0.80**	0.56/0.11	**0.69/0.35**	**0.64/0.27**	**0.60/0.30**
C-IV					0.50/0.11	0.58/0.25	0.57/0.22	0.57/0.25
A-I						**0.83/0.60**	**0.69/0.53**	**0.64/0.42**
A-II							**0.72/0.70**	**0.69/0.66**
A-III								**0.79/0.88**
A-IV								

a) there are typical situations (like A-I) where the ring growth depends more on environmental conditions (altitude and soil depth) than on climatic parameters;
b) the ring growth seems to depend primarily on WT of the same year, SP and to a lesser extent ST of the preceding year;
c) the correlation with WT is more pronounced in shallow soil at high altitude; with ST and SP it is more evident at low altitude and in shallow soil at high altitude only.

The results obtained in this work encourage additional research where the goals are to ascertain and quantify the effect of environmental factors on the growth of silver fir with a better knowledge of the climatic parameters in the examined area.

Finally, there is the chance to face the problem using meteorological parameters more directly linked to biological growth processes like potential evapotranspiration and the water balance or the Palmer Drought Severity Index (PDSI).

5. REFERENCES

Becher M., Can. J. For. Res. 19 (1989) 1110-1117.
Braker U., F. H. Schweingruber, FDK 561.24: 101: (450): (44) (1989).
Calistri, L'It. For. E Mont. 4 (1962) 148-160.
Ciampi C., L'It. For. E Mont. 6 (1954) 303-312.
Corona E., Ann. Acc. Ital. Sci. For. 32 (1983) 149-163.

Ferri C., Ann. Acc. Ital. Sci. For. 32 (1955) 135-158.
Gandolfo C., M. Sulli, Ann. Ist. Sper. Selv. 21 (1990) 147-18 I.
Gindel J., Monti e Boschi 6 (1959) 157-164.
Lo Vecchio G., T. Nanni, Dendrochronologia Il (1993) 165-168.
Romagnoli M., B. Schirone, Ann. Acc. Ital. Sci. For. 41 (1992) 3-29.
Santini, N. Martinelli, Giorn. Bot. Ital. 125 (1991) 895-906.

Trends in High Frequency Precipitation Variability in Some Northern Italy Secular Stations

MICHELE BRUNETTI (1), LETIZIA BUFFONI (2), MAURIZIO MAUGERI (3), TERESA NANNI (1)

(1) ISAO-CNR - via Gobetti, 101 - I40129, Bologna, ITALY

(2) Osservatorio Astronomico di Brera - via Brera, 28 - I20121, Milan, ITALY

(3) Istituto di Fisica Generale Applicata - via Brera, 28 - I20121, Milan, ITALY

Key words: Daily Precipitation, Heavy Precipitation events, Precipitation trend, Northern Italy

Abstract: Recent studies on changes in precipitation intensity have found for some areas evidence of an increase in the proportion of total precipitation contributed by heavy and extreme rainfall events in the last 80 years. The purpose of the paper is to verify whether such a signal can also be detected in Northern Italy where daily precipitation (DP) is available starting from the beginning of the 19' century. The analysis is performed on 5 stations: Genoa (1833-1998), Milan (1858-1998), Mantova (1868-1997), Bologna (1879-1998) and Ferrara (1879-1996). It gives evidence that in Northern Italy there is a positive trend in the proportion of total precipitation contributed by heavy precipitation events (i.e. DP > 25 mm and DP > 50 mm). The trend is mainly caused by the last 60-80 years and is particularly evident in the periods 1930-1960 and 1975-1995. It is more evident in the western than in the eastern area.

1. INTRODUCTION

The analysis of daily precipitation series shows for some areas a trend in precipitation intensity and a tendency toward higher frequencies of heavy and extreme rainfalls both in the last decades and in the last century [1], [2]. Within this context the purpose of our research was to set up ultra-secular daily precipitation series for five stations uniformly distributed over Northern Italy and to analyse them in order to verify whether an increase in the proportion of precipitation contributed by heavy rainfall events can be

G. Visconti et al. (eds.), Global Change and Protected Areas, 29–36.

detected in this area in the last 100/150 years. This item is particularly important for Northern Italy as it is an area where heavy rainfall events are rather frequent and where many disastrous floods have been reported in the last 100/150 years [3].

2. DATA

The analysis has been performed on 5 North Italian stations: Genoa, Milan, Mantova, Bologna and Ferrara.

The Genoa series begins in 1833; it has been digitised for the period 1833-1980 by [4] and updated to 1998 by the authors; it has no missing data. It is worth noticing that the application of homogeneity tests to monthly Italian precipitation series [5] allowed the Genoa series to be classified as one of the most homogeneous in Italy.

The Milan series begins in 1835, but till 1857 sometimes cumulative precipitation over some days has been reported. In the last years the series has been revised and digitised [6] and it is now available, with only a few missing data, for the period 1858-1998. Moreover a complete and accurate historical research on archive documents has been performed in order to give information on the quality and on the homogeneity of the data [7].

The Mantova series begins in 1840. It has been recently recovered by Bellumè et al. [8]. The application of homogeneity tests to monthly data showed that there were some problems before 1868, due to incorrect management of snow [8], [5]. The Mantova series is not complete and 2.3 % of the values are missing in the period 1868-1997. The Bologna data were collected at the Astronomical Observatory from 1813 to 1988, a year in which the meteorological monitoring was interrupted, so we updated the series to 1998 using the data from a near (600 meters) station of the Servizio Idrografico. At present the revision of the series is in progress and till now only the data of the period 1879-1998 have been studied and tested for homogeneity. The 1879-1998 series has only a few missing data. The Ferrara series begins in 1879. The data have been recovered from the UCEA database (1879-1974), completed in some of its missing data with the archive of the observatory and updated to 1996 by the authors. As the Mantova series also the Ferrara one is not complete and 9.3 % of the values are missing. Even if all located in Northern Italy, the stations are representative of quite different geographical areas. As a consequence they have rather different pluviometric regimes: yearly precipitation ranges from around 1300 mm in Genoa to 600/700 mm in Bologna, Ferrara and Mantova; the precipitation pattern throughout the year exhibit two maxima (May and October) and two minima (February and August) in the Po plain stations,

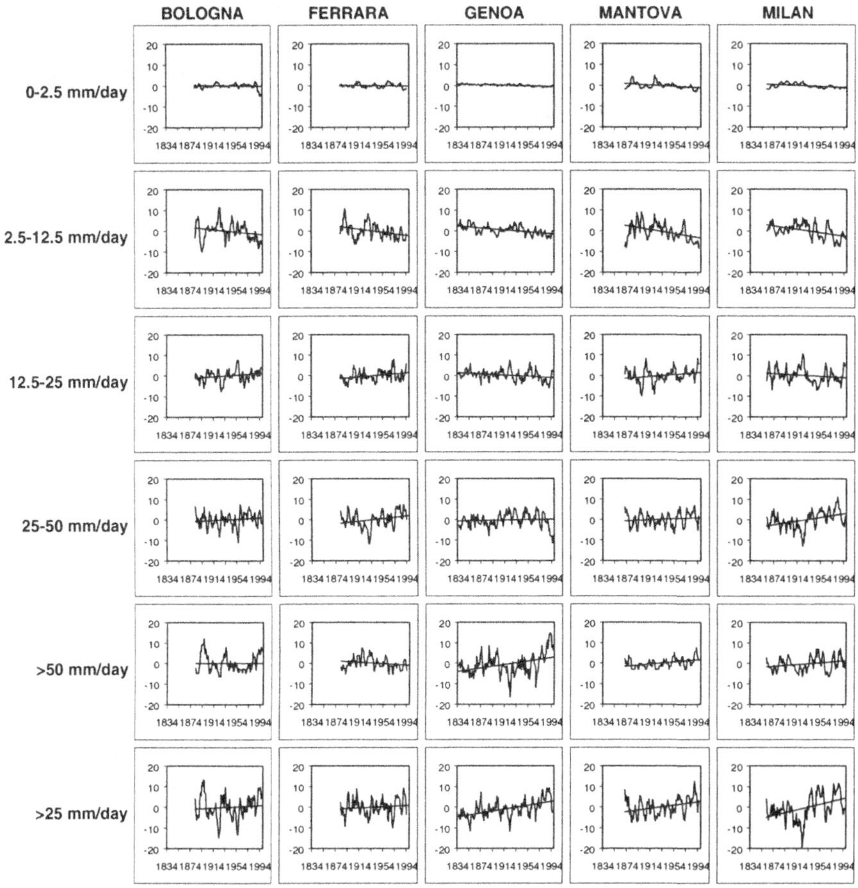

Figure 1. 5 year moving average anomalies of yearly relative contributions of precipitation falling in class intervals 1-6 for the five stations. The straight lines indicate the linear fitting of the series.

whereas Genoa has a more Mediterranean behaviour with only one maximum in autumn (October) and one minimum in summer (July) [9]. The stations exhibit a different behaviour also in relation to heavy precipitation events with differences both in the frequency of the events and in their seasonal distribution [10].

3. METHODS

To prevent missing data from introducing any bias we used a procedure described by Karl et al. [11] and by Karl et al. [2] to estimate them. Basically, a gamma function is fit to each station's daily data for each month of the year. To determine if precipitation occurs on any missing day, a random number generator is used such that the probability of precipitation is set equal to the empirical one on that day. if precipitation occurs, then the gamma distribution is used to determine the amount that fails for that day, again using a random generator.

After completing the data, we calculated for each station anomalies for the proportion of daily precipitation (DP) falling in 5 precipitation class intervals in the year (CI I, ..., Cl5) compared

with the corresponding total precipitation [11].

As class intervals we used: $0 < DP \leq 2.5$ mm; $2.5 < DP \leq 12.5$ non; $12.5 < DP \leq 25.0$ mm; $25.0 < DP \leq 50.0$ mm and $DP > 50$ mm. In order to better study the proportion of daily precipitation in the upper part of the daily rainfall value distribution, we introduced another class interval (C16) simply defined as the sum of C14 and CIS ($DP > 25$ mm). Yearly values were conventionally made to correspond to the period from December 1st to November 30th and dated by the year in which January occurred. All the anomalies are differences between yearly values and the corresponding means calculated over the common period of the 5 series (1880-1996).

The statistics on the proportion of daily precipitation in different class intervals provide information about changes in precipitation intensity being unrelated to changes in precipitation amount as a consequence of the normalization with total precipitation. The Cl series were analysed with the Mann-Kendall non parametric test as described by Sneyers [12] to look for a trend. The slopes of the trends were calculated by least square linear fitting. The Mann-Kendall test was also used for a progressive analysis of the series, consisting of the application of the test to all the series starting with the first term and ending with the i-th and to those starting with the i-th one and ending with the last [12]. In the absence of any trend the graphical representation of the direct (u_i) and the backward (u_i') series obtained with this method gives curves which overlap several times, whereas in the case of significant trend the intersection of the curves enables the start of the phenomenon to be located approximately [12].

4. RESULTS AND DISCUSSION

Figure 1 shows 5 year moving averages of Cl1, Cl6 of the 5 stations, whereas table 1 contains the results of the application of the Mann Kendall test and of casa squares linear fitting to Cl series.

In order to allow comparison among the 5 stations, only the period 1880-1996 is considered for trend analysis. The figure gives evidence of a tendency in Northern Italy toward decreasing trends in the relative

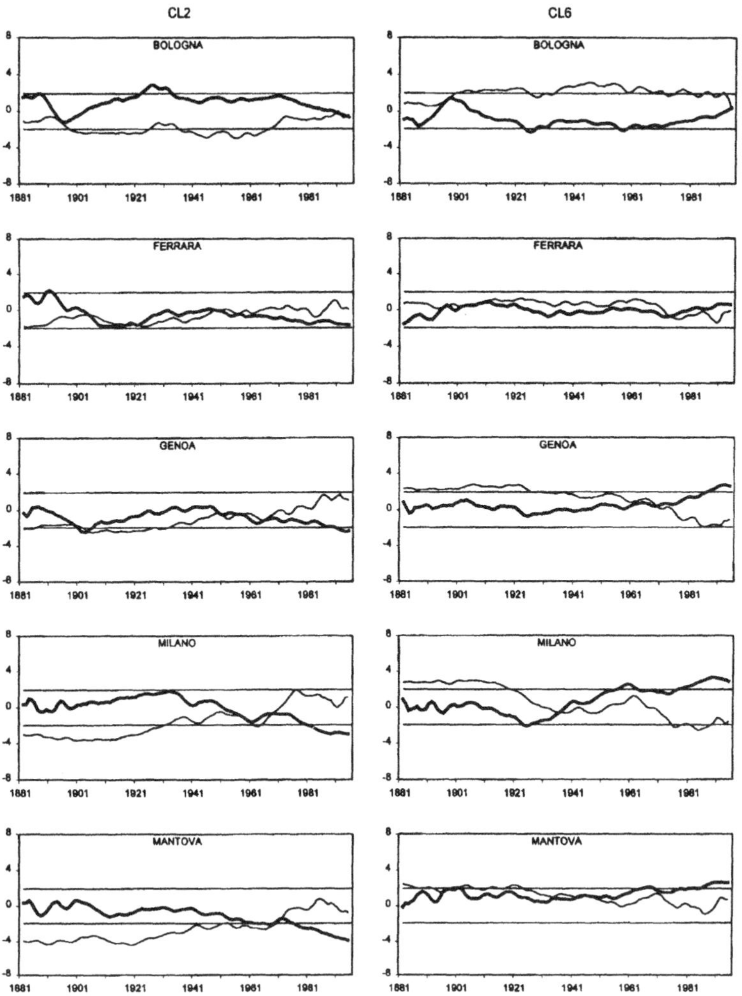

Figure 2. Progressive values of u_i (thick line) and u_i' (thin line) for the relative contributions of yearly precipitation falling in class intervals 2 and 6 for the five stations. Straight lines indicate a confidence level of 95%.

contributions of the lower class intervals (1 and 2 - DP < 12.5 mm) and increasing trends in the one of class intervals 4, 5 and 6 (DP > 25 mm).

The class intervals with the most clear results are Cl1, Cl2 and Cl6: the two lower class interval contributions have a negative trend in all the stations whereas C16 has a positive trend everywhere. For all the three class interval contributions the trends have a confidence greater than 99% in Genoa, Milan and Mantova. Other significant trends are Milan Cl4 (confidence > 99%) and Genoa and Mantova Cl5 (confidence > 95%). Normalizing the linear trends by the correspondent mean class interval contributions the relative variation of each class interval contribution can be estimated. This means for example that the -8%/100y trend of Mantova C12 corresponds to a -19%/100y relative variation, being normalized by Mantova Cl2 mean contribution (42%).

In Genoa, Milan and Mantova these relative variations are around 20% in 100 years both for the lower (Cl1+Cl2) and the higher (Cl6) class intervals whereas in Bologna and Ferrara they are around 5%.

A more detailed analysis of the behaviour of the class interval contributions can be obtained with the progressive application of the Mann Kendall test. The results are shown in figure 2 that gives the graphical representation of the direct $(u_i)$and backward u_i' series for yearly Cl2 and Cl6 of the 5 stations.

In Genoa, Milan and Mantova - where significant trends are present - both for Cl2 and Cl6 u and u' curves have a single intersection in the period 1940-1970 and generally u begins to assume significant values only around 1980. In these stations Cl6 direct curves have a monotone increase after 1920 with only a short period (1960-1970) with a less evident growth. For Cl2 the behaviour is apposite: the direct curves have a monotone decrease after 1920 because their slopes are particularly high in the same periods as the Cl6 curves.

In Bologna and Ferrara the curves do not give evidence of a clear trend. It is however worth noticing that also in these station in the last 20÷30 years Cl6 direct curves show a clear increase whereas the Cl2 ones have an opposite behaviour pattern.

5. CONCLUSIONS

The analysis of 5 secular Northern Italy precipitation series gives evidence of an increase in the proportion of daily precipitation falling in higher precipitation class intervals (25 mm < DP ≤ 50 mm and DP > 50 mm) and a decrease in the proportion falling in the lower ones (O mm < DP ≤ 2.5 mm and 2.5 mm < DP ≤ 12.5 mm). The progressive application of the Mann Kendall test shows that the trend is mainly due to the last 60- 80 years. The results are in agreement with the ones of Karl et al. (1 995): both for Northern Italy and for the USA the percentage of total annual precipitation occurring in heavy rainfall events (DP > 2 inches for the USA; 25 mm < DP ≤ 50 mm, DP > 50 mm and DP > 25 mm for Northern Italy) there is a significant positive trend in the period 1910-1996.

The increase in the proportion of daily precipitation falling in high class intervals has not a uniform spatial pattern over Northern Italy, showing that

Table I. Results of the application of the Mann-Kendall test and of least squares linear fitting to the relative contributions of yearly precipitation falling in class intervals 1-6 for the five stations: period 1880-1996. b is the linear regression coefficient, a and r are the associated error and the associated linear correlation coefficient respectively; u is the Mann-Kendall test value. (* = significance level greater then 95%; * * = significance level greater then 99%).

	BOLOGNA			FERRARA			GENOA			MANTOVA			MILAN		
	b±σ	r	u	b±σ	r	u	b±σ	r	u	b±σ	r	u	b±σ	r	u
0-2.5	-0.3±0.6	0.04	-0.5	-0.3±0.6	0.05	-0.5	-1.1±0.3	0.35	-4.3**	-2.4±0.7	0.32	-3.6**	-2.5±0.3	0.58	-6.5**
2.5-12-5	-3±2	0.09	-0.8	-3±2	0.13	-1.6	-2.4±1.1	0.20	-2.2*	-8±2	0.35	-4.0**	-5±2	0.27	-2.9**
12.5-25	3±2	0.08	1.0	3±2	0.12	1.6	-2±2	0.09	-1.2	4±2	0.15	1.4	-1±2	0.04	-0.6
25-50	2±2	0.01	0.3	2±2	0.10	1.2	-1±2	0.04	-0.2	2±2	0.10	1.4	6±2	0.26	2.6**
>50	-1.5±2	0.02	0.1	-1.5±2	0.07	-1.0	6±3	0.17	2.0*	4±1	0.27	1.9	2±2	0.08	0.6
>25	1±3	0.02	0.6	1±3	0.03	0.5	5±2	0.20	2.4*	6±2	0.24	2.6**	8±3	0.25	2.7**

the western area has a greater trend than the eastern one. This result is very interesting as heavy precipitation events in Genoa and in Milan are associated with the same pressure patterns - low pressure systems that determine southerly flow over Northwest Italy with warm and wet Mediterranean air masses forced upwards by the Alps and by the Apennines - that can cause extreme precipitation events over wide areas of the Po basin [10], [13], [3]. Therefore the strong trend in precipitation intensity in the last 60- 80 years could be associated with an increase of the flood risk over this region.

6. REFERENCES

Bellumè M., M. Maugeri, M. Mazucchelli, (Eds.), Due secoli di osservazioni meteorologiche a Mantova, Edizioni CUSL, Milan 1998.

Buffoni L., F. Chlistovsky, (Eds.), Precipitazioni giornaliere rilevate all'Osservatorio Astronomico di Brera in Milano dal 1835 al 1990, EdiErmes, Milan, 1992.

Buffoni L., M. Maugeri, T. Nanni. Theor. Appl. Climatol. in press (1999)

Flocchini G., C. Palau, I. Repetto, M. P. Rogantin, (Eds.), 1 dati pluviometrici della serie storica di (1 833-1980) di Genova, CNR Report AQ/5139, Rome, 1982.

Gazzola, (Ed.), Distribuzione ed evoluzione delle temperature e delle precipitazioni in Italia in relazione alla situazione meteorologica, CNRIIFA Report, Rome, 1978.
Geneve, 1990.

Giacobello N. and G. Todisco, Riv. Met. Aer. 3 9 (1979) 13 9-15 I.

IPCC, 1996: J. T. Houghton, L. G. Meira Filho, B. A. Callander, N. Harris, A. Kattenberg and Maskell K. (Eds.), Climate Change. The IPCC Second Assessment Report, Cambridge University Press, N.Y., 1996.

Karl T. R., R. W. Knight, Bull. Am. Met. Soc. 79 (1998) 231-241

Karl T. R., R. W. Knight, N. Plummer, Nature 377 (1995) 217-220.

Maugeri M., L. Buffoni, F. Chlistovsky, Acqua & Aria 5 (1995) 549-560.

Maugeri M., P. Bacci, R. Barbiero, M. Bellumè, Physics and Chemistry of the Earth in press 1999.

Mennella C., (Ed.), Il Clima d'Italia, Fratelli Conti Editori, Napoli, 1967.
Sneyers R., On the statistical analysis of series of observation, WMO, Technical Note N. 143,

Climate Change Experiments on a Glacier Foreland in the Central Alps

BRIGITTA ERSCHBAMER
Institute of Botany, Sternwartestr. 15, A-6020 Innsbruck, Austria

Key words: global change, growth, leaf number, ramet groups, temperature enhancement

Abstract: An experiment is being carried out on the glacier foreland of the Rotmoosferner (Oetztal Alps, Tyrol, Austria) to study the effect of enhanced temperatures on the vegetation. The main aim of study is to analyse the growth and biomass production of an early and a late-successional species under the present natural conditions and under experimentally-altered microclimatic conditions. In 1996, 10 open top chambers (OTC's) and 10 control plots were established on the moraine of the 1971 glacier stage at 2400 m above sea level. 5 open top chambers and 5 control plots were planted with seedlings of *Trifolium pallescens* (= an early successional species). Another 5 open top chambers and 5 control plots were planted with ramet groups *of Carex curvula* (= a late-successional species). Leaf and ramet growth have been monitored in each subsequent growing season. Preliminary results show that *Trifolium pallescens* develops significantly more leaves under enhanced temperature conditions. The *Carex curvula* ramet groups decreased in size in the OTC's as well as in the controls. The final results of the five-year study are expected in August 2000.

1. INTRODUCTION

Glaciers are highly sensitive to changes in temperature and they can be regarded as indicators of global warming. The retreat of the glaciers during more than one century is a striking phenomenon all over the Central Alps. According to Haeberli (1) the glaciers of the Alps have lost about 30 – 40 % of their surface area and around 50 % of their ice volume. The spectacular discovery of the Ice-man in the Oetztal Alps shows clearly that the alpine glaciers are more reduced today than during the past 5000 years.

G. Visconti et al. (eds.), Global Change and Protected Areas, 37–44.

Fig. 1 OTC's and control plots on the glacier foreland of the Rotmoosferner (Oetztal Alps, Austria), at 2400 m above sea level.

According to the IPCC Second Scientific Assessment Report 1995, air temperatures are expected to increase continuously: increases of 2°- 4°C by the middle of the next century, or even 0.7° - 5.2°C according to other models are suggested (2).

These scenarios have stimulated manipulative experiments on vegetation world-wide, the main concentration of such studies being mainly in the Arctics (3). In the Central Alps such experiments have been carried out by Körner (4, 5) on alpine grasslands and by Stenström et al. (6) on *Saxifraga oppositifolia*.

The main aim of the present study was to establish an in situ experiment on a glacial retreat area in the Central Alps. The ambient temperature is increased by use of International Tundra Experiment (ITEX) shelters (7). Together with air temperatures, several other microclimatic conditions are altered. However, the ITEX-shelters (open top chambers) do not change either the CO_2 – or the UV-B conditions.

The importance of air temperature as a factor controlling growth is being studied on an early successional species (*Trifolium pallescens*) and on a late-successional species (*Carex curvula*). Leaving these shelters in place during the winter, also allows the effects of changes in the duration of the snow cover and length of the growing season to be analysed.

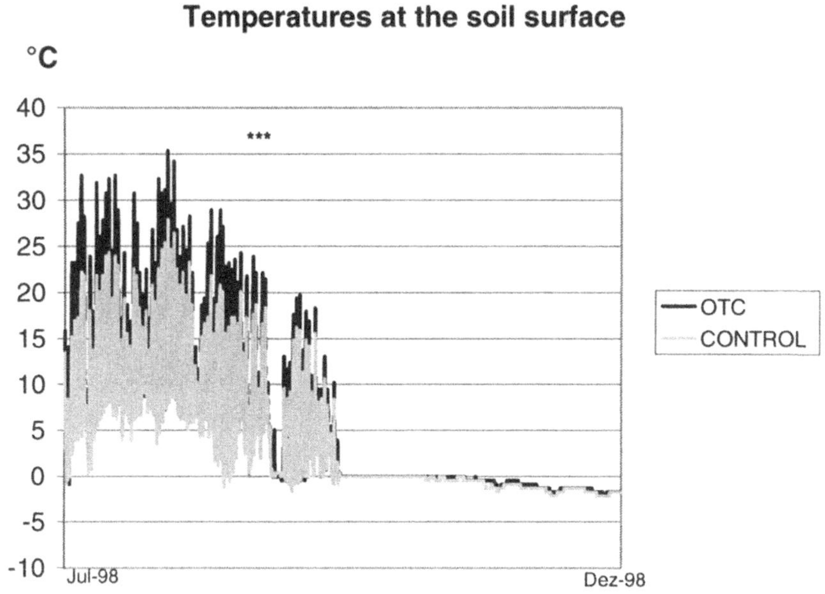

Fig. 2 Temperatures measured at the soil surface of an OTC and a control plot from July 1998 till May 1999.

2. STUDY AREA

The study area is located within the glacier foreland of the Rotmoosferner (Oetztal Alps, Tyrol, Austria; 46°49'N, 11°02'E) at an altitude of 2400 m above sea level. On the moraine of the 1971 glacier stage a research area of 20x30 m was fenced off in 1996. A pioneer vegetation, dominated by *Saxifraga* species (*S. oppositifolia* and *S. aizoides*), prevails there. Moreover, scattered patches of the mid-successional species *Trifolium pallescens*, *Poa alpina*, *Artemisia genipi* are already established. A mean plant cover of 15 % - 30% was estimated.

3. METHODS

Within the research area, 10 conical open top chambers (= OTC's, made of polycarbonate, height = 30 cm, upper diameter = 50 cm, lower diameter = 84,6 cm, built according to the ITEX-guidelines given by Molau (7), Fig. 1) were established.

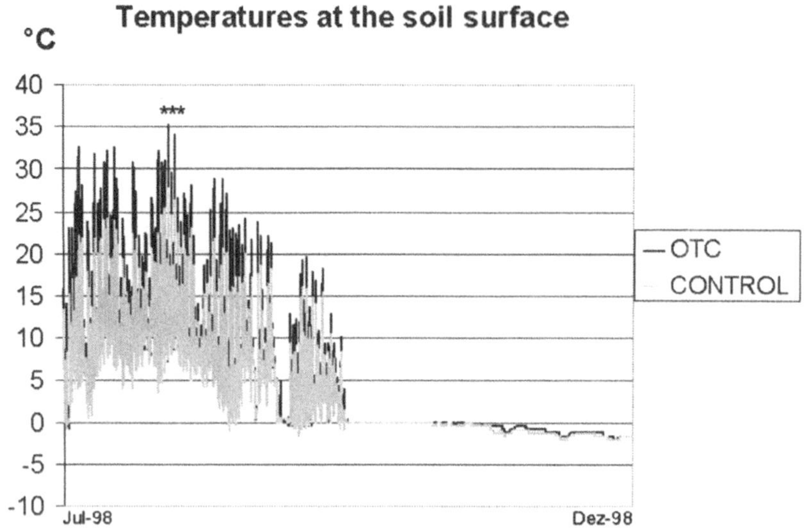

Fig. 3 : Temperatures measured at 3 cm soil depth of an OTC and a control plot from July 1998 till May 1999.

Ten permanent plots of the same extension were marked off as controls (Fig. 1). The seedlings of *Trifolium pallescens* were collected in July 1996 on the 1956/57 glacier stage and transplanted immediately on to the 5 OTC plots and the 5 control plots (10 individuals on each plot).

The other 5 OTC plots and 5 control plots were planted with ramet groups of *Carex curvula*. Tussocks of this species were dug up beyond the glacier foreland on an altitude of 2300 m above sea level. They were divided into ramet groups (5-14 connected ramets) and planted (10 ramet groups on each plot).

Leaf and ramet growth have been monitored during every growing season. In August 2000 all the plants will be harvested and their leaf areas and dry weights will be determined.

5 of the OTC's and 5 of the control plots are equipped with data loggers to record the above-ground and soil temperatures. The soil moisture content is also being recorded at 3 cm depth on 2 OTC's and 2 control plots, using soil moisture sensors SMS 3 Cyclobios (output signal range: 0-1 Volt).

Statistical analyses have been carried out using the programmes Spss (Mann-Whitney-test for comparisons of the growth results) and Excel 6.0 (t-test for comparisons of the temperature conditions).

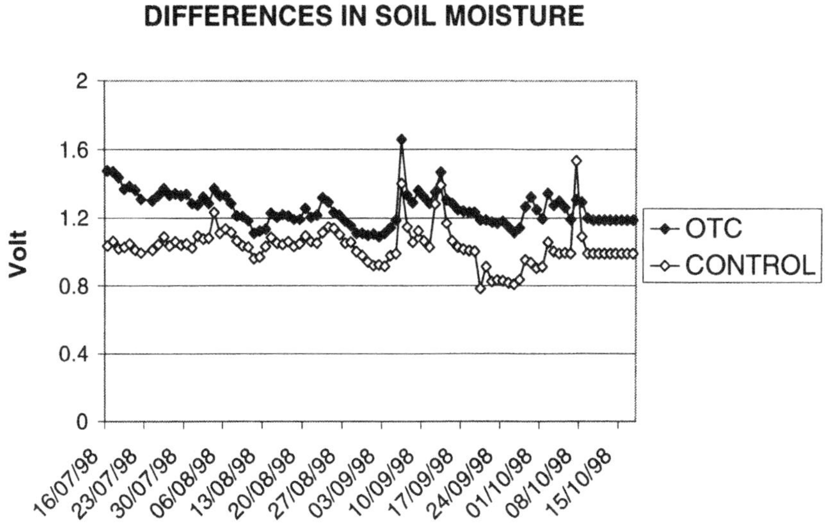

Fig. 4 Differences of soil moisture between an OTC and a control plot from July 1998 till October 1998.

Since the project is intended to continue until the year 2000, only preliminary results can be shown here.

4. RESULTS

The above-ground temperatures of the OTC's were at least 1.5°C higher than those of the control plots (Fig. 2). The maxima were more than 7.5 °C higher. Also the soil temperatures at 3 cm depth were 0.3°C higher within the OTC's compared to the control plots, the maxima being 1.6°C higher (Fig. 3). All these differences are statistically significant.

The differences between the soil moisture of the OTC's and the control plots are shown in Fig. 4. Within the OTC, the soil would seem to remain moister than in the control plot, however the differences are not statistically significant. A significantly higher increase ($p = 0.01$) in leaf number was detected for the *Trifolium pallescens* individuals in the OTC's, compared to those within the control plots (Fig. 5).

For the *Carex curvula* ramet groups, a decrease in shoot number was observed within all the OTC's and all the controls, the decreases in 1998 being slightly higher in the OTC's (Fig. 6). However, the differences are not statistically significant.

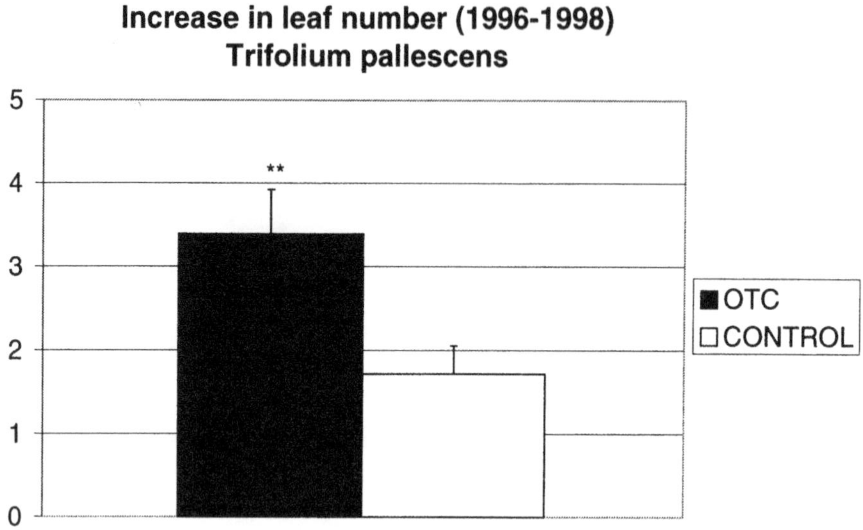

Fig. 5 Increase in leaf numbers of *Trifolium pallescens* in the OTC's and in the control plots. The differences are highly significant (p = 0.01).

5. DISCUSSION

A vast number of studies have been published during the last ten years concerning the effects of global temperature change (see the review by Guisan (2)). Most of the studies, however, have been of short duration and relatively small effects were observed (6, 8, 9). According to Körner (10) and Theurillat (11), a temperature increase of 1- 2°C would not greatly affect the alpine flora. However, the preliminary results of this study show that at least *Trifolium pallescens* reacts sensitively to enhanced temperatures by developing significantly more leaves than under ambient conditions. In general, favourable temperatures seem to be more important for phenological development and reproduction than for biomass increase (6, 12, 13, 14, 15). The long-term study presented here is expected to provide more information about biomass effects under enhanced temperatures and it is hoped that the results will lead to new hypothesis.

6. ACKNOWLEDGEMENTS

I would like to thank all the people who have helped in establishing the

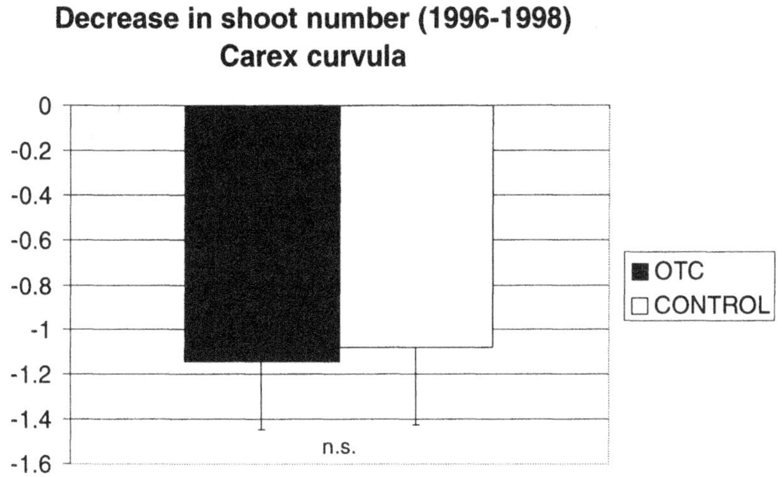

Fig. 6: Decrease in ramet numbers of *Carex curvula* in the OTC's and in the control plots.

fences and the experiment and in repairing the equipment on the glacier foreland of the Rotmoosferner. I am particularly grateful to Bertram Piest, Manuela Hunn, Josef Schlag, Ruth Niederfriniger Schlag, Elisabeth Kneringer, Corinna Raffl, Helmut Scherer, Klaus Vorhauser, Dirk Lederbogen, Martin Mallaun, Meini Strobl, Max Kirchmair, Walter Steger, Rüdiger Kaufmann, Erwin Meyer, to the participants of the botanical course of the University of Innsbruck held at Obergurgl in 1996 and to the participants of the botanical course of the University of Essen (Prof. Dr. Maren Jochimsen) held at Obergurgl in 1997.

7. REFERENCES

Alatalo J.M. and. O. Totland, Global Change Biol. 3, Suppl. 1 (1997) 74-79.

Guisan A., J.I.Holten, W. Haeberli and M. Baumgartner, In: A. Guisan, J.I.Holten, R. Spichiger and L. Tessier (Eds.), Potential Ecological Impacts of Climate Change in the Alps and Fennoscandian Mountains, Genève, 1995, pp.15-38.

Haeberli W., In: A. Guisan, J.I.Holten, R. Spichiger and L. Tessier (Eds.), Potential Ecological Impacts of Climate Change in the Alps and Fennoscandian Mountains, Genève, 1995, pp. 97-104.

Havström M., T.V. Callaghan and S. Jonasson, Oikos 6 (1993) 389-402.

Henry G.H.R. and U. Molau, Global Change Biol. 3, Suppl. 1(1997) 1-9.

Jones M.H., C. Bay and U. Nordenhäll, Global Change Biol. 3, Suppl. 1 (1997) 55-60.

Körner C., Catena 22, Suppl. (1992) 85-96.

Körner C., In: F.S. Chapin III and C. Körner (Eds.) Arctic and Alpine Biodiversity, Ecol. Studies 113, Springer, Heidelberg, 1995, pp. 45-62.

Körner C., In: M. Beniston (Ed.), Mountain environments in changing climates, Routledge, London and New York, 1994, pp. 155-166.

Molau U., International Tundra Experiment: ITEX-Manual. Danish Polar Center, Copenhagen, 1993.

Parsons A.N., J.M. Welker, P.A. Wookey, M.C. Press, T.V. Callaghan and L.A. Lee, J. Ecol. 82 (1994) 307-318.

Stenström M., F. Gugerli and G.H.R. Henry, Global Change Biol. 3, Suppl. 1 (1997) 44-54.

Theurillat J.-P., In: In: A. Guisan, J.I.Holten, R. Spichiger and L. Tessier (Eds.), Potential Ecological Impacts of Climate Change in the Alps and Fennoscandian Mountains, Genève, 1995, pp. 121-127.

Welker J.M., U. Molau, A.N. Parsons, C.H. Robinson and P.A. Wookey, Global Change Biol. 3, Suppl. 1 (1997) 61-73.

Wookey P. A. , A.N. Parsons, J.M. Welker, J.A. Potter, T.V. Callaghan, J.A. Lee and M.C. Press, Oikos 65 (1994) 490-502.

High Mountain Summits as Sensitive Indicators of Climate Change Effects on Vegetation Patterns: The "Multi Summit-Approach" of GLORIA (Global Observation Research Initiative in Alpine Environments)

HARALD PAULI, MICHAEL GOTTFRIED, KARL REITER & GEORG GRABHERR

Department of Vegetation Ecology and Conservation Biology, Institute of Plant Physiology at the University of Vienna, Althanstr. 14, A-1090 Wien, Austria

Key words: Altitudinal gradients, climate change, high mountain ecology, observation network, vegetation sampling

Abstract: GLORIA, a Global Observation Research Initiative in Alpine Environments, aims to establish an urgently required global indicator network to detect climate-induced changes in high mountain regions. High mountains appear to be particularly appropriate for such a global initiative, because they still comprise low-temperature determined, natural ecosystems in a world wide distribution. Evidence of upward migrations of vascular plants was found at high mountain peaks in the Alps – most likely resulting from the climate warming since the 19th century. GLORIA is aligned to "target regions" in alpine or nival environments of all principal vegetation zones from polar to tropical latitudes. The "Multi Summit-Approach", part of the proposed GLORIA-network, shall provide an effective method to compare mountain ecosystems and their climate-induced changes by using summits of different altitude in each target region. Reasons why high mountain summits can be particularly beneficial as indicator environments are pointed out. The sampling design and the method – already tested in field – are outlined along with first results. The final part gives some notes on the implementation of the research initiative. A close co-operation among international research co-ordinators and high mountain ecologists appears to be crucial for a globally active and successful indicator network.

G. Visconti et al. (eds.), Global Change and Protected Areas, 45–51.

1. INTRODUCTION

High mountains can be considered as particularly appropriate environments to detect effects of climate change on natural biocoenoses in a global scale for the following reasons: Firstly, ecosystems at the low-temperature limits of plant life are generally thought to be especially sensitive to climate change [1][2][3]. An already ongoing upward shift of vascular plants at high summits in the Alps, determined by the Austrian IGBP-research [4][5][6][7][8], is most likely a response to the atmospheric warming since the 19th century. Secondly, high mountains still comprise the most natural ecosystems in many countries, being largely untouched by human settlements and agricultural influences. Therefore, climatic effects on ecosystems can be studied without masking effects from human land use. Thirdly, high mountain ranges are present in virtually every major zonobiome of the earth.

The research initiative GLORIA aims to establish an urgently needed global monitoring network, by using high mountain ecosystems as sensitive indicators, as required in the "IGBP-Mountain Workplan" [9]. Moreover, a deeper understanding of assemblage mechanisms and assemblage processes in vegetation patterns as a contribution to ecological theory can be expected.

This paper gives a short general overview about GLORIA and a first outline about the concept, method, and some few results of the "Multi Summit-Approach", one of the basic intentions within the proposed network. It aims to encourage the involvement of high mountain researchers and research co-ordinators in a detailed discussion of the proposed research activities and in a co-operation within the planned global monitoring network.

2. THE PROPOSED NETWORK

GLORIA is geographically aligned to "target regions" within high mountain systems of the principal vegetation zones along the latitudinal scale from the polar to the humid tropical biomes (zonobiomes after Walter [10]). The distribution of these regions should be geographically balanced in latitude and longitude. The term "high mountains" is defined here by the following characteristics (according to Troll [11]): The uppermost altitudinal level exceeds the upper, cold-determined tree line; the landscape is shaped by glaciers (glaciation was present at least in the Pleistocene); frost is still an important factor for pedogenesis and soil structure.

The research area in each target region is focused on these high mountain areas (corresponding with the alpine and nival vegetation belt), from the tree line upwards. Permanent plots for a long term monitoring of climate-induced effects on the vegetation should be established at sites with no or low pressure form local human land use. Requirements like accessibility and the availability of local research stations and personal scientific resources are crucial to keep the network feasible.

Two different, synergistic approaches are suggested: 1) The "Single Mountain-Approach", which investigates one mountain per target region with special emphasis on transect studies across sensitive ecotones, and 2) the "Multi Summit-Approach", investigating summits of different altitude in each target region with the focus on an altitudinal comparison of ecological patterns. For further details about GLORIA and the two approaches see also [12] and the website: http://www.pph.univie.ac.at/gloria/gloria.html

3. THE MULTI SUMMIT-APPROACH

The Multi Summit-Approach within the proposed GLORIA-network shall provide a feasible and cost-efficient method to compare high mountain biocoenoses and their climate-induced changes along altitudinal and latitudinal gradients – the two fundamentally climatic gradients. This approach focuses on several summit areas at different elevation levels within each target region to detect vegetation patterns and species richness. Summit areas appear to be most relevant and beneficial for such an altitudinal comparison, because: 1) They represent at best the average local climate of a distinct elevation level (shading effects are absent or low); 2) Summits have habitats in all directions within a small area and changes of vegetation patterns occur over short distances; 3) Disturbances from avalanches and debris falls are minimised; 4) Mountain peaks may act as traps for upward migrating species; 5) They are pronounced landmarks which can easily be found again for revisitations without special marking. On the other hand, summits have to be carefully selected to avoid possible disadvantages, e.g. from frequent visits by tourists or from intensive animal grazing. Further, very steep or very flat summit terrain situations will not be beneficial for an efficient field work and for the comparison of data.

4. SAMPLING DESIGN AND METHOD

Fig. 1 shows a summit area with the sampling design for the Multi Summit-Approach, as used at 8 summits from two European Mountain systems with different climate (North-eastern Alps and Sierra Nevada of Spain).

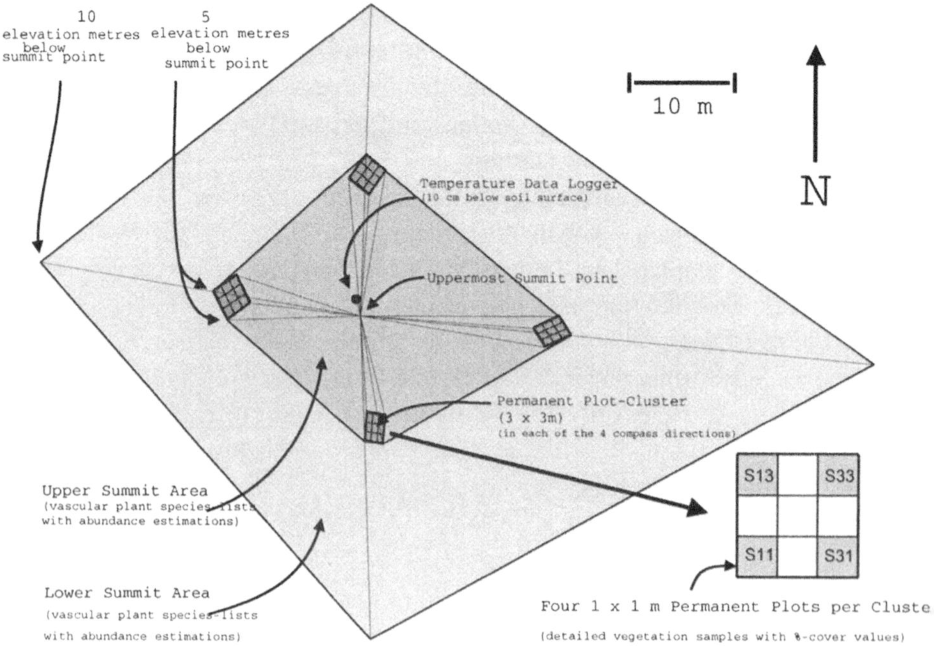

Fig. 1: Proposed sampling design for the Multi Summit-Approach (vertical view) with 4 permanent plot-clusters and an upper and a lower summit area with exact boundaries for species lists with abundance estimations

Permanent plot-clusters are positioned in all 4 compass directions with the lower boundary at the 5 m isoline below the highest summit point. A deviation from the exact compass direction was accepted in cases where the terrain was to steep or composed of bare rocks without micro habitats for plants to establish. Each cluster consists of a sampling grid of 3 x 3 m, subdivided with flexible measuring tapes into nine 1 x 1 m plots. Only the 4 plots at the corners were investigated, to avoid trampling impacts within the permanent plots, caused by the investigators.

A total of 16 plots of 1 x 1 m were recorded at each summit in the following manner: The percent-cover of vascular plants and cryptogams (bryophytes and lichenes) as well as the percent-cover of the abiotic surface categories (solid rock, scree, open soil) were recorded. Cover-percentages of species were investigated for the vascular plants. Photographs were taken from all plots for future revisitations.

A second investigation focuses on the uppermost summit area (the uppermost 5 elevation metres), where all vascular plant species were recorded with their abundance given in verbal terms (e.g. dominant, common, scattered, rare, very rare). The lower boundary of the uppermost summit area is marked by the 4 clusters and by straight lines between the lower corners of the clusters (see. Fig. 1). The investigation area reaches the 5 metre level (below the highest summit point) only at the 4 clusters. This helps to keep the area in reasonable size – particularly at summits with long ridges. On the other hand, an exact marking along the 5 m isoline would multiply the measuring work without enhancing the value of data for comparison. In the same manner, a belt between the lower boundary of the uppermost summit area and 4 points at the 10 m level below summit was investigated. The four 10 m points – again connected with straight lines - are located at the extended lines between the highest summit point and one of the lower corners of each cluster (see Fig. 1). Measurements were made with flexible measuring tapes, an electronic spirit-level, and with GPS (differential GPS with sub-metre accuracy).

Finally, a miniature data logger for temperature (StowAway Tidbit) was positioned close to the summit point in 10 cm below soil surface. Temperature is measured by an interval of 1.5 hours for a 5 year period.

Two full working days for two investigators should be calculated for each summit. For the future monitoring reinvestigation intervals of 5 to 10 years are proposed.

5. SOME RESULTS FROM THE FIRST TEST-SUMMITS

The 4 summits within the target region "North-eastern Alps" are located in different altitudinal levels between 1855 and 2255 m a.s.l. The obvious gradient in vegetation structure and density (with the lowest summit at the tree line and the highest summit composed of alpine grassland, scree and rock patterns) was well expressed by a stepwise decrease of plant cover within the permanent plots. It was 2.2 times as high at the lowest summit compared with the highest summit. Species richness (vascular plants) clearly dropped with increasing altitude, from 142, 119, 80 to 58 species within the summit areas, and from 107, 98, 70 to 47 species within the permanent plots. On the other hand, the relative share of endemic vascular plant species (endemics of the Eastern Alps), compared with the number of all species per summit, increased remarkably with altitude, from 5.6, 10.9 to 18.8% at the second highest summit, dropping only slightly to 17.2% at the highest summit. These first values of the Multi Summit-Approach, found along an altitudinal range of 400 m in now permanently established observation sites, show some fundamental climatically caused gradients in plant composition, which can be of particular interest when plant migration processes increase in changing climates.

6. THE CURRENT POSITION AND THE IMPLEMENTATION OF THE INITIATIVE

A preliminary outline of the research initiative was first presented at the "European Conference on Environmental and Societal Change in Mountain Regions" in Oxford in December 1997 [12], where it was highly recommended as a contribution towards the realisation of the "IGBP Mountain Workplan" [13]. GLORIA is further linked to the newly emerging "Global Mountain Biodiversity Assessment" (MBA) within DIVERSITAS (an international programme of biodiversity science). MBA is supported by the Swiss Academy of Sciences to create an international mountain biodiversity network.

The generation and the testing of appropriate field methods for GLORIA have been so far conducted at the Department of Vegetation Ecology and Conservation Biology of the University of Vienna, and are supported by the Austrian Academy of Sciences from the national IGBP/GCTE budget. For the Multi Summit-Approach, field method was already tested in two potential target regions: 1) The "North-eastern Alps" in Austria in zonobiome VI, with temperate-nemoral climate, and 2) The "Sierra Nevada" in southern Spain in zonobiome IV, with mediterranean climate. Four summits per region with different altitude were investigated.

A feasibility study of the research initiative, financed by the Austrian Federal Ministry of Science and Transport with a half-year contract, was started in early summer 1999. It includes the call for potential contributors (principal investigators, institutions, mountain ranges, potential target regions) as a first step towards the implementation of GLORIA. In a second step, a workshop is planned in the near future for further discussion of work plans, observation manuals, target regions, the organisation and funding. The start of a globally active GLORIA-network will be, hopefully, within the next few years – preferably in 2002 which was declared as "Year of the Mountains" by the United Nations.

7. REFERENCES

Becker A. and H. Bugmann, (Eds.), Predicting global change impacts on mountain hydrology and ecology: integrated catchment hydrology/altitudinal gradient studies, IGBP Report 43, Stockholm, 1997.

Beniston M. (Ed.), Mountain Environments in Changing Climates, Routledge, London, New York, 1994.

Beniston M. and D. G. Fox, In: R. T. Watson, M. C. Zinyowera and R. H. Moss, (Eds.), Climate change 1995 - Impacts, adaptations and mitigation of climate change: scientific-technical analysis, Cambridge University Press, Cambridge, 1996, pp. 191-213.

Gottfried M., H. Pauli and G. Grabherr, Jahrbuch d. Vereins zum Schutz d. Bergwelt 59 (1994) 13-27.

Grabherr G., M. Gottfried and H. Pauli, In: C. Burga and A. Kratochwil (Eds.), Vegetation monitoring/Global Change, Tasks for Vegetation Science, Kluwer, Dordrecht, (in press).

Grabherr G., M. Gottfried and H. Pauli, Nature 369 (1994) 448.

Grabherr G., M. Gottfried, A. Gruber and H. Pauli, In: F. S. Chapin III and C. Körner, (Eds.), Arctic and Alpine Biodiversity: Patterns, Causes and Ecosystem Consequences, Ecological Studies 113, Springer-Verlag, Berlin, 1995, pp. 167-181.

Pauli H., M. Gottfried and G. Grabherr, In: M. F. Price, T. H. Mather and E. C. Robertson, (Eds.), Global Change in the Mountains, Parthenon Publishing, New York, 1999, pp. 25-28.

Pauli H., M. Gottfried and G. Grabherr, World Resource Review 8 (1996) 382-390.

Price M. F. and R. G. Barry, In: B. Messerli and J. D. Ives, (Eds.), Mountains of the World, Parthenon Publishing, New York, 1997, pp. 409-445.

Price M. F., In: M. F. Price, T. H. Mather and E. C. Robertson, (Eds.), Global Change in the Mountains, Parthenon Publishing, New York, 1999, pp. 10-11.

Troll C., Ökologische Landschaftsforschung und vergleichende Hochgebirgsforschung, Erdkundliches Wissen - Schriftenreihe für Forschung und Praxis 11, Franz Steiner Verlag, Wiesbaden, 1966.

Walter H., Vegetation of the earth and ecological systems of the geo-biosphere - 3rd edition, Springer Verlag, Berlin, 1985.

Temperature and Precipitation Trends in Italy During the Last Century

E.PIERVITALI [(1)] AND M.COLACINO [(2)]

(1) Consorzio Ricerche "CRATI" - Università della Calabria - Rende (CS); tel.06-49934318; fax 06-20660291;

(2) Istituto di Fisica dell'Atmosfera - CNR (Roma); tel 06-49934320; fax 06-20660291;

Key words: climatology, temperature, precipitation, trends, secular series

Abstract: In the study of the climate change, probably related to the anthropic enhancement of greenhouse effect, an important topic is represented by the reconstruction of past climate, because the study of the past variations can help in understanding the present day trends. In this regard a prominent role is due to the analysis of long term data series, that can give quantitative information about climate change, going back from now to XVIII century. In a preceding paper, the investigation on the main climatic parameters in the period 1951-1995 in the Central-Western Mediterranean basin indicates a noticeable climatic evolution. In the present work we have examined only the Italian territory, where are available secular series of temperature and precipitation relative to about 50 years. The obtained results generally indicate a temperature increase and a rainfall reduction. In addition we have started to reconstruct some long-term data sets (Venezia, Taranto, Foggia, Catania), for the development of a data base of secular series.

1. INTRODUCTION

Several studies are devoted to the problem of climatic change, because the anthropic enhancement of greenhouse effect could have an impact on climatic parameters.

Special attention is concerned to the analysis of long-range data series, since the examination of the past patterns on one hand points out possible

G. Visconti et al. (eds.), Global Change and Protected Areas, 53–60.

Table 1 - Temperature stations

Station	Observation beginning year	Missing years
Alessandria	1866	-
Arezzo	1877	1922-23; 1944-45; 1947-49
Belluno	1873	1918-20; 1922-23
Bologna	1866	-
Cagliari	1879	1943-44
Cuneo	1877	1943-46
Ferrara	1879	1943-44
L'Aquila	1874	1915-24; 1972
Livorno	1866	1944-50
Mantova	1866	1971
Messina	1881	1944
Milano	1866	1937-40
Napoli	1865	1925-31
Padova	1866	-
Palermo	1866	-
Parma	1866	-
Pavia	1866	1935; 1975
Perugia	1866	-
Pesaro	1871	1944
Piacenza	1872	1944-45
Reggio Calabria	1879	1909-24
Reggio Emilia	1876	1824; 1928
Roma	1811	1931-40
Rovigo	1879	1951
Sassari	1876	1881-1882
Siracusa	1878	1913; 1943-50; 1956
Torino	1866	1927-34; 1942-50

climatic evolution in time, on the other it can give information about the present day trends. In particular, the presence of cycles and periodicities in the secular series could allows us to investigate whether the actual climate variations can be attributed to the natural variability either to the anthropic contribution. With reference to the temperature many researches [1, 2, 3] indicate an increase of 0.5 °C/100 years at the global scale in the past century, although this increase appears not linear. Precipitation trends have also been analysed [3, 4, 5] and in the Northern Hemisphere they indicate a decrease at latitudes lower than 50°N and a rainfall increase at higher latitudes. At the regional scale, in the study of climate evolution by the analysis of the trends, very important is the availability of several long-period records of meteorological variables. In Europe some quantitative series go back until XVIII century and more information is found from XIX century. Great Britain, for example possesses a great number of long data sets and many papers have been published about the examination of such series, specially of precipitation [6, 7, 8, 9].

Table 2 - Precipitation stations, having yearly data in the period 1951-1995. In the station marked by * also monthly totals are available

Alghero *	Firenze	Pisa *
Amendola	Genova *	Potenza
Ancona	Ginosa	Prizzi
Arezzo	Grosseto *	Ravenna
Bellavista	Leuca	Reggio Calabria *
Bologna	Messina	Roma *
Brindisi	Milano	Torino
Cagliari *	Napoli *	Trapani *
Campobasso	Palinuro	Trieste
Catania *	Perugia	Venezia
Cozzo	Pescara	Verona
Crotone	Piacenza	Viterbo

Also in Italy some works have been produced, but they mainly refer to single secular series [10, 11]. In the last 50 years period, an investigation on climate evolution in the Central-Western Mediterranean basin has been performed [12], in which the main climatic parameters have been analysed and in particular a systematic study of precipitation has been developed [13]. A climate evolving in a consistent way has been found, however definite results can not be drawn because the examined period is not so long. In the present work the analysis is mainly concentrated on temperature and precipitation trends in Italy. While precipitation data refer to the period 1951-1995, for temperature secular series are available, that can give more reliable indications. A decreasing pattern of -3.4 mm/year (-20%) has been obtained for precipitation and an increasing trend of +0.49°C/100 years for temperature.

In addition, in order to enrich the temperature data set and to extend back in time precipitation records, the reconstruction of some series of the U.C.E.A. (Ufficio Centrale di Ecologia Agraria) network is in progress.

2. DATA SET

Temperature data, starting from past century, are relative to the stations shown in table 1, in which also the gaps for each series are indicated. Precipitation records are available for the stations reported in table 2, in the period 1951-1995 and do not have missing data.

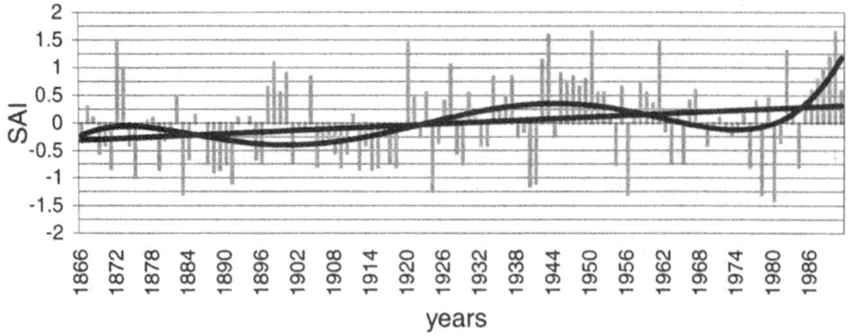

Fig.1 - SAI for temperature in Italy. The linear trend and the 5-order polynomial smoothing are reported.

Since all these series derive from the network of the Meteorological Service of the Italian Air Force, quality controls have been performed before data diffusion.

3. ANALYSIS OF TEMPERATURE

To analyse the temperature pattern in Italy the Standardised Anomaly Index (SAI) has been calculated, using the stations reported in table 1. It is a regional index, given by:

$$I_j = \frac{1}{N_j} \cdot \sum_{i=1}^{N} \frac{\left(T_{ij} - T_i\right)}{\sigma_i}$$

where:
I_j is the index for the J th year
T_{ij} is the temperature in the year j for the station i
T_i is the mean temperature in the ith station
σ_i is standard deviation of the temperature in the i th station
N_j is the number of stations available in the year j

When the index is lower than -0.25 or higher than +0.25, it indicates an anomaly statistically significant [14]. The SAI allows us to have a regional series also when there are gaps in the records of some stations, because, for

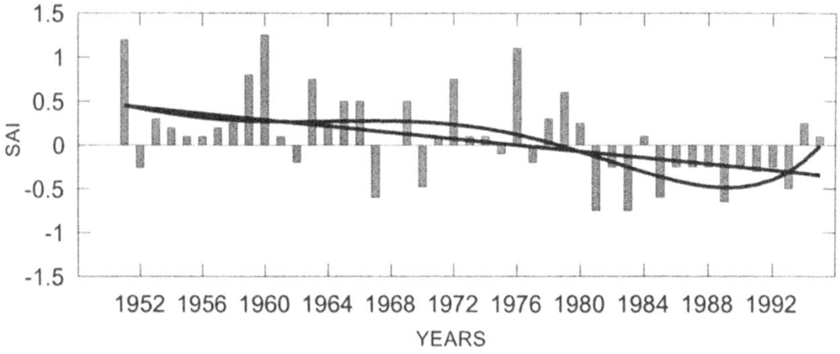

Fig.2 - SAI for precipitation in Italy. The linear trend and the 5-order polynomial smoothing are reported.

each year, the calculation is effected using the available stations. In fig.1 the index relative to the Italian temperature is shown, together with the linear trend and the 5 order polynomial smoothing. It indicates an increasing pattern statistically significant, since the linear trend, equal to 0.49°C/100 years, goes outside the range -0.25 ÷ +0.25. However, observing the polynomial smoothing, it appears that this increase is not constant: temperature is almost stationary until the end of 1800, increases until the forties, shows a reduction in the successive period and a new positive trend starting from the eighties. These results seem in agreement to the studies at the global scale, that indicate a temperature increase of 0.5°C/100 years in the past century.

4. ANALYSIS OF PRECIPITATION

Precipitation pattern in Italy has been examined using the stations of table 2, for which data are available in the period 1951-1995. The same statistical technique than temperature has been applied, the Standardised Anomaly Index. Fig.2 shows the rainfall SAI index with the relative trend and the 5-order polynomial smoothing. It is clear a precipitation decrease, that is statistically significant. The reduction rate is -3.4 mm/year that corresponds for the whole period to -150.8 mm, equivalent to -20%. It is also evident that negative values of the SAI occurred frequently after 1980. The seasonal analysis indicate that the observed reduction is higher during the cold season, in which rainfall is mainly concentrated in the Mediterranean climate. In winter it arrives to 32.9%, as reported in table 3. The observed negative trend could be related to the increase of the anticyclonic activity in the last 50 years, over the Central Western Mediterranean, although it could have not influence on the rainfall amount but only on the number of the rai-

Table 3 - Seasonal precipitation variation in the period 1951-1995

	Trend	ΔP	ΔP%
Autumn	-0.5	-23.7	-8.5
Winter	-2.0	-89.1	-32.9
Spring	-0.3	-14.6	-8.9
Summer	-0.03	-1.6	-2.5

ny days [12]. This investigation indicates a variation in precipitation field, however since it has been performed over a not very long period, about 50 years, the results cannot be considered definite.

5. OUTLINE OF RECONSTRUCTION OF THE PAST SERIES

In order to extend the performed analysis, we are working to reconstruct secular data series in some Italian stations. In Italy, in fact, many observatories began their activity since last century so that quantitative information is available to examine climatic evolution. In the frame of a research project, supported by CNR, actually a data base of secular series is in development, concerning a very large number of observatories. We have started to analyse data relative to Venezia, Taranto, Foggia, Catania, belonging to the network of the U.C.E.A. (Ufficio Centrale di Ecologia Agraria). Data before 1960, were recorded daily and collected on paper registers. The first step of the work is to transfer temperature and precipitation data from paper to informatic support. It has been already performed for the above stations, in which measurements start from 1901. Data from 1961 are available in the U.C.E.A. database and we have received precipitation records of Taranto and Venezia from 1961 to 1992, until now. Data quality controls have to be done on these data series, in particular the continuity and the homogeneity has to be tested. With reference to the continuity the gaps in the data sets are reported in table 4. Few missing values are found in the records of the stations of Catania, Foggia, Taranto, with a longer break of 4,5 years (from August 1943 to December 1947) in Foggia, corresponding to the second war world, while the stations of Venezia presents a lot of missing periods. The homogeneity is another important element in the study of climatic evolution, because systematic errors in the series can induce to wrong conclusions.

The displacement of the instrument location constitutes the most common reason of inhomogeneity.

Table 4 - Reconstructed climatic series, belonging to the U.C.E.A. network

Station	Precipitation		Temperature	
	Observation period	missing years	Observation period	missing years
Catania	1901-1960	-	1901-1960	1936 - 1937 (Sept.)
Foggia	1901-1960	1907; 1943 (Aug.) - 1947 (Jan); 1956 (Apr.-May); 1957 (Apr., Nov.; Dec.)	1901-1960	1907; 1943 (Aug.) - 1947 (Jan); 1956 (Apr.-May); 1957 (Apr., Nov.; Dec.)
Taranto	1901-1992	1913 (Jan.- Feb.); 1914 (Mar.-Dec.); 1944 (Aug.-Dec.); 1953 (Oct.); 1975	1901-1960	1913 (Jan.; Feb.); 1914 (Mar.-Dec.); 1944 (Aug.-Dec.); 1953 (Oct.)
Venezia	1901-1992	1901-1902; 1904;1911-1914; 1916-1920; 1926-1937; 1945; 1947 (Aug.-Dec.); 1948 (June); 1952 (June-Sept.); 1953 (Aug.-Sept.); 1954 (Mar.)-1960; 1989; 1992	1901-1960	1901-1902; 1904;1911-1914; 1916-1920; 1926-1937; 1945; 1947 (Aug.-Dec.); 1948 (June); 1952 (June-Sept.); 1953 (Aug.-Sept.); 1954 (Mar.)-1960

A preliminary analysis is the reconstruction of the history of the observatories. For the stations of Foggia, Taranto, Venezia, we have examined the archives of the U.C.E.A., in Rome, where historical documents and mails between the head of the observatory and the director of the UCEA are gathered for each station. A displacement of the observation location occurred in Foggia and in Venezia. In the first station all the instruments of the Observatory "Nigri" were moved to the town hall, during the second war world and were then installed in a new observatory from 1956. In the second one, measurements were performed by the Observatory of Seminar until 1954, when they were interrupted due to the sickness and the death of the responsible of the observatory. From 1962 data received by the U.C.E.A. in Rome were collected by the private Observatory "Cavanis", that assured to send also measurements relative to the preceding missing years. This change can explain gaps in the series found in this period.

6. CONCLUSIONS

On the basis of the performed analysis the following conclusions can be drawn:

a) temperature pattern in Italy, examined in the last 100 year period, shows an increasing trend of 0.49 °C/100 years;
b) yearly precipitation amount seem to be reduced in the last 50 years by about 20%
c) the seasonal analysis of rainfall points out that the higher decrease occurred in the cold season, particularly in winter.

These results seem to be in agreement with those at the global scale, that indicate an increase of temperature of 0.5°C/100 years and a decrease of precipitation at the latitudes lower than 50°, in the Northern hemisphere. With reference to rainfall, however, the analysed period in Italy is too short (50 years about) to draw a definite conclusion.

A reconstruction of secular series relative to temperature and precipitation is in progress to obtain further information.

7. REFERENCES

Bradley R.S., Groisman P.Ya., In: Proceedings of the International Conference on "Precipitation Measurements", WMO, Geneva, 1989, pp.168-184.
Colacino M., Purini R., Theor. Appl. Climatol., 37 (1986) 90-96.
Colacino M., Rovelli A., Tellus, 35A (1983) 389-397.
Diaz H.F., Bradley R.S., Eischeid J.K., J. Geoph. Res. 94 (1989) 1195-1210.
Gregory J.M., Jones P.D., Wigley T.M.L., Int. J. Climatol., 11 (1991) 331-345.
Hansen, J., Lebedeff, S., J. Geoph. Res. (D11) 29 (1987) 13345 - 13372.
Jones P.D., Conway D., Int. J. Climatol., 17 (1997) 427-438.
Jones, P.D., Wigley, T.M.L., Wright, P.B., Nature 322 (1986) 430-434.
Nicholson S.E., Monthly Weather Rev. 111 (1983) 1646-1654.
Piervitali E., Colacino M., Conte, M., Il Nuovo Cimento, 21C, 3 (1998) 331-344.
Piervitali E., Colacino M., Conte, M., Theor. Appl. Climatol. 58 (1997) 211-219.
Vinnikov K.Ya., Groismann P.Ya., Lugina, K.M., Jour. of Climate 3 (1990) 662-677.
Wigley T.M.L., Jones P.D., J. Climatol., 7 (1987) 231-246.
Woodley M.R., Int. J. Climatol., 16 (1996) 677-687.

Climate and other Sources of Change in the St. Elias Region

D. SCOTT SLOCOMBE
Geography & Environmental Studies, and Cold Regions Research Centre, Wilfrid Laurier University 75 University Ave. W. Waterloo, ON, CANADA, N2L 3C5

Key words: National parks, St. Elias region, Land use planning, Comanagement, Disturbance, Global change

Abstract: Climate change is hypothesized to have both a greater effect and/or to be more visible in high latitudes and/or elevations. This is significant for both the peoples who live in these regions and scientists seeking evidence about the nature and magnitude of climate change. The Kluane National Park region of southwest Yukon and adjoining parks in Alaska and British Columbia, is one high-latitude, mountainous region well-suited to such studies. Many factors cause change and disturbance in mountainous regions. Thus important questions for research and management are distinguishing and understanding the interaction of climate related changes and changes due to other factors. This paper reviews the literature on the interaction of climate-related and other environmental change. It provides an initial assessment of the causes, nature and magnitude of environmental change in the broader St. Elias region as a basis for distinguishing climate-related changes. Key sources of change include resource management policies and practices, land use change, wildlife population fluctuations, long-range transport of pollutants, forestry and mining, and tourism activities.

1. INTRODUCTION

Global climate change is an increasingly recognized and studied phenomenon. Issues surrounding it include whether it exists, the magnitude and distribution of its effects, and appropriate responses to reduce its extent and effects. While most scientists now agree on the existence of anthropogenic climate change, and that effects at high latitudes are expected to be greater, many questions remain about the extent and distribution of

G. Visconti et al. (eds.), Global Change and Protected Areas, 61–69.

effects. There are also questions about the interaction of climate change with other causes of environmental change. Indeed, one summary of potential climate change impacts on northern resource management highlighted "the need to assess the relative importance of climate in the context of other factors with implications for resource management decisions" [1, p.1]. This paper examines the Greater Kluane region of southwest Yukon, and surrounding regions of Alaska and BC, to provide an initial identification of key questions, priorities, and disturbance interactions.

2. THE ST. ELIAS REGION

The St. Elias region includes well over 100,000 km^2 in southwest Yukon, southeast Alaska, and northwest British Columbia. The area has been home to native peoples for millennia; the Tlingit on the coast, the southern Tutchone in Yukon, and the Ahtna in the Copper River Basin. European exploration and settlement began with the Russians in the Copper River Valley in the late 18th century; but didn't really begin until mineral exploration in the late 19th century, and especially the Klondike gold rush. The Kluane region was not readily accessible until the completion of the Alaska Highway in 1942. Still, today, total regional population is under 10,000 if one excludes the cities of Valdez, Juneau, and Whitehorse on its edges.

The region's ecosystems range from the temperate rainforests of the Gulf of Alaska up through the high ice fields and mountains of the St. Elias ranges at 4000-6000 m.a.s.l., to the boreal spruce forests of interior Yukon and Alaska. In the interior the forests and wetlands are found only in the valley bottoms, rapidly giving way to alpine meadows around 2000 m.a.s.l. The landscape is tectonically and geomorphologically dynamic, subject to extensive periglacial and permafrost processes. Substantial large mammal, bird of prey, and migratory bird populations are found in the region. The watersheds are short in distance, if sometimes large in runoff volume, very silty, and often steep, running out of the mountains. The most famous of the region's rivers are the Alsek and Tatshenshini of rafting fame; others like the White and Copper are famous for their mining and other history.

At the heart of the region are several major national and provincial parks: Kluane National Park and Park Reserve (22,000 km^2), Wrangell-St. Elias National Park and Preserve (49,000 km^2), Tatshenshini-Alsek Provincial Wilderness Park (12,000 km^2), and Glacier Bay National Park and Preserve (13,000 km^2). The Tongass National Forest, the Kluane Game Sanctuary, and Tetlin National Wildlife Refuge add more protected area. While Glacier Bay was first protected in 1925; and Kluane in 1942; most of these protected areas date from the late-1970s or early 1980s. The Tatshenshini-Alsek wasn't established until the early 1990s. Big-game hunting and guiding are long-standing activities, now limited to outside the parks. Mining, particularly placer gold, but also hard rock copper, nickel, and a range of related metals are or could be mined. Tourism activities are growing, from hiking, wilderness B&B's, to kayaking and river rafting. Cruise ships are a major source of activity in coastal towns such as Juneau and Skagway; now with associated bus tours going north on the Haines and Alaska Highways to Kluane, Whitehorse, and into Alaska. Oil development touches the eastern edge of the region in Alaska where the Alyeska pipeline passes Glennallen, and terminates in Valdez. Science has been a major activity in the region since at least the late 19th century: from ethnography to geology, glaciology and geomorphology to high-altitude physiology and wildlife ecology.

Land claims in the Alaska part of the region were settled in the early 1970s; those in Kluane are largely settled in the southeast, but still being negotiated in the northwest; those in BC are unsettled. There are or have been regional planning exercises and/or jurisdictions in the Alaska and Yukon parts of the region. A range of issues and activities from poaching control in the contiguous wilderness parks, to the threat of large-scale mine development, to the potential for coordinated and mutual tourism promotion have demonstrated the need for some level of regional management. See [2,3,4,5,6] for more information.

3. CHANGE AND DISTURBANCE

Natural disturbances in the Kluane region include forest fire, insect pests, geomorphological and periglacial processes, and storm flooding events, and natural animal population cycles. Forest fire intervals are long in Kluane, perhaps 200-300 years, with size varying greatly [7]; while insect infestations occur more regularly, on a scale of one to several decades. The current severe spruce beetle infestation is the largest post-1922 in Kluane region [8]. Geomorphological processes largely affect vegetation establishment and maintenance in dynamic areas, and interact with fire and insects, as well as synergistically with each other [9].

Anthropogenic causes of disturbance include sport and subsistence hunting, wildlife management, trapping, mining, LRTAP, transportation infrastructure and settlement/tourism building and activities. Subsistence hunting pressures have certainly affected population levels, and even brought about large-scale efforts to alter species equilibria through predator control. Mining is largely small-scale and placer, although the possibility of larger-scale developments exists. Building and development has followed the highway infrastructure. This has likely had limited effect although cumulative effects deserve attention, particularly in regard to animal movements. Tourism's impact is in infrastructure and, to some degree, on bear and other large mammal management, and perhaps vegetation in some places.

Policies and processes also have impacts on the region, through assessment and mitigation requirements, changes in harvesting practices and limits, park activities and development, and tourism development plans.

Disturbance impacts are primarily felt in water quality, landuse change, wildlife populations and habitat use, forest and other vegetation change, and culturally on lifestyles.

4. CLIMATE CHANGE AND KLUANE

There are two key dimensions to assessing the effects of climate change in the southwest Yukon. First is the identification of their effects on the biophysical systems of the region; second is their effect on human activities and cultures in their region. Each will be discussed briefly in turn.

There has been little work on climate change specific to the Kluane region, although a loose coalition of researchers is beginning to address the topic. Some probabilities can be identified based on expert assessments of likely impacts on a range of biophysical systems in western mountains and the southwest Yukon generally. A major complication in the Kluane region is the complexity of the topography and microclimates. As in other mountainous regions the vegetational impacts of climate warming will likely be felt altitudinally. Small, sharply defined habitats; short, highly variable rivers and streams; and complex patterns of ecological and physical organization, e.g. in permafrost and habitat distribution all complicate prediction of climate change effects. At a middle scale assessment is weakened by the difficulties of applying GCM results at regional [10].

First, there are indicative general statements about climate change and the north. For example, we might expect longer growing seasons [11]; increased precipitation; decreased permafrost, glaciers, and snow cover; and changed forest stability including fire and pest cycles, and species composition [1]. The Mackenzie Basin Impact Study reached similar conclusions of impacts related to permafrost thaw, increased landslides, lower annual minimum lake and river levels, more forest fires, and lower softwood yields [12]. Then there are assessments of climate change impacts on mountain environments. Common predictions include short-term increase in run-off due to glacier recession, but longer-term decreases in summer river flows due to a rise in snowline [13].

It is a major conclusion of recent GCTE work that "wherever human activities have a direct, significant impact on water and nutrient cycles and on disturbance regimes, this impact will override any direct CO2 effects on ecosystem functioning" [14, p. 4]. They also note the importance of unfragmented ecosystems to permit plant migration [15], the potential for increased alien invasions, greater disturbance and dieback as environmental conditions change, and shifts to more early-successional state systems, and the significance of shifts toward agricultural/rangeland uses. All these will likely be seen in Kluane, and are consistent with the results of the more system and region-specific studies below.

Somewhat more specifically, erosional processes in mountainous regions have been assessed, including specific comments on the Coastal/St. Elias mountains. Substantial impacts of summer warming are expected, particularly on glacial dynamism, treeline, permafrost, and on processes on northeast facing slopes [16]. A review of the Canadian Rockies reached similar conclusions for similar distributions of events such as debris flows, snow avalanches, floods, and rockfalls, but potentially increased magnitude and frequency, depending on changes in triggering events such as storms, snowfall, freeze-thaw cycles, and other human activities such as forest clearing [17]. Climate change effects on geomorphological processes and their impact on humans have also been examined. Relevant impacts include increased slide, debris flow, flood, and avalanche dangers; reduced long-term water storage capacity, a rise in timberline, and changes in scenery in mountain parks [18, 19].

The first more region-specific study looked at permafrost and tectonics and climate change at a regional scale in Yukon, based on an assessment of GCMs and the actual current physical environment and constraints on climate. Current models do not represent topography and its effects well, but rather speculatively it is likely that precipitation would increase, particularly outside of winter; warmer temperatures and greater snow cover in winter, and permafrost possibly disappear, if slowly [20].

Other general forecasts for the mountain regions of the St. Elias include earlier spring freshet, greater winter and spring runoff, and slight summer flow increase [21]; and continuation of the current glacier advances (increased precipitation offsetting greater summer melt) [22].

Biologically it is likely that species at the northern limits expand, those at southern limits retract, wetlands will be especially hard to protect [23]; there will be species composition changes above and below treeline; black spruce may well decrease, white spruce and lodgepole pine increase, with possibility of substantial change in southwest Yukon to a cold dry steppe without modern analogue [24, 45]. Changes in disturbance regimes could well affect the current relative homogeneity of the forest and landscape that is a function of the generally slow disturbance cycles [9]. Homogeneity also increases with elevation and climate change could particularly affect alpine communities. On fisheries, trout, char and grayling could do better in the north as long as periods and centrarchids are not introduced [26].

In summary, the most likely and significant impacts appear to be on the parameters of the biophysical system:

- Temperature and growing season changes (increase)
- Precipitation and snow cover changes (increase)
- Changes in extent and depth of permafrost (decrease)
- Changes in fire and pest frequencies and magnitudes (increase)
- Changes in water flows and levels (decrease)

None of these is simple or certain alone; their combined effects are highly uncertain at this point. And there are possibilities of a more catastrophic nature such as an Alsek Glacier surge, or summer storms and widespread debris flows as have happened in the past.

Effects of global change on human activities are even harder to foresee, depending both on climate and biophysical system changes. Only agriculture has seen specific study to date. One study exists of soil changes and impacts on agriculture due to climate change [27, 28]. A temperature increase could remove the risk of mid-August killing frost; although available water is also a major limitation [29]. A very significant question will be the impacts of change on wildlife, which in turn could substantially affect subsistence and sport hunting. The main avenues for impact would appear to be:

- Increased agricultural activity feasibility
- Changes in wildlife populations
- Changes in forest cover and productivity
- Increased infrastructure/commercial/tourism development

It is critical however to bear in mind Walker & Steffen's point about the effects of other human activities on systems also affected by climate change. Even in Kluane this is surely true: most notably now in the context of responses to spruce beetle infestation, forestry activities, tourism development, mining development, and wildlife management.

5. CONCLUSIONS

A range of possible direct and indirect effects of climate change can be identified in the St. Elias. Given their uncertainty, yet possible significance, it is important to pay attention to larger issues and planning and management contexts and frameworks.

A first crosscutting issue is surely the relationship between the major subregions of the St. Elias [cf. 2]. Climate change could conceivably affect interactions and connections in terms of, for example, wildlife movements and corridors, species distributions, and hydrology. A related issue is the differences in potential effects between the interior and coastal parts of the greater St. Elias. This paper has focused on the interior; the other side of the mountains could see quite different changes, such as greater streamflow and storm related changes, and sea level changes? In a related vein it is necessary to explore how the details of physiography and ecology in the region will affect the expression of global change; and to seek a better understanding of vegetation and wildlife population responses to physical changes.

The strong protected areas network in the region also has implications. Park managers have been concerned with climate change for some time. Similar potential issues have been identified for Great Smoky and Glacier N.P.s in Montana's Rockies: e.g. microhabitat, ecotone, biodiversity, stream chemistry and flow changes. Responses include natural and human resource sensitivity monitoring, developing regional landscape analyses, and a framework for monitoring at several scales, and strategic plans for species conservation and cultural resources [30]. Monitoring will be critical, perhaps facilitated by an Ecological Integrity Statement now being developed for Kluane N.P.

Many of Cohen's [12] five themes of response also apply to Kluane: interjurisdictional water management, ecosystem sustainability, economic development, infrastructure maintenance, and sustainability of native lifestyles. Key lessons include coordinating the scientists, communicating with stakeholders and integrating TEK and scientific research on climate change. Much might be done here building on Julie Cruikshank's early work [31, 32] which touched on disturbance and landuse. The region has undergone complex social and ecological transformations in the past, and undoubtedly will again [3]. A fuller understanding of environmental history and the relationships of environment and societies is essential in the region.

Equally, it is important that goals and concerns at a regional scale are identified and integrated into planning and assessment activities [e.g. 33] and that stronger links with sustainability planning and assessments are fostered [34]. This would contribute to assessing the cumulative effects of current human activities in relation to the magnitude of possible climate change-induced effects.

The bottom line is to be found in the question of what should be done. Perhaps the key is not to let climate change dominate other management concerns and processes. What is needed is to begin focused research and monitoring, communication, and integration of information around climate change and other activities, within existing consultation and decision-making vehicles. This is especially true in the context of the new resource and land management institutions developing in Yukon, from the Kluane Park

Management Board to the regional Alsek Renewable Resources Council and the territory-wide Development Assessment Process and Fish and Wildlife Management Board.

6. REFERENCES

Barry R.G., Mountain Research and Development 10(1990) 161-70.

Beamish R.J., M. Henderson, & H.A. Regier. In: E. Taylor, and B. Taylor, (Eds.), Responding to Global Climate Change in British Columbia and Yukon. Environment Canada, BC & Yukon Region, Vancouver, 1997.

Brugman M.M., P. Raistrick, & A. Pietroniro. Taylor, B. In: E. Taylor, and B. Taylor, (Eds.), Responding to Global Climate Change in British Columbia and Yukon. Environment Canada, BC & Yukon Region, Vancouver, 1997.

Burn C.R., Can. J. Earth Sci. 31(1994) 182-91.

Cohen S., D. Demeritt, J. Robinson, and D. Rothman, Global Environmental Change 8(1998) 341-71.

Cohen S.J., Arctic 50(1997) 293-307.

Coulson, H. In: E. Taylor, and B. Taylor, (Eds.), Responding to Global Climate Change in British Columbia and Yukon. Environment Canada, BC & Yukon Region, Vancouver, 1997.

Cruikshank J., Arctic Anthropology 18(1981) 67-93.

Cruikshank J.M., Through the Eyes of Strangers: A Preliminary Survey of Land Use History in the Yukon During the late 19th Century, Report to the Yukon Territorial Government and the Yukon Archives, Whitehorse, 1974.

Danby R., Regional Ecology of the St. Elias Mountain Parks: A Synthesis with Management Implications. MES thesis, Geography, Wilfrid Laurier University, 1999.

Evans S.G. and J.J. Clague, In: E. Taylor, and B. Taylor, (Eds.), Responding to Global Climate Change in British Columbia and Yukon. Environment Canada, BC & Yukon Region, Vancouver, 1997.

Ferris R., History of Important Forest Pests in the Yukon Territory, 1952-1990. FIDS Report 91-13. Canadian Forest Service, Pacific and Yukon Region, Victoria, BC, 1991.

Goos T. and G. Wall, Impacts of Climate Change on Resource Management of the North: Symposium Summary. Climate Change Digest 94-02, Environment Canada, Ottawa, 1994.

Harding L.E. and E. McCullum. In: E. Taylor, and B. Taylor, (Eds.), Responding to Global Climate Change in British Columbia and Yukon. Environment Canada, BC & Yukon Region, Vancouver, 1997.

Hawkes B.C., In: Wein, R.W., R.R. Riewe, and I.R. Methven, (Eds.), Resources and Dynamics of the Boreal Zone. Association of Canadian Universities for Northern Studies, Ottawa, 1983.

Hebda R.J., In: E. Taylor, and B. Taylor, (Eds.), Responding to Global Climate Change in British Columbia and Yukon. Environment Canada, BC & Yukon Region, Vancouver, 1997.

Krannitz P.G. and S. Kesting. In: E. Taylor, and B. Taylor, (Eds.), Responding to Global Climate Change in British Columbia and Yukon. Environment Canada, BC & Yukon Region, Vancouver, 1997.

Luckman B.H., Mountain Research and Development 10(1990) 183-95.

Mills P.F., In: C.A. Scott Smith, (Ed.), Proc's 1st Circumpolar Agriculture Conference, Agriculture Canada, Research Branch, Centre for Land and Biological Resources Research, Ottawa, 1994, pp. 195-204.

Peine J.D. and C.J. Martinka, In: J.C. Pernetta, et al., (Eds.), Impacts of Climate Change on Ecosystems and Species: Implications for Protected Areas. IUCN, Gland, 1994, pp. 55-75.

Pitelka L.F. and Plant Migration Working Group, Amer. Scientist 85(1997) 464-73.

Ryder J.M., Geomorphological Processes in the Alpine Area of Canada: The Effects of Climate Change and their impacts on Human Activities. Geological Survey of Canada Bull. 524, Ottawa,1998.

Slaymaker O., Mountain Research and Development 10(1990) 171-82.

Slocombe D.S., and R. Danby, Toward Collaborative Bioregional Management in the St. Elias Region, Yukon, Alaska, B.C. Ecostewardship Session, American Association of Geographers, Boston, Mass, March 26, 1998

Slocombe D.S., Environmental Review, 13(1989) 1-13.

Slocombe D.S., Mountain Research and Development, 12(1992) 87-96.

Smith C.A.S., In: C.A.S. Smith, ed., Proc's of the 1st Circumpolar Agriculture Conference, Whitehorse, Yukon, Canada, September 1992. Agriculture Canada, Research Branch, Centre for Land and Biological Resources Research.Ottawa, 1994, pp. 217-21.

Solomon A.M., In: J.C. Pernetta, et al., (Eds.), Impacts of Climate Change on Ecosystems and Species: Implications for Protected Areas. IUCN, Gland, 1994, pp. 1-12.

Taylor B., In: E. Taylor, and B. Taylor, (Eds.), Responding to Global Climate Change in British Columbia and Yukon. Environment Canada, BC & Yukon Region, Vancouver, 1997.

Theberge J.B., (Ed.), Kluane: Pinnacle of the Yukon. Doubleday Canada, Toronto, 1980.

Walker B., and W. Steffen, Conservation Ecology [online] 1 (1997): 2. URL: http://www.consecol.org/vol1/iss2/art2

Wurtele B. and D.S. Slocombe, In: P. Jonker, et al., (Eds.), Caring for Home Place: Protected Areas and Landscape Ecology, Extension Press, Univ. of Saskatchewan, Saskatoon & Cdn Plains Research Center, Univ. of Regina, 1997, pp. 227-39

Yin Y. and S.J. Cohen, Global Environmental Change 4(1994) 246-60.

Zebarth B., Caprio, J. K. Broersma, P. Mills and S. Smith. In: E. Taylor, and B. Taylor, (Eds.), Responding to Global Climate Change in British Columbia and Yukon. Environment Canada, BC & Yukon Region, Vancouver, 1997.

Permafrost and Climate in Europe. Climate Change, Mountain Permafrost Degradation and Geotechnical Hazard

CHARLES HARRIS[1] AND DANIEL VONDER MUHLL[2]
[1]*Department of Earth Sciences, Cardiff University, P.O. Box 914, Cardiff CF1 3YE, UK Tel: +44 1222 874336, Fax +44 1222 874326.*
[2]*Laboratory of Hydraulics, Hydrology and Glaciology (VAW), Swiss Federal Institute of Technology (ETH); Gloriastr. 37/39, CH-8092 Zürich, Switzerland. Tel: +41 1 632 41 13.*

Key words: Permafrost, Climate Changes Global Warming.

Abstract: Mountain permafrost is highly vulnerable to present and future climate warming, since ground temperatures are generally only a few degrees below zero. The European Union PACE Project (Permafrost and Climate in Europe), which includes partners from Norway, Sweden, U.K., Germany, Switzerland, Italy and Spain, was established to develop new methods of assessing the potential impact of warming climate on mountain permafrost slopes in Europe. The research programme is first described and then progress in selected topics is summarised

1. INTRODUCTION

The PACE project commenced in December 1997 and in this paper we report preliminary results from the first 18 months of the programme. The project objectives are as follows:

a) To establish a framework for monitoring global climate change by detecting changes in permafrost ground temperatures in the mountains of Europe;
b) To develop methods of mapping and modelling the distribution of thermally- sensitive mountain permafrost, and predicting climatically-induced changes in this distribution;

G. Visconti et al. (eds.), Global Change and Protected Areas, 71–82.

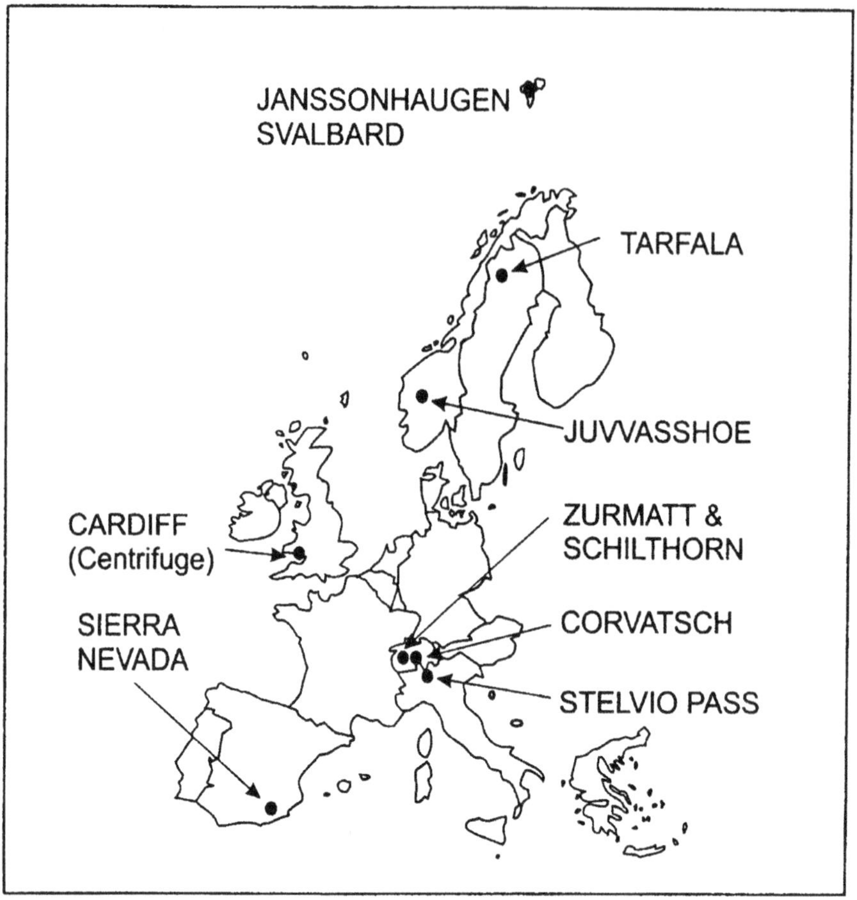

Figure 1: Location of PACE project test sites, plus centrifuge laboratory

c) To provide new, process-based methods for assessing environmental and geotechnical hazards associated with mountain permafrost degradation.

The scientific rational and research structure of the project will first be outlined and then results from three aspects of the research (permafrost thermal monitoring, geophysical mapping and geotechnical centrifuge modelling) will be reviewed.

2. RESEARCH STRATEGY

European mountains are particularly sensitive to climate warming since they are characterised by the presence of permafrost (permanently frozen ground) at higher altitudes [1].

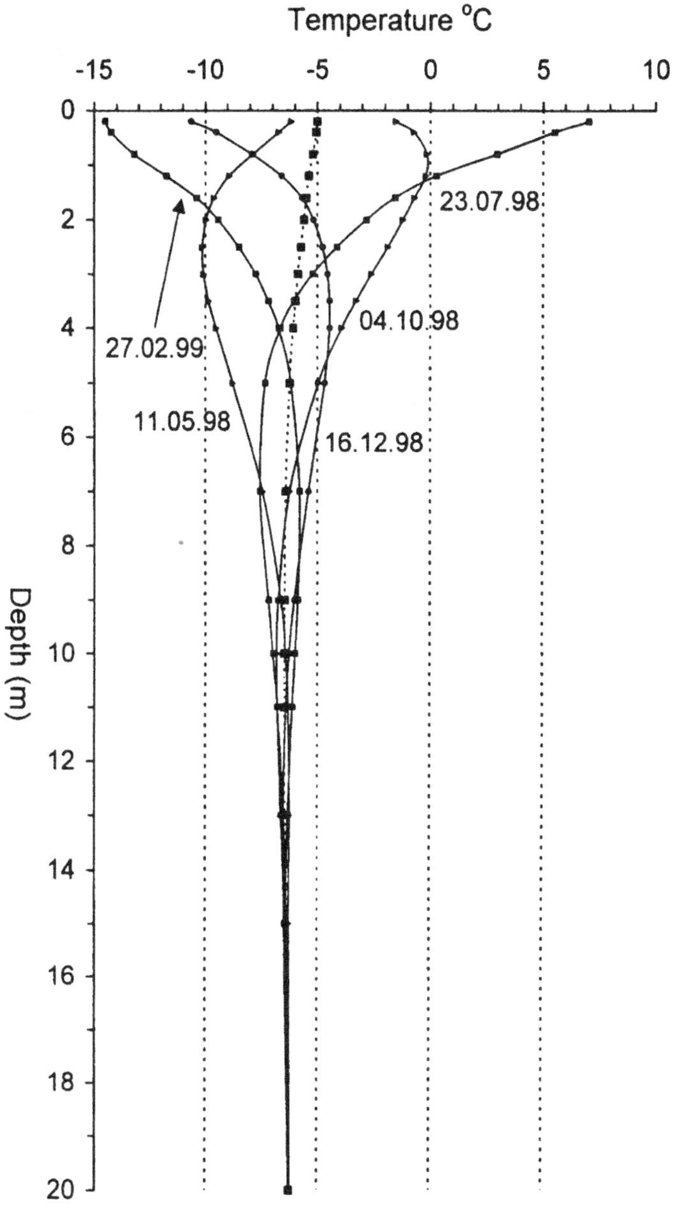

Figure 2: Temperature profiles in the upper 20 m of the Janssonhaugen borehole, May 1998 to June 1999. Dotted line shows the mean annual temperature profile.

Permafrost temperatures are generally only a few degrees Celsius below freezing, so that even slight warming may lead to significant permafrost thaw. Evidence from Switzerland suggests that warming of mountain permafrost has taken place over the past decade [2].

The presence of frozen ground is a vital factor in the stability of mountain slopes, since, in most cases, thawing leads to a rapid loss of strength. The combination of ground temperatures only slightly below zero, high ice contents and steep gradients, makes mountain slopes vulnerable to the de-stabilising effects of permafrost degradation.

The PACE project, therefore, not only seeks to monitor future changes in permafrost temperatures, but also to predict resulting changes in permafrost distribution, and the environmental and geotechnical impact of these changes in terms of mountain slope instability.

The major objectives of the PACE project will be achieved through six interrelated work packages based on mountain field sites in Scandinavia, the Swiss and Italian Alps and Spain. Field monitoring sites have been established in Svalbard and the Jotunheimen in Norway, Tarfala in Sweden, at Piz Corvatsch, Schilthorn and the Zermatt area of Switzerland, the Stelvio Pass and Foscagno area of Italy, and on Valetta Peak in the Spanish Sierra Nevada (Figure 1).

In Work Package 1 a series of new boreholes will be drilled in a transect from Svalbard in the north to Spain (Sierra Nevada) in the south (Figure 1). Boreholes will be instrumented for automatic logging of permafrost ground temperatures. Work Package 2 is testing new geophysical techniques to provide reliable and efficient methods for mapping and characterisation of mountain permafrost. Work Package 3 is in process of compiling GIS-format maps of permafrost distribution, ground and environmental conditions, and current processes. Vegetation mapping as an indicator both of permafrost and near-surface mass movements is also included. In Work Package 4, new approaches to the numerical modelling of mountain permafrost distribution are being developed, based on microclimatological data collected at a series of field stations. Advanced numerical modelling will combine energy flux between atmosphere active layer and permafrost with digital elevation models to provide improved prediction of permafrost distribution patterns in different mountain regions and for various climatic scenarios. Work Package 5 uses scaled centrifuge modelling of thawing slopes in which detailed process monitoring is possible.

Thresholds for slope instability will be determined and process/intensity relations explored. Finally, Work Package 6 will integrate the previous five work packages in the context of geotechnical and environmental hazard prediction, to provide new practical guidelines for risk assessment in the mountains of Europe.

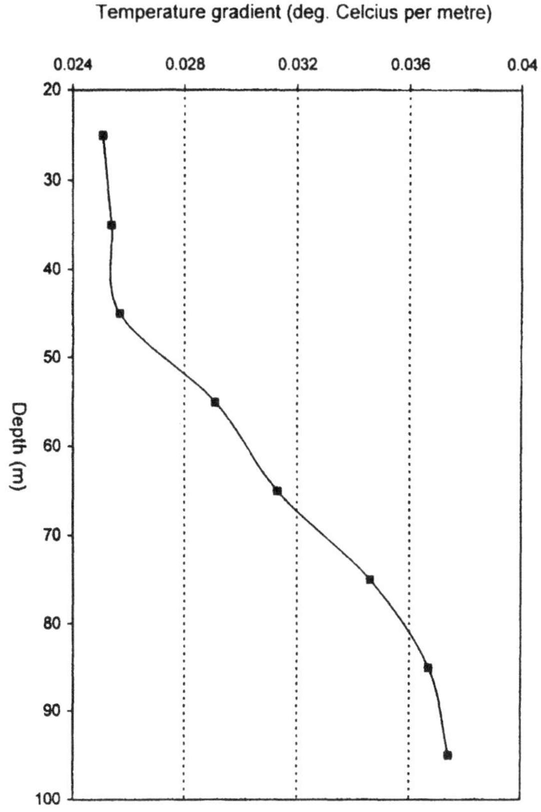

Figure 3: Temperature gradient in the Janssonhaugen borehole below the depth of seasonal temperature variation.

3. PRELIMINARY RESULTS FROM SELECTED WORK PACKAGES

3.1 Permafrost Drilling

Deep (at least 100 m) permafrost boreholes were drilled and instrumented at Janssonhaugen, Svalbard, Norway and in the Stelvio Pass, Italian Alps in spring 1998. A shallow (14 m) borehole was also installed on Schilthorn in Switzerland in October 1998. A third deep borehole was drilled at Juvvasshoe, Jotunheirnen, Norway in August 1999, where, as in Svalbard, an additional 20m shallow borehole has been installed. Further shallow boreholes will be drilled on the Valetta Peak, Sierra

Nevada, Spain in September 1999. Remaining deep boreholes at Tarfala (Sweden) and in Switzerland will be drilled in 2000.

CASE STUDY: JANSSONHAUGEN 78"12'N, 161128'E, 275 M A.S.L.

The Janssonhaugen drill site is located on a low sandstone hill in Adventdalen, some 15 km cast of Longyearbayen. A detailed analysis of the first year of ground temperature recording is given by Isaksen *et al.* [3] and a summary is presented here. Permafrost is continuous in the arctic islands of Svalbard, with thickness varying from 200 to 400m (41. The nearest meteorological station is at Longyearbyen (28 m a.s.l.) where the mean annual air temperature was -6.1 'C during the period 1976-1998.

Only the months of June, July and August have average temperatures above zero. It is estimated that at the altitude of Janssonhaugen, mean annual air temperature is about -8.0 °C.

The borchole reached a depth of 102 m and a plastic lining tube was installed into which a 100 m long string of type YSI 44006 thermistors was lowered. The string included 30 thermistors from 0.2 m to 100 m. Thermistor installation is designed to allow periodic removal and re-calibration. A Campbell CR2 1 X logger records temperature in the uppermost 5m every 6 hours, and at greater depths every 24 hours. Results from the first year of monitoring are presented in Figures 2 and 3. The maximum depth of seasonal ground temperature fluctuation was approximately 17 m, and active layer depth in the first summer was -1.5 m. Extrapolation of the mean thermal gradient indicates an estimated permafrost thickness of roughly 220 m.

At Janssonhaugen, the thermal gradient below 17m is approximately 0.024°C per metre, but increases below 50 m to around 0.038°C per metre (Fig. 3), suggesting recent warming of the ground surface. Analysis of the thermal profile using an inversion procedure suggests that warming began about 60-80 years ago, with a maximum in the 1960s [31. The magnitude of surface warming in this period was 1.5° to 2.5°C. The mean surface temperature at the borehole in the first year of measurement was -5.0°C, but extrapolation of the thermal profile from below the depth of seasonal temperature fluctuations suggests the "equilibrium surface temperature" is -6.8°C.

Clearly the establishment of long-term permafrost temperature monitoring stations across the mountains of Europe offers not only the potential for early warning the impact of future climate change, but also the prospect of reconstructing recent trends in atmospheric temperatures from observed geothermal gradients.

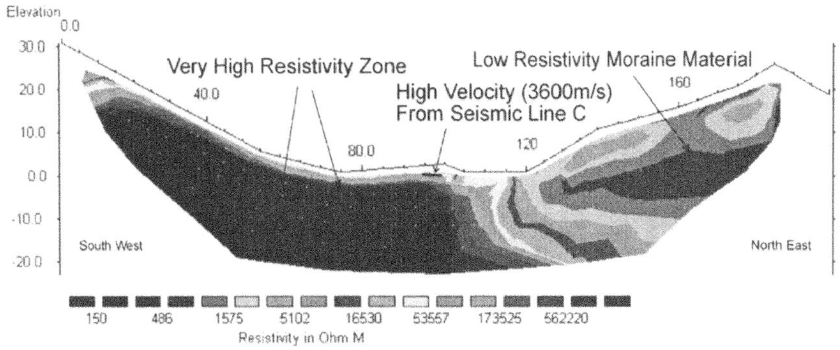

Figure 4: Resistivity and seismic survey, Line D, Valeta Peak, Sierra Nevada, Spain.

3.2 Geophysical Survey

Refraction seismic and DC resistivity soundings are the most established methods of mapping permafrost. In addition, at sites with ice-supersaturated sediments, gravimetry has been shown to be effective in the few cases where this method has been applied [5].

Methodological improvement and application of methods rarely used in the difficult terrain of mountain permafrost is the real challenge of Work Package 2. Within these categories, ground penetrating radar (GPR) in winter and in summer, the method of spontaneous potential SP, two-dimensional resistivity imaging (tomography), two-dimensional refraction seismic, the EM-31 measurements and radiometry have been tested and continue to be developed [6].

The principal aim has been to assess permafrost distribution and character within potential drill sites in the Alps and the Sierra Nevada. Surveys have been undertaken at the Stelvio Pass (Italy), Schilthom and the Zermatt areas (Switzerland), in the Sierra Nevada (Spain), in Trafala (Sweden) as well as in Jotunheimen and Svalbard (Norway).

In Jotunheimen, the transition from permafrost to no-permafrost was detected by applying EM-3 1, DC resistivity tomography, refraction seismic and BTS measurements.

At the Sierra Nevada, the southern most site, identification of a drill site within frozen ground was the initial priority.

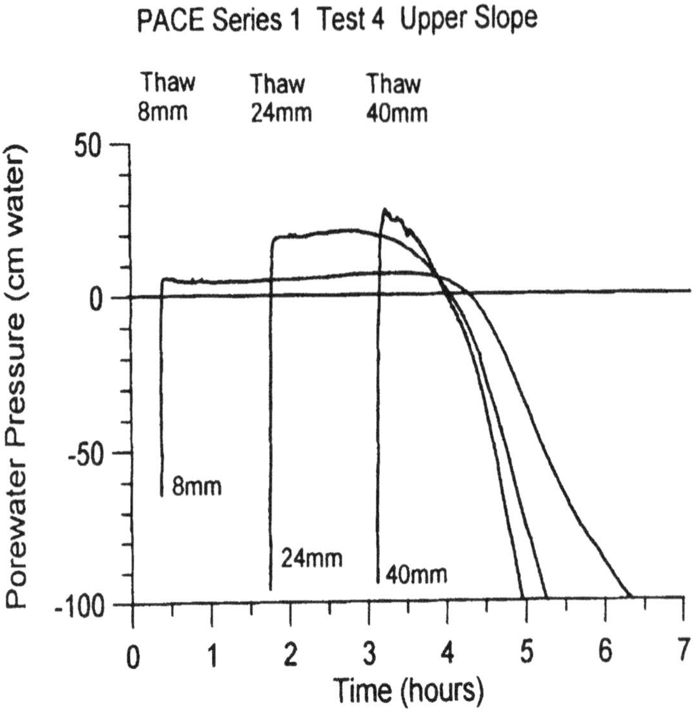

Figure 5: Porewater pressures during thaw cycle 4, Test Series 1 (gelifluction).

Results of the survey at this site illustrate well the effectiveness of geophysical techniques in detecting permafrost. High resolution seismic and two-dimensional resistivity surveys were undertaken in an attempt to prove the presence of permafrost on the Veleta Peak (3394m), which is the second highest in the Sierra Nevada. The survey area consists of mica schist bedrock with a mantle of weathering products and moraine.

An array of regularly-spaced electrodes along a survey line is deployed for resistivity tomography survey. Resistivity data are then recorded via complex combinations of current and potential electrode pairs, to build up a pseudo cross-section through the underlying soil and rock of apparent resistivity.

The sub-surface resistivity model is then derived from iterative finite-difference forward calculation. Seismic refraction surveying involves the observation of seismic waves that has been refracted at a geological boundary.

Table 1. Principles adopted in the interpretation of resistivity and seismic data.

Resistivity Value of Layer	Seismic Value of Layer	Likely Nature of Layer
HIGH	LOW	Coarse grained unfrozen material that is deficient in clay minerals
LOW to HIGH	HIGH	Bedrock
VERY HIGH	HIGH	Permafrost or shallow dry Bedrock deficient in clay

Shots are deployed at the surface and recordings made via a linear array of sensors (geophones) in contact with the ground surface. Interpretation of the resistivity and seismic results in the context of permafrost detection is based on the following principles.

Five resistivity and four seismic profiles were completed during assessment of the potential Sierra Nevada drill site. One such line is illustrated in Figure 4, surveyed close to the headwall of the north-facing cirque below the Veleta Peak. In Figure 4 the section is plotted as depth below ground level incorporating surface geometry calculated from topographical maps. At very shallow depth «2m) a high seismic velocity layer (360Om/s) is identified which also has a very high resistivity (>50,000 ohm.m). Thus, applying the principles outlined in Table 1, the observed zone is most likely to consist of frozen sediments (permafrost). This interpretation is reinforced by data from adjacent survey lines that show local bedrock resistivities to be much lower, and the bedrock to lie at a greater depth than the observed high resistivity/high seismic velocity zone in line C.

This example of results from the Sierra Nevada geophysical survey illustrates well the effectiveness of these techniques in detecting ice-rich frozen ground.

3.3 Centrifuge Modelling of Thaw -Induced Instability

Mass movement processes associated with thawing mountain permafrost are a major geotechnical hazard in the high mountains of Europe. Such processes include slow soil movements (solifluction) and rapid failures such as shallow landslides, mudflows, debris flows and rock falls). A programme of tests currently underway at the Cardiff University Geotechnical Centrifuge Centre aim to model processes of thaw-related instability and identify trigger levels and movement mechanisms. This research provides a new approach to mountain permafrost hazard assessment.

Stress/strain behaviour of granular soils is stress level dependent and accurate scale modelling therefore requires both similitude between material properties in prototype and model and the correct stress distribution within the model. if a model is constructed at I/N scale using soil from the prototype (similitude in soil properties), and tested at N gravities in the centrifuge, stress similitude in model and prototype may be demonstrated as follows:

For prototype: $$\sigma_p = h \cdot \rho \cdot g$$

For model: $$\sigma_m = (h / N)\rho \cdot (N \cdot g)$$

Therefore $$\sigma_p = \sigma_m$$

Where h = soil thickness, ρ = soil density, g = gravity and σ = stress.

Specific aims of the first series of centrifuge tests were to monitor mass movement rates and mechanisms and to measure changes in porewater pressure through the transition from slow gelifluction to rapid mudflow. Model slope gradients of 12', 18' and 24' were selected in successive test series, maintaining all other parameters constant. Models were constructed at 1110th scale and tested at 10 gravities [7]. A 7 cm thick slope (scaling to 70 cm) was formed above a basal sand drainage unit. Six miniature Druck pore pressure transducers were installed together with ten thermocouples in two vertical strings. Plastic markers were placed on the soil surface to allow video recording and measurement of movement during thaw. Vertical columns of plastic beads were inserted into the soil, and excavated following each test series, to reveal profiles of accumulated soil displacements (Fig. 6). The soil was saturated and consolidated prior to freezing from the surface downwards. The amount of frost heaving was recorded. The 12' model underwent four freeze-thaw cycles, with each thawing phase taking place in the centrifuge at 10g. The 1 8' model was subjected to two cycles and the 24' model one cycle.

Pore pressures in the thawing soils during each test typically rose following thaw, but fell slowly as the soil subsequently drained (Fig. 5). Surface displacements increased significantly in each test series, with classical slow gelifluction recorded on the 12' model, rapid mudflow on the 24' model, and a transition between the two on the 18' model.

Excavated soil displacement profiles following four cycles of freezing and thawing (Test 1, gelifluction), two cycles of freezing and thawing (Test 2, transition) and one cycle (Test 3, mudflow) revealed soil shear strain associated with these processes (Fig. 6).

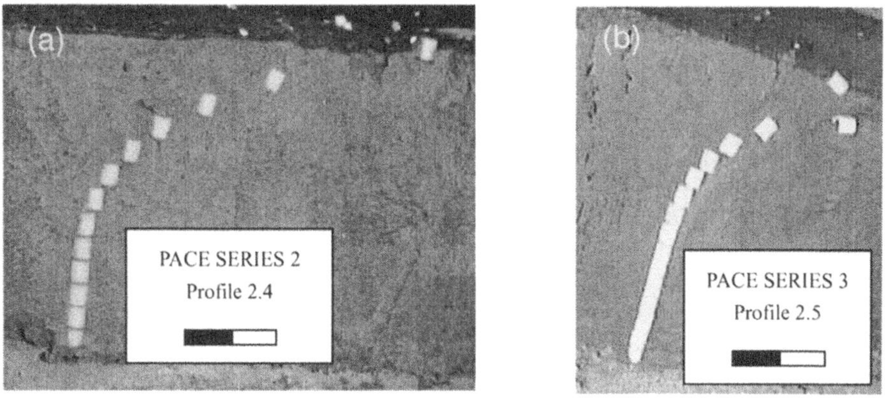

Figure 6: Profiles of soil movements (a) following four freeze/thaw cycles, Test Series 1 (12' slope gradient) and (b) following two freeze/thaw cycles, Test Series 2 (18' slope gradient).

Slope stability analysis based on effective stresses and the cycle (Test 3, mudflow) revealed soil shear strain associated with these processes (Fig. 6). Slope stability analysis based on effective stresses and the thaw consolidation ratio will be supported by analysis of displacement rates, displacement profiles and porewater pressures to allow styles of shear strain to be determined, and mechanisms of failure to be better understood.

4. CONCLUSION

In this paper three examples of work undertaken during the first year of the PACE Project are presented. Data analysis is as yet preliminary, but already significant progress has been made. The final mountain slope hazard assessment recommendations will be based on integration of the diverse scientific research undertaken in each work package. In addition, a permafrost monitoring network will be established, providing a long-term early warning of the impact of future climate warming to permanently frozen slopes of the mountains of Europe.

5. ACKNOWLEDGEMENTS

This research was supported by the "Environment and Climate Programme" under contract ENV4-CT97-0492 and the Swiss Government (97.0054).

6. REFERENCES

Harris, C, Rea, B and Davies, M.C.R (in press). Annals of Glaciology.

Isaksen, K, Vonder Muhll, D., Gubler, H, Kohl, T, Sollid, J.L. (in press) Annals of Glaciology.

King, L., Gorbunov, A.P. and Evin, M. Permafrost and Periglacial Processes. 3 (1992) 73-81

Liestol, 0. Frost i Jord, 21 (1980) 23-28.

Vonder Muhll, D.S., Hauck, C. and Lehmann, F. (in press) Annals of Glaciology.

Vonder Muhll, D.S., Stucki, Th. and Haeberli, W. 7th International Conference on Permafrost, (1998) Yellowknife, Canada, pp. 1089-1095.

Vonder MW, D.S. and Klingelé. E.E. (1994): Gravimetrical investigation of ice rich permafrost within the rock glacier Murtèl-Corvatsch. Permafrost and Periglacial Processes, 5(1). 13-24.

Thermal Variations of Mountain Permafrost: an Example of Measurements Since 1987 in the Swiss Alps.

DANIEL VONDER MÜHLL
Laboratory of Hydraulics, Hydrology and Glaciology (VAW), Swiss Federal Institute of Technology (ETH)

Key words: Permafrost, thermal state, temperature, monitoring, rock glacier, Swiss Alps

Abstract: Alpine permafrost is particularly sensitive to climate change, since it's temperature is often close to the melting point of ice. In summer 1987, several hundred debris flows caused considerable damage and several victims in the Swiss Alps. Analysis showed that one out of three debris flows started at the lower boundary of mountain permafrost. A 58m deep borehole through creeping permafrost was drilled in 1987 near Piz Corvatsch (Upper Engadine, Swiss Alps). Temperatures have been measured regularly since then. Comparisons of two permafrost boreholes some 20km apart, where temperatures were measured once a year, indicated at least the regional character of the signal. Between 1987 and 1994, the uppermost 25m warmed rapidly. Surface temperature is estimated to have increased from -3.3°C (1988) to -2.3°C (1994), thereby probably exceeding previous peak temperatures during the 20th century. In the two-year period from 1994 to 1996, when winter snowfall was low, intensive cooling of the ground occurred, the temperatures reaching values similar to those in 1987. Since 1996, permafrost temperatures have once again been raising, followed by a cooling last winter. The variability of the observed permafrost temperatures is caused by several processes, including: (1) a reduced period of negative temperatures within the active layer due to long-lasting zero-curtains in autumn; (2) global radiation and air temperature changes influencing ground temperatures mainly in summer; and (3) variations in the duration of winter snow-cover. If the observed warming trend in alpine mountain permafrost temperatures continues into the foreseeable future, widespread permafrost degradation is likely, with potentially serious consequences with regard to mountain slope instability.

G. Visconti et al. (eds.), Global Change and Protected Areas, 83–95.

1. INTRODUCTION

The distribution of permafrost in low-latitude mountain areas is quite different from circumpolar regions: It is mainly controlled by radiation and altitude ([1], [2]) and, hence, there are important differences within distances of some hundred metres only. The most typical and reliable indication for mountain permafrost are so-called active rock glaciers, a geomorphologic feature which is formed due to creep of frozen debris-ice mixture. Since permafrost is defined by temperature, the investigation of the rock-glacier thermal regime is fundamental.

In 1987, a 60m deep drilling through the active Murtèl-Corvatsch rock glacier (2670m a.s.l.) created the opportunity to investigate the thermal regime in a creeping permafrost body ([3], [4]). The cores, borehole logging, instruments for long-term monitoring (borehole deformation, temperature) and a number of geophysical surveys contributed to a better understanding of the internal structure and ongoing processes. Moreover, the probable evolution and development of an active rock glacier can be reconstructed ([5]). Temperatures were measured, in principle, twice every month. Since 1993, a logger stores one value every day.

In 1990, two permafrost drillings were completed at Pontresina-Schafberg (around 2740m a.s.l.) within an avalanche protection project ([6]). Combined with Murtèl-Corvatsch, this allows the comparison of two sets of drill sites, which are at a distance of some 20km apart. Because of the difficult access, only one temperature reading per year was originally foreseen at Pontresina-Schafberg.

Nevertheless, between December 1991 and September 1994 a datalogger furnished additional daily temperature data for every thermistor in each borehole. Mainly due to lightning problems the logging was removed afterwards. There are several papers reporting borehole temperatures, especially in circumpolar permafrost (e.g. [7], [8], [9]). However, articles about time series of permafrost temperatures are quite rare ([10], [11], [12]). This paper summarises temperature measurements in permafrost of the rock glacier drilling at Murtèl-Corvatsch between 1987 and 1999. Analysis of the most important effects and comparison to the Pontresina-Schafberg drill site are also presented.

2. TEMPERATURES WITH DEPTH

The thermal regimes of the three borehole sites are quite different (Figure 1 and Table 1), although elevation, surface conditions and lithology are similar.

Table 1: Overview of some temperature key data from three boreholes in Alpine permafrost.

	Murtèl-Corvatsch 2/1987	Schafberg 1/1990	Schafberg 2/1990
Elevation [m a.s.l.]:	2671	2755	2733
Surface temperature [°C]:	-2.5 to -3.0	~ -1.5	~ -0.5
10m temperature, 1992 [°C]:	-2.15	-1.5	-0.6
10m temperature, 1993 [°C]:	-1.73	-1.0	-0.4
10m temperature, 1994 [°C]:	-1.34	-0.7	-0.2
10m temperature, 1995 [°C]:	-1.98	-1.0	
10m temperature, 1996 [°C]:	-2.60	-1.2	-0.4
10m temperature, 1997 [°C]:	-1.82	-0.8	-0.3
Temperature gradient between 20m and 30m (1991) [°C/m]:	0.050	0.022	0.015

Surface temperature estimated by a linear extrapolation of the gradient below the zero annual amplitude (ZAA) to the surface is about 1 to 2°C colder at Murtèl-Corvatsch than at the Pontresina-Schafberg localities. A big difference exists in the temperature gradients of the two drill sites as well: the high value of about 0.5°C/10m at Murtèl-Corvatsch is caused by the two boundary conditions: one at the surface (surface temperature: -2.5°C), the other at about 50 m depth where an intra-permafrost aquifer is active (talk, see below: 0°C). According to the Swiss heat flow map, values of about 0.2°C/10m at Pontresina-Schafberg can be expected for this area ([13]). Thermal conductivity of cores from the Murtèl-Corvatsch drilling, determined in a cold laboratory, is between 2.3 and 3.0W/m°C. Values calculated from amplitude attenuation and phase lag with depth scatter slightly more but most are between 2.0 and 3.0W/m°C as well ([4]).

3. INTRAPERMAFROST TALIK

A special feature is observed at Murtèl-Corvatsch as described by [14]: seasonal temperature variations occur not only down to a depth of roughly 20m but also within a layer between 51 and 57m depth. Every year, at the end of June or the beginning of July, the temperatures rise within a few days from -0.05°C to about +0.15°C. Temperatures remain positive until late September and drop within a short period of time towards -0.1°C during winter and spring. Above 51m and underneath 57m, the values do not vary and have been negative since 1987. The measured maximum temperature in summer increased slightly between 1989 and 1993. Since then the warming has even accelerated (0.1°C in 1994; 0.3°C in 1997).

4. RECENT EVOLUTION OF THE TEMPERATURE AT VARIOUS DEPTHS

A fundamental question for the assessment of the permafrost distribution in mountain regions concerns the representative ness of the permafrost temperatures. Do the measurements show the particular thermal regime at the drill sites (is it a local signal?) or are they characteristic for the region or even a larger area? Therefore, three comparisons between the two localities were made:

a) temperature records just below the permafrost table (boreholes 2/1987 Murtèl-Corvatsch and 2/1990 Pontresina-Schafberg from 1992 to 1994, when a datalogger was used at Pontresina-Schafberg; [15]);
b) the 10m temperature in August/September (when measurements are available from all three boreholes; see Table 1);
c) the temperature data in the uppermost 20m in general.

All comparisons confirm that the evolution of the temperatures is synchronous in all three boreholes although there are differences in absolute values as well as in the temperature gradient. In the following, only data from Murtèl-Corvatsch are discussed, because the highest temporal resolution is available here.

Figure 2 shows the temperature measurements at the most interesting depths between July 1987 and July 1999. In addition to the raw data, a running mean of a one year interval is plotted.

4.1 Surface and active layer

The uppermost thermistor at 0.6m furnishes reference value for the bottom temperature of the winter snow cover, which is used as indicator whether permafrost is present of not (BTS, [16]).

The variation of the winter temperature from one year to the other is remarkably high (between -3° and -9°C). However, in every year, temperatures are below –3°C, indicating 'permafrost probable' in terms of BTS categories.

The amplitude of the temperature signal (i.e. half difference between minimum and maximum value) at the lowermost thermistor in the active layer (at 2.6m, 4°C) is about half of the uppermost (9°C). A so-called zero curtain effect, basically an isothermal situation of 0°C over a period of time due to latent heat processes, can be observed in various years (e.g. 1993, see also Figure 3).

Sometimes it hardly ever occurred, especially when the first snowfall came late (autumn) or when the snow melted quickly (spring). The running mean with a one year interval shows variations between -2°C and +1°C.

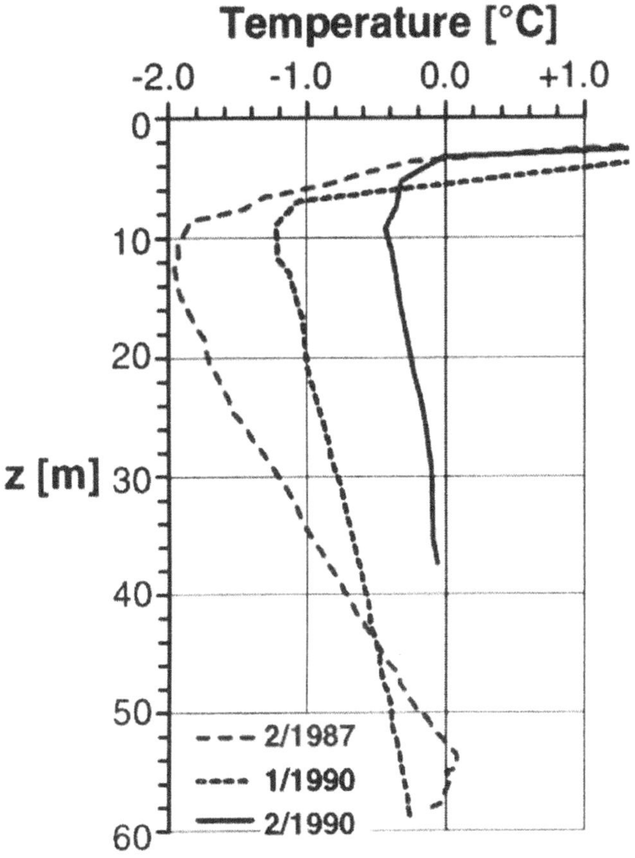

Figure 1: Permafrost temperatures in the Upper Engadine in September 1991 from the drillings Murtèl-Corvatsch (2/1987) and Pontresina-Schafberg (1/1990 and 2/1990).

The uppermost thermistor even reveals a positive annual mean for periods of several months. The periods 1991 to 1995 and 1997 to 1998 were particularly warm.

4.2 3.6 m depth

The uppermost thermistor in permafrost reveals that the warmest temperatures are almost constant (between -0.3°C and -0.1°C) throughout the whole observation period while the coldest values are governed by the active layer temperatures. Consequently the running mean remains negative and its behaviour is similar to that of the thermistors above.

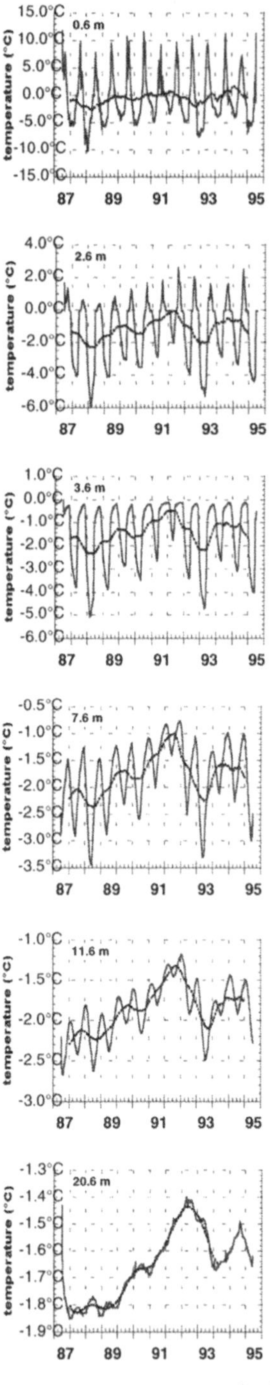

Figure 2: Selected time series of borehole temperatures in Alpine permafrost of Murtèl-Corvatsch 1987-1999 including the annual running mean (dashed line). Note the varying scale.

The warming temperature in summer is very typical: after a fast warming in late spring to roughly -0.5°C only a small temperature increase is observed during the following few summer months because of the latent heat processes. Hence, the shape of the summer peak is asymmetrical (a slow increase in temperature followed by a fast cooling). In addition, the shape is quite different from one year to the next: in 1991 the peak is quite sharp, whereas in 1993 temperature remained near the maximum temperature for almost half a year. This means that a large of heat amount penetrated into the frozen ground, which of course strongly influences the running mean.

4.3 7.6 m depth

The signal is still slightly asymmetric but more closely resembles a sine curve superposed by an amplitude variation and by a long-term fluctuation. The amplitude ranges between 0.3°C and 1.2°C, phase lag is of the order of some 4 to 5 months. The running mean ranges from -2.4°C (1989) to -1.0°C (1994).

4.4 11.6 m depth

A characteristic depth. The shape of the temperature signal is symmetrical, in particular the maximum peak. Amplitude (0.1°C to 0.6°C), phase lag (about half a year) and fluctuation of the running mean (-2.3°C in 1989 to -1.3°C in 1994) are easily detectable because the absolute accuracy of the used sensors is on the order of +/- 0.05°C.

4.5 20.6 m depth

After dissipation of the drilling heat lasting about half a year, seasonal temperature variations are visible although they are smaller than 0.1°C. The signal shows a temperature trend integrated over about one year (the annual running mean corresponds to the measured values). However, temperature at 20.6m depth rose by 0.4°C within 4 years (1991 - 1995).

5. ANALYSIS

5.1 Running means

Figure 4 shows the running means with a running interval of 365 days for temperature readings at various depths in the uppermost 20m.

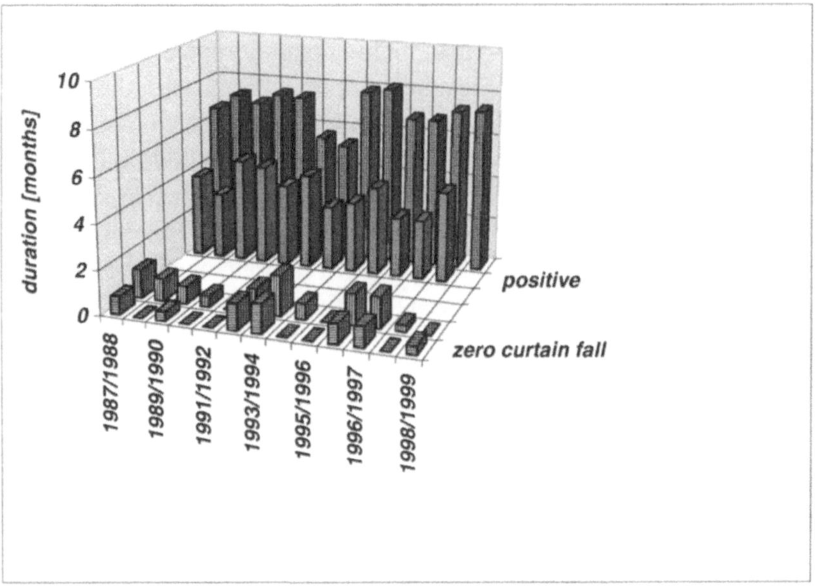

Figure 3: Analysis of the time series from the uppermost thermistor (0.6m below the top of the slope tube; actually just within the debris). Periods of positive and negative temperatures and duration of an observed zero curtain in fall and in spring separated for each year (fall 1987 to spring 1999).

Characteristic features of heat conduction are present: phase lag, amplitude attenuation and filtering of high frequencies with increasing depth. At the depths of 7.6m and 20.6m, mean temperatures are about the same (-1.7°C). In general, mean temperatures cool down with depth in the uppermost 15m, indicating an on going warming trend. In fact, in steady state conditions one would assume the contrary (the deeper the warmer).

Another interesting fact is the distance between the curves. In the active layer, effects of advection and convection can be expected. Therefore, large differences (jump) in mean temperatures above and below the permafrost table would be a surprise. However, a jump of more than 0.5°C can be observed between 1.6m and 2.6m. Just above and below the permafrost table the maximal difference is less than 0.4°C.

5.2 Active layer

The surface of mountain permafrost often consists of coarse blocks and boulders of different size, generating self protecting micro-climate conditions: in early winter and especially when the snow cover is thin, cold entering through natural funnels circulates in the voids between the scree ([17], [18], [19]) and cools down the mean annual surface temperature (MAST).

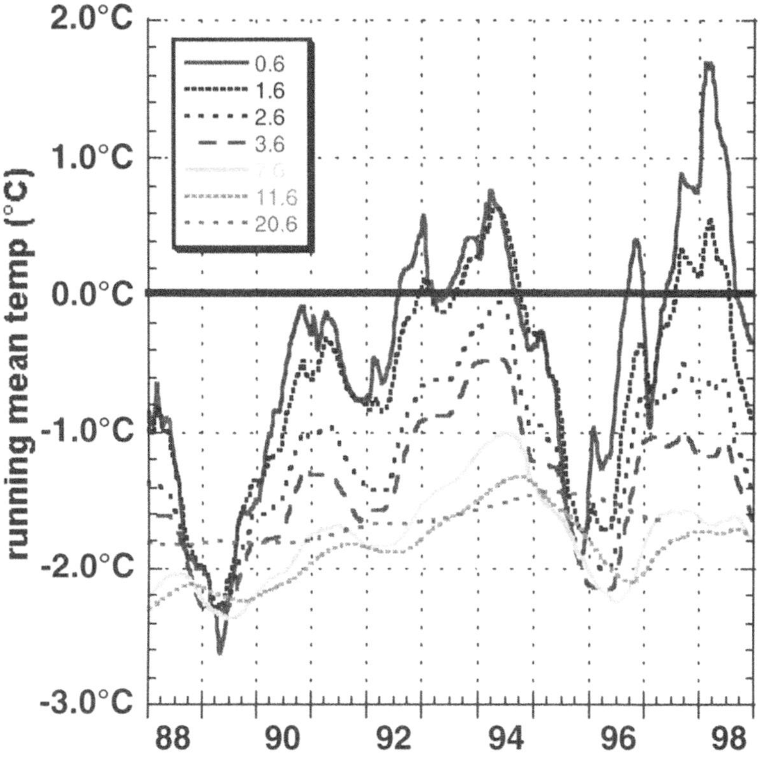

Figure 4: Running means (interval one year) of borehole temperature with time at various depths [m] (Murtèl-Corvatsch).

Big boulders are free of snow for a long time conducting cold temperatures into the ground. Moreover, in spring, the snow lasts much longer between the boulders than at vegetated sites of comparable elevation and aspect.

The influence of advection (by air and/or water) in the active layer is obvious. Nevertheless, the temperature signal from the active layer is one boundary condition for the measured permafrost temperatures farther down, where heat conduction is the dominant process.

Several processes are to be observed, which are important for the thermal regime of Alpine permafrost:

First major snowfall. Besides the overall thickness of the snow cover, the date of the first major snowfall is most important.

In 1988/1989 for instance, after a first small snowfall in early December, no precipitation was registered until the end of February. At the end of April, snow-cover thickness was higher than the long-term average. The thin snow cover during the first part of the winter allowed the winter cold to penetrate into the ground. Consequently, this winter was the coldest since measurements began. In October 1993, snow-cover thickness had already reached almost 1.0m. Between New Year and April, more than 1.5m of snow protected the ground from cooling. In addition, the snow fell on warm ground, as indicated by a long-lasting zero curtain. The heat stored in the active layer during summer could not escape because of the insulating snow cover. In early November 1996 a heavy snowfall of more than 1.5m had a similar effect. The winters of 1993/1994 and 1996/1997 are - in terms of permafrost temperature - the warmest, although the mean annual air temperatures were not particularly warm.

b) Duration of positive and negative temperatures (Fig. 3). After a thick snow cover in winter, generally warming the permafrost temperature, a late melting of the snow follows. As long as snow covers the ground, temperature is below or at 0°C. This in turn reduces the time of positive temperature and hence the heat amount introduced into the ground in summer. The same is true if the first snowfall occurs early. This extends the duration of negative temperature. Average durations over the last ten years are 3.8 months for the positive temperatures and 7.0 months for the negative.

c) Zero curtain. As mentioned above, a long-lasting zero curtain can be observed only under special circumstances. In principle, a zero curtain in fall shortens the duration of negative temperature in the following winter and in spring, the zero curtain causes a shorter positive temperature time in the following summer mainly. The latter effect is less pronounced than the first one as Fig. 3 indicates: The shorter the duration of the zero curtain in fall, the longer the period with negative temperatures. In contrast, a zero curtain in spring does not necessarily cause a shorter period of positive temperature values.

6. CORRELATION BETWEEN RADIATION AND GROUND TEMPERATURE

The relation between monthly means of the global radiation and the temperature in the active layer (0.6m depth) was also investigated. The values from January to December are scattered and do not show any significant correlation. The separation into summer (July and August) and winter (November to June) reflects the influence of the above-discussed snow cover. The correlation for July and August is 0.8, for November to June 0.4. Especially in summer, radiation is an important factor for the permafrost temperature.

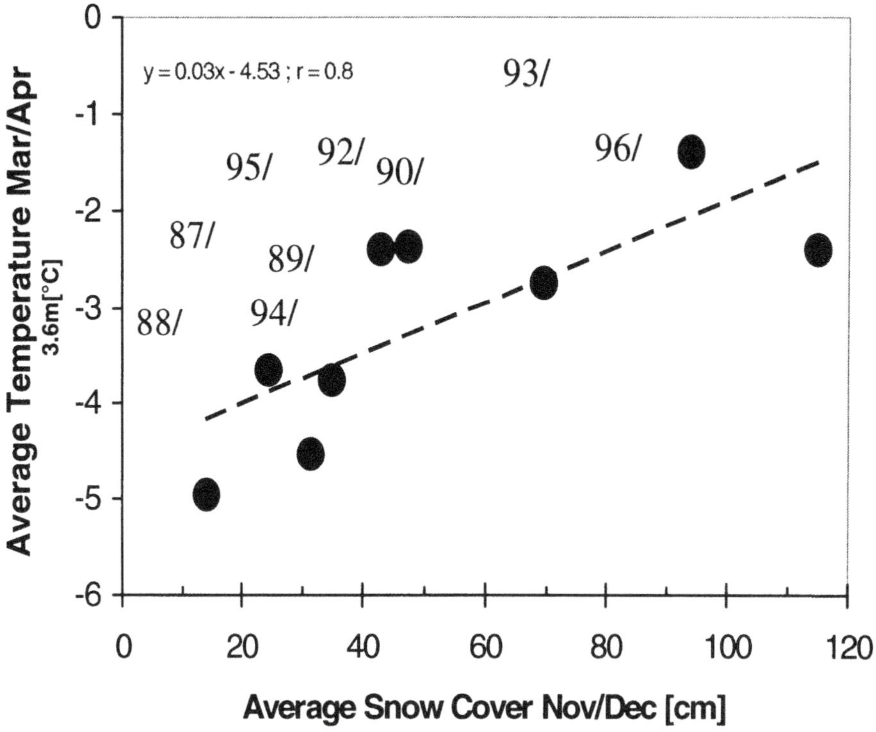

Figure 5: Correlation between the averages of snow height in November / December (x) and temperature at 3.6m depth in March / April (y).

7. CORRELATION BETWEEN SNOW-COVER THICKNESS IN NOVEMBER AND DECEMBER AND THE PERMAFROST TEMPERATURE IN MARCH AND APRIL

As shown before, the snow cover is an important factor for the evolution of permafrost temperatures. A snow cover with a thickness of more than about 80cm acts as insulation. It preserves the heat introduced in summer and protects the permafrost from cold winter air temperature. In contrast, a thin (5 to 15 cm) snow cover in late autumn is most efficient in allowing cooling of the ground ([17]).

The correlation coefficient r for the relation between the mean snow-cover thickness in November and December and the mean permafrost temperature at 3.6m depth in March and April is 0.8 (Fig. 5). A decrease of snow-cover thickness by 10cm causes a cooling in permafrost temperature by 0.3°C [20] calculated a similar value by correlating the mean snow-cover thickness in November and February with the mean permafrost temperature at 3m depth in February and May at Gruben rock glacier (6 years, r=0.97).

These relations statistically confirm the influence of the snow-cover thickness in early winter to permafrost temperatures. Local effects such as variations of snow cover distribution as a function of boulder size or local climate cause particular conditions for every site.

8. CONCLUSIONS

The analysis of borehole temperatures within the permafrost of the active Murtèl-Corvatsch rock glacier revealed that

- temperatures in the uppermost 20m showed remarkable interannual variations;
- a trend of rapid warming by about 1°C/decade until 1994 was largely compensated by rapid cooling in 1994/1995 and 1995/1996waring up afterwards again;
- snow conditions - especially in early winter - exert an important influence on ground temperatures; and
- the documented ground thermal signals probably reflect conditions and evolutions characteristic of regional rather than local scales.

The measurements will continue into the future and serve as a basis for a permafrost observation network to detect effects and impacts of climate change to mountain permafrost.

9. REFERENCES

Balobaev, V.T., Devyatkin, V.N. and Kutasov, I.M. (1983): Contemporary geothermal conditions of the existance and development of permafrost. Fourth International Conference on Permafrost, Fairbanks. Final Proceedings. 8-12.

Bernhard, L., Sutter, F., Haeberli, W. and Keller F. (1998): Processes of snow/permafrost-interactions at a high-mountain site, Murtèl-Corvatsch, Eastern Swiss Alps. Proceedings of the Seventh International Conference on Permafrost, Yellowknife., Canada. Collection Nordicana, 57. 35-41.

Bodmer, Ph. and Rybach, L. (1984): Geothermal map of Switzerland (heat flow density). Commission Suisse de Géophysique. Materiaux pour la géologie de la Suisse, 22. 47.

Haeberli, W. (1973): Die Basis Temperatur der winterlichen Schneedecke als möglicher Indikator für die Verbreitung von Permafrost. Zeitschrift für Gletscherkunde und Glazialgeologie, 9 (1-2). 221-227.

Haeberli, W. (1985): Creep of mountain permafrost: Internal structure and flow of alpine rock glaciers. Mitteilung der VAW-ETH Zürich, 77. 142.

Haeberli, W., Hoelzle, M., Keller, F., Vonder Mühll, D. and Wagner, S. (1998): Ten years after the drilling through the permafrost of the active rock glacier Murtèl, eastern Swiss

Alps: answered questions and new perspectives. Proceedings of the Seventh International Conference on Permafrost, Yellowknife., Canada. Collection Nordicana, 57. 403-410.

Haeberli, W., Huder, J., Keusen, H.-R., Pika, J. and Röthlisberger, H. (1988): Core drilling through rock glacier-permafrost. Fifth International Conference on Permafrost, Trondheim N. Proceedings, 2. 937-942.

Hoelzle, M. (1994): Permafrost und Gletscher im Oberengadin - Grundlagen und Anwendungsbeispiele für automatisiserte Schätzverfahren. Mitteilung der VAW-ETH Zürich, 132. 121.

Keller, F. (1994): Interaktion zwischen Schnee und Permafrost: Eine Grundlagenstudie im Oberengadin. Mitteilung der VAW-ETH Zürich, 127.145.

Keller, F. and Gubler, H.U. (1993): Interaction between snow cover and high mountain permafrost Murtèl-Corvatsch, Swiss Alps. Sixth International Conference on Permafrost, Beijing. Proceedings 1. 332-337.

Lachenbruch, A.H., Brewer, M.C., Greene, G.W. and Marshall, B.V. (1962): Temperatures in permafrost. Temperature - its measurement and control in science and industry, 3 (1). 791-802.

Lachenbruch, A.H., Cladouhos, T.T. and Saltus, R.W. (1988): Permafrost temperature and the changing climate. Fifth International Conference on Permafrost, Trondheim N. Proceedings, 3. 9-17.

Lachenbruch, A.H., Greene, G.W. and Marshall, B.V. (1966): Permafrost and the geothermal regimes. Environment of the Cape Thompson region, Alaska. USAEC Division of Technical Information. 149-165.

Lachenbruch, A.H., Sass, J.H., Marshall, B.V. and Moses Jr, T.H. (1982): Permafrost, heat flow, and geothermal regime at Prudhoe Bay, Alaska. Journal of Geophysical Research, 87 (B11). 9301-9316.

Osterkamp, T.E and Romanovsky, V.E. (1996): Characteristics of changing permafrost temperatures in the Alaskan Arctic, USA. Arctic and Alpine Research, 28 (3). 267-273.

Stucki, T. (1995, unpubl.): Permafrosttemperaturen im Oberengadin. Masters thesis, VAW-ETH Zürich, Department of Earth Sciences. 110.

Sutter, F. (1996, unpubl.): Untersuchungen von Schloten in der Schneedecke des Blockgletschers Murtèl-Corvatsch. Masters thesis, Geographical Institute, University of Zürich.

Vonder Mühll, D. (1992): Evidence of intrapermafrost groundwater flow beneath an active rock glacier in the Swiss Alps. Permafrost and Periglacial Processes. 3 (2). 169-173.

Vonder Mühll, D. and Haeberli, W. (1990): Thermal characteristics of the permafrost within an active rock glacier (Murtèl/Corvatsch, Grisons, Swiss Alps). Journal of Glaciology, 36 (123). 151-158.

Vonder Mühll, D. and Holub, P. (1992): Borehole logging in Alpine permafrost, Upper Engadin, Swiss Alps. Permafrost and Periglacial Processes, 3 (2). 125-132.

Climate Change and Air Quality Assessment in Canadian National Parks

DAVID WELCH
Physical Sciences Advisor, Parks Canada 25 Eddy Street, 4th floor, Hull, Québec K1A 0M5, Canada

Key words: Climate change, air quality, air issues, national parks, threats, Canada

Abstract: At each national park in Canada, a panel of experts assessed threats to ecological integrity, and a park resource manager completed a structured questionnaire on air studies, air issues, local air pollution and air quality related values. These surveys, the general literature, and the findings of a Canada/US park air issues workshop give an overview of the issues facing national parks. Seasonal average temperature and precipitation values generated by four global circulation models under a doubling CO_2 scenario were interpolated for each park. 1994-1996 average annual wet sulphate and nitrogen deposition and precipitation pH were interpolated from the national air pollution monitoring system. These data place national parks within the context of continental scale climate change scenarios and national pollution levels. In Canada, the air issues threatening national park ecosystems are, in order of importance, 1) acidification, 2) climate change, 3) toxics, especially persistent organochlorines, 4) UV-B, 5) the interacting and cumulative effects of several air issues, 6) enrichment from airborne nitrates and CO_2 increases and 7) ground level ozone. The air issues affecting park visitors are 1) particulate matter, 2) ground level ozone, 3) UV-B, 4) noise from aircraft and traffic, and 5) light from towns obscuring the night sky. Canada's boreal and Arctic national parks are severely threatened by climate change due to the relatively high levels of warming predicted at higher latitudes, coupled with drought prone, fire dependent boreal forests or widespread permafrost in the Arctic and sub-Arctic, and wildlife dependent on particular snow and ice conditions. Despite recent reductions in Canada's sulphate emissions, the national parks in south-eastern Canada remain at risk of acidification due to continued high levels of nitrate emissions from automobiles, and loss of buffering capacity in soils and lakes after decades of acidification.

G. Visconti et al. (eds.), Global Change and Protected Areas, 97–107.

1. THREATS TO ECOLOGICAL INTEGRITY

Many of Canada's 39 national parks inherit the legacies of prior occupation, such as town sites, fields, orchards, forest plantations, and of decades of park management oriented to tourism facilities. Nonconforming uses may also continue under park establishment agreements. Examples include domestic wood cutting and subsistence hunting by native peoples. Many through roads and railways continue in use. All parks are subject to local, regional, continental and global stresses, such as urbanization, loss of habitats for wide ranging species, and a variety of threats like airborne toxics, regional haze and visibility impairment, global warming and stratospheric ozone depletion. Twenty-nine significant stresses to national parks' ecological integrity have been identified (Fig.1, Table 1) [1,2]. A stressor is significant if it has a definite ecological impact, affects more than 1 km^2, and is not diminishing over time. Stresses range in frequency from two parks reporting heavy metal pollution to 24 reporting stress from visitor and tourism facilities, and average three to four per park. The higher levels of amalgamation shown in Table 1 are of roughly equal frequency. Until the 1980s, park management practices allowed or even encouraged some town sites to expand, golf courses, ski runs and roads to be built, natural fire to be suppressed, predatory wildlife to be extirpated, and charismatic mammals and sport fish to be introduced. Since then, Parks Canada has started to turn the tide on some of the *in situ* stressors. It has begun to restore natural fire and has capped the development of park towns and roads. The future is less certain, however, for regional developments that destroy and fragment habitats of wide ranging species, including widespread logging, encroachment of agricultural livestock, urbanization and rural road building. Many of these regional problems impact through habitat destruction and fragmentation. Sometimes they are also responsible for pollution, the transmission of exotic species, and wildlife disturbance and mortality. If they are of local or regional origin, then regional actions and national policies can combat them. However, pollutants carried by air, and their effects upon climate, soil and water chemistry, wildlife health and reproduction, are continental and global scale phenomena and require international solutions. The balance of this paper addresses these air issues, their impacts on park values, and ways in which a park agency can help to solve the issues.

2. EXAMPLES OF AIR ISSUES

Acid deposition. Like the rest of Atlantic Canada, Kejimkujik National

Figure 1. Canada's national parks.
Provincial and territorial boundaries, and coordinates referred to in the text are also shown.

Park receives the full brunt of acid precipitation blowing east from the major urban and industrial regions of North America. Its low pH levels decrease the reproductive success of brook trout, reduce angling success and contribute to the disappearance of Atlantic salmon. Reductions of fish biomass lead to decreased reproduction of loons. The leaching of minerals from wetlands causes fen plants like sedges and shrubs to be replaced by bog species such as Sphagnum and Kalmia [3]. Acid deposition in national parks is discussed in more detail below.

Climate change scenarios for Canada feature more total precipitation, more Winter rain at the expense of snow, earlier Spring runoff, more intense and prolonged droughts, and increases in sea level [4]. Warmer temperatures will raise the summer snow line, so there will be accelerated loss of glaciers and permafrost in alpine and Arctic environments.

Earlier Springs and more drought in Summer will increase the prevalence of wild fire, and many areas of Canada may change from boreal forest to grasslands and aspen parkland. Climate change and national parks is discussed in more detail below.

Mercury is released from combustion and manufacturing processes, sewage treatment, exposed soils, and decomposing and burning vegetation. It is the only metal that can be liquid or gaseous at atmospheric temperatures and pressures, and so mixes easily with air and is transported globally. It combines easily with carbon and hydrogen compounds and bio accumulates. Its highest levels are in piscivorous birds, marine mammals and native people in remote rural or natural areas. Cape Breton Highlands National Park has the highest known concentration of mercury in lake water in Atlantic Canada [5]. Kejimkujik National Park has the highest known mercury levels in loons in North America, where it reduces nesting and hatching success.

Organochlorines. Most organochlorine pesticides were banned two decades ago in developed countries, but are still in use around the world. They are easily transported in the atmosphere and fall with snow and rain. They evaporate less at colder temperatures and so concentrate at high latitudes and altitudes. Research at Bow Lake, Banff National Park, shows that toxaphene is taken up by some zooplankton and bio accumulates in trout at up to 10-20 times the concentration in other fish, and up to 1000 times the concentration in fish at low elevation lakes in the park [6].

In Point Pelee National Park DDT has been found at significant levels in sediments where it was once handled and stored. High DDT levels have been blamed for reducing frog populations in several parks and wildlife reserves along the northern edge of Lake Erie. Only five frog species remain at Point Pelee.

Ozone forms when sunlight acts on nitrates and volatile organic compounds released mainly from internal combustion engines. It takes several hours to build in concentration, so levels are typically higher in downwind rural areas than in urban source regions.

High ozone levels are dangerous to active children and people with respiratory problems. The Canadian health standard for the maximum acceptable one hour average for ozone is 82 parts per billion (ppb).

From 1986 to 1993 this was exceeded at Kejimkujik National Park on 24 days [7]. In 1994, Fundy National Park recorded the highest mean concentration of ozone recorded that year in Canada, 36 ppb.

Particulate. Some park visitors seek isolation in the back country, but for many a drive-in campground is home for the night, and an open fire is an essential part of their experience. Campfires produce smoke particles small enough to enter the respiratory tract, exacerbating pulmonary and cardiovascular diseases in sensitive people, the young and the elderly. In Jasper National Park, a study of smoke in the main campground measured total suspended particles (TSP) to determine whether they were above the

Table 1. Significant stresses affecting Canadian national parks

Stresses originating within the park
1. Infrastructure Visitor and tourism facilities
Park infrastructure Roads, railways and utility corridors
2. Activities Park management practices Human disturbance of wildlife
Sport fishing Vehicle/animal collisions Poaching
Stresses originating outside the park and acting directly on park values
3. Exotic species Vegetation Mammals Birds
Fish Invertebrates Microorganisms
4. Pollution Solid waste Petrochemical pollution Pesticides
Sewag Climate change Acid precipitation
Heavy metals Ground level ozone
Regional stresses affecting the greater ecosystem of which the park is part
5. Infrastructure Urbanization Dams Mining
6. Resource use Forestry Agriculture Hunting
Commercial fishing

During one October week, TSP peaked at 170 $\mu g/m^3$ and exceeded 120 on half the days, even though only 413 of 781 camping sites were occupied. A full campground might push the level over 400 $\mu g/m^3$, at which point the federal government is supposed to take immediate action to protect human health.

Comment. Natural areas are often more exposed to air pollution than cities. Acid precipitation and ozone levels increase downwind from source areas, canopy plants cannot seek shade from UV-B, predators cannot stop themselves from ingesting mercury and organochlorines, and natural ecosystems will not adapted readily to climate change.

3. A SURVEY OF AIR ISSUES

Air issues are air or airborne phenomena of unnatural origin that degrade the integrity of ecosystems or the enjoyment and health of visitors. They are enumerated in Table 2, a ranking that emerged from an air issues workshop [9], a literature review and a questionnaire sent to each national park. For example, while high levels of ozone damage leaf tissue on seasonal time scales, there is much less evidence of multi year harm to plant populations. Acid deposition, on the other hand, has been widely documented to cause forest productivity declines, increase the prevalence of tree diseases, and prevent reproduction of some fish species.

High concentrations of ozone cause respiratory stress in active children and in adults with respiratory ailments.

Table 2. Air issues and examples of national park concerns. In each group, the greater concerns are higher in the list

Ecosystem health	
Acid deposition	Leaching, low pH of water, fish reproduction
Climate change	Heat stress, permafrost, altered water regimes, drought, climate variability, extreme events, sea level rise, reduced sea ice regime
Hazardous air pollutants	Mercury and organochlorines, causing various reproductive problems in wildlife
UV-B	Photosynthesis in aquatic organisms, herptile deformities
Compounding effects	Acidification, UV-B and pesticides working together
Enrichment	Nitrogen, CO_2 altering plant successions
Ground level ozone	Leaf damage
Individual human health and enjoyment	
Particulate matter	Reduced visibility, respiratory ailments
Ground level ozone	Damage to lung function, breakdown of many materials
UV-B	Cataracts, skin cancer, immune system deficiency
Traffic and aircraft noise	Impaired wilderness experience
Light from towns, cities	Reduced night sky contrast and appreciation of starscapes

Particulate matter and ozone combine as smog, to impair the enjoyment of natural vistas by reducing visual range and scene contrast.

Air issue related values are the things put at risk by air issues. Abiotic examples include surface water regimes affected by climate change and soils saturated with nitrogen from pollutant, fertilizer and biogenic ammonia deposition. Biotic examples include fish species reproduction impaired by acidification and organisms that bioaccumulate toxic substances that affect their health and reproductive success. Cultural examples are limestone buildings and tombstones, and natural exposures bearing pictographs that may be corroded by acid rain. Human amenity values include vistas unimpaired by regional haze and recreational activities dependent upon some aspect of climate. Human health values include protection from melanomas caused by excessive UV-B and freedom from respiratory stress due to excessive ozone.

Local pollution sources. Most air issues stem from regional, continental and global pollution related to manufacturing, urban transportation, domestic heating, air conditioning and agriculture. However, thirty parks have *in situ* or nearby air pollution sources. Most are insignificant, such as diesel electricity generators for small, remote communities, or smoke and odour from landfill sites. Some are significant but not common, like pesticides drifting in from adjacent forestry land. The leading local air pollution sources are commercial and visitors' vehicles, agricultural pesticides, smelters, saw mills and refineries.

4. PARK INITIATIVES TO RESOLVE AIR ISSUES

Parks should work to improve air quality because they have 1) a legal obligation to take reasonable measures to safeguard the health of visitors, 2) a legal obligation to protect valued ecosystem components, 3) a moral duty to inform citizens about environmental issues that affect their health and enjoyment of protected areas, and 4) a policy duty to help meet international obligations concerning air quality. The direct contribution of park efforts to improve air quality is trivial on a global scale, but as hosts to millions of visitors, parks can play an important role in demonstrating best practices and broadcasting air issues and solutions. Here are some of Parks Canada's initiatives in this respect.

Smoke management. Natural fire is an important ecosystem process, and Parks Canada conducts prescribed burning to meet fire restoration goals. Because burning wild land fuels release large quantities of smoke, particularly during periods of high fuel moisture, smoke management is considered in planning burns. Therefore planned ignition prescribed fire is preferred over wildfire or lightning ignited prescribed fire, since the selection of appropriate fuel moisture and atmospheric conditions, ignition technique and pattern reduces the amount of smoke emitted.

Green operations. Parks Canada is reducing the use of chemical pesticides by assessing the need to control unwanted organisms, and using alternative pest control methods. Energy conservation, reduction of air emissions, and the reduced use of ozone depleting substances are government priorities. Actions include minimizing the consumption of gasoline in favour of alternative fuels. Unfortunately, Parks Canada has no direct sway over the main sources of greenhouse gases emitted from within national parks, namely through traffic, railway operations and buses. A greater contribution to greenhouse gas emission reduction may come from educating visitors about air quality, global change and ecosystem responses, and demonstrating best practices.

Campfires, ecosystem management and health risk. At campgrounds where firewood is free, wood consumption is about ten times greater than where it is purchased. In 1994 Kouchibouguac National Park switched from giving to selling firewood to visitors, a change that reduced the exposure of visitors to inhaling particulate matter and volatile organic compounds. During some periods without inversions, for example, Benzene (a) Pyrenees levels exceeded health standards. The park is monitoring vegetation around the campground to assess the impact of visitors who branches and woody debris for their recreational combustion.

Northeast Regional Air Quality Committee. Parks Canada co-chairs the Northeast Regional Air Quality Committee, a partnership of federal, state

Table 3. Acid deposition data, 1994-1996 average[1]

Park[2]	precipitation pH	SO_4 kg/ha/yr[3]	Park	precipitation pH	SO_4 kg/ha/yr[3]
Bruce Peninsula	4.54	21.77	Mingan Archipelago	4.74	8.60
Cape Breton Highlands	4.79	8.97	Point Pelee	4.36	23.82
Forillon	4.75	9.28	Prince Albert	5.09	3.02
Fundy	4.68	11.12	Prince Edward Island	4.76	8.91
Georgian Bay Islands	4.48	19.73	Pukaskwa	4.71	11.59
Grasslands	5.17	2.21	Riding Mountain	5.26	4.27
Gros Morne	4.80	9.00	Saint Lawrence Islands	4.35	20.78
Kejimkujik	4.70	10.46	Terra Nova	4.83	7.08
Kouchibouguac	4.67	9.46	Wapusk	4.91	3.44
La Mauricie	4.44	18.23	Wood Buffalo	5.02	1.99

1 Data courtesy of Environment Canada.
2 Parks within the region of confident interpolation, east of 110°W and south of 60°N.
3 Wet sulphate, sea salt corrected.

and provincial protected area and air quality agencies in New England (USA) and Atlantic Canada. It exchanges information between member agencies, and provides a link between land management and air quality agencies across jurisdictions. The partners cooperate to understand air issues and document air quality improvements, increase public and employee understanding of the issues and opportunities, and develop support for air quality improvement goals from other agencies.

5. FOCUS ON ACIDIFICATION

Precipitation pH and sulphate deposition for Canadian national parks east of 110°W and south of 60°N are shown in Table 3. All Canadian national parks for which pH can be interpolated with confidence have precipitation pH averages less than 5.3 (Riding Mountain) and as low as 4.35 (Saint Lawrence Islands), about 10 to 50 times more acid than should be the case. Clearly, acid rain, acid snow and acid fog continue in eastern Canada despite recent sulphate emission reductions. As well, recovery is stalled due to the loss of buffering capacity after decades of acidification [10], exacerbated by increasing nitrate emissions from the ever growing and more powerful North

American vehicle fleet. The Canadian target load for wet sulphate deposition is 20 kg/ha/yr. However, many researchers consider this too high to protect sensitive forest ecosystems, and 8 kg/ha/yr has been proposed as a target. Calcareous soils can neutralize acid better than acidic soils, so the wet sulphate critical load for forest damage depends on soil type. The national parks of southern Ontario and La Mauricie suffer deposition around 20 kg/ha/yr, and it is not until one travels as far east as Terra Nova that loads fall below 8 kg/ha/yr. It is clear that in many eastern parks, sulphate deposition exceeds the critical load. The precipitation interpolations are corroborated by *in situ* surface water pH measurements in Atlantic parks. In 1994 the lowest surface water pH in Cape Breton Highlands was 4.6, Gros Morne 4.8, Kejimkujik 4.2 and Terra Nova 5.1. With a pH over 5.5 to 6 there is a good chance of maintaining aquatic biodiversity, but 75% of fish species are lost as pH declines to 5. Some sport fishes can be lost at pH of 5.6, while Atlantic salmon and brook trout are usually present until pH goes below 5.1. Among benthic macroinvertebrates, acid sensitive species include mayflies, caddisflies, stoneflies, amphipods, crayfish, snails, clams and leeches.

6. FOCUS ON CLIMATE CHANGE

National park seasonal temperature and precipitation values for $2xCO_2$ were interpolated from four general circulation models (GCMs), the Canadian Climate Centre's GCM II (CCCII) and Coupled GCM I (CGCMI), Princeton University's Geophysical Fluid Dynamics Laboratory model (GFDL), and NASA's Goddard Institute for Space Studies model. The results show that 1) there will be warming, 2) there will be more warming during the winter, and 3) that this effect increases poleward [11]. What is striking about the interpolations is the extreme amount of winter warming expected for many parks, e.g. GISS showing +11.5°C for Aulavik, CCCII showing +8.0°C at Grasslands, and CGCMI showing +8.2°C in Wapusk. The models also reveal much variation at regional to local scales, even before micro and meso climatic phenomena are taken into account.

Precipitation scenarios show a great range of dryer to wetter conditions. Aulavik and Tuktut Nogait, for example, might experience Winter precipitation from 20% dryer to 30% wetter, whereas Prince Edward Island may be wetter by 10% to 15%. Some consistencies emerge at regional scales. Most areas will be distinctly wetter in Winter.

The same is true in Spring, but with more exceptions such as Ellesmere Island and Pacific Rim. Fall precipitation values reveal much greater uncertainty between models, the extreme being Point Pelee that could be as much as 35% dryer or 25% wetter. Summer scenarios are again more consistent, projecting an overall dryer climate in boreal and southerly areas and wetter in the Arctic.

These data reflect the conclusions of the general literature, typically expressed as continental averages. They also underscore the great variations possible at regional to local scales and from season to season. Winters will be wetter but warmer, so that there will be more rain as opposed to snow, and so that the snow pack will form later and melt sooner. An earlier Spring, coupled with less snow to be melted, will reduce Spring flooding, although there may be more local storm related flood events. In many regions, the earlier and reduced snow melt will lead to dryer soil conditions for longer periods, even without considering the warming in Summer. We can expect chronic and occasional severe drought in Summer. Wind erosion will increase over the Great Lakes dune shorelines and the Prairies, especially at Point Pelee, Elk Island and Grasslands National Parks. In the Arctic, permafrost melting will accelerate and combine with runoff and ice melt to increase erosion.

7. CONCLUSION

Acidification remains a significant threat to Canadian national parks everywhere east of Manitoba. It is the leading air issue, but toxics, climate change, ground level ozone and particulate matter are crowding in. There is a need for enforceable regulations to protect all ecosystems and species from these threats. Such measures are exemplified by the acid deposition critical load concept, secondary standards and regional rules to supplement point source emission controls. A common cause of many air quality and climate change problems is the burning of fossil fuels. Acid gases, acid aerosols, particulate matter, nitrates, carbon dioxide, volatile organic compounds and heavy metals are all released in abundance by this one basic process. While governments and industry can and do encourage, force and implement many changes to improve the picture, it will take societal shifts in lifestyles, consumption choices and urban design to achieve radical improvements in the global ecosystem.

8. ACKNOWLEDGMENTS

Thanks to Bob Vet and Chul Un Ro, Environment Canada, for the acid deposition data, to Neil Munro, Parks Canada, for the surface water pH data, to Daniel Scott and Bonnie Hui, University of Waterloo, for climate change scenarios, and to numerous colleagues in Parks Canada and other protected area agencies in Canada and the United States for answering questionnaires, attending meetings and providing relevant park documents and data bases.

9. REFERENCES

Bailey R. and A. Stendie, Should campgrounds have campfires? Research Links 1(1) (1993) 3,7.

Campbell L., Research Links 4(3) (1996) 11,6.

Commission for Environmental Cooperation, Long range transport of ground level ozone and its precursors: assessment of methods to quantify transboundary transport within the northeastern United States and eastern Canada, Montréal, 1997.

Environment Canada, Canada United States Air Quality Agreement 1998 Progress Report, Ottawa, Supplies and Services Canada, 1998.

Environment Canada, Canada's second national report on climate change, Environment Canada on-line document at http://www1.ec.gc.ca/climate/index.html, 1997.

Hauge E. and D. Welch (Eds), International Air Issues Workshop, Waterton Lakes National Park 5-8 June 1995, United States National Park Service, publication no. NPS D-1116/May, 1996.

Hengeveld H., Understanding atmospheric change: a survey of the background science and implications of climate change and ozone depletion, Supply and Services Canada, State of the Environment Report No.92-2, 1995.

Kerekes J. et al, In: T.B. Herman et al, Ecosystem Monitoring and Protected Areas, Proceedings of the Second International Conference on Science and the Management of Protected Areas, Science and Management of Protected Areas Association, 1995, pp.326-331.

Northeast States for Coordinated Air Use Management et al, Northeast States/Eastern Canadian Provinces Mercury Study, Boston, Massachusetts, on-line document at http://www.cciw.ca/eman, 1998.

Parks Canada, State of the parks 1994 report, Ottawa, Canada, Supply and Services Canada, 1995.

Parks Canada, State of the parks 1997 report, Ottawa, Canada, Supply and Services Canada, 1998.

Regional Clean Air Partnerships and the ETEAM

ERIK R. HAUGE
Head, ETEAM, 30378 Appaloosa Drive, Evergreen, CO 80439-8635, USA

Key words: RCAP - Regional Clean Air Partnership; ETEAM - Ecoteam PSD - Prevention of Significant Deterioration; FLM - Federal Land Manager; AQRV - Air Quality Related Value

Abstract: As the planner for the US National Park Service Air Resources Division, I conceived of regional clean air partnerships (RCAPs) as a tool to address air pollution effects on national park and other protected areas' resources on a regional basis. These RCAPs are voluntary associations of land managing and air regulatory agencies, Indian tribes, industries, and environmental groups. They share the costs of monitoring, research, and outreach programs, and cooperate in regulatory reviews. There are several RCAPs in North America. (See attached map.) In September 1998, I presented the RCAP concept at the 11th World Clean Air Congress in Durban, South Africa. While in South Africa, I also presented the concept at an East Cape province symposium. I returned to the province in March to establish an RCAP organizing committee. The committee has already adopted a charter, and has begun a regional air monitoring program. I returned once again last month, where I was keynote speaker at the African Energy and Environment Conference, and also assisted the partnership committee in selecting a visibility monitoring site at Addo Elephant National Park and installing the first of several US donated monitoring instruments. As a result of my African experience, I conceived the ecoteam (ETEAM), a group of internationally experienced experts who could travel to developing countries and help them deal with environmental and economic problems. The first ETEAM effort will be to facilitate an air quality training course in South Africa in November. In September 1996, the National Park Service and the Environmental Protection Agency began a long term program to monitor environmental stresses on park ecosystems, including establishing a UV-B monitoring program to determine changes in irradiation that may effect human health and ecosystem processes. UV-B monitoring will be recommended as part of most RCAP monitoring programs throughout the world.

G. Visconti et al. (eds.), Global Change and Protected Areas, 109–117.

1. PREFACE

Air pollution is a regional if not a global problem. It requires cooperation between various organizations to solve the problem.

The concept of preserving portions of nations' cultural and natural resources for posterity has spread from the United States to more than 160 other countries.

Tourism is the world's largest industry. In 1998, it generated $3.4 TRILLION in revenue world wide. It employs 204 million people – 11% of the world's workforce. [1]

2. INTRODUCTION

The recent forest fires in Amazonia, Indonesia, and Mexico graphically illustrate that air pollution is a regional if not a global problem. It respects no boundaries.

Congress amended the Clean Air Act in 1967 to require the establishment of air quality control regions throughout the United States and called for intergovernmental cooperation in dealing with the problem. [2]

In 1977, Congress further amended the Clean Air Act to establish a program to prevent the significant deterioration of air quality in clean air areas of the country. A major purpose of this prevention of significant deterioration (PSD) program is to preserve, protect, and enhance air quality in nationally or regionally significant lands such as national parks, wildernesses and wildlife refuges. The federal land managers (FLMs), including the National Park Service, the Fish and Wildlife Service, and the Forest Service, were given an "affirmative responsibility" to protect air pollution sensitive resources or "air quality related values" (AQRVs) in 158 national parks and wilderness areas (class 1 areas) from adverse impacts of air pollution. [3]

In response to these Clean Air Act mandates, the NPS and the other FLMs have established air quality programs. In the NPS case, it was also in response to the mandate of the NPS Organic Act, which requires the Service to "conserve the scenery and the natural and historic objects and the wildlife therein...unimpaired for the enjoyment of future generations". [4] These air quality programs include monitoring, effects research, regulatory review, and outreach activities.

One of the major elements of the NPS and other FLMs' air programs is permit review. [5] The Clean Air Act provides FLMs' the opportunity to review applications for permits to construct major new stationary air pollution sources near class 1 areas. If the FLM review indicates that the new source could have adverse impacts on one or more AQRVs, then it can recommend that the permit be denied. However, if approved, these sources would be coming on line using Best Available Control Technology, assuring that their emissions would be minimal. The major cause of current air pollution problems in class 1 areas is emissions from existing sources not subject to the permit review requirements of the Clean Air Act. Thus the problem facing the FLMs trying to protect their resources from air pollution is how to deal with those many existing sources.

Related to this situation is the fact that the permit review process is done on a case-by-case, unit-by-unit basis. Since air pollution is a regional problem which respects no boundaries, how can the FLM agencies deal with the problem of regional air pollution from multitudes of sources when they are primarily geared to a program which has limited them to dealing with individual new sources which may impact single parks or wildernesses?

3. ONE ANSWER - REGIONAL CLEAN AIR PARTNERSHIPS

What is a regional clean air partnership (RCAP) and what is its significance? An RCAP is a voluntary cooperative association of land managing and air regulatory agencies, Indian tribes, industry and environmental organizations in an ecosystem with similar sensitive resources which deals with the existing or potential impacts of air pollution on those resources on a regional basis. The significance is the focus on the region, rather than individual units, and the sharing of monitoring, research, regulatory review, and outreach programs. It avoids duplication of effort, and saves money. It allows the development of consensus positions. Because much of the activities of the partnership can be conducted electronically, face to face meetings can be kept to a minimum. Travel costs are reduced considerably.

What does this mean for dealing with regional air pollution? The Clean Air Act was amended again in 1990, and among other things, reemphasized the regional aspect, implications, and impacts of air pollution. [6]

How have the FLMs coped with the problems of regional air pollution and its impacts on protected areas? In 1990, a group of NPS, Bureau of Land Management, and Forest Service employees in various units in the Sierra Nevada mountains of California got together and formed the Sierra Federal Clean Air Partnership, the first RCAP. [7]

At the same time, the NPS published its first Air Quality Management Plan, for Colonial National Historical Park. [8] That park, which includes the first English settlement in the United States, as well as the final battleground of the Revolutionary War, is surrounded by major sources of air pollution. A survey of vegetation at the park yielded several species sensitive to air pollution, and visible symptoms were observed. The park's monuments and statuary were deteriorating. The park's vistas were obscured. The superintendent declared that air pollution was the park's number one resources management problem, and wanted the problem focussed on in a separate document. The document, when published, became a prototype throughout the NPS. Several other parks now have AQMPs.

In 1991, I began to develop the first NPS regional AQMP for Shenandoah and Great Smoky Mountains National Parks and the Blue Ridge Parkway. These three parks are all part of the Southern Appalachian Biosphere Reserve, and have documented air pollution problems, including visibility degradation, vegetation damage, and acidified ponds and streams. [9] The Forest Service, managers of eight class 1 wildernesses in the Southern Appalachians, asked to participate. Soon NPS and FS funds were sponsoring a contract to prepare an interagency plan. In 1992, this effort was superseded by the Southern Appalachian Mountains Initiative (SAMI), made up of the FLMs plus the Environmental Protection Agency, state air regulatory agencies, industry and environmental organizations. The original contract was modified, and the final document, an air quality status report for the Southern Appalachians, was published in 1996 and submitted to SAMI. [10]

The RCAP approach has proven popular and effective. Additional RCAPs have been and are being established. Others have been proposed. New partners are participating. Just before I retired from the National Park Service in December 1998, I edited the Clean Air Partnership Guidelines, a document to be used by new partnerships in developing their organizations. [11]

4. US – CANADA RCAPS

In 1991, the US-Canada Air Quality Agreement was signed. [12] It called for increased international cooperation regarding transboundary air quality issues. One major provision of the Agreement deals specifically with protected areas – Annex 1-4. This provision calls for the establishment in Canada of a program similar to the US PSD and visibility protection programs, and for the two countries to cooperate in dealing with PSD related issues.

Shortly after the US-Canada Air Quality Agreement was signed, I attended the first Science and the Management of Protected Areas (SAMPA) conference in Canada, where I made a presentation about the NPS air quality program. [13] A dialogue was begun there with representatives from Parks Canada regarding potential cooperative efforts in transboundary areas experiencing similar air pollution problems. This led to the first International Air Issues Workshop at Roosevelt Campobello International Park (New Brunswick) in June 1993. The workshop participants recommended continuing and expanded cooperative efforts between the land managing and air regulatory agencies regarding transboundary issues. The participants also recommended a second workshop be held in the West, with invitations extended to all FLMs, air regulatory agencies, Indian tribes, and the International Joint Commission on the Great Lakes. [14]

Another direct result of the first workshop was a meeting held in September 1994 at Roosevelt Campobello to explore the establishment of the first US-Canadian RCAP for the New England – Atlantic Canada region. AT subsequent meetings, the partnership – the Northeast Regional Air Quality Committee (NERAQC) – has adopted a charter and prepared a regional air quality assessment. [15] It is currently finalizing an information brochure on the partnership and its activities which can be distributed at any participating partner's facilities.

The Second International Air Issues Workshop was held in June 1995 in western Canada at Waterton Lakes National Park and Lethbridge, Alberta. Sixty US and Canadian officials participated. The participants recommended further cooperation on transboundary issues. They also recommended that a follow up session be held at the third SAMPA conference, in Calgary, Alberta in May 1997. [16] A day long concurrent session on air quality in protected areas was included in the SAMPA III Conference. [17]

5. THE GREAT LAKES AIR QUALITY PARTNERSHIP

Although there are several organizations already established to deal with environmental problems in the Great Lakes, the Great Lakes RCAP will focus on significant air pollution impacts on the resources of protected lands in the region.

The first organizing meeting was held in Sault Sainte Marie, Ontario in December 1996 after several exploratory meetings. Representatives of 15 agencies and Indian tribes established an organizing committee to review possible partnership objectives, mission, membership, procedures, responsibilities, and other elements. A second organizing meeting was held in Ann Arbor, Michigan in July 1997 to draft the charter for the partnership. A third meeting was held in July 1998 in Windsor, Ontario to finalize the draft charter and establish a permanent steering committee. The committee will be meeting once again this month to select US and Canadian co-chairs and begin the process of obtaining organizational endorsement of the charter on both sides of the border.

6. ADDITIONAL RCAPS WITH CANADA

The US and Canada signed several joint environmental initiatives on April 7, 1997 in Washington, DC. The two nations agreed to a renewed effort to promote exchange of research findings and technical data related to prevention, monitoring, and control of transboundary air pollution. [18] These agreements provide additional stimuli for establishing other RCAPs along the border including the Northern Great Plains, the Crown of the Continent (the crest of the Rocky Mountains), and the Pacific Northwest. An exploratory meeting to discuss a Pacific Northwest partnership was held at Mount Rainier National Park in Washington state in March 1998. A second meeting was held in October 1998 in Seattle, Washington, at which time the participants decided to continue their informal, international cooperative efforts and not develop a formal partnership yet.

7. PARTNERSHIPS WITH MEXICO

The US and Mexico have signed a number of bi-national agreements over the years regarding environmental pollution control. In October 1996 the two governments released the Border XX I Framework for public review. This program is directed toward conserving natural resources, protecting the environment and environmental health, and promoting the transition to sustainable development in the border region. [19] Also, Mexico has established two new Biosphere Reserves near Big Bend National Park in Texas in addition to the several existing national parks and biosphere reserves near the border.

A US-Mexico air quality monitoring program has been established to assess sources of the transboundary visibility problem in the Big Bend area, the first step in establishing a formal partnership.

8. SOUTH AFRICA

In September 1998 I presented the RCAP concept at the 11th World Clean Air Congress in Durban, South Africa. [20] It was well received, and among other things, has led to invitations to present the concept at several additional international conferences.

While in South Africa, I was also invited to present the RCAP concept at an ecotourism and environment symposium in East London, a city in the East Cape province. The participants endorsed the concept, and I was invited to return to the province to help organize the first RCAP outside of North America. This I did in March, and the East Cape Province RCAP has subsequently drafted a charter and initiated a regional air monitoring and emissions inventory program.

9. ETEAM

After returning from South Africa in September 1998, I realized that although I would continue to provide expertise to various nations regarding the establishment of RCAPs, these nations needed more and varied expertise in order to deal with environmental problems in their haste to industrialize.

Remember, more than 160 nations have established national parks and other protected areas, and are interested in tapping into the tourism and ecotourism industry. However, at the same time, many of these nations have allowed the construction and operation of major industrial facilities without proper environmental controls. This has led not only to major public health problems, but as visitors to Greece's Parthenon or India's Taj Mahal can attest, to impacts on cultural and natural resources of shrines set aside for the enjoyment of future generations, including tourists.

In order to help these developing nations deal with the problem, I conceived the ecoteam or ETEAM (Ecological and Technical Experts Advising Mankind). This would be a group of internationally experienced experts who could travel to those countries and assist them by facilitating conferences, establishing monitoring , research and outreach programs, developing planning and regulatory programs, conducting training courses, and providing advice on the development of sustainable economic alternatives such as ecotourism.

Several of my colleagues endorsed the concept, and some sent me their resumes/cv's, hoping to participate on the ETEAM. I now have 25 resumes from world class experts who will participate on the team. As more of these experts learn of the ETEAM and its future activities, I anticipate receiving additional resumes.

I have begun to formally organize the ETEAM and seek funding from a variety of sources, including individuals, foundations, and corporations, to support its activities. Even though the team has not yet met for the first time to develop its mission, goals, objectives, and priorities, it has already been invited to participate in various international activities, such as environmental conferences in Africa and Australia. The first official ETEAM activity will be to facilitate and instruct an air quality training course in South Africa in November.

While the ETEAM will at first focus its efforts on air quality and ecotourism, I believe it will broaden its base to include experts who can help developing countries with other environmental problems such as water quality, solid waste management, transportation planning, and remedial recovery programs, among other things.

10. RETURN TO SOUTH AFRICA

I returned to the East Cape Province yet again last month. I was the keynote speaker at the African Energy and Environment Conference in Port Elizabeth, where I presented my RCAP and ETEAM concepts to a very receptive audience from 26 countries. [21]

I also joined members of the East Cape RCAP committee in selecting a site at nearby Addo Elephant National Park for the installation and operation of the first of several US donated visibility monitoring instruments (an automated camera). This site will be part of the regional air monitoring network being established by the RCAP committee.

As a result of my efforts, I have been invited to discuss the establishment of a second RCAP in South Africa, this time in KwaZulu-Natal Province centered on the Durban metropolitan region. Also, a number of other nations expressed interest in the RCAP concept, and I expect invitations to discuss establishment of such partnerships in the future. In addition, the Norwegian Air Research Institute and other similar organizations expressed interest in and commitments to providing monitoring instruments and technical expertise, among other things, to the South African RCAPs. If additional RCAPs are established elsewhere, I will work to establish such working relationships with similar support organizations.

11. REFERENCES

Agreement Between the Government of the United States of America and the Government of Canada on Air Quality, Ottawa, Ontario, Canada, March 13, 1991

Air Quality Act of 1967 – Public Law 90-148 (November 21, 1967) (81 STAT. 485)

Air Quality Management Plan, Colonial National Historical Park, July 1991

An Act to Establish a National Park Service – (39 STAT. 35) (August 25, 1916) (16 U.S.C. 1)

Clean Air Act Amendments of 1977 – Public Law 95-95 (August 7, 1977) (42 U.S.C. 7401 et seq.)

Clean Air Act Amendments of 1977 – Section 165(d)(2)

Clean Air Act Amendments of 1990 – Public Law 101-549 (November 15, 1990) (104 STAT. 2399) – Sections 169B (f), 176A

Clean Air Partnership Guidelines, Erik R. Hauge, Editor, National Park Service, Denver, Colorado, December 1998

Ecosource Website (www.podi.com/ecosource/), Ecotourism Society, 1999

Federal Clean Air Partnership Charter, June 1990

Hauge, E.R., "Regional Air Quality Partnerships – International Implications and Applications,"

Hauge, E.R., "Regional Clean Air Partnerships and the ETEAM," keynote address in Africa Energy and Environment Conference, Port Elizabeth, South Africa, August 4-6, 1999

Hauge, E.R., "The National Park Service Air Quality Program: The Cutting Edge of Science in Resource Protection," in Science and the Management of Protected Areas, Proceedings of an International Conference, Acadia University, Wolfville,Nova Scotia, Canada, May 1991

In Papers of the 11th World Clean Air and Environment Congress, Durban, South Africa, September 1999

International Air Issues Workshop (Proceedings), Waterton Lakes National Park, Canada, June 5-8, 1995

Preliminary Notice of Adverse Impact on Shenandoah National Park, 55 F.R.38403ff., September 18, 1990; Preliminary Notice of Adverse Impact on Great Smoky Mountains National Park, 57 F.R. 4465ff., February 5, 1992

Program to Develop Joint Plan of Action for Addressing Transboundary Air Pollution, Washington, D.C., April 1, 1997

SAMPA III, the 3rd International Conference of Science and the Management of Protected Areas, University of Calgary, Alberta, Canada, May 1997

Southern Appalachian Clean Air Partnership, Technical Publication RS-TP-30, USDA Forest Service and USDI National Park Service, Atlanta, Georgia, September 1996

Terms of Agreement (Draft) for the Northeast Regional Air Quality Committee (NERAQC), September 1996

U.S.-Mexico Border XXI Program Framework Document, October 1996, Washington, D.C., October 7, 1996, and Mexico City, Mexico, October 15, 1996

United States/Canada Air Quality Workshop (Proceedings), Roosevelt Campobello International Park, June 8-10, 1993

Land-Atmosphere Interactions

R.A. PIELKE SR.*, T.N. CHASE*, J. EASTMAN*, L. LU*, G.E. LISTON*, M.B. COUGHENOUR**, D. OJIMA**, W.J. PARTON*** AND T.G.F. KITTEL****

**Department of Atmospheric Science, Colorado State University, Ft. Collins, Colorado 80523*
***Natural Resource Ecology Lab, Colorado State University, Ft. Collins, Colorado 80523*
****Rangeland Ecosystem Science, Colorado State University, Ft. Collins, Colorado 80523*
*****National Center for Atmospheric Research, Boulder, Colorado 80307*

Key words: Land-use, Climate Changes, Climate Models.

Abstract: There is substantial evidence that land use change and vegetation/soil/snow dynamics processes have a significant influence on climate on regional and global scales. The effect of these influences on the global scale has been found to be comparable in magnitude to the radiative effect of carbon dioxide. On the regional scale, these influences appear to be more important, and act on time scales of months and less. The paper presents evidence of the importance of landscape processes, including land use change, on climate. Results are shown on a regional scale comparing the relative importance of the radiative and biological effects of a doubling of CO_2, and of landuse change from the natural to current landscapes. On the global change scale, results are presented comparing the sensitivity of the Earth's climate to anthropogenic increases of CO2, and from human-caused landuse change. The modelling tools used in these studies include the Regional Atmospheric Modeling System (RAMS) and the National Center for Atmospheric Research (NCAR) CCM_3 global model. Global data sets including the National Center for Environmental Prediction (NCEP) Reanalysis and regional data from the National Climate Data Center's (NCDC) historical data are presented to in order to provide confirmation of the modeling results. [1]

[1] *This contribution was prepared for the American Meteorological Society 8th Conference on Climate Variations, 13-17 September 1999, and is reproduced here for the 8-13 September Global Change and Protected Areas Meeting in L'Aquila, Italy.

G. Visconti et al. (eds.), Global Change and Protected Areas, 119–126.

1. INTRODUCTION*

Two coupled atmosphere-land surface models (RAMS/CENTURY and RAMSIGEMTM) have been applied to investigate the feedbacks between weather, and vegetation and soil on seasonal and longer time scales. The RAMS version used is de- scribed in Liston and Pielke (1999), and has been applied in Pielke et al. (1999). The coupled modeling systems were validated against observed atmospheric and vegetation evolution during the growing season. The coupled models were then used to investigate, for example, the relative importance of landuse change and of radiative and biological effects of current and doubled CO_2 on seasonal weather. Global model results using the NCAR CCM3 model have also been completed to explore whether these feed-backs teleconnect over long distances and can alter the global Atmospheric circulation. A major conclusion of this study is that climate models must include dynamic, interactive land-surface parameterisations for the assessment of seasonal and longer-term atmospheric predictability.

2. RAMS/CENTURY COUPLING

Land-surface characteristics play a key role in partitioning energy received at the earth's surface. Vegetation, through transpiration and evaporation, modifies Atmospheric and land-surface hydrological processes. Both observational and modeling studies have shown that two-way atmosphere and biosphere interactions are very important components of both Atmospheric and ecosystem dynamics.

A coupled RAMSICENTURY modeling system has been developed to study regional-scale two-way interactions between the atmosphere and biosphere (Lu 1999; Lu et al. 1999). Both Atmospheric forcings and ecological parameters (LAI, etc.) axe prognostic variables in the linked system. The atmospheric and ecosystem models exchange information on a weekly timestep. The ecosystem model CEN- TURY receives as input: air temperature, precipitation, radiation, wind speed, and relative humidity simulated by the regional Atmospheric model RAM- S. From CENTURY-produced outputs, variables including leaf area index (LAI), vegetation transmissivity, vegetation fractional coverage, displacement height, roughness length, rooting profile, and albedo can be computed and returned to RAMS. In this way, biogeochemical-constrained vegetation responses to weekly and seasonal Atmospheric changes axe simulated and fed back to the atmospheric/land-surface hydrology model.

The coupled model has been used to simulate the two-way interactive biosphere and atmosphere feedbacks from 1 January through 31 December for 1988, 1989, and 1993, which represent dry, average, and wet years, respectively, focusing on the central United States. In these experiments, the CENTURY- produced outputs of LAI and vegetation transmissivity are input into RAMS. Validation is performed for the Atmospheric portion of the model by comparing with over 3,800 meteorological-station observations over the entire domain, and for the ecological component by comparison to AVHRR remote-sensing ND-VI data sets.

A series of sensitivity experiments have been conducted to highlight interactions and feedback- s between Atmospheric and land-surface processes. The coupled control run's Atmospheric lateral boundary conditions have been perturbed to create both dry and wet springs. The model's ability to represent the interannual and seasonal variations in both climate and biomass has been examined. The results show that seasonal and interannual climate pat- terns axe significant influences on land-water energy exchange. The coupled model captures key aspects of weekly, seasonal, and annual feedbacks between the atmosphere and ecological systems. This demonstrates the coupled model's usefulness as a research tool for studying complex interactions between the atmosphere, biosphere, and hydrosphere. In the modeling system, vegetation is permitted to grow in response to the simulated weather with the weather feeding back to influence subsequent plant and biogeochemical dynamics over the central Great Plains of the United States (e.g., see Lu 1999). In addition, plant development feeds back to the evolution of weather. As a result of the feedback, fine-grid domain-averaged temperatures axe up to 2°C cooler with associated increases in precipitation. More details of this study are reported in Lu et al. (1999).

3. RAMS/GEMTM COUPLING

The plant model, the General Energy and Mass Transfer Model (GEMTM), was coupled to the meteorological model, the Regional Atmospheric Modeling System (RAMS) over the same domain (Eastman 1999; Eastman et al. 1999). The modeling system was then used to investigate the effects of when landcover is changed from current to potential vegetation, radiative forcing is changed from 1 x CO_2 to 2 x CO_2, and biological effects of doubled CO_2 are included. On the domain average, both landuse change from natural to the current landscape, and the biological effect of doubled CO_2 resulted in significant cooling.

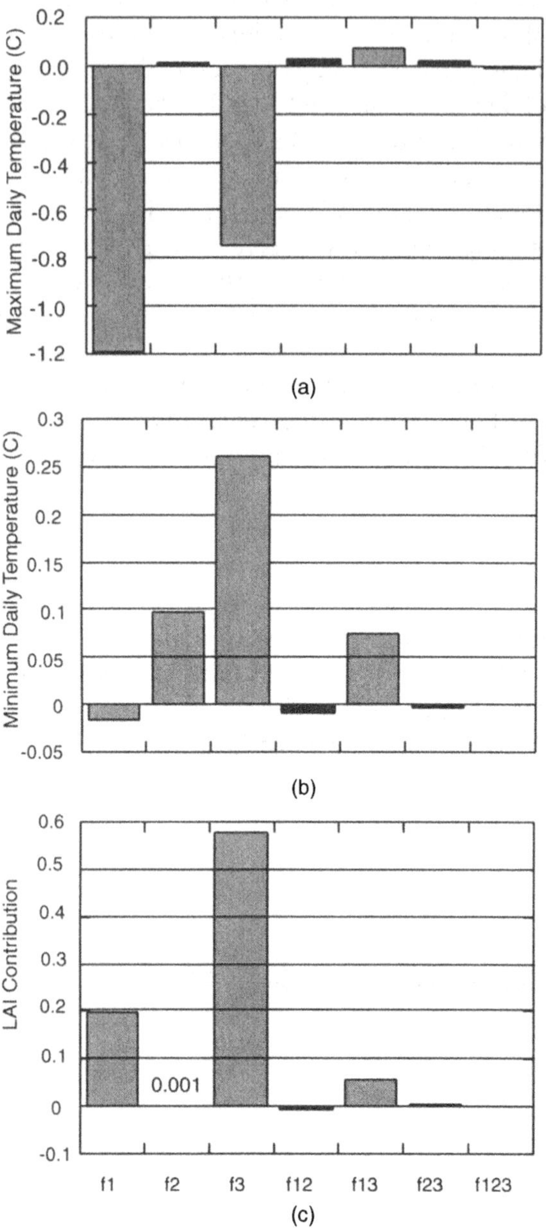

Figure1.RAMS/GEMTM coupled model results - the seasonal domain-averaged contributions to maximum daily temperature due to: fl = natural vegetation, f2 = 2xCO2 radiation, f3 = 2xCO2 biology, f12 = interaction of natural vegetation and 2xCO2 biology, f13 = interaction of natural vegetation an 2xCO9 biology, f23 = interaction of 2xCO2 for radiation and biology, f123 = the interaction of all three factors. (b) Same as (a) except for minimum daily temperature contribution. (c) Same as (a) except for LAI (from Eastman et al. 1999).

Figure lc shows the increase in vegetation cover due to the enrichment of CO_2 in the atmosphere. The model results indicate that 2 x CO_2 felt by the biology, and landuse change exhibit dominant effects on meteorological and biological fields (Figure la-e). This was found at daily to seasonal temporal scales, and grid to regional spatial scales. The radiation impacts of 2 x CO_2 are found to be minimal, with interactive effects between the three areas of investigation as large as the radiational impact. More details of the study are found in Eastman et al. (1999).

4. CCM3 GLOBAL SIMULATION

Two ten-year general circulation model experiments were performed to compare a simulation where land-surface boundary conditions were represented by observed, present-day landcover with a simulation where the surface was represented by natural, potential landcover conditions assumed to be representative of the pre-settlement landcover distribution (Chase 1999; Chase et al. 1999). As a result of these estimated changes in historical landcover, significant temperature and hydrology changes affected tropical land surfaces, where some of the largest historical disruptions in total vegetation biomass have occurred. Also of considerable interest, because of their broad scope and magnitude, were changes in high latitude Northern Hemisphere winter climate which resulted from changes in tropical convection, upper-level tropical outflow, and the generation of low-frequency tropical waves which propagated to the extratropies. These effects combined to move the Northern Hemisphere zonally-averaged westerly jet to higher latitudes, broaden it, and reduce its maximum intensity (Figure 2a and b). Low-level easterlies were also reduced over much of the tropical Pacific basin while positive anomalies in convective precipitation occurred in the central Pacific. There were large simulated ten-year average changes in near-surface temperature (Figure 3), although globally-averaged changes were small.

5. CONCLUSION

Land-atmosphere interactions clearly have importance on regional and global climate.

Indeed, climate cannot be adequately understood if these interactions are not considered. Pielke (1998) and Pielke et al. (1999) discuss the implications of this conclusion in the context of climate prediction. Other researchers (e.g., Claussen 1994, 1998; Claussen et al. 1999; Foley 1994;

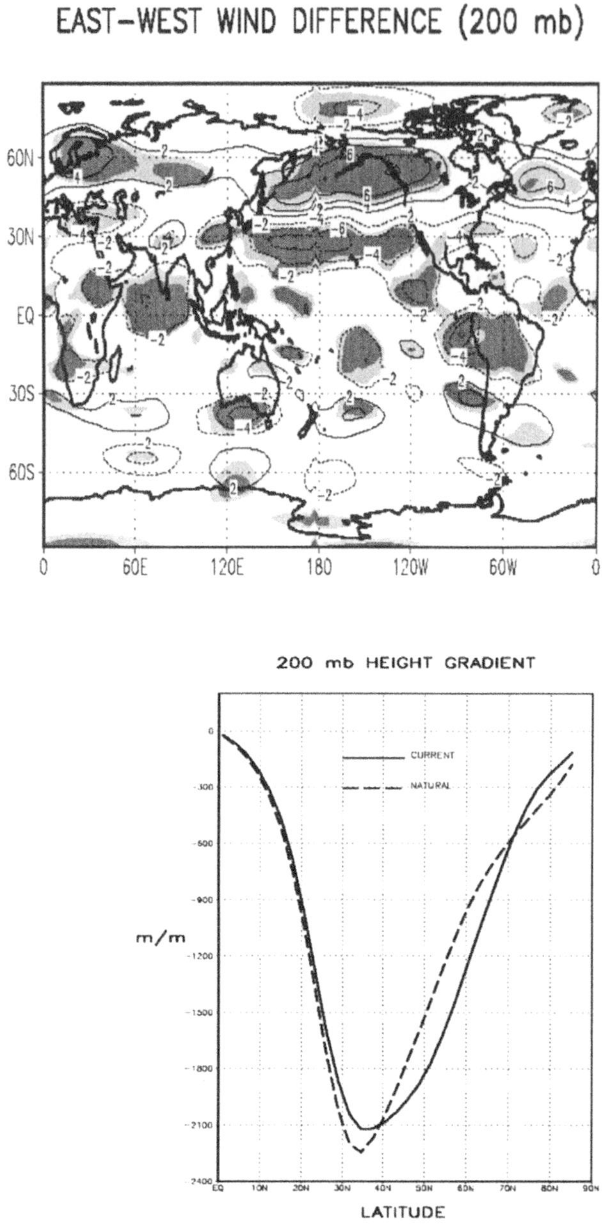

Figure 2: CCM simulations - (a) difference in 200 mb east-west wind (current-natural landcover). Contour by 2 m s^{-1} Light shading represents the 90% significance level for a 1-sidedt-test. Dark shading rep- resents the 95% significance level. (b) Comparison of north-south derivative of zonally-averaged 200 m- b heights (d(Z200)Idy) in the Northern Hemisphere (from Chase et al. 1999).

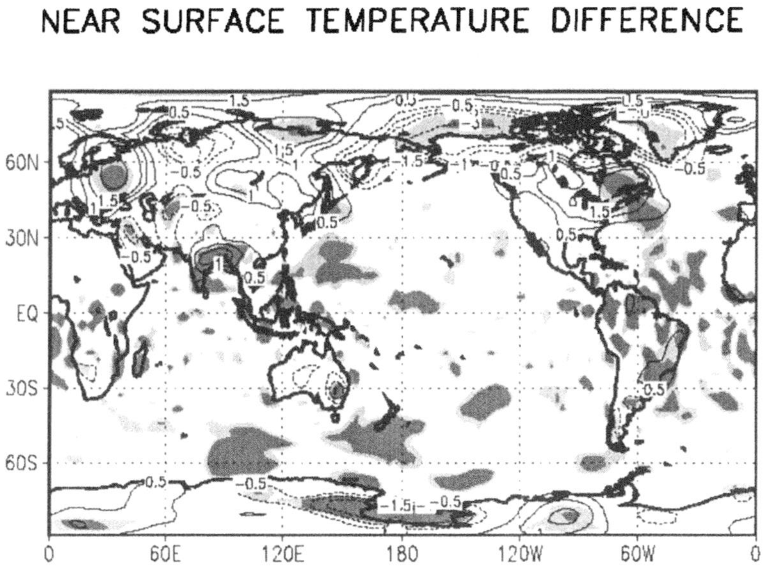

Figure 3: Difference in near-surface air temperature (current-natural) using a 9-point spatial filter for easier visibility. Contour by 0.5, 1.0, 1.5, and 3.0 C. Shaded regions as in Figure 2 (from Chase et al. 1999).

Texier et al. 1997; Dirmeyer 1995, 1999; U.S. National Research Council 1994) provide results which support this conclusion.

6. ACKNOWLEDGMENTS

Support for this research was provided by NASA Grant No. NAG8-1511, NPS Contract No. CA 12682-9004 CEGR-R92-0193, NPS Contract No. CA 12682-9004 COLR-R92-0204, EPA Grant No. R82499301-0, and NSF Grant No. DEB-9011659.

7. REFERENCES

Chase, T.N., 1999: The role of historical land-cover changes as a mechanism for global and regional climate change. Ph.D. Dissertation, Department of Atmospheric Science, Colorado State University, Fort Collins, CO 80523.

Chase, T.N., R.A. Pielke, T.G.F. Kittel, R.R. Nemani, and S.W. Running, 1999. Simulated impacts of historical landcover changes on global climate. Climate Dynamics, submitted.

Claussen, M., 1994: On coupling global biome models with climate model. Climate *Res.,* 4, 203-221.

Claussen, M., 1998: On multiple solutions of the atmosphere-vegetation system in present-day climate. *Global Change Biology,* 4, 549-560.

Claussen, M., V. Brovkin, A. Ganopolski, C. Kubatzki, and V. Petoukhov, 1999. Modeling global terrestrial vegetation climate interaction. Phil. *Trans. Roy. Soc.,* in press.

Dirmeyer, P.A., 1995: Meeting on problems in initializing soil wetness. BUIL Amer. *Meteor. Soc.,* 76, 2234-2240.

Dirmeyer, P.A., 1999: Assessing GCM sensitivity to soil wetness using GSWP data. *J. Meteor. Soc.* Japan, in press.

Eastman, J.L., M.B. Coughenour, and R.A. Pielke, 1999: The effects Of C02 and landscape change using a coupled plant and meteorological model. *Global Change* Biology, submitted.

Foley, J.A., 1994: The sensitivity of the terrestrial biosphere to climate change: a simulation of the middle Holocene. Global *Biogeochemical Cycles,* 8, 505-525.

Liston, G.E. and R.A. Pielke, 1999. A climate version of the Regional Atmospheric Modeling System. J. Climate submitted.

Lu, L., 1999. Implementation of a two-way interactive atmospheric and ecological model and its application to the central United States. Ph.D. Dissertation, Department of Atmospheric Science, Colorado State University, Fort Collins, CO, 134 pp-

Lu, L., R.A. Pielke, G.E.Liston, W.J. Paxton, D. Ojima, and M. Hartman, 1999: Implementation of a two-way interactive atmospheric and ecological model and its application to the central United States. J. Climate submitted.

Pielke, R.A. Sr., G.E. Liston, L. Lu, and R. Avissax, 1999: Land-surface influences on atmospheric dynamics and precipitation. Chapter 6 in Integrating *Hydrology, Ecosystem* Dynamic, 9, and *Bio-* geochemistry in *Complex Landscapes,* J.D. Tenhunen and P. Kabat, Eds., John Wiley and Sons Ltd., 105-116.

Pielke, R.A., 1998: Climate prediction as an initial value problem. Bull. Amer. *Meteor. Soc.,* 79, 2743-2746.

Texier, D., N. de Noblet, S.P. Harrison, A Haxeltine, D. Jolly, S. Joussaume, F. Laarif, I.C.]Prentice, and P. Tarasov, 1997: Quantifying the role of biosphere-atmosphere feedbacks in climate change: Coupled model simulations for 6000 years BP and comparison with paleodata for northern Eurasia and northern Africa. Climate Dynamics, 13, 865-882.

U.S. National Research Council, 1994: *GOALS Glob*al Ocean-Atmosphere-Land System for Predicting Seasonal-to-Interannual Climate. National Academy Press, Washington, DC, 103 pp. [Available from National Academy Press, Box 285, Washington, DC 20055.]

Uncertainties in the Prediction of Regional Climate Change

FILIPPO GIORGI AND RAQUEL FRANCISCO
Abdus Salam International Centre for Theoretical Physics, Physics of Weather and Climate Group, P.O. Box 586, 34100 Trieste, Italy

Key words: Regional Climate, General Circulation Models.

Abstract: Uncertainties in regional climate change predictions for the 21st century by five coupled atmosphere-ocean General Circulation Models (AOGCMs) (two of them including ensembles of simulations), for different anthropogenic forcing scenarios and 23 regions in the World, are examined. The variables considered are seasonally and regionally averaged precipitation and surface air temperature for the future period of [2071--2100] as compared to the present day period of [1961--1990]. We find that the dominant source of uncertainty in the simulation of regional climatic changes by AOGCMs is due to inter-model variability with inter-scenario and internal model variability playing secondary roles. For both models including ensemble simulations, the spread of predicted average climatic changes by different realizations of the same ensemble is small. In addition, simulated regional climatic changes exhibit a high level of coherency among different forcing scenarios. Overall, uncertainties in predicted regional changes by the five AOGCMs are of the order of 3 K or greater for surface air temperature and 25% of present day values or greater for precipitation. These uncertainties would be transmitted to any regionalization technique used to enhance the regional information of AOGCMs. Differences between AOGCMs and their effects on the model simulations need to be better understood in order to increase confidence in the prediction of regional climate change due to anthropogenic forcings.

1. INTRODUCTION

During the last decade, different coupled atmosphere-ocean General Circulation Models (AOGCMs) have been used to simulate transient

G. Visconti et al. (eds.), Global Change and Protected Areas, 127–139.

climatic changes for the 21st century due to anthropogenic forcings such as greenhouse gas (GHG) and sulfate aerosol concentration (IPCC 1996). Coupled AOGCMs are the most powerful tools presently used for climate change prediction. In order to evaluate the possible impacts of climatic changes and thus implement policy actions to best face these impacts, climate change information is needed at the regional scale (i.e. up to 10^7 km^2) or at the country level. To date, most impact research has employed regional climate change information from AOGCM simulations (IPCC 1998). However, such information has limitations related to the models' coarse horizontal resolution (300--500 km). This has prompted the development of regionalization techniques aimed at enhancing the AOGCM information through the use of high resolution and variable resolution global atmospheric models, limited area climate models and statistical methods (e.g. Giorgi and Mearns 1991). With these premises, a comprehensive assessment of regional climatic changes for the 21st century due to anthropogenic forcings requires a four step strategy of: 1) A range of forcing scenarios related to different assumptions of future economic and technological development; 2) A range of different AOGCMs, since different models have different representations of physical processes with their respective strengths and weaknesses; 3) Ensemble simulations for each forcing scenario and AOGCM, as non-linearities in the climate system render a model simulation dependent on initial conditions; and 4) use of different regionalization techniques to enhance the regional information produced by the AOGCMs. Within the framework of this strategy, uncertainties in anthropogenic climate change predictions at the regional scale stem from a hierarchy of sources: 1) Uncertainty related to the forcing scenarios (e.g. GHG and aerosol forcings), hereafter referred to as ``inter-scenario variability"; 2) Uncertainty related to simulations by different AOGCMs for the same forcing scenario, i.e. due to "inter-model (AOGCM) variability", 3) Uncertainty related to predictions by different realizations of a given scenario with a given AOGCM, i.e. due to "internal model (AOGCM) variability"; 4) Uncertainty related to sub-GCM grid scale forcings and processes. Past work has dealt separately with these different sources of uncertainty (e.g. Whetton et al. 1996; Lal et al. 1998; Kittel et al. 1998; Hulme and Brown 1998; Giorgi and Francisco 1999). However, a new suite of AOGCM experiments has recently become available which allows a comparative study of uncertainties due to the sources 1) - 3) above. We present here such a study based on a regional analysis of AOGCM simulations of transient climate change for the 21st century. We do not address here the issue of sub-GCM grid scale processes, which requires an entirely separate work, but stress that uncertainties in AOGCM regional predictions would be transmitted to any regionalization technique used to

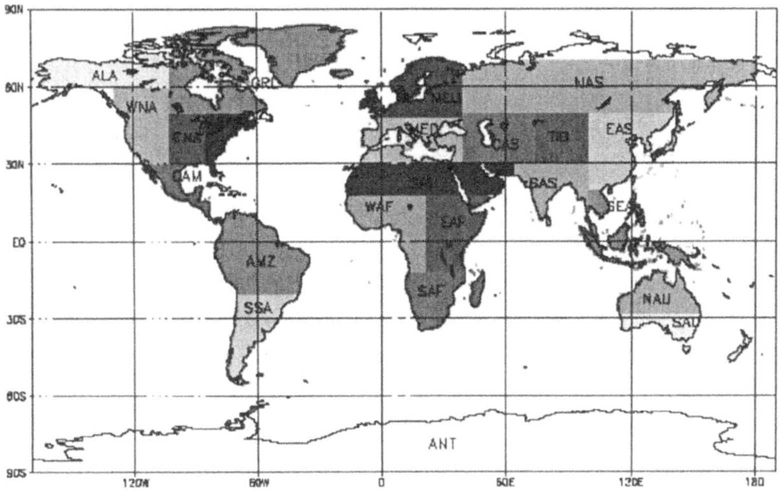

Figure 1. Regions used in the analysis presented in this work.
SAU = Southern Australia; NAU = Northern Australia;
AMZ = Amazon Basin; SSA = Southern South America;
CAM = Central America; WNA = Western North America; CNA = Central North America; ENA = eastern North America; ALA = Alaska; GRL = Greenland (and Northern Territories); MED = Mediterranean Basin; NEU = Northern Europe;
WAF = Western Africa; EAF = Eastern Africa; = SAF = Southern Africa;
SAH = Sahara; SEA = Southeast Asia; EAS = East Asia; SAS = South Asia;
CAS = Central Asia; TIB = Tibet; NAS = North Asia; ANT = Antarctica.
Giorgi and Francisco (1999) provide the regions' definition in terms of latitude and longitude.

enhance such predictions.

2. MODELS AND EXPERIMENTS

For our analysis we divided all land areas in the World into 23 regions using a base 0.5 degree grid (see Fig. 1). The size of the regions (a few thousand km or greater) is such that significant skill can be expected at the resolution of current AOGCMs. We analyzed output from transient climate change simulations with five AOGCMs available in the Data Distribution Center (DDC), a data archive recently created under the auspices of the Intergovermental Panel for Climate Change (IPCC) (information on the DDC can be found by accessing the website http://ipcc-ddc.cru.uea.ac.uk). This output was interpolated onto the 0.5 degree grid and it was then regionally averaged over the regions of Fig. 1.

The interpolation procedure is described in Giorgi and Francisco (1999). A third order polynomial interpolation scheme in latitude and longitude is utilized which, given the smoothness of the AOGCM fields, produces minimal modification of the original fields, both for surface air temperature and precipitation. Note that, since the AOGCMs have different land masks and resolutions, and since only AOGCM land points are used in the interpolation procedure, the actual shape of the regions encompassing complex coastal areas is somewhat different between different AOGCMs. Admittedly, both the use of a particular interpolation scheme and the different coastlines in the models may add a factor of uncertainty whose importance, however, is reduced by taking averages over broad regions such as in Fig. 1. Only land points common to the AOGCM land mask and to the base grid are considered in the regional averaging.

All AOGCM runs covered the period of 1860 to 2100 with transient forcing due to GHG and sulfate aerosol effects. For our analysis, we extracted the 30-year period of [1961-1990] to represent present day climate conditions and the period of [2071-2100] to represent future climate conditions. The variables examined are average surface air temperature and precipitation for December-January-February (DJF) and June-July-August (JJA), i.e. variables used in previous studies (e.g. IPCC 1996; Kittel et al. 1998; Giorgi and Francisco 1999) and of importance for climate change impact work. Here we limit our analysis to averages over the 30-year periods. We indicate with the term "sensitivity" the difference between the averages for the [2071-2100] and [1961--1990] periods and with the term ``bias" the difference between simulated and observed averages for the [1961--1990] period. Throughout this paper, units for precipitation sensitivity and bias are percent of present day precipitation and percent of observed precipitation, respectively. Climate variability is not examined.

Results from five AOGCMs were analyzed, referred to as HADCM2 (Mitchell and Johns 1997; Johns et al. 1997; horizontal resolution in the atmosphere of 3.75° x 2.5° lon x lat or approximately 400 x 270 km), CCC (McFarlane et al. 1992, Boer et al. 1999a,b; horizontal resolution in the atmosphere of 3.7° x 3.7° lon x lat or approximately 400 x 400 km), CSIRO (Gordon and O'Farrell 1997; horizontal resolution in the atmosphere of 5.6° x 3.2° lon x lat or approximately 600 x 350 km), CCSR (Hasumi and Suginohara 1998; horizontal resolution in the atmosphere of 5.6° x 5.6° lon x lat or approximately 600 x 600 km), MPI (Bacher et al. 1998; horizontal resolution in the atmosphere of 2.89° x 2.89° lon x lat or approximately 300 x 300 km). All these coupled AOGCMs include ocean flux corrections for heat and moisture.

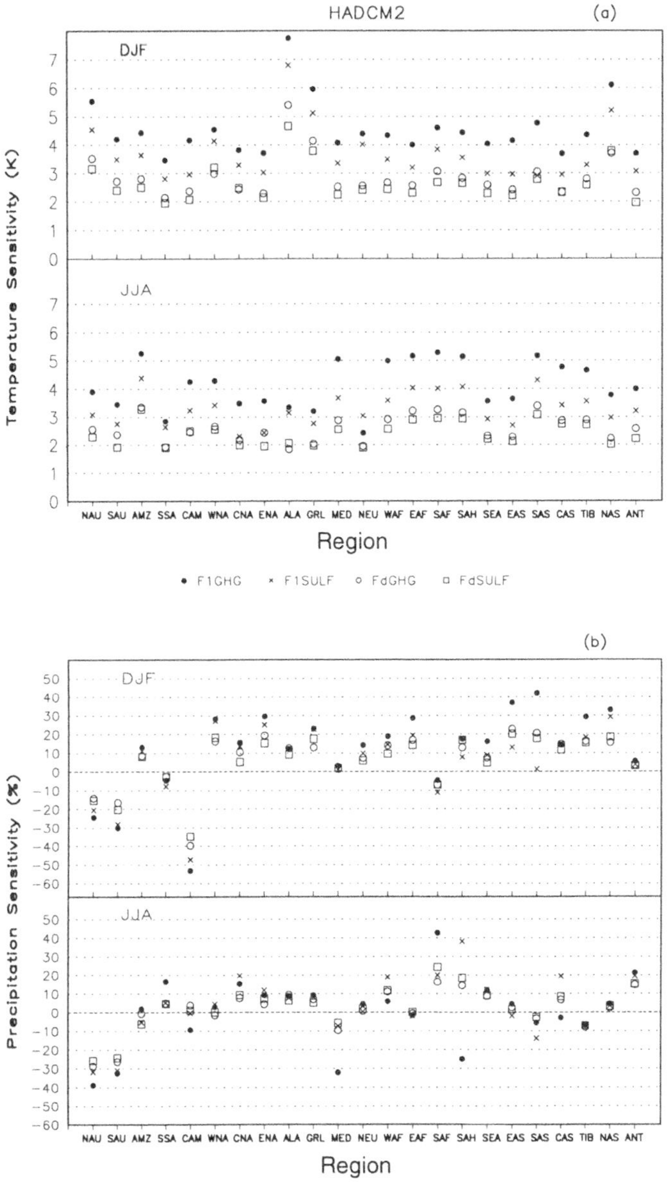

Figure 2. Ensemble average sensitivities ([2071-2100] minus [1961-1990] values) for the HADCM2 simulations of four different forcing scenarios (see text) and for the regions of Figure 1.
a) Surface air temperature (Units are K);
b) Precipitation (Units are percentage of [1961-1990] value).

The following future climate (i.e. past 1990) forcing scenarios are considered: F1GHG: 1% year^{-1} compounded GHG increase (after 1990) and no aerosol effects; F1SULF: 1% year^{-1} compounded GHG increase and inclusion of aerosol effects; FdGHG: 0.5% year^{-1} compounded GHG increase and no aerosol effects; FdSULF: 0.5% year^{-1} compounded GHG increase and inclusion of aerosol effects.

Historical GHG and aerosol forcing and future aerosol forcing are from the IS92a scenario of IPCC (1992). The aerosol forcing is included as a perturbation to the surface albedo to represent direct aerosol effects and a perturbation to the cloud albedo to represent indirect aerosol effects.

The following simulations were available: ensembles of four realizations of all four scenarios for the HADCM2; one realization of the F1GHG and F1SULF scenarios for the CSIRO, CCS and MPI; one realization of the F1GHG scenario and an ensemble of three realizations of the F1SULF scenario for the CCC.

(The CSIRO experiments actually used a 0.9% year^{-1} compounded GHG increase, while the MPI F1SULF experiment only extended to year 2050.) Different realizations of the same transient scenario start at different times in corresponding control experiments (i.e. experiments with pre-industrial levels of GHG), and therefore employ different initial conditions of atmosphere, ocean and land surface (e.g soil moisture).

3. RESULTS

Figures 2a,b first illustrate the effect of inter-scenario variability by showing the ensemble average surface air temperature and precipitation sensitivities in the HADCM2 runs for the four scenarios. The scenarios vary from one of relatively strong forcing (F1GHG) to one of weak forcing (FdSULF) and represent a range of plausible forcing conditions for the next century. The simulated regional warmings mostly vary in the range of 2-5 K, with the greatest warming occurring in the F1GHG scenario. Exceptions are the cold climate regions of Alaska, Greenland and North Asia, where the warming is up to 6--8 K due to the snow/ice feedback mechanism. The range of surface air temperature sensitivities due to varying scenarios (i.e. due to inter-scenario variability) is 1.5 - 2.5 K, with the greatest contribution to this range deriving from the F1GHG case.

For precipitation, the sensitivities are mostly positive, resulting from an intensified hydrologic cycle in future climate conditions, with a few exceptions (e.g. Australia) likely related to regional shifts in storm tracks.

The magnitude of the sensitivities is mostly in the range of a few percent to 30% of present day precipitation with the exceptions of Central America in DJF and Southern Africa and Sahara in JJA. In addition, there is a high level of coherency, at least in sign, between regional precipitation sensitivities associated with different forcing scenarios. The spread of precipitation sensitivities associated with inter-scenario variability is variable from region to region, from a few percent to 20-30%, but it is mostly of the order of 15% or less.

The effects of inter-model and internal model variability are illustrated in Figs. 3a,d, which present surface air temperature and precipitation sensitivities for different models and realizations of the F1GHG and F1SULF scenarios. Concerning surface air temperature (Figs. 3a,b), it is evident that the effect of internal model variability is much less pronounced than that of inter-model variability. The temperature sensitivities of different realizations of the same ensemble are close to each other, both for the HADCM2 and CCC models. In fact, a t-test showed that the average temperatures of two realizations of the same ensemble did not differ from each other at the 5% confidence level in the vast majority of cases. The effect of internal model variability on the temperature sensitivities is generally less than 1 K. By contrast, the effect of inter-model variability is generally greater than 2 K, with many instances of up to 4-7 K, in both the F1GHG and the F1SULF scenarios. (It should be noted, however, that in several cases the inter-model range of sensitivities is dominated by results from one "outlier" model).

Similar conclusions are found for precipitation, although the internal model variability appears more pronounced for precipitation than for surface air temperature. Both for the HADCM2 and CCC experiments, in most cases the precipitation sensitivities associated with different realizations of the same ensemble have the same sign, be it positive or negative, and the spread associated with internal model variability is mainly of the order of 10-20%. By comparison, the spread associated with inter-model variability is in the range of 20--80% with many instances for which the sensitivities produced by different models are of opposite sign. A t-test showed that all the temperature sensitivities in Figs. 2 and 3 were statistically significant at the 5% confidence level, while the number of statistically significant precipitation sensitivities varied among models, regions and season. If we consider the spread of the sensitivities in Figs. 2, 3 as a measure of the uncertainty in the simulation of regional climate change, it is evident that the largest source of uncertainty is due to inter-model variability. At the other end, both the HADCM and CCC experiments indicate that the smallest contribution to uncertainty is due to internal model variability and that a small number of realizations is sufficient to characterize the average multi-decadal regional climatology of an AOGCM.

The results of Fig. 2 also show that, at least for the HADCM2 model, the regional sensitivities for different scenarios show similar trends.

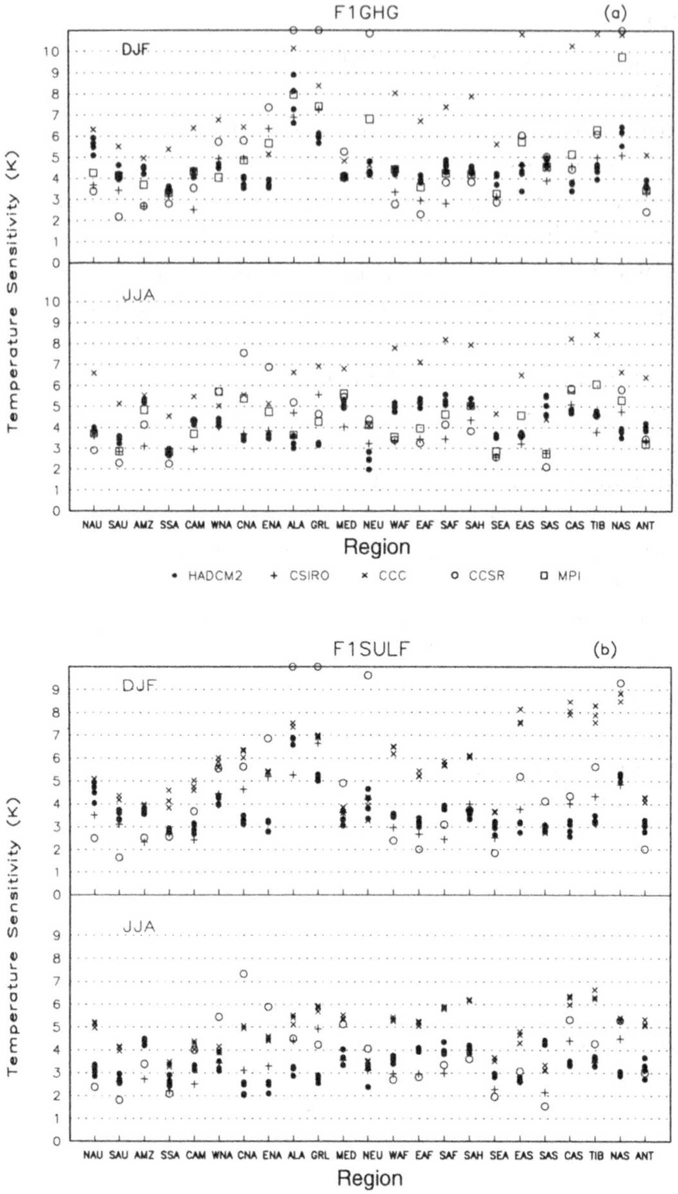

Figure 3. Sensitivities ([2071-2100] minus [1961-1990] values) for different model simulations (individual realizations) of two forcing scenarios (see text) and for the regions of Fig. 1.

a) Surface air temperature (Units are K), F1GHG scenario;

b) Surface air temperature (Units are K), F1SULF scenario;

This is somewhat surprising, at least for precipitation, given the different nature of GHG (global) and aerosol (regional) forcings.

Our analysis also gives some tentative indications for specific regions: 1) Surface air temperature sensitivities are highest for the cold climate regions, likely as an effect of the snow/ice albedo feedback mechanism. The CCC and CCSR models show the largest sensitivities (up to 10 K or more) over cold climate regions; 2) The inter-model spread in surface air temperature sensitivities is largest over the Asian and African regions, Alaska, Greenland and Northern Europe; 2) Aerosol effects tend to reduce the surface air temperature sensitivities over most regions but do not strongly affect the precipitation sensitivities; 3) In only a few cases all models agree in the sign of the precipitation sensitivities (noticeable cases where most models indicate negative precipitation changes are Northern and Southern Australia, Central America and the Mediterranean in JJA); 4) Regional precipitation sensitivities are mostly positive, especially during DJF. As a second measure of uncertainty we can take the models' ability to reproduce present day climate, as measured for example by the model bias. Figures 4a,b present the regional surface air temperature and precipitation biases for all F1SULF simulations with respect to the observed dataset developed at the Climatic Research Unit (CRU) of the East Anglia University for the 30-year period of 1961--1990 (New et al. 1999). This dataset is defined on the 0.5 degree grid of Fig. 1. The biases, as well as their inter-model ranges, are of the order of a few to 4-6 degrees for surface air temperature and several tens of percent to over 100% of observed values for precipitation. The biases associated with different realizations of the same ensembles are close to each other, both for the HADCM2 and CCC models, which is again an indication of low internal model variability. The performance of the models in reproducing observed averages is highly variable from region to region and there is no model that ubiquitously shows lower biases than the others. Comparison of Figs. 2, 3, and 4 shows that, as was also concluded by analyses of a previous generation of AOGCM experiments (Kittel et al. 1998), the models tend to agree with each other better in simulating sensitivities than in reproducing present day climate.

4. CONCLUSIONS

Our analysis indicates that the current uncertainty in simulated regional and seasonal sensitivities by coupled AOGCMs for the future climate period of [2071-2100] is of 3 K or greater for surface air temperature and 25% (of present day values) or greater for precipitation.

This uncertainty is highly variable from region to region. Our results clearly show that inter-model variability represents the primary source of uncertainty in the prediction of regional climate change by AOGCMs, with

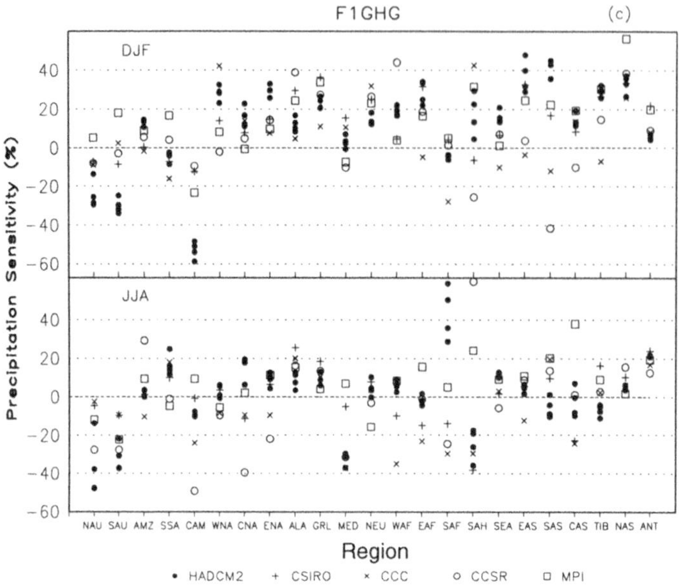

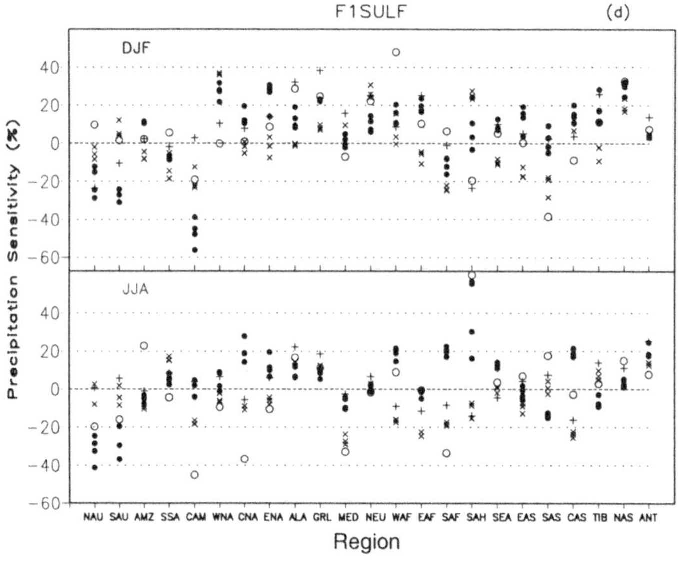

Figure 3. Sensitivities ([2071-2100] minus [1961-1990] values) for different model simulations (individual realizations) of two forcing scenarios (see text) and for the regions of Fig. 1.

c) Precipitation (Units are percentage of [1961-1990] value), F1GHG scenario;

d) Precipitation (Units are percentage of [1961-1990] value), F1SULF scenario.

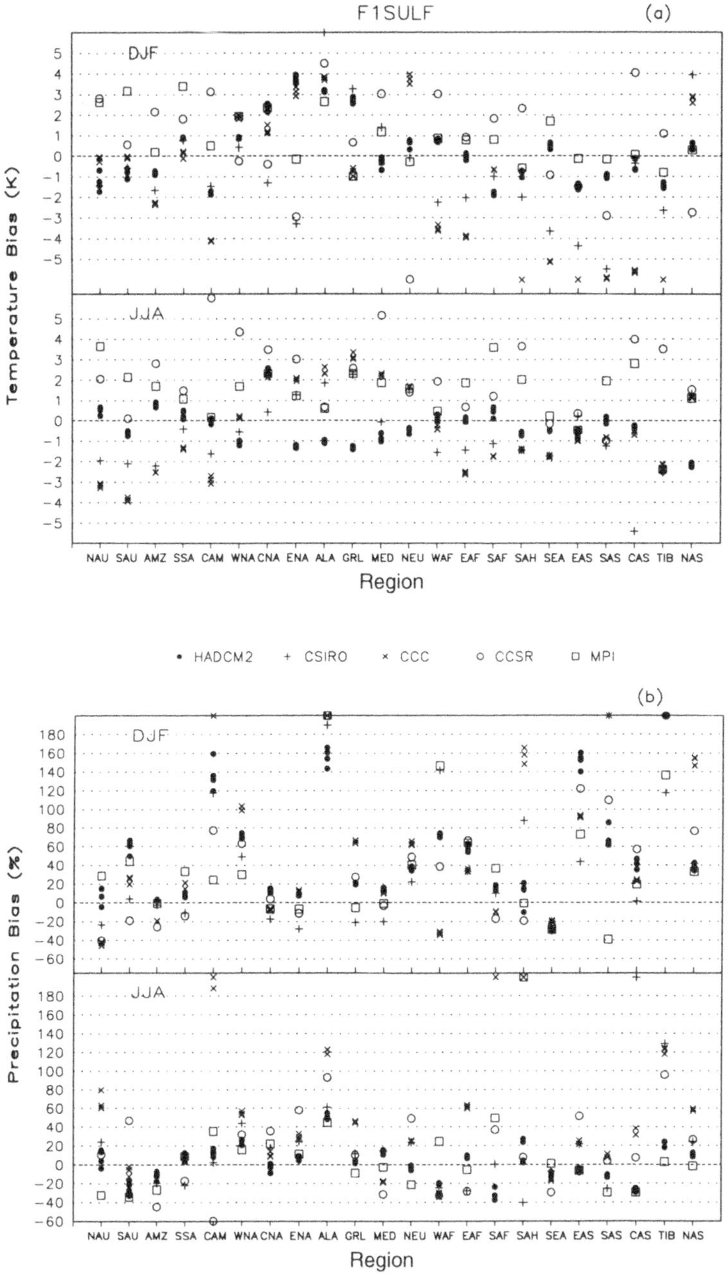

Figure 4. Biases (simulated minus observed values for the [1961-1990] period) for different model simulations of the F1SULF scenario and for the regions of Fig. 1.
a) Surface air temperature (Units are K);
b) Precipitation (Units are percentage of observed value). Observations are from the Climatic Research Unit's dataset (New et al. 1999). Biases for Antarctica are not included because data was not available for this region in the observation dataset.

inter-scenario and internal model variability playing less prominent roles.

Similar conclusions were also found for sensitivities relative to the future climate period of [2041-2070] (not shown for brevity). In addition, the models still show deficiencies in reproducing present day climate conditions, with biases widely varying among models and regions. Therefore, in order to increase our confidence in the model simulations of climate change it is of foremost importance to 1) understand fundamental differences between models and their effects on the model performance; and 2) improve the model performance in simulating present day climatology.

In this respect, it appears that the model sensitivity to the treatment of ice and snow processes greatly affects inter-model variability, at least over cold climate regions.

Some caveats need to be considered in the evaluation of these results. First, all models analyzed here employed ocean flux correction. This correction is primarily included to prevent model drift and obviously influences the model simulations of climate change. In addition, our conclusions are limited to climatic averages. For many impact applications, climate variability can be more important than climate averages, so that more work is needed to analyze model simulations of climate variability at different temporal scales.

5. ACKNOWLEDGMENTS

We thank the Hadley Centre for Climate Prediction and Research, the Canadian Climate Center, the Commonwealth Scientific and Industrial Research Organization, the Max Plank Institute for Meteorology and the Centre for Climate Study Research for making available the results of their AOGCM simulations.

We also thank the Climatic Research Unit of the University of East Anglia for making available the observation datasets.

6. REFERENCES

Bacher, A., J.M. Oberhuber, E. Roeckner, ENSO dynamics and seasonal cycle in thetropical Pacific as simulated by the ECHAM4/OPYC3 coupled general circulation model, Clim. Dyn., 14, 431, 1998.

Boer, G.J., G.M. Flato, D. Ramsden, 1999a, A transient climate change simulation with historical and projected greenhouse gas and aerosol forcing: projected climate for the 21st century. Submitted to Clim. Dyn., 1999b.

Boer, G.J., G.M. Flato, M.C. Reader, D. Ramsden, A transient climate change simulation with historical and projected greenhouse gas and aerosol forcing: experimental design and comparison with the instrumental record for the 20th century. Submitted to Clim. Dyn., 1999a.

Giorgi F. and L.O. Mearns, Approaches to regional climate change simulation: a review, Rev. Geo., 29, 191-216, 1991.

Giorgi, F. and R. Francisco, Uncertainties in regional climate change predictions. A regional analysis of ensemble simulations with the HADCM2 GCM, Clim. Dyn., in press (1999).

Gordon H.B., and S.P O'Farrell, Transient climate change in the CSIRO coupled model with dynamic sea ice, Mon. Wea. Rev., 125, 875, 1997.

Hasumi H., and N. Suginohara, Effects of the seasonal variation on forming the steady state of an atmosphere-ocean coupled system, Clim. Dyn., 14, 803, 1998.

Hulme, M., O. Brown, Portraying climate scenario uncertainties in relation to tolerable regional climate change, Climate Research, 10, 1-14, 1998.

IPCC, Climate change 1992: the supplementary report to the IPCC scientific assessment. IPCC Working Group I. Houghton JT, B.A. Callander and S.K. Varney (eds). Cambridge University Press, Cambridge, UK, 200 pp., 1992.

IPCC, Climate change 1995: The science of climate change, Contribution of Working Group I to the Second Assessment Report of the Intergovernmental Panel on Climate Change. J.T. Houghton, L.G. Meira Filho, B.A. Callander, N. Harris, A. Kattenberg, K. Maskell (eds), Cambridge University Press, New York, 572 pp., 1996.

IPCC, The Regional Impacts of Climate Change, An Assessment of Vulnerability, A Special Report of IPCC Working Group II, R.T. Watson, M.F. Zinyowera, R.H. Moss, and D.J. Dokken (eds), Cambridge University Press, Cambridge, U.K., 517 pp., 1998.

Johns T.C., R.E. Carnell, J.F. Crossley, J.M. Gregory, J.F.B. Mitchell, C.A. Senior, S.F.B. Tett, R.A. Wood, The second Hadley Centre coupled ocean-atmosphere GCM: model description, spin up and validation, Clim. Dyn., 13, 103-134, 1997.

Kittel T.G.F., F. Giorgi, G.A. Meehl, Intercomparison of regional biases and doubled CO2 sensitivity of coupled atmosphere-ocean general circulation model experiments, Clim. Dyn., 14, 1-15, 1998.

Lal M., U. Cubasch, R. Voss, and J. Waszkewitz, Effects of transient increase in greenhouse gases and sulfate aerosols in monsoon climate, Current Science, 69, 752-763, 1998.

McFarlane, N.A., G.J. Boer, J.-P. Blanchet, and M. Lazare, The Canadian Climate Center second generation general circulation model and its equilibrium climate, J. Climate, 5, 1013, 1992.

Mitchell J.F.B., and T.C. Johns, On modification of global warming by sulfate aerosols, J. Climate, 10, 245-267, 1997.

New, A.M., M. Hulme, P.D. Jones, Representing twentieth-century space time climate variability. Part I: Development of a 1961-1990 mean monthly terrestrial climatology. J. Climate, 12, 829, 1999.

Whetton, P.H., M.H. England, S.P. O'Farrell, I.G. Watterson, and A.B. Pittock, Global comparison of the regional rainfall results of enhanced greenhouse coupled and mixed layer ocean experiments: implications for climate change scenario development. Climatic Change, 33, 497-519.

Gamma-Ray Spectrometer for "In Situ" Measurements on Glaciers and Snowfields

ANTONELLA BALERNA, ENRICO BERNIERI, MAURIZIO CHITI, UBALDO DENNI, ADOLFO ESPOSITO, ANTONIETTA FRANI
INFN - Laboratori Nazionali di Frascati Via E. Fermi 40, 00044 Frascati Italy

Key words: Radioactivity, Fallout, Gamma-ray detectors, Glaciology, Time-markers, Cs-137

Abstract: A gamma-ray spectrometer based on a NaI(Tl) scintillator crystal coupled with a photomultiplier for measurements on glacier and snowfields is being developed. The whole instrument (detector+electronics+computer+power supply) will be portable, temperature controlled and able to work in harsh environmental conditions. One of the main purposes is the detection of the radioactive gamma-peak of Cs-137, due to nuclear fallout, for *"in situ"* determination of absolute time markers in ice layers. The instrument is even able to detect a wide range of natural and artificial radioactive isotopes, allowing the determination of the kind of radioactive contamination in remote areas where sampling is difficult or impossible.

1. INTRODUCTION

It is well known that radioactive fallout can be used as time markers in snow layers and in ice bore-holes. A clear example is the beta activity measured in ice samples in Antarctica [1] that shows clearly, as a function of the depth, peaks linked to the nuclear explosions in fifties and sixties and to the Chernobyl accident (Fig. 1).

Usually this method requires samples from snow or ice cores to be gathered and returned to the laboratory.

However these samples must often be taken from remote places were handling and shipping can be difficult, and sometimes those places are of difficult access.

G. Visconti et al. (eds.), Global Change and Protected Areas, 141–146.

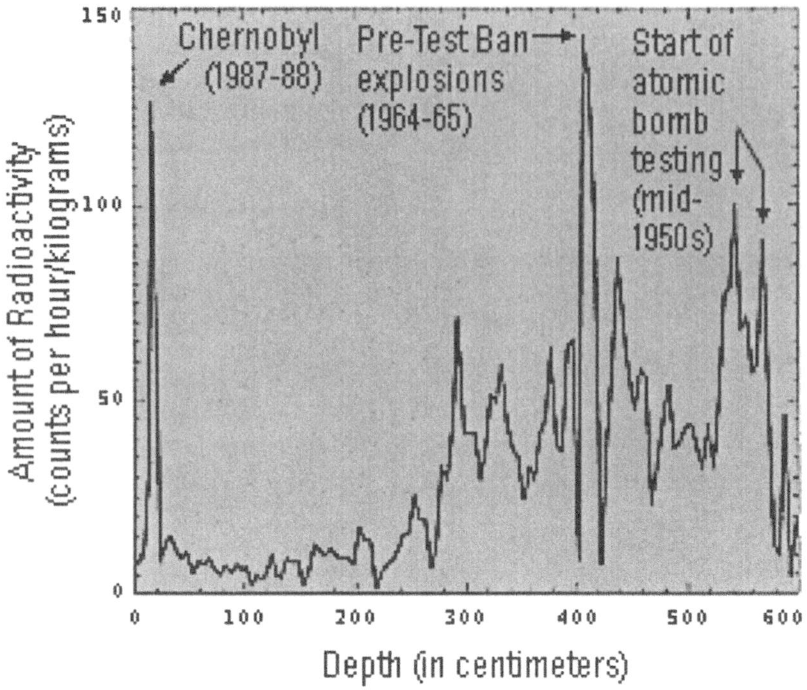

*Fig.*1. Beta activity vs.depth in ice core (from ref. [1])

This is particularly true in the higher parts of mountain regions which are particularly sensitive to small amounts of radioactive contaminants and offer the unique opportunity to detect in advance small changes in the atmospheric composition.

In 1981, Pinglot and Pourchet [2] proposed a method to measure "in situ" radioactive fallout monitoring the emission at 662 keV of Cs-137 by using a NaI(Tl) scintillator detector coupled with a photomultiplier tube and a multichannel analyser. Their measurements showed that the gamma activity can be detected "in situ" and is strongly correlated with the radioactivity measured in samples.

The limit of their system was the portability, being the weight of the whole detector system of about 250 kg !

In 1994, Dunphy, Dibb and Chupp [3] proposed a lighter portable system based on a NaI(Tl) scintillator detector computer controlled. Their measurements were done near a permanent base by using a standard AC power source which is not usually available - or portable - in some remote areas, like high altitude glaciers.

For this reason we are developing a completely portable instrument that should be easily carried and handled by a small team without any external support.

Table 1. Main fission products from thermonuclear tests or accidents

Fission products	Gamma-energy (keV)	Half-life
Nb-95	765	35 d
Zr-95	756-724	65 d
Ce-144	133	285 d
Ru-106	511	372 d
Sb-125	428	2.8 y
Ce-137	662	30.17 y

2. THE TECHNIQUE

The aim of the proposed detector, developed for "in situ" measurements, is to detect the 662 keV gamma-rays emitted by Cs-137. This energy is sufficiently high to penetrate the snow with an attenuation length of about 20 cm. Among the main fission products from nuclear tests or accidents, Cs-137 is the gamma-emitter with the longest half-life, 30.17 y (Tab. 1). Two radioisotopes show energies higher than Cs-137, Nb-95 and Zr-95, but their relatively short half-life leads to weak activity several years after the deposition. The only other isotope with gamma emission at 609 keV, near the Cs gamma peak, is the Bi-214 which is of natural origin. However these two lines can be easily resolved even with a detector of moderate energy resolution.

The "in situ" measurements of Cs-137, in bore-holes or along trench profiles, is made in the presence of cosmic rays, and the detected signal contains even this contribution. However it has been observed that, at the reference levels, the Cs-137 signal is up to twenty times higher than the cosmic ray background [2].

The gamma-rays "in situ" measurement technique requires that the detector, the associated electronics and the power sources must operate in the cold, wet and harsh environment where the measurements are made.

3. THE PROPOSED SYSTEM

The entire instrument, showed in Fig. 2, must satisfy the following characteristics:

- High efficiency in the Cs-137 emission region
- Low temperature working conditions
- Low power consumption
- Portability
- Work in wet conditions and resistance to mechanical shocks

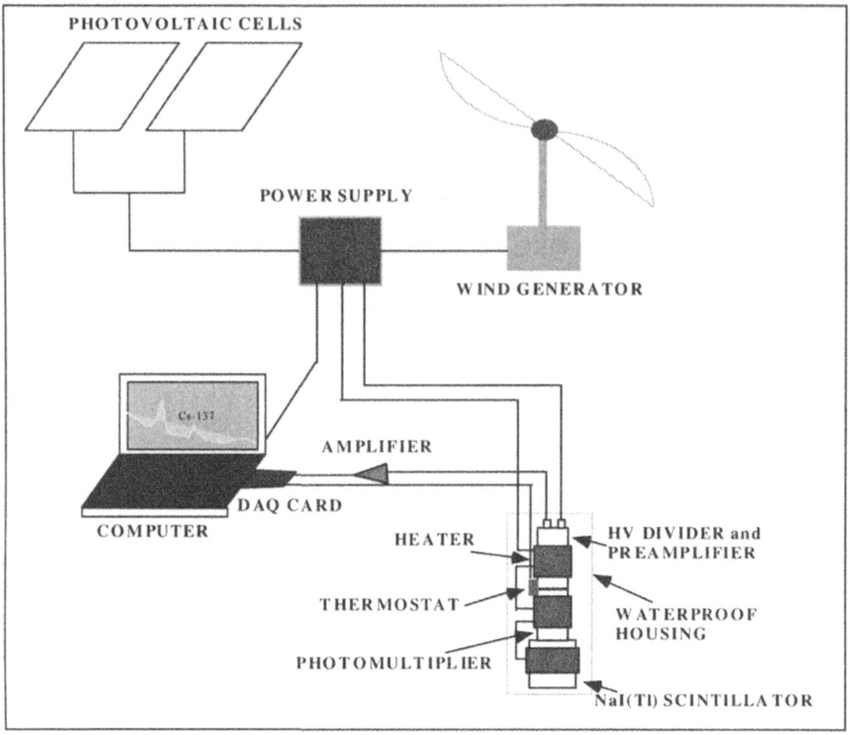

Fig. 2. The gamma-ray spectrometer for "in situ" measurements

As detector a NaI(Tl) scintillator coupled with a photomultiplier has been chosen. NaI(Tl), shows a high efficiency for the Cs-137 gamma-ray region and enough resolution to discriminate the Bi-214 peak. A great scintillator volume is needed in order to obtain a very low minimum detectable signal, and since the bore-holes in snow have a minimum diameter of 100 mm, a cylindrical scintillator crystal with 76 mm diameter and length of about 100 mm was selected. The scintillator, the photomultiplier, the high-voltage divider and the preamplifier will be assembled and housed in a thin stainless steel water-resistant cylinder.

The detector can be exposed to temperatures as low as –50°C. Since the temperature gradient can damage the detector and affects the detector response changing the position and the width of the detected peak, the stainless steel cylinder will be wrapped with a flexible heater controlled by a thermostat. To protect from water and mechanical shocks, the all system (detector, heaters and thermostat) is housed in an acrylic tube. The temperature is monitored (inside and outside the housing tube and in the environment) by using semiconductor temperature sensors.

All the detector system must be carried with backpacks by a team of two or tree persons.

So all its parts must be accurately selected specially as a function of weight in order to obtain an overall weight not higher than 15-20 kg. For this reason the use of a portable generator as power supply is excluded.

We are planning to use as power source a mixed system composed by batteries, photovoltaic cells and wind generators, taking into account the possible different environmental conditions. A wind generator of 6 kg weight – working with a windspeed of about 10 m/s - can generate a power of about 300 W. The same power can even be obtained by using about 3 m^2 of photovoltaic cells. This power is assumed to be the maximum consumption limit of our system. This requirement can be satisfied considering that the power required by the computer and the detector should be lower than 100 W and that 100 W is the evaluation of the power necessary to warm up the detector when working in very low temperature conditions.

4. DATA ACQUISITION

Data acquisition and temperature controls will be performed by using a Panasonic (Toughbook CF 27) portable computer - protected against water, dust and mechanical shocks - and a National Instruments DAQ Card (AI-16E-4) that allows the input/output of analogue and digital signals. A stretching circuit has been developed to match the preamplifier output signal with the input requirements of the card.

All the signals will be processed by means of LabVIEW (5.1) software able to reproduce a multichannel analyser system.

5. PRELIMINARY TESTS

Some preliminary tests have been done on the field, on the Calderone glacier, by using a commercial portable detector. The tests have shown that even in mild climate conditions it is necessary to work with a suitable instrument since the commercial one does not satisfy the requirements for "in situ" measurements. Laboratory tests to analyse the temperature dependence of the signal and the performances of various power supply sources are in progress in order to optimise the final configuration of the entire system.

6. REFERENCES

Dibb J.E., P.A. Mayewski, C.S. Buck and S.M. Drummey, Nature 345 (1990) 25

Dunphy P.P., J.E. Dibb, E.L. Chupp, Nucl. Instr. Meth. A 353 (1994) 482-485

Pinglot J.F. and M. Pourchet, In: Methods of low-level counting and spectrometry, IAEA, Vien (1981), pp 161-172

Cs-137 Gamma Peak Detection in Snow Layers on Calderone Glacier

*ANTONELLA BALERNA, *ENRICO BERNIERI, *ADOLFO ESPOSITO, **MASSIMO PECCI AND ***CLAUDIO SMIRAGLIA
**INFN - Laboratori Nazionali di Frascati, Via E. Fermi 40, 00044 Frascati Italy;*
*** ISPESL - Dipartimento Insediamenti Produttivi e Interazione con l'Ambiente, Via Urbana 167, 00184 Roma Italy;*
**** Universita' degli Studi di Milano - Dipartimento Scienze della Terra, Via Mangiagalli 34, 20133 Milano Italy*

Key words: Calderone glacier, Gran Sasso d'Italia, climate change, pollution, radioactivity, Cs-137

Abstract: The Calderone glacier, located in the Gran Sasso d'Italia mountain group (Abruzzo, Italy), is the most southern glacier in Europe. The reduced dimension and the general conditions of the glacier make it a powerful environmental indicator in evaluating global change processes including the radio-chemical pollution induced by human activity. Two high altitude samples of snow, collected during spring 1999, and summer 1995, have been analysed using a high purity Germanium solid state gamma ray detector. The analysis of these samples revealed the presence of the radioactive isotope Cs-137. The possible origin of this contamination and differences between different samples are analysed. The physical features of the snow, where the spring 1999 samples were gathered, are also presented.

1. INTRODUCTION

The Calderone glacier is a small glacier of about 50.000 m^2, located in a deep northward valley forming from the top of Corno Grande d'Italia (2912 m asl) in the center of the Gran Sasso d'Italia mountain group (Central Italy). The glacier, characterised by a strong reduction phase, is at present the most southern glacier in Europe (42 28' 15" N), since the Picado de Veleta glacier in the Sierra Nevada has completely melted.

G. Visconti et al. (eds.), Global Change and Protected Areas, 147–152.

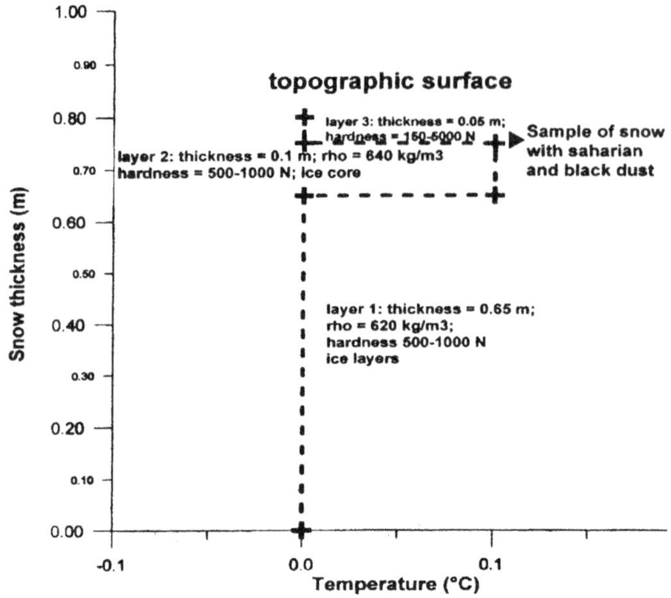

Fig.1 Trench 1995. Stratigraphy of the snow. "rho" indicates the snow density. The arrow indicates the sampled layer.

Since 1994, a set of multidisciplinary studies were started in order to evaluate the actual role of the glacier as an environmental indicator in climate change processes and in the effects induced by industrial activity in the Mediterranean area [1]. Those studies include historical survey, geomorphologic analysis, continuous climate parameters monitoring, analysis of the annual snow layer, and, recently, radioactive monitoring.

The aim of radioactive monitoring is to look for the presence of radioisotopes produced by human activity and sent in the atmosphere, that can cause severe environmental pollution. High altitude glaciers and snowfields are among the most sensitive indicators of such a pollution and, in particular, the Calderone glacier, due to its particular position in the Mediterranean area, can be considered one of the most representative indicators in this region.

2. RADIOACTIVITY MEASUREMENTS

Among the main fission products from thermonuclear tests or accidents, Cs-137 and Sr-90 are the two isotopes with the longest half-life.

Sr-90, with a half-life of 28.6 years, is a beta emitter. Cs-137, with a half-life of 29.5 years, is a gamma-emitter at 662 keV and 33 keV, and beta-emitter at 514 keV.

To detect radioactivity from beta-emitters it is necessary that samples be brought back to the laboratory, melted and filtered. Gamma emission, instead, can be in principle measured in situ, allowing the possibility to determinate faster, in the field, absolute time markers in snow or in ice bore-holes [2], [3].

Radioactive isotopes of natural origin do not interfere with the measurement of the Cs-137 gamma peak at 662 keV. The nearest gamma-peak, coming from Bi-214, located at 609 keV - and easily discriminated by a solid-state detector - can be resolved even by using a NaI(Tl) scintillator of moderate 10% energy resolution.

Due to is high efficiency in this energy region this kind of detector is the most suitable for measurements in situ.

3. FIELD SAMPLING ON SNOW COVERAGE

The field surveys and sampling have been carried out in July 1995 and June 1999. The field operations included:

- general measurements of thickness of snow and choice of the best site
- digging of a trench in the whole thickness of the whole snow layer, deposed above the glacier body and the superficial debris;
- stratigraphy of the snow thickness and, in the first sampling, the characterisation of physical and mechanical properties (Fig 1 and 2), according to the international classification of snow [4].
- Sampling of the most representative layers of the trench.

The July 1995 snow coverage showed a still wide distribution on the whole glacier surface and was characterised by a thin but evident superficial deposit of Saharan and black dust. The trench was realised at about 2800 m a.s.l. with a slope angle of about 35 °, in cloudy weather condition with an air temperature of about +3°C. The sampling was performed on this characteristic deposit. The June 1999 snow coverage was extended to the whole glacier surface, even if with a minor thickness than the standard situation of the month [5]. The snow coverage (Fig. 2) showed a complex stratification, due to the relatively initial condition in summer season. In fact in the snow stratigraphy the single atmospheric event is still well detectable in a single snow layer. The trench was realised at about 2700 m a.s.l. with a slope angle of about 25 °, in good weather condition with an air temperature of about + 8.5 °C. The sampling was performed at the bottom and at the top of the snow thickness.

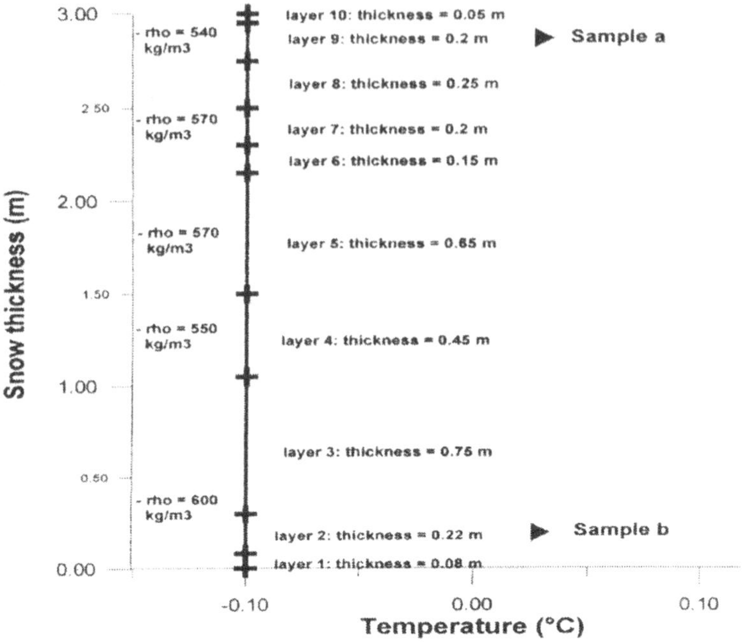

Fig.2 Trench 1999. Stratigraphy of the snow. "rho" indicates the snow density. The arrows indicate the two sampled layers.

A first attempt of "in situ" detection, in the 1999 trench, of the gamma emission of radioactive contaminants, by using a standard (50x50 mm) NaI(Tl) scintillator detector was unsuccessful, probably due to the short acquisition time. Two samples of snow, belonging to the layers 2 and 9, were gathered to perform laboratory measurements.

4. RADIOACTIVITY MEASUREMENTS ON SAMPLES

The measurements on the snow samples were performed by using a low-background laboratory gamma spectrometer composed by:

- an HPGe - high purity Germanium - solid state gamma ray detector
- (PGT) with 16% efficiency and a resolution (FWHM) of 1.9 keV/1.33 MeV.
- a low noise amplifier for gamma spectrometry
- a 8000 channel AD converter
- a PC for data acquisition and processing
- a software for automatic analysis of the spectra

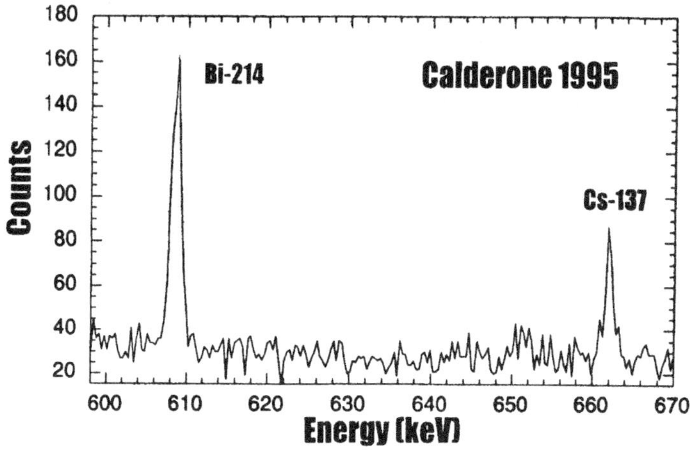

Fig. 3. The gamma-ray spectra of the 1995 sample. The peaks from 662 keV Cs-137 and 609 Bi-214 gamma-rays are indicated.

The non-filtered samples were put in Marinelli's bekers (500 cm^3 capacity) and, in order to reduce the environmental radioactivity contribution, were well shielded by a 10 cm lead layer and two layers of copper and cadmium, of 1 mm each. The acquisition time was about 170000 seconds.

Fig. 3 and 4 show the spectra obtained for the 1995 and the 1999 samples respectively. (No difference has been found between the spectra of the two 1999 layers and only the layer 2 spectra is reported.) The spectra show clearly the Cs-137 gamma peak, at the energy of 662 keV. The other bigger peak at 609 keV is due to the natural isotope Bi-214. The measured activity of the 1995 and 1999 samples were 0.67 Bq/kg and 0.22 Bq/kg, respectively.

5. CONCLUSIONS AND PERSPECTIVES

Strong quantities of Cs-137 were sent in the atmosphere during thermonuclear tests (for example in the years 1954 and 1962) and nuclear accidents, like Chernobyl in 1986. Traces of this last event are still present in the environment and were even found in the soil on the Gran Sasso mountain [6]. Our measurements show that small quantities of Cs-137 are still present in the atmosphere and can be detected in snow layers associated to single atmospheric events. The development of a suitable detector, able to work in situ in high sensitive mountain regions, could allow the real-time detection of small amounts of gamma-emitters radioactive isotopes in the atmosphere.

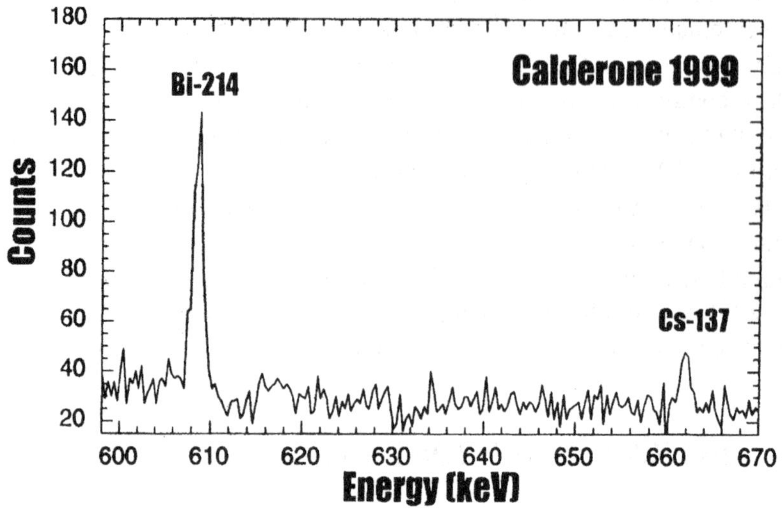

Fig. 4. The gamma-ray spectra of the 1999 sample (layer 2). The peaks from 662 keV Cs-137 and 609 Bi-214 gamma-rays are indicated.

The difference in Caesium content between the two samples can, or reflect a change in the atmospheric composition between 1995 and 1999, or a different concentration of dust particles in the snow due to the melting. This latter hypothesis will be tested with further measurements during summer 1999.

6. REFERENCES

Arpesella C. and G. Schirippa Spagnolo, *Monitoraggio del Radon e della Ionizzazione*, a cura del Consorzio di Ricerca del Gran Sasso (1993).

Balerna A., E. Bernieri, M. Chiti, U. Denni, A. Esposito, A. Frani, *Gamma-ray spectrometer for in situ measurements on glaciers and snowfields*, these Proceedings.

D'Alessandro L, M. D'Orefice, A. Marino, M. Pecci, C. Smiraglia, *The Calderone Glacier (Gran Sasso D'italia Mountain Group): Knowledge And Geo-Environmental Issues*. VI Congr. Giov. Ric Geol. Appl. Ottobre 1998. Mem. Soc. Geol. It. (in press).

Dunphy P. P., J. E. Dibb, E. L. Chupp, Nucl. Instr. Meth. A, 353 (1994) 482-485.

ICSI-IASH-IGS. Clasificazione internazionale della neve stagionale presente al suolo. Traduzione italiana. Gruppo di lavoro dei previsori AINEVA, Neve e Valanghe, 19 (1993).

Pecci M., C. Smiraglia, M. D'Orefice, Neve e Valanghe, 32 (1997) 46 –57.

Section 2

IMPACT ON THE BIOSPHERE AND HYDROLOGY

The Effects of Global Warming on Mountain Regions: a Summary of the 1995 Report of the Intergovernmental Panel on Climate Change

M. BENISTON
Department of Geography, University of Fribourg, Switzerland

Key words: Climate change, climate impacts, environmental systems, socio-economic systems, policy

Abstract: This paper focuses on the impacts of climatic change in mountain regions, based on the material published in the 1996 Second Assessment Report of the Intergovernmental Panel on Climate Change (IPCC). A particular focus is provided for selected sensitive environmental and socio-economic sectors, in particular water, snow and ice, ecosystems, mountain agriculture, tourism, and energy. Some approaches and recommendations for policy response to the overall problem of global warming on mountain regions is also given.

1. INTRODUCTION

Few assessments of the impacts of environmental change in general, and climatic change in particular, have been conducted in mountain regions, as opposed to other biomes such as tropical rainforests, coastal zones, high-latitude, or arid areas.

This is mainly because mountain orography is often too poorly represented in global or regional climate models for meaningful projections to be applied to the impacts sector, although recent progress in high-resolution models is leading to an improved situation.

There is also a significant lack of comprehensive multi-disciplinary data for impact studies, which is one of the pre-requisites for case studies of

G. Visconti et al. (eds.), Global Change and Protected Areas, 155–185.

impacts on natural or socio-economic systems (Barry, 1992; Kates *et al.*, 1985; Riebsame, 1989; Parry *et al.*, 1992).

In addition, the complexity of physical, ecological and social systems in mountains, and their mutual inter-dependency, pose significant problems for assessment, particularly because there is often little or no quantification of the value of mountains in monetary terms, as Price (1990) has pointed out. The complexity of mountain systems presents major problems for assessing the potential impacts of environmental change.

This applies to assessments of changes in both biophysical systems (e.g., Rizzo and Wiken, 1992; Halpin, 1994) and societal systems, particularly because many of the most valuable products of mountain regions are not easily quantified in monetary terms (Price, 1990). Tourism, which is an increasingly important component of mountain economies around the world, represents an economic sector which, in contrast to agriculture or forestry, is closely linked to other aspects of mountain economies and is not a "stand-alone" sector. A further consideration is that few mountain regions can be described comprehensively, because of the heterogeneity of the available data for climatic, biological, or socio-economic impact assessments. Because of the diversity of mountain economies, from the exclusively tourist-based ones to those characterized by subsistence agriculture, no single impacts study will adequately represent the range of potential socio-economic responses to climate change.

A case-by-case approach is therefore essential to understand how, for example, mountain agriculture may change in Bolivia, tourism may change in Switzerland, or hydro-power resources may be impacted in New Zealand. Despite these severe constraints, it is possible to assess to a certain extent the manner in which environmental change may impact various biogeophysical systems, either based on climate model scenarios or on a what if approach, which determines the sensitivity and vulnerability of a system to a hypothetical change or range of plausible changes.

2. IMPACTS OF GLOBAL WARMING ON NATURAL SYSTEMS IN MOUNTAINS

2.1 Impacts on Hydrology

Because mountains are the source region for over 50% of the globe's rivers, the impacts of climatic change on hydrology are likely to have significant repercussions, not only in the mountains themselves but also in populated lowland regions which depend on mountain water resources for domestic, agricultural and industrial purposes. Hydrological systems are controlled by soil moisture, which also largely determines the distribution of ecosystems, groundwater recharge, and runoff ; the latter two factors sustain river flow and can lead to floods.

These controls are themselves governed by climate, and hence any shifts in temperature and precipitation will have significant impacts on water. Increased temperatures will lead to higher rates of evaporation, a greater proportion of liquid precipitation compared to solid precipitation; these physical mechanisms, associated with potential changes in precipitation amount and seasonality, will affect soil moisture, groundwater reserves, and the frequency of flood or drought episodes.

According to Shiklomanov (1993), the global annual water demand is likely to increase from 4,130 km^3 in 1990 to 5,190 km^3 in 2020, if present consumption patterns are sustained. Because of increasing population, the additional demand will be accompanied by a sharp decline in water availability per capita. While a consumption of 1,000 m^3 of water per year and per capita is considered a standard for well-being in the industrialized world, projections of annual water availability per capita by the early 21st century for North Africa are 210 m^3, 700 m^3 for Central Asia and Kazakhstan, and 1,100 m^3 for southern Asia. This trend is declining in all parts of the world, including those that are considered to have ample water resources. This in turn has prompted the need for a more rational approach to the conservation and use of what is probably the most vital single resource for humankind. An additional source of concern is that mountains have long been considered as an exclusive source of water for the lowland populations. New initiatives are aimed at conservation and distribution of water within the mountains of the developing world so that mountain people, in particular women, can avoid spending a large part of their working lives merely carrying drinking water for their families.

Against this backdrop of social problems, it is obvious that water resources will come under increasing pressure in a changing global environment. Significant changes in climatic conditions will affect demand, supply and water quality. In regions which are currently sensitive to water stress (arid and semi-arid mountain regions), any shortfalls in water supply will enhance competition for water use for a wide range of economic, social, and environmental applications. In the future, such competition will be sharpened as a result of larger populations, which will lead to heightened demand for irrigation and perhaps also industrialization, at the expense of drinking water (Noble and Gitay, 1998).

It would be hazardous to assume that present-day water-supply and consumption patterns will continue in the face of increasing population pressures, water pollution, land degradation and climatic change. Events in recent history may provide useful guidelines for developing such strategies (Glantz, 1988).

Projections of changes in precipitation patterns are tenuous in General Circulation Models (GCMs), even those operating at high spatial resolution, because rainfall or snowfall are difficult variables to simulate, compared to temperature. A number of assessments of the potential impacts of climate change on water resources, including snowfall and storage, have been conducted at a variety of spatial scales for most mountain regions, as reported by Oerlemans (1989), Rupke and Boer (1989), Lins et al. (1990), Slaymaker (1990), Street and Melnikov (1990), Nash and Gleick (1991), Aguado et al. (1992), Bultot et al. (1992), and Leavesley (1994). Riebsame et al. (1995) has found that in many cases, it is difficult to find changes in annual river flows in response to climatic change, but that seasonality changes were often detected. In Latin America, which accounts for 35% of global non-cryosphere freshwater, the impacts of climate change are expected to occur in the more arid regions of the continent, which are often associated with the rain-shadow influences of the Andes ranges. Shifts in water demands will depend on population growth, industrial expansion, and agricultural potential. In many countries of the region, water availability is expected to decline, which is likely to generate potential for international conflicts. The IPCC (1998) estimates that water availability per capita and per annum will decrease from 4,750 m^3 in 1990 to 2,100 m^3 in Mexico by 2025, without any change in climate, i.e., due to population and economic growth. Based on several GCM simulations, projected shifts in precipitation in a warmer climate yield a range of 1,740 - 2,010 m^3. For Peru, the respective set of figures are 1,860 m^3, 880 m^3 resulting from demography alone, and 690 - 1,020 m^3 with climatic change, i.e., close to or below the minimum requirements for well-being. Water resources in tropical Asia are very sensitive to tropical cyclones and fluctuations in their trajectories and intensity. The dominant effect of the Monsoon may be perturbed in a changing climate. Runoff in the Ganges, for example, is more than 6 times that of the dry season. As elsewhere in the world, water resources will become increasingly vulnerable to increasing population growth, urbanization, industrial development and agriculture, as shown by Schreir and Shah (1996). An impacts assessment study by Mirza (1997) for a number of Himalayan basins contributing to the Ganges has shown that changes in the mean runoff in different sub-basins ranged from 27 to 116% in a climate forced by a doubling of CO_2 concentrations relative to their pre-industrial levels. The sensitivity of basins to climate change was seen to be greater in the drier catchments than in the wetter ones. However, water demand is greatest in the dry season in India, and demand cannot be met by supply in this season. Shifts in the timing and intensity of the Monsoon, and the manner in which the Himalayan range intercepts the available precipitable water content of the atmosphere, will have major impacts on the timing and amount of runoff in river basins such as the Ganges, the Brahmaputra or the Irrawaddy. Divya and Mehrotra (1995) and Mirza and Dixit (1997) have shown that amplification or weakening of the Monsoon circulations would indirectly impact upon agriculture and fisheries, freshwater supply, storage capacity, and salinity control.

2.2 Impacts on Mountain Cryosphere

Changes in the mountain cryosphere will have a number of indirect consequences; in terms of water supply, changes in seasonal snow pack and glacier melt will influence discharge rates and timing in rivers which originate in mountains. In terms of tourism, the negative impacts of lack of snow in winter, and the perception of landscape changes in the absence of glaciers and snow, may deter tourists from coming to certain mountain regions.

In most temperate mountain regions, the snow-pack is close to its melting point, so that it is very sensitive to changes in temperature. As warming progresses in the future, current regions of snow precipitation will increasingly experience precipitation in the form of rain. For every °C increase in temperature, the snowline rises by about 150 m; as a result, less snow will accumulate at low elevations than today, while there could be greater snow accumulation above the freezing level because of increased precipitation in some regions. Shifts in snow-pack duration and amount as a consequence of sustained changes in climate will be crucial to water availability for hydrological basins, as Steinhauser (1970) has shown.

Glaciers are possibly the most sensitive system to climatic change, because any changes in the ratio of accumulation to ablation of snow and ice, which are dependent on temperature and precipitation, will trigger glacier mass wasting. Glacier behavior thus provides some of the clearest signals of ongoing warming trends related to the enhanced greenhouse effect (Haeberli, 1990; Wood, 1990; WGMS, 1993). Haeberli (1994) suggests that current glacier retreat is now beyond the range of natural variability as recorded during the Holocene.

The effects of temperature and precipitation changes on glaciers are complex and vary by location. In polar latitudes and at very high altitudes of mid-latitudes, atmospheric warming does not directly lead to mass loss through melting/runoff but to ice warming (Robin, 1983). In areas of temperate ice which predominate at lower latitudes or altitudes, atmospheric warming can directly impact the mass and geometry of glaciers (Haeberli, 1994).

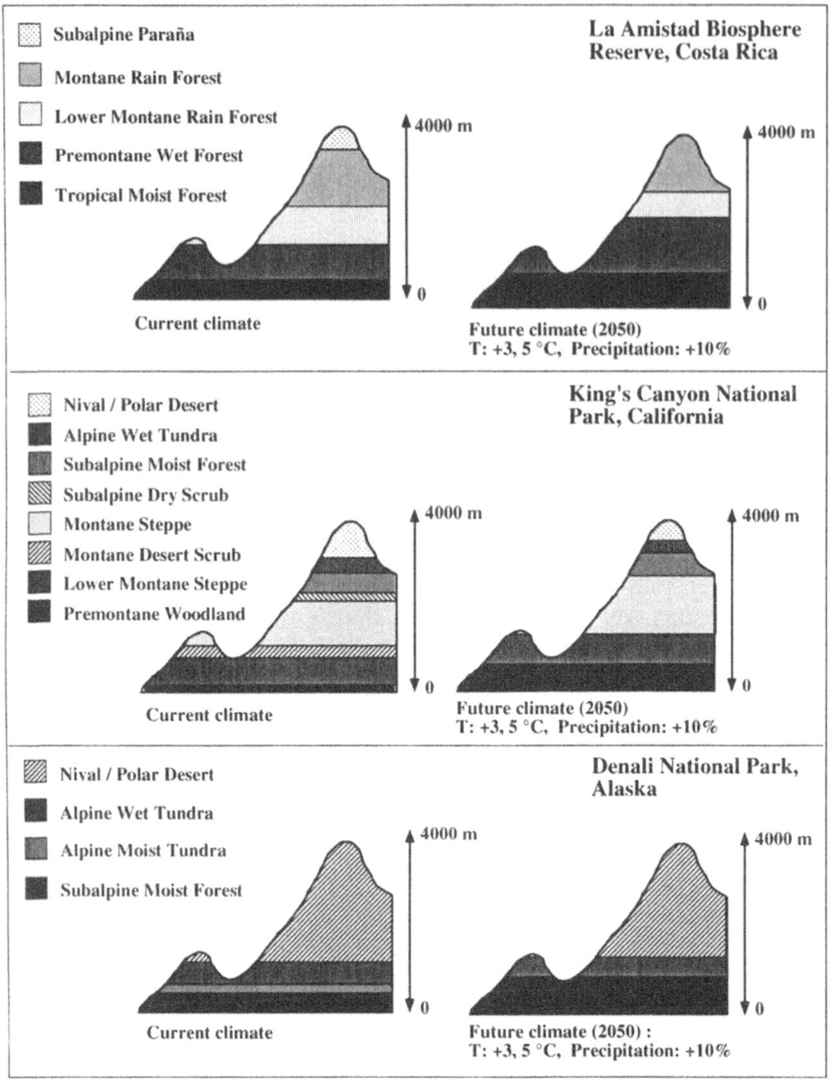

Figure 1. An example of vegetation changes under different climatic conditions, as a function of the movement of climatic ranges along altitudinal gradients and intercompetition between vegetation communities (after Halpin, 1994).

Empirical and energy-balance models indicate that 30 – 50% of existing mountain glacier mass could disappear by the year 2100 if global warming scenarios indeed occur (Fitzharris et al. 1996; Haeberli, 1995; Haeberli and Beniston, 1998; Kuhn, 1993; Oelemans and Fortuin, 1992). The smaller the glacier, the faster it will respond to changes in climate. With an upward shift of 200–300 m in the equilibrium line between net ablation and net accumulation, the reduction in ice thickness of temperate glaciers could reach 1-2 m per year. As a result, many glaciers in temperate mountain regions would lose most of their mass within decades (Maisch, 1992).

Impacts studies for different mountain regions of the globe confirm these projections. In East Africa, for example, Hastenrath and Rostom (1990) have shown the Lewis and Gregory glaciers on Mount Kenya to be in constant regression since the late 19th century. Should this trend continue as expected in the 21st century, there would be very little permanent ice left in this region within the next 100 years. Similarly, Schubert (1992) has used photographic evidence from the last century to show that the snowline has risen from 4,100 m to more than 4,700 m today. These changes in ice extent and in the snowline altitude have had important geoecological effects, leading to shifts in vegetation belts and to the fragmentation of previously continuous forest formations. Enhancement of the warming signal in these regions would lead to the disappearance of significant snow and ice surfaces. In northwest China, projected changes in precipitation and temperature over the next century is likely to lead to the disappearance of one-fifth of glacier surfaces in this region (Wang, 1993).

A consequence of increased glacier mass ablation in coming decades is that there will be enhanced runoff as the ice disappears; this extra runoff could persist for a few decades to a few centuries, according to the size of the melting glaciers. Use could be made of this resource in terms of hydropower or irrigation in the interim period until the glaciers disappear completely. Yoshino et al. (1988), for example, indicate that runoff from glaciers in Asia could triple in volume by the middle of the next century in response to global warming as projected by the IPCC (1996).

Permafrost will also respond to climatic change, although investigations of mountain permafrost are not extensive and monitoring is confined essentially to the European Alps. Evidence from borehole profiles in permafrost helps to determine the rate and magnitude of temperature changes (Mackay, 1990, 1992; Vonder Mühll and Holub, 1995). Vonder Mühll et al. (1994) have shown that permafrost temperatures are increasing at an annual rate of 0.1 °C; this reflects the particularly strong warming signal in the Alps since the 1980s. It is difficult to interpret changes in permafrost, as the amount of permafrost in any particular region depends on regional geomorphologic characteristics, soil and geology, exposure to weather elements, seasonal precipitation and temperature, and the amount and duration of the snow-pack. This latter parameter significantly controls perennially frozen surfaces, because snow insulates the ground surface and suppresses the propagation of cold temperatures into the soil which are favorable to permafrost. Changes in just one of the controls on permafrost, such as temperature, need to be viewed in the larger context of the other controls on this element of the cryosphere.

2.3 Impacts on Ecological Systems

Because temperature decreases with altitude by 0.5-1.0 °C/km, a first-order approximation concerning the response of vegetation to climate change is that species will migrate upwards to find climatic conditions in tomorrow's climate which are similar to today's (e.g., MacArthur, 1972; Peters and Darling, 1985). According to this hypothesis, the expected impacts of climate change in mountainous nature reserves would include the loss of the coolest climatic zones at the peaks of the mountains and the linear shift of all remaining vegetation belts upslope. Because mountain tops are smaller than bases, the present belts at high elevations would occupy smaller and smaller areas, and the corresponding species would have smaller populations and may thus become more vulnerable to genetic and environmental pressure (Peters and Darling, 1985; Hansen-Bristow *et al.*, 1988; Bortenschlager, 1993).

While temperature is indeed a major controlling factor on vegetation, it is by no means the only one, nor is it necessarily the dominant factor for some species. Certain types of vegetation have a large tolerance range to temperature, while others may suffer severe damage in response to a relatively modest change in climatic conditions. Some plants are capable of migrating by various seeding mechanisms, while others can adapt within their original environment. Plant communities will experience competition, with the most robust species adapting or migrating at the expense of the less robust ones. Furthermore, species capable of migrating upwards may not find other adequate environmental conditions, in particular soil types and soil moisture, such that even if climate is appropriate at a particular site in the future, the plant's migration will be slowed or suppressed by other factors. In populated mountain regions of the world, additional stress factors include the fragmentation of biomes and the obstacles to migration, such as roads and settlements. The success of colonization may also depend on regions where erosion and overland flows may increase under changing climatic conditions. Whatever the response of ecosystems, it must be borne in mind that they are exceedingly complex dynamic systems whose response is likely to be non-stationary and stochastic (Hengeveld, 1990). The ability for species to maintain viable populations at ecoclines or ecotones will be affected by numerous interactions between existing populations and site-specific factors (Halpin, 1994). These interactions will be contained in a complex cascade of environmental and ecological feedbacks.

Huntley (1991) suggests that there are three responses which can be distinguished at the species level, namely genetic adaptations, biological invasions through species inter-competition, and species extinction. Street and Semenov (1990) show that these responses may take one of the following adaptation pathways in the face changing environmental conditions. A first scenario is that the currently dominant species would progressively be replaced by a more thermophilous (heat-loving) species. A second plausible mechanism is that the dominant species is replaced by pioneer species of the same community that have enhanced adaptation capability. A third possibility is that environmental change may favor less dominant species, which then replace the dominant species through competition.

These scenarios are based on the assumption that other limiting factors such as soil type or moisture will remain relatively unaffected by a changing environment.

Halpin (1994) has used Geographical Information Systems (GIS) techniques to map the changes in vegetation that could take place under hypothetical climate change scenarios. By mapping the potential zones of shifts in thermal belts with altitude, and knowing the potential for one species to dominate over another, he has applied the concept to three mountain regions of the western hemisphere, namely Costa Rica, the Sierra Nevadas of California, and the Alaskan ranges. In Figure 6.3, an example is given of the shifts in vegetation in a changed climate in the Californian mountains. It is seen that extinction is projected to occur for the montane desert scrub and the subalpine dry scrub as a result of species inter-competition. While reduced in area, the alpine tundra zone located towards the top of the mountains is able to survive under this particular scenario. Results for the other ranges suggest species extinction at the mountain tops in Costa Rica, with little change occurring at height in Alaska. Examples of past extinctions attributed to upward shifts are found in Central and South America, where vegetation zones have moved upward by 1000 - 1500 m since the last glacial maximum (Flenley, 1979; van der Hammen, 1974). Romme and Turner (1991), in their study on possible implications of climate change for ecosystems in Yellowstone National Park (USA), project species extinctions as a result of fragmentation and shrinking mountain-top habitats.

The examples provided by the Halpin (1994) study emphasize the caution which is needed when interpreting such results; even a rather simple conceptual model suggests that changes in ecoclimatic zones are not linear functions of altitudinal climatic gradients.

It is expected that, on a general level, the response of ecosystems in mountain regions will be most important at ecoclines (the ecosystem boundaries if these are gradual), or ecotones (where step-like changes in vegetation types occur). Guisan et al. (1995) note that ecological changes at ecoclines or ecotones will be amplified because changes within adjacent ecosystems are juxtaposed. In steep and rugged topography, ecotones and ecoclines increase in quantity but decrease in area and tend to become more fragmented as local site conditions determine the nature of individual ecosystems. Even though the timberline is in many regions not a perfect ecocline, it is an example of a visible ecological boundary which will be subject to change in coming decades.

This change will either take place in response to a warmer climate, or as a result of recolonization of pastures that have been cleared in the past for pastoral activities. McNeely (1990) has suggested that the most vulnerable species at the interface between two ecosystems will be those which are genetically poorly adapted to rapid environmental change. Those reproduce slowly and disperse poorly, which are isolated, or which are highly specialized will therefore be highly sensitive to seemingly minor stresses.

Numerous investigations have attempted to show that negative effects of climatic change may be offset by the enhanced atmospheric concentrations of CO_2 in the future. Körner (1995) has substantially reviewed this problem and notes that on average, high-altitude plants are able to fix more CO_2 per unit leaf area than lowland plants.

However, the net uptake is about the same for both high and low-elevation plants because of the lower CO_2 at height. In an atmosphere with enhanced CO_2, therefore, high-altitude plants are expected to improve their primary productivity compared to today. Körner et al. (1995) have used experimental plots near the Furka Pass in Central Switzerland to detect changes in alpine plant productivity as a result of CO_2 enrichment. The plants subjected to enrichment were seen to have a higher photosynthetic rate per unit area compared to those in the control plants at current CO_2 levels. Other studies have also shown this acclimation effect, which in certain species diminishes over time (e.g., Gunderson and Wullschleger, 1994). However, these experiments under carefully-controlled conditions are probably far removed from changes which may take place in the free environment in coming decades; even if photosynthetic uptake of carbon by alpine plants were to be the rule, the consequences for plant growth and development are still uncertain.

The length and depth of snow cover, often correlated with mean temperature and precipitation, is one of the key climatic factors in alpine ecosystems (Barry and Van Wie, 1974; Aulitzky *et al.* 1982; Ozenda, 1985; Burrows, 1990). Snow cover provides frost protection for plants in winter, and water supply in spring. Alpine plant communities are characterized by a very short growing season (i.e., the snow-free period) and require water to commence growth. Ozenda and Borel (1991) have shown that predicted that vegetation communities which live in snow beds and in hollows will be the most vulnerable to change, because they will be subject to summer desiccation.

There are currently a number of ecosystem models available which can be used to test the sensitivity of a particular system to processes such as nutrient cycling (e.g., CENTURY), for investigating species composition under changed environmental conditions (e.g., BIOME, DOLY, MAPSS ; e.g., Woodward et al., 1995), or for assessing forest health (e.g., FORET ; Innes, 1998). The problem here is that many of these models are nominally capable of operating at higher spatial resolution than the GCMs which are providing part of the essential input data, i.e., spatially-distributed climate scenario data. A supplementary problem is that these models are designed to operate over continental to global scales, with little possibility of emphasis on the details of altitudinal vegetation belt typology and dynamics typical of mountain regions. Ecosystems are expected to shift upwards and polewards in response to global warming, but expansion will be determined by edaphic factors and dispersal rates, which will determine colonization rates. Additional constraints include land-use competition by human activities in many parts of the world, including the Middle East where significant biodiversity is being lost as a result of human encroachment (settlements ; industry ; flooding of valleys by hydro-power facilities ; etc.), as reported by Bie and Imevbore (1995) and Kharin (1995). In the Himalayas, Yoshino et al. (1998) suggest that weedy species with a wide ecological tolerance will be able to adapt more than other species to shifts in temperature and precipitation. Mountain forests in some regions of Southeast Asia, Africa and Central and South America are presently under greater threat from direct human interference (deforestation ; slash-and-burn practices) than from climatic change *per se* ; unless there is a reversal of current deforestation trends, this situation is unlikely to change in the near future. Sinha and Swaminathan (1991) believe that the combined effects of deforestation and climatic change may impact heavily on sustainable food security, and of course on native populations who still dwell in isolated forested mountain regions.

It should be emphasized that there are considerable limitations in present-day simulation techniques for assessing ecosystem response to climate change, in particular the temporal changes of these responses. In general, increases in atmospheric temperature will affect the structure and function of vegetation, and also species composition where time may not be sufficient to allow species to migrate to suitable habitats (Kienast, 1991; Bugmann, 1994; Klötzli, 1994). According to the detail of biogeographical models, including for example the response of vegetation not only to temperature but also to the CO_2 fertilization effect, results can be very different and sometimes even contradictory. Shriner et al. (1998) have shown that without the CO_2 effect and with a moderate increase in mean temperatures, forest responds by increased growth, while the reverse is frequently observed in simulations where the fertilization effect is absent. On the other hand, even when CO_2 influences are taken into account, forest dieback becomes significant when global warming is towards the upper range of IPCC scenarios. A long-term consequence of global change could comprise two phases, one in which enhanced growth occurs as a first response to warming, followed by dieback when warming exceeds a particular threshold beyond which particular forest species can no longer survive.

The impact of climatic change on altitudinal distribution of vegetation cannot be analysed without taking into account interference with latitudinal distribution. Especially at low altitudes, Mediterranean tree species can substitute for sub-montane belt species. In the southern French Alps, Ozenda and Borel (1991) predict a northward progression of mediterranean ecotypes (steppification of ecosystems) under higher temperatures and lowe rainfall amounts. Kienast et al. (1998) have applied a spatially-explicit static vegetation model to alpine vegetation communities. The model suggests that forests which are distributed in regions with low precipitation and on soils with low water storage capacity are highly sensitive to shifts in climate. Under conditions of global warming, the northward progression of Mediterranean influences would probably be important, and it is estimated that 2-5% of currently forested areas of Switzerland could undergo steppification, particularly on the Italian side of the Alps and in the intra-alpine dry valleys. A similar change is less likely to take place in the southeastern part of the range (Julian and Carnic Alps), where the climate is much more humid. In boreal latitudes, migration of the treeline polewards into previously barren regions would significantly modify the surface characteristics and local climates, in particular through changes in albedo and surface energy balance (Fitzharris et al., 1996). With the expansion of boreal forest zones in both mountain and lowland regions, new assemblages of plant and animal species can be expected in regions such as the northern Alaskan ranges and the eastern Siberian mountains.

Fire is an element which is of particular importance in many ecological systems; it can be devastatingly destructive in certain circumstances, but it plays a vital role in the recycling of organic material and the regrowth of vegetation. Changing climatic conditions are likely to modify the frequency of fire outbreaks and intensity, but other factors may also play a major role; for example, changes in fire-management practices and forest dieback leads to a weakening of the trees in response to external stress factors (Fosberg, 1990; King and Neilson, 1992). In North America in particular, fire management was in favor of suppression of forest fires in recent decades, and as a consequence, there has been a substantial increase in biomass compared to natural levels. Under such circumstances, Stocks (1993) and Neilsson et al. (1992) have shown that forests tend to transpire most of the available soil moisture, so that catastrophic fires can occur as a result of the greater sensitivity of trees to seemingly minor changes in environmental conditions. One example of the combination of deadwood accumulation resulting from fire-suppression policies and a prolonged drought, is the long and spectacular fire outbreak which occured in Yellowstone National Park in the United States during the summer of 1988. Fires can thus occur in the absence of significant climate change.

With climatic change as projected by the IPCC (1996), prolonged periods of summer drought would transform areas already sensitive to fire into regions of sustained fire hazard. The coastal ranges of California, the Blue Mountains of New South Wales (Australia), Mt. Kenya, and mountains on the fringes of the Mediterranean Sea, already subject to frequent fire episodes, would be severely affected. Fires are also expected to occur in regions which are currently relatively unaffected, as critical climatic, environmental and biological thresholds for fire outbreaks are exceeded.

Because many regions sensitive to fires are located close to major population centers, there could be considerable damage to infrastructure and disturbances to economic activities at the boundaries of cities such as Los Angeles and the San Francisco Bay Area in California, Sydney in Australia, coastal resorts close to the mountains in Spain, Italy, and southern France. This has already occured in the past and is likely to become more frequent in the future as fire hazards increase and urban centers expand as a result of population growth.

3. IMPACTS ON SOCIO-ECONOMIC SYSTEMS: INTRODUCTORY REMARKS

In view of the fact that humans have influenced mountain ecosystems in many different ways throughout history, anthropogenic impacts generally cannot be dissociated from climate change impacts., as illustrated schematically in Figure 2. Climatic influences are often obscured by the impacts of change in land use. An example is the fragmentation of the forest and natural vegetation cover. Because of persistent anthropogenic influences in the past, timberline in mountains such as the Alps has dropped between 150 and 400 m compared to its uppermost position during the post-glacial optimum (Holtmeier, 1994).

At present, the climatic limit of tree growth in the Alps is situated above the actual forest limit (Thinon, 1992; Tessier et al., 1993). By reducing species diversity and even intra-species genetic variability of some species, humans have reduced the ability of alpine vegetation to respond to environmental change (David, 1993; Peterson, 1994).

In general terms, it should be emphasized that many sectors of global environmental change will impact upon the poorest members of society, who are clearly the most vulnerable to changes which will affect mountains in coming decades. The potential impacts of global change will probably exacerbate hunger and poverty around the world. New and fluctuating conditions could have a strongly negative impact on economic activities, particularly in the resources sector. People who are highly-dependent on farming and forestry might see their livelihood severely disrupted by changes in rainfall patterns, impoverished forests, and degraded soils. The poor would suffer the most because they have fewer options for responding to environmental change, in terms of technological and financial resources. For example, they would find it more difficult to change over to new crops requiring less water, to pump water for irrigation, or to extend their cultivatable land. Such solutions typically require extensive inputs such as machinery or fossil-fuel energy, which are beyond the financial capabilities of the populations concerned. If global change were to have severe local or regional impacts in certain mountains and uplands, then waves of refugees and immigrants would be likely to move from rural to urban areas within national borders, and from the South to the North across national boundaries. Such migrations from non-urban populations would probably become an additional source of social and political conflict. Displaced and impoverished populations would suffer an erosion of their cultural identity. The resulting disruption to their culture might create social and political problems just as intractable as the environmental problems that generated them in the first place (IUC, 1997).

3.1 Mountain agriculture

Upland regions contribute a significant proportion of the world's agricultural production in terms of economic value. While mountainous areas in the middle and high latitudes are often marginal for agricultural production compared with the warmer lowland areas, the converse may be true at lower latitudes, where highland zones frequently offer a more temperate climate. In addition, because mountains and uplands are the source region for many of the world's major rivers, changes in environmental conditions may modify the seasonal character and the amount of discharge in hydrological basins. This would in turn disrupt the lowland agriculture that is dependent on the availability of water in these rivers.

Upland regions are characterized by climatic gradients that can lead to rapid altitudinal changes in agricultural potential over comparatively short horizontal distances. Where elevations are high enough, a level will eventually be reached where agricultural production ceases to be viable, in terms of either economic profit or subsistence. Upland crop production, practised close to the margins of viable production, can be highly sensitive to variations in climate. The nature of that sensitivity varies according to the region, crop and agricultural system of interest. In some cases, the limits to crop cultivation appear to be closely related to levels of economic return. Yield variability often increases at higher elevations, so that climate change may mean a greater risk of yield shortfall, rather than a change in mean yield (Carter and Parry, 1994).

The effects of climate on agriculture in individual countries cannot be considered in isolation. Agricultural trade has grown in recent decades and now provides significant increments of national food supplies to major importing nations and substantial income for major exporting ones. There are therefore close links between agriculture and climate, the international nature of food trade and food security, and the need to consider the impacts of climate change in a global context. Despite technological advances such as improved crop varieties and irrigation systems, weather and climate are still key factors in agricultural productivity. For example, weak monsoon rains in 1987 caused large shortfalls in crop production in India, Bangladesh, and Pakistan, which forced these countries to import corn (World Food Institute, 1988).

Agricultural production will be affected by the severity and pace of climate change. If change is gradual, there will be time for political and social institutions to adjust. Slow change also may enable natural biota to adapt. Many untested assumptions lie behind efforts to project global warming's potential influence on crops. In addition to the magnitude and rate of change, the stage of growth during which a crop is exposed to drought or heat is important. When a crop is flowering or fruiting, it is extremely sensitive to changes in temperature and moisture; during other stages of the growth cycle, plants are more resiliant. Moreover, temperature and seasonal rainfall patterns vary from year to year and region to region, regardless of long-term trends in climate. Temperature and rainfall changes induced by climate change will likely interact with atmospheric gases, fertilizers, insects, plant pathogens, weeds, and the soil's organic matter to produce unanticipated responses. Rainfall is the major limiting factor in the growth and production of crops worldwide. Adequate moisture is critical for plants, especially during germination and fruit development.

Projections of future agricultural production stem from both «controlled-conditions» experiments and from initializing crop simulation models with climate scenario data. Quantitative results of simulations are therefore highly dependent on the type of climate scenario used, especially in terms of shifts in precipitation regimes. Furthermore, most simulation and experimental studies have so far used expected fluctuations of mean values for climate variables, but the emphasis is increasingly shifting to the possible consequences of a more variable interannual and intra-annual climate and extremes.

Several authors have predicted that currently viable areas of crop production will change as a result of climate change in the Alps (Balteanu et al., 1987), Japan (Yoshino et al., 1988), New Zealand (Salinger et al., 1989), and Kenya (Downing, 1992). These projections do not consider other constraints such as soil types which may no longer be suitable for agriculture at higher elevations. In-depth studies of the effects of climatic change in Ecuador's Central Sierra (Parry, 1978; Bravo et al., 1988) and Papua New Guinea (Allen et al., 1989) have shown that crop growth and yield are controlled by complex interactions between various climatic factors. Specific methods of cultivation may permit crop survival in sites where the microclimates would otherwise be unsuitable. Future climate scenarios suggest both positive and negative impacts, such as decreasing frost risks in the Mexican highlands (Liverman and O'Brien, 1991) and less productive upland agriculture in Asian mountains, where impacts would depend on various factors, particularly types of cultivars and the availability of irrigation (Parry et al., 1992).

Given the wide range of microclimates already existing in mountain areas and which have been exploited through cultivation of diverse crops, the direct negative effects of climate change on crop yields may be relatively small. While crop yields may rise if moisture is not limiting, increases in the number of extreme events may offset any potential benefits. In addition, increases in both crop and animal yields may be negated by greater populations of pests and disease-causing organisms, many of which have distributions which are climatically-controlled. Global warming will favor conditions for insects to multiply and prosper. Rising temperatures will lengthen the breeding season and increase the reproductive rate. This in turn will raise the total number of insects attacking a crop and subsequently increase crop losses. In addition, some insects will be able to extend their range northward as a result of the warming trend (Chippendale, 1979).

At some sites near the high-latitude and high altitude boundaries of current agricultural production, increased temperatures can benefit crops otherwise limited by cold temperatures and short growing seasons, although the extent of soil suitable for expanded agricultural production in these regions may not be appropriate for viable commercial agriculture (Rosenzweig et al., 1993). Increases in crop yields at high elevations will be the result of the positive physiological effects of CO_2, the lengthened growing season and the amelioration of cold temperature effects on growth. It can be surmised that at the upper limit of current agricultural production, increased temperatures will extend the frost-free growing season and provide regimes more favorable to crop productivity. However, in lower-latitude mountain regions, there could be causes for reductions in crop yields, in particular through a shorter growing period and a decrease in water availability. Higher temperatures during the growing season speed annual crops through their development, allowing less grain to be produced. Problems of water availability result from a combination of higher evapotranspiration rates in the warmer climate, enhanced losses of soil moisture and, in some cases, decreases in precipitation.

If current climate variability remains the same, adaptive strategies such as a change in sowing dates, in genotypes, and in crop rotation could counteract expected production losses. In economic terms, most agricultural model studies suggest that mid-latitude mountain regions would on average benefit from climate change, because of an overall increase in crop yields leading to lower consumer prices (IPCC, 1998). Other adaptation options include both technological advances and socio-economic options, such as land-use planning, watershed management, improved distribution infrastructure, adequate trade policy, and national agricultural programs.

It should be noted, when considering adaptation options in the agricultural sector, that the ultimate impacts of climate on the agricultural sector may be determined fact by non-climate factors that control the system. In Europe, for example, the main driving force is the Common Agricultural Policy (CAP) for the countries in the European Union, that also affects other countries in Europe. As a result, the consequences of climatic variations in agronomic systems that are highly regulated, such as the systems in the countries of the European Union (EU), are difficult to predict, since the crops are highly subsidized and therefore the crop prices are artificially high (IPCC, 1998). In mountain regions of Europe, particularly in Switzerland and to a lesser extent in Austria, subsidies to mountain agriculture encourage farmers to remain in these regions, thereby acting as caretakers of the mountain scenery. Reductions of subsidies to such farmers, as planned following the ratification of the Uruguay Round of negotiations of the World Trade Organization, could lead to a collapse of mountain agriculture in these countries. The impacts of direct economic interference would be far more fundamental than those of expected environmental change in coming decades. More important in certain parts of the developing world is the potential for complete disruption of the life pattern of mountain communities which climate change may represent in terms of food production and water management. People in the more remote regions of the Himalayas or Andes have for centuries managed to strike a delicate balance with fragile mountain environments.

This balance would likely be disrupted by climate change and it would take a long time for a new equilibrium to be established.

In cases such as this, positive impacts of climate change (e.g., increased agricultural production and/or increased potential of water resources) are unlikely because the combined stressors, including negative effects of tourism, would overwhelm any adaptation capacity of the environment.

3.2 Tourism

Tourism is of great economic importance and is one of the fastest growing economic sectors in the world. It accounts for 10% of the world's real net financial output, but many countries and in particular those in the developing world are dependent on tourism to a far greater degree than in the industrialized world. In the developing countries, tourism of all types contributes roughly US$ 50 billion annually (Perry, 1999). Even in the current period of widespread economic recession and depression, tourism has remained surprisingly strong. Figures released by the World Tourism Organisation indicate that the number of international tourists has increased 25-fold in the second half of the 20th Century. It is estimated that if current trends continue, international tourism will double every 20 years. Furthermore, tourism patters have become more diversified: new activities have joined traditional recreational patterns. As a consequence, even remote and so-far untouched natural areas, in particular mountain regions in the Himalayas, the Andes and East Africa, are being visited by tourists more frequently.

The major trends in international tourism are characterized by higher demands for air travel to remote destinations, and an increasing trend towards various forms of tourism in natural areas, such as climbing, kayaking, diving, hang-gliding or snow- boarding.

These activities often take place in highly-sensitive mountain regions which would warrant particular protection. In order to keep up with the increasing number of tourists in mountain regions of the world, there is a parallel boom in the development of tourism infrastructure and construction in attractive cultural and natural landscapes which are can be detrimental to those landscapes and the sensitive ecosystems which they support.

Climate change is likely to have both direct and indirect impacts on tourism in mountain areas. Direct changes refer to changes in the atmospheric resources necessary for specific activities. Indirect changes may result from both changes in mountain landscapes which Krippendorf (1984) refers to as the "capital of nature", and wider-scale socio-economic changes such as patterns of demand for specific activities or destinations and for fuel prices.

The marked seasonality of mountain climates implies that their attractions for tourists vary greatly through the year.

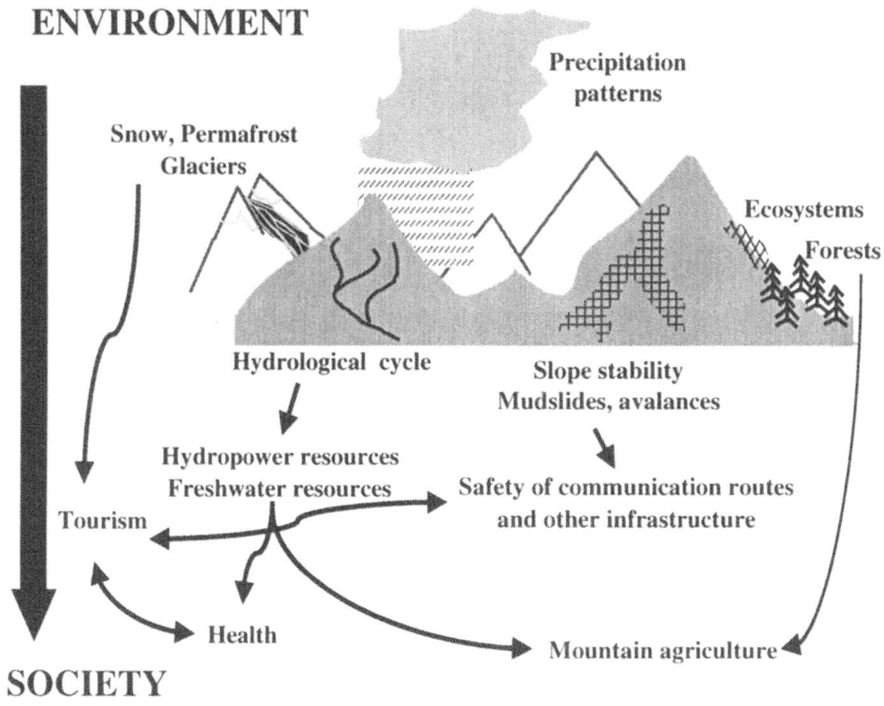

Figure 2: Schematic illustration of the interlinks between environmental change impacts on the natural environment and the repercussions of these on socio-economic systems

Various methodologies have been developed to assess the suitability of regions for specific activities in different seasons. While such approaches are based on long-term averages, others have been developed to assess the economic implications of historical climate variability over short periods, mainly for the skiing industry (Lynch et al., 1981; Perry, 1971).

In the European Alps and the North American Rockies in particular, the ski industry is for some resorts by far the greatest single source of income. In many mountain communities, there is no alternative to skiing capable of generating such major financial resources. Capital investment for cable-cars, ski-lifts and chairlifts in countries such as Austria, Switzerland, France and Italy need 20 – 30 years for a positive return on investment. Lower revenues would put these investments at risk, which would impact negatively on the financial revenue of mountain ski resorts that are often the major shareholders in the cable-car and ski-lift companies. In an attempt to alleviate problems related to the lack of snow at lower elevations, recent investments such as snow-making equipment have been made in Europe and the United States. However, in a warmer climate as projected for the 21st Century, these snow-making machines would be of little use as temperatures need to be well below freezing for them to be effective. Given the prospects of a periodic snow, and deficient winters in many mountain resorts, it is likely that tourists will either no longer come during the peak seasons, or may delay booking accommodation until snow is secure. In both cases, this will lead to reduced sales and rental of ski gear, and a consequent increase in partial or full unemployment in sectors dependent on skiing, namely hotels, restaurants, sports stores, etc.

Using scenarios derived from GCMs, a number of investigations have been carried out to examine the possible implications of climate change for skiing in Australia, eastern Canada, and Switzerland (Abegg and Froesch, 1994; Galloway, 1988; McBoyle and Wall, 1987; Lamothe and Périard, 1988). Abegg and Froesch (1994) have shown in their study of the Swiss ski industry that if temperatures were to rise by about 2-3°C by the year 2050, the low to medium elevation resorts located below 1,200 – 1,500 m above sea level would be adversely affected. Warmer winters bring less snow, and the probability of snow lying on the ground at peak vacation periods (Christmas, February and Easter) would decline. A general rule for the viability of the ski season in Europe is a continuous snow cover of over 30 cm depth for at least 100 days. Based on these figures, Abegg and Froesch (1994) have shown that while towards the latter part of the 20th Century, 85% of ski resorts have reliable amounts of snow for skiing, a 2°C warming would bring this figure down to 63%. Regions such as the Jura Mountains to the west of the country, whose average altitude lies between 900 – 1,200 m, would seldomly experience significant periods of snow-cover, whereas the elevated ski resorts in the central and southern Alps would be less severely affected.

The economic impacts on ski resorts of changing patterns of snowfall in a changing climate may appear to be far removed from the preoccupations of communities in the mountains of certain developing countries. Yet ski resorts are also found in the Andes (Fuentes and Castro, 1982; Solbrig, 1984) and the Himalayas, and changes in the length of the snow-free season would be of critical importance for most mountain communities. In South Asia, another important potential change for communities which increasingly depend on tourism concerns the monsoon, whose timing may well change (IPCC, 1996). This could have substantial effects on countries, such as Nepal and Bhutan, for whom tourists are the principal source of foreign exchange (Richter, 1989).

To some extent, such impacts might be offset by new opportunities in the summer season and also by investment in new technology, such as snow making equipment, as long as climatic conditions remain within appropriate bounds. Mountaineering and hiking may provide compensation for reduced skiing, and thus certain mountain regions would remain attractive destinations. However, global climate change has wider implications for traditional holiday breaks, with destinations other than mountains in winter becoming far more competitive. Higher temperatures imply longer summer seasons in mid-latitude countries, and Perry and Smith (1996) have noted that these may well result in a new range of outdoor activities. In the Mediterranean region, much warmer temperatures will have negative health impacts related to heat stress and skin cancers, and mountains in the vicinity of many Mediterranean beaches are likely to offer a cooler alternative to the hot beaches. Mountainous islands such as Corsica, Sardinia or Cyprus, or the coastal ranges in Spain, Italy and Greece may continue attracting tourists who traditionally spend their vacation at the seashore.

In addition to these potential direct impacts of climate change on tourism, a critical indirect impact needs to be emphasized. One of the most likely types of policy response to climate change will be the imposition of "carbon taxes" on fossil fuels (Bryner, 1991). These will increase the costs of fuels, a major component of the cost of tourism, and in particular to mountain regions which are generally not readily accessible. Other indirect impacts might include decreasing attractiveness of landscapes, and new competition from other tourist locations as climate changes.

3.3 Hydro-power and other commercial activities

An important socio-economic consequence of global warming on the hydrological cycle is linked to potential changes in runoff extremes. Not only the mountain population but also the people in the plains downstream (a large proportion of the world population) presently depend on unregulated river systems and thus are particularly vulnerable to climate-driven hydrological change. Current difficulties in implementing water resource development projects will be compounded by uncertainties related to hydrologic responses to possible climatic change. Among these, possible increases in sediment loading would perturb the functioning of power generating infrastructure. Thermal, nuclear and hydropower stations rely on the supply of water for cooling or for the direct generation of electricity.

Many of the more important hydropower dams are located in mountains, where the head of water can reach considerable heights in Switzerland, Austria, Norway, Russia, the United States, and New Zealand. Changes in flow regimes, induced either by changes in total precipitation, the amount of snowmelt, or a combination of both, would affect hydropower potential. Citing pan-European research examples, Arnell (1999) mentions that a shift in peak discharge rates from spring to winter in Norway would reduce power-generation potential in spring, but would increase this in winter during the peak demand season. This would not necessarily be the case in other regions, such as Greece, where the reduction of spring and autumn flows could shift the seasonality of power generation and lead to reductions of 5 – 25% by the decade of the 2050s. There is a technical problem here, since electricity cannot be stored in any significant amount ; it is thus critical that electricity be generated precisely when required.

The sensitivity of some thermal and nuclear power generating stations to shortfalls in water for cooling purposes may increase in the future, particularly during the summer months. Lack of water can lead to reducing or halting energy production, for obvious security reasons. There may be a real risk of increases in such reductions in energy production in coming decades, particularly in those areas which are likely to become drier in the future.

Sensitivity of mountain hydrology to climate change is a key factor that needs to be considered when planning hydro-power infrastructure. In the future, a warmer and perhaps wetter greenhouse climate needs to be considered. The impact of climate on water resources in alpine areas has been examined by Gleick (1986, 1987a, 1987b) and Martinec and Rango (1989). Similar studies have related electricity demand to climate (Warren and LeDuc, 1981; Maunder, 1986; Downton et al., 1988). However, few have attempted to integrate these impacts of climate change by considering both electricity supply and electricity consumption (Jaeger, 1983).

Mountain runoff (electricity supply) and electricity consumption (demand) are both sensitive to changes in precipitation and temperature. Long-term changes in future climate will have a significant impact on the seasonal distribution of snow storage, runoff from hydro-electric catchments and aggregated electricity consumption. On the basis of a study made in the Southern Alps of New Zealand, Garr and Fitzharris (1994) have concluded that according to future climate scenarios used (New Zealand Ministry for the Environment, 1990), the seasonal variation of electricity consumption will be less pronounced than at present, with largest changes in winter which corresponds to the time of peak heating requirements. There will also be less seasonal variation in runoff and more opportunity to generate power from existing hydropower stations. The electricity system will be less vulnerable to climate variability in that water supply will increase, but demand will be reduced. These conclusions suggest that climate change will have important implications for hydro-electricity systems in other mountain areas as well.

The countries of the Hindu Kush-Himalaya region are currently undergoing rapid economic transition in order to meet their overall requirements for development purposes. This includes energy demand, which until recently depended almost entirely upon on fuel wood, which has been a critical factor for deforestation in the region. However, forests are no longer considered to be the only source of energy and water is now a principal source for economic development (Verghese and Ramaswamy, 1993). The obstacles to harnessing hydropower in countries such as Nepal or Bhutan are of an economic, technical, and political nature. So far, poverty appears to be the main limiting factor for developing hydropower potential (Chalise, 1997). The complexity of the problems associated with the development of water resources in the region could be further complicated by the potential impacts of enhanced Monsoon rainfall and intensity due to global warming. The implications of such an increase in precipitation amounts in the geologically active high mountain environments of the Hindu Kush-Himalaya may be quite significant, and increased sediment loading could severely damage turbines and dam infrastructure, leading to prohibitive maintenance costs for the countries concerned.

Commercial utilization of mountain forests can be affected directly and indirectly by climate change. Direct effects include loss of viability of commercial species, including problems in regeneration and lower seedling survival. Indirect effects relate to disturbances such as fire, insect and disease losses. These indirect effects depend on the influence of climate on the disturbance agents themselves. Many of the commercially-viable mineral deposits in the world are located in mountain regions. While climate has only a minor direct influence on exploitation of these resources, it may exert a significant indirect influence. Mining causes a surface disruption and requires roads and other infrastructure. Changes in climate that lead to increases in precipitation frequency and/or intensity may exacerbate the potential for mass wasting and erosion associated with these developments. Furthermore, the economics of mineral exploitation often requires in situ processing of the extracted ore, for example smeltering and hydrochemical processing. In the latter case, climate, especially precipitation and temperature are critical factors in process design.

4. POLICY RESPONSE

In facing up to environmental change, human beings are going to have to think in terms of decades and centuries. Many of the impacts of these profound changes may not become unambiguously apparent for two or three generations. Perhaps the key to success is through long-term economic thinking, based on concepts of sustainable development. Although sustainability is a much-flaunted term today, the common-sense basis for sustainability (i.e., environmental conservation and careful resource use to improve living standards worldwide today, and to provide these resources for future generations) should be seen as the only long-term alternative to current economic trends. The search for sustainability in any form of development presumes that the thresholds of the environmental carrying capacity for a given region are known or can be established on the basis of existing information. For the present time, sustainable economic development can be observed after the fact; in addition, sustainability is a notion which is not necessarily valid for an infinite time, but may change over time as population, technology, or the environment shift in response to sustainable policies. The establishment of the goals of sustainable development are essentially social decisions related to the desirability of establishing a dual environmental-economic system which can survive as long as possible. The real problem here is not to define the goals of sustainability *per se*, but rather to determine the policy implications of what will lead to the establishment of a sustainable system. These considerations, and the large uncertainties associated with them, can only be alleviated to some extent by a consistent application of the precautionary principle mentioned earlier in this chapter.

Many of the policies and decisions related to pollution abatement, climatic change, deforestation or desertification would provide opportunities and challenges for the private and public sectors. A carefully selected set of national and international responses aimed at mitigation, adaptation and improvement of knowledge can reduce the risks posed by environmental change to ecosystems, food security, water resources, human health and other natural and socio-economic systems.

There are large differences in the cost of attempting to address crucial global environmental problems among countries due to their state of economic development, infrastructure choices and natural resource base. International cooperation in a framework of bilateral, regional or international agreements could significantly reduce the global costs severe environmental stress. If carried out with care, these responses would help to meet the challenge of climate change and enhance the prospects for sustainable economic development for all peoples and nations.

When progress has been made towards attaining some of these global objectives, and the positive effects of implemented policies begin to be perceived, mountain and upland environments will also benefit from these measures. Mountains are unique features of the Earth system in terms of their scenery, their climates, their ecosystems; they provide key resources for human activities well beyond their natural boundaries; and they harbor extremely diverse cultures in both the developing and the industrialized world. The protection of mountain environments against the adverse effects of economic development should be a priority for both today's generation and the generations to come.

5. REFERENCES

Abegg, B., and R. Froesch, 1994: Climate change and winter tourism: impact on transport companies in the Swiss Canton of Graubünden. In: Beniston, M. (ed.), Mountain environments in changing climates. Routledge Publishing Company, London and New York, pp. 328-340

Aguado, E., et al., 1992: Changes in the timing of runoff from West Coast streams and their relationships to climatic influences, Swiss Climate Abstracts, Special Issue: International Conference on Mountain Environments in Changing Climates

Allen, L. H., Jr., Boote, K. J., Jones, J. W., Jones, P. H., Valle, R. R., Acock, B., Rogers, H. H. and Dahlman, R. C., 1989: Response of vegetation to rising carbon dioxide: Photosynthesis, biomass and seed yield of soybean. Global Biogeochemical Cycles 1: 1-14.

Arnell, N., 1999 : The effect of climate change on hydrological regimes in Europe. Global Environmental Change, 9, 5-23

Aulitzky, H., H. Turner, and H. Mayer, 1982: Bioklimatische Grundlagen einer standortgemässen Bewirtschaftung des subalpinen Lärchen – Arvenwaldes, Eidgenössische Anstalt für forstliches Versuchswesen Mitteilungen, 58, 325-580.

Balteanu, D., et al, 1987: Impact analysis of climatic change in the central European mountain ranges, European Workshop on Interrelated Bioclimatic and Land Use Changes, Volume G, Noordwijkerhout, The Netherlands.

Barry, R.G., 1992: Mountain weather and climate. Routledge Publishing Company, London and New York

Barry, R.G., and C.C. Van Wie, 1974: Topo- and microclimatology in alpine areas, Arctic and Alpine Environment, Methuen, London, 73-83.

Bie, S.W. and A.MA. Imevbore, 1995 : Executive summary. In : Biological Diversity in the Drylands of the World [Bie, S.W. and A.M.A. Imevbore (eds.)]. United Nations, Washington, DC, USA, pp. 5-21.

Bortenschlager, S., 1993: Das höchst gelegene Moor der Ostalpen "Moor am Rofenberg" 2760 m. Festschrift Zoller, Diss. Bot., 196, 329-334.

Bravo, R.E. et al., 1988: The effects of climatic variations on agriculture in the Central Sierra of Ecuador, The Impact of Climate Variations on Agriculture: Volume 2, Assessments in Semi-Arid Regions, M.L. Parry, T.R. Carter, and N.T. Konijn, (eds.), Kluwer, Dordrecht, 381-493.

Bryner, G., 1991 : Implementing global environmental agreements. Policy Studies Journal, 19, 103-114

Bugmann, H., 1994: On the ecology of mountain forests in a changing climate: A simulation study. PhD Thesis No. 10638, Swiss Federal Institute of Technology (ETH), Zürich, Switzerland, 268.

Bultot, F., D. Gellens, B. Schädler, and M. Spreafico, 1992: Impact of climatic change, induced by the doubling of the atmospheric CO_2 concentration, on snow cover characteristics - Case of the Broye drainage basin in Switzerland.Swiss Climate Abstracts, Special Issue: International Conference on Mountain Environments in Changing Climates

Burrows, C.J, 1990: Processes of Vegetation Change, Unwin Hyman Publishing, London, 551.

Carter, T.R., and M.F. Parry, 1994: Evaluating the effects of climatic change on marginal agriculture in upland areas, Mountain Environments in Changing Climates, M. Beniston, (ed.), Routledge Publishing Company, London and New York, 405-421.

Chalise, S., 1997 : Mountain tourism in Nepal. Mountain Forum Internet Publication on http://www.mtnforum.org/

Chippendale, G. M., 1979 : The Southwestern Corn Borer, *Diatraea grandiosella*: Case History of an Invading Insect, Research Bulletin 1031. University of Missouri Agricultural Experiment Station, Columbia

David, F., 1993: Altitudinal variation in the response of vegetation to late-glacial climatic events in the northern French Alps, New Phytol., 125, 203-220.

Divya, and R. Mehrotra, 1995 : Climate change and hydrology with emphasis on the Indian subcontinent. Hydrological Sciences Journal, 40, 231-241.

Downing, T.E., 1992: Climate Change and Vulnerable Places: Global Food Security and Country Studies in Zimbabwe, Kenya, Senegal, and Chile, Environmental Change Unit, Oxford.

Downton, M.W., T.R. Stewart, and K.A. Miller, 1988: Estimating historical heating and cooling needs: per capita degree days, J. Appl. Meteorol., 27, 84-90.

Fitzharris, B. B., Allison, I., Braithwaite, R.J., Brown, J., Foehn, P., Haeberli, W., Higuchi, K., Kotlyakov, V.M., Prowse, T.D., Rinaldi, C.A., Wadhams, P., Woo, M.K., Youyu Xie, 1996: The Cryosphere: Changes and their impacts. In: Second Assessment Report of the Intergovernmental Panel on Climate Change (IPCC), Chapter 5, Cambridge University Press, pp. 241-265

Flenley, J.R., 1979: The late quaternary vegetational history of the equatorial mountains, Progress in Physical Geography, 3, 488-509.

Fosberg, M.A., 1990 : Global change – a challenge to modeling. In : Process modeling of forest growth responses to environmental stress [Dixon, R.K., R.S. Meldahl, G.A. Ruark, and W.G. Warren (eds.)]. Timber Press, Inc., Portland, OR, USA, pp. 3-8.

Fuentes, E.R., and Castro, M., 1982 : Problems of resource management and land use in two mountain regions of Chile. In : di Castri, F., Baker, G., and Hadley, M. (eds.), Ecology in Practice, Tycooly Press, Dublin, pp. 315-330

Galloway, R.W., 1988: The potential impact of climate changes on Australian ski fields, Greenhouse Planning for Climate Change, G.I. Pearman, (ed.), CSIRO, Aspendale, Australia, 428-437.

Garr, C.E., and B.B. Fitzharris, 1994: Sensitivity of mountain runoff and hydro-electricity to changing climate, Mountain Environments in Changing Climates, M. Beniston, (ed.), Routledge Publishing Company, London and New York, 366-381.

Glantz, M.H., (ed.), 1988: Societal Responses to Regional Climatic Change, Westview Press, Boulder, Colorado.

Gleick, P.H., 1986: Methods for evaluating the regional hydrologic impacts of global climatic changes, J. Hydrology, 88, 97-116.

Gleick, P.H., 1987a: Regional hydrologic consequences of increases in atmospheric CO_2 and other trace gases, Clim. Change, 10, 137-161.

Gleick, P.H., 1987b: The development and testing of a water balance model for climate impact assessment: modelling the Sacramento Basin, Water Resources Research, 23, 1049-1061.

Guisan, A., J. Holten, R. Spichiger, and L. Tessier, (eds.), 1995: Potential Impacts of Climate Change on Ecosystems in the Alps and Fennoscandian Mountains. Annex Report to the IPCC Working Group II Second Assessment Report, Publication Series of the Geneva Conservatory and Botanical Gardens, University of Geneva, Switzerland, 194 pp.

Gunderson, C.A. and S.D. Wullschleger, 1994 : Photosynthetic acclimation in trees to rising atmospheric CO_2 : a broader perspective. Photosynthetic Research, 39, 369-388.

Haeberli, W., 1990: Glacier and permafrost signals of 20th century warming, Ann. Glaciol., 14, 99-101.

Haeberli, W., 1994: Accelerated glacier and permafrost changes in the Alps, Mountain Environments in Changing Climates, M. Beniston, (ed.), Routledge Publishing Company, London and New York, 91-107.

Haeberli, W., 1995: Glacier fluctuations and climate change detection - operational elements of a worldwide monitoring strategy. WMO Bulletin 44, 1, 23 - 31.

Haeberli, W., and Beniston, M., 1998: Climate change and its impacts on glaciers and permafrost in the Alps. Ambio, 27, 258 – 265

Halpin, P.N., 1994: Latitudinal variation in montane ecosystem response to potential climatic change, M. Beniston, (ed.), Mountain Ecosystems in Changing Climates, Routledge Publishing Company, London and New York, 180-203.

Hansen-Bristow, K.J., J.D. Ives, and J.P. Wilson, 1988: Climatic variability and tree response within the forest-alpine tundra ecotone, Annals of the Association of American Geographers, 78, 505-519.

Hastenrath, S., and R. Rostom, 1990: Variations of the Lewis and Gregory Glaciers, Mount Kenya, 1978- 86-90, Erdkunde, 44, 313-317.

Hengeveld, R., 1990: Theories on species responses to variable climates, Landscape-Ecological Impact of Climatic Change, M.M. Boer, and R.S. de Groot, (eds.), IOS Press, Amsterdam, 274-293.

Holtmeier, F.-K., 1994: Ecological aspects of climaticaly caused timberline fluctuations: review and outlook, Mountain Environments in Changing Climates, M. Beniston, (ed.), Routledge Publishing Company, London and New York, 220-233.

Huntley, B. 1991: How plants respond to climate change: migration rates, individualism and the consequences for plant communities, Annals of Botany, 67, 15-22.

Innes, J. L., 1998 : The impacts of climatic extremes on forests : An introduction. In : Beniston, M. and Innes, J.L. (eds.), The Impacts of Climate Variability on Forests, Springer-Verlag, Heidelberg and New York, pp. 1-18

IPCC, 1996: Climate Change. The IPCC Second Assessment Report. Cambridge University Press, Cambridge and New York. Volumes I (Science), II (Impacts) and III (Socio-economic implications)

IPCC, 1998: The regional impacts of climate change. Cambridge University Press, Cambridge and New York, 517 pp.

IUC, 1997 : International Unit on Conventions, Fact Sheets, United Nations Environment Program, Geneva

Jaeger, J., and W. W. Kellogg, 1983, Anomalies in temperature and rainfall during warm Arctic seasons, Climatic Change, 34-60

Kates, R.W., J.H. Ausubel, and M. Berberian, (eds.), 1985: Climate Impact Assessment, SCOPE 27, John Wiley and Sons, Chichester.

Kharin, N.G., 1995 : Change of biodiversity in ecosystems of central Asia under the impacts of desertification. In : Biological Diversity in the Drylands of the World [Bie, S.W. and A.M.A. Imevbore (eds.)]. United Nations, Washington, DC, USA, pp. 23-31.

Kienast, F., 1991: Simulated effects of increasing atmospheric CO_2 and changing climate on the successional characteristics of Alpine forest ecosystems, Landscape Ecology, 5, 225-235.

Kienast, F., Wildi, O., Brzeziecki, B., Zimmermann, N., and Lemm, R., 1998 : Klimaänderung und mögliche langfristige Auswirkungen auf die Vegetation der Schweiz. VdF Hochschulverlag, Zurich, 71 pp.

King, G.A. and R.P. Neilson, 1992 : The transient response of vegetation to climate change : a potential source of CO_2 to the atmosphere. Water, Air and Soil Pollution, 64, 365-383.

Klötzli, F., 1994: Vegetation als Spielball naturgegebener Bauherren, Phytocoenologia, 24, 667-675.

Körner, C., 1995: Impact of atmospheric changes on Alpine vegetation; the ecophysiological perspective. In: Guisan, A., J. Holten, R. Spichiger, and L. Tessier, (eds.), 1995: Potential Impacts of Climate Change on Ecosystems in the Alps and Fennoscandian Mountains. Annex Report to the IPCC Working Group II Second Assessment Report, Publication Series of the Geneva Conservatory and Botanical Gardens, University of Geneva, Switzerland, pp. 113-120

Körner, C., Diemer, M., Schäppi, B., and Zimmermann, L., 1995: The response of Alpine vegetation to elevated CO_2. In: Koch, G.W., and Mooney, H.A. (Eds.), Terestrial Ecosystem Response to Elevated CO_2. Academic Press, New York.

Krippendorf, J., 1984: The capital of tourism in danger, The Transformation of Swiss Mountain Regions, E.A. Brugger, et al., (eds.), Haupt Publishers, Bern, 427-450.

Kuhn, M., 1993: Possible future contribution to sea-level change from small glacier. In: Warrick, R.A., Barrow, E.M., and Wigley, T.M.L., (Eds.), Climate and sea-level change: Observations, projections and implications. Cambridge University Press, Cambrisge, UK, pp. 134-143

Lamothe, M., and D. Périard, 1988: Implications of climate change for downhill skiing in Quebec, Climate Change Digest, Atmospheric Environment Service, Downsview, 88-103.

Leavesley, G.H., 1994: Modeling the effects of climate change on water resources - A review, Assessing the Impacts of Climate Change on Natural Resource Systems, K.D. Frederick, and N. Rosenberg, (eds.), Kluwer Academic Publishers, Dordrecht, 179-208.

Lins, H., et al., 1990: Hydrology and water resources, Climate Change: The IPCC Impacts Assessment, W.J.McG. Tegart, G.W. Sheldon, and D.C. Griffiths, (eds.), Australian Government Publishing Service, Canberra, Chapter 4.

Liverman, D.M., and K.L. O'Brien, 1991: Global warming and climate change in Mexico, Global Environmental Change: Human and Policy Dimensions, 1, 351-364.

Lynch, P., G.R. McBoyle, and G. Wall, 1981: A ski season without snow, Canadian Climate In Review-1980, D.W. Phillips, and G.A. McKay, (eds.), Atmospheric Environment Service, Toronto, 42-50.

MacArthur, R.H., 1972: Geographical Ecology, Harper and Row, New York.

Mackay, J.R., 1990: Seasonal growth bands in pingo ice, Canadian Journal of Earth Science, 27, 1115-1125.

Mackay, J.R., 1992: Frequency of ice-wedge cracking 1967-1987 at Garry Island, Western Arctic Coast, Canada, Canadian Journal of Earth Science, 29, 11-17.

Maisch, M., 1992: Die Gletcher Graubündens - Rekonstruktion und Auswertung der Gletscher und deren Veränderungen seit dem Hochstand von 1850 im Gebiet der östlichen Schweizer Alpen (Bündnerland und angrenzende Regionen). Publication Series of the Department of Geography of the University of Zurich, Switzerland.

Martinec, J., and A. Rango, 1989: Effects of climate change on snow melt runoff patterns, Remote Sensing and Large-Scale Global Processes, Proceedings of the IAHS Third Int. Assembly, Baltimore, Md, May 1989, IAHS Publ. No. 186, 31-38.

Maunder, W.J., 1986: The Uncertainty Business, Methuen and Co. Ltd, London, 420.

McBoyle, G.R., and G. Wall, 1987: The impact of CO_2-induced warming on downhill skiing in the Laurentians, Cahiers de Géographie de Québec, 31, 39-50.

McNeely, J.A., 1990: Climate change and biological diversity: policy implications, Landscape-Ecological Impact of Climatic Change, M.M. Boer, and R.S. de Groot, (eds.), IOS Press, Amsterdam.

Mirza, M.Q., 1997 : The runoff sensitivity of the Ganges river basin to climate change and its implications. Journal of Environmental Hydrology, 5, 1-13.

Mirza, M.Q. and A. Dixit, 1997 : Climate change and water management in the GBM basins. Water Nepal, 5(1), 71-100.

Nash, L.L., and P.H. Gleick, 1991: Sensitivity of streamflow in the Colorado Basin to cliamtic changes, Hydrology, 125, 221-241.

Neilsson, R.P., G.A. King, and G. Koerper, 1992 : Toward a rule-based biome model. Landscape Ecology, 7, 27-43.

New Zealand Ministry for the Environment, 1990: Climatic Change, a Review of Impacts on New Zealand, DSIR Publishing, Wellington, New Zealand.

Noble, I., and Gitay, H. : Climate change in desert regions. In : IPCC, 1998, The Regional Impacts of Climate Change, Watson, R. T., Zinyowera, M., and Moss, R. (eds.), Cambridge University Press, pp. 191 – 217

Oerlemans, J., (ed.), 1989: Glacier Fluctuations and Climate Change, Kluwer Academic Publishers, Dordrecht, The Netherlands.

Oerlemans, J., and Fortuin, J. P. F., 1992: Sensitivity of glaciers and small ice caps to greenhouse warming. Science, 258, 115-117.

Ozenda, P., 1985: La Végétation de la Chaine Alpine dans l'Espace Montagnard Européen, Masson, Paris, 344.

Ozenda, P., and J.-L. Borel, 1991: Les Conséquences Ecologiques Possibles des Changements Climatiques dans l'Arc Alpin. Rapport Futuralp No. 1, International Centre for Alpine Environment (ICALP), Le Bourget-du-lac, France.

Parry, M.L., 1978: Climatic Change, Agriculture and Settlement. Dawson-Archon Books, Folkestone, England, 214pp.

Parry, M. L., de Rozari, M. B., Chong, A. L., and Panich, S., eds. 1992. The Potential Socio-Economic Effects of Climate Change in South-East Asia. Nairobi: UN Environmental Programme.

Perry, A.H., 1971: Climatic influences on the Scottish ski industry, Scottish Geographical Magazine, 87, 197-201.

Perry, A.H., 1999: Impacts of climate change on tourism, energy and transportation. In : IPCC Third Assessment Report, Impacts of Climate Change on Europe, Cambridge University Press, Cambridge (in preparation)

Perry, A.H., and Smith, K., 1996 : Recreation and Tourism. Climate Change Impacts Report, UK Department of the Environment, London

Peters, R.L., and Darling, J.D.S., 1985: The greenhouse effect and nature reserves: global warming would diminish biological diversity by causing extinctions among reserve species, Bioscience, 35, 707-717

Peterson, D.L., 1994: Recent changes in the growth and establishment of subalpine conifers in Western North America, Mountain Environments in Changing Climates, M. Beniston, (ed.), Routledge Publishing Company, London and New York, 234-243.

Price, M.F., 1990: Temperate mountain forests: common-pool resources with changing, multiple outputs for changing communities, Natural Resources Journal, 30, 685-707.

Richter, L.K., 1989: The Politics of Tourism in Asia, University of Hawaii Press, Honolulu

Riebsame, W.E., 1989: Assessing the Social Implications of Climate Fluctuations, United Nations Environment Programme, Nairobi.

Riebsame, W.E et al., 1995 : Complex river basins. In : As Climate Changes : International Impacts and Implications [Strzepek, K.M. and J.B. Smith (eds)]. Cambridge University Press, Cambridge, United Kingdom, pp. 57-91.

Rizzo, B., and E. Wiken, 1992: Assessing the sensitivity of Canada's ecosystems to climatic change, Clim. Change, 21, 37-54.

Robin, G. de Q., 1983: The Climatic Record in Polar Ice Sheets, Cambridge University Press, Cambridge, 212.

Romme, W.H., and M.G. Turner, 1991: Implications of global climate change for biogeographic patterns in the greater Yellowstone ecosystem, Conserv. Biol., 5, 373-386

Rosenzweig, C., et al., 1993 : Climate Change and World Food Supply. Oxford University Press, Oxford

Rupke, J., and M.M. Boer, (eds.), 1989: Landscape Ecological Impact of Climatic Change on Alpine Regions, with Emphasis on the Alps, Discussion report prepared for European conference on landscape ecological impact of climatic change, Agricultural University of Wageningen and Universities of Utrecht and Amsterdam, Wageningen, Utrecht and AmsterdamSalinger, M. J., 1983 : New Zealand climate : From ice age to present. Environmental Monitoring in New Zealand, 32-40

Salinger, M.J., J.M. Williams, and W.M. Williams, 1989: CO_2 and Climate Change: Impacts on Agriculture, New Zealand Meteorological Service, Wellington.

Schreir, H. and P.B. Shah, 1996 : Water dynamics and population pressure in the Nepal Himalayas. Geojournal, 40(1-2), 45-51.

Schubert, C. 1992: The glaciers of the Sierra Nevada de Mérida (Venezuela): a photographic comparison of recent deglaciation. Erdkunde, 46, 58-64.

Shiklomanov, I., 1993: World freshwater resource. In: Gleick, P. (Ed.), Wtare in crisis: A guide to the World's freshwater resources. Oxford University Press, Oxford, UK, pp. 13-24

Shriner, D.S. and R.B. Street, 1998 : North America. In : The Regional Impacts of Climate Change. Cambridge, UK, pp. 253-330.

Sinha, S.K., and M.S. Swaminathan, 1991 : Deforestation, climate change and sustainable nutrition security. Climatic Change, 19. 201-209.

Slaymaker, O., 1990: Climate change and erosion processes in mountain regions of Western Canada, Mountain Research and Development, 10, 171-182.

Solbrig, O., 1984 : Tourism. Mountain Research and Development, 4, 181-185

Steinhauser, F., 1970: Die säkularen Änderungen der Schneedeckenverhältnisse in Oesterreich. 66-67 Jahresbericht des Sonnblick-Vereines, 1970-1971, Vienna, 1-19.

Stocks, B.J., 1993 : Global warming and forest fires in Canada. The Forestry Chronicle, 69(3), 290-293.

Street, R.B., and Melnikov, P.I., 1990: Seasonal snow, cover, ice and permafrost, Climate Change: The IPCC Impacts Assessment, W.J.McG. Tegart, G.W. Sheldon, and D.C. Griffiths, (eds.), Australian Government Publishing Service, Canberra, Chapter 7.

Street, R.B., and Semenov, S.M., 1990: Natural terrestrial ecosystems. In: Tegart, W.J. KcG., Sheldon, G.W., and Griffiths, D.C. (Eds.), Climate Change: The First Impacts Assessment Report. Australian Government Publishing Service, Chapter 3.

Tessier, L., de Beaulieu, J.-L., Couteaux, M., Edouard, J.-L., Ponel, Ph., Rolando, Ch., Thinon, M., Thomas, A., and Tobolski, K., 1993: Holocene palaeoenvironments at the timberline in the French Alps - A multidisciplinary approach, Boreas, 22, 244-254.

Thinon, M., 1992: L'analyse pédoanthracologique. Aspects méthodologiques et applications, PhD Dissertation, University of Aix-Marseille III, France, 317.

van der Hammen, T., 1984: Datos eco-climáticos de la transecta Buritaca y alrededores (Sierra Nevada de Santa Marta), La Sierra Nevada de Santa Marta (Colombia), Transecta Buritaca-La Cumbre, T. Van der Hammen, and P. Ruiz, (eds.), J. Cramer, Berlin, 45-66

Verghese, B.G., and Ramaswamy, I., 1993 : Harnessing the Eastern Rivers. Regional Cooperation in South Asia, Konark Publishers, Kathmandu, 286 pp.

Vonder Mühll, D., and Holub, P., 1995: Borehole logging in Alpine permafrost, Upper Engadine, Swiss Alps, Permafrost and Periglacial Processes, 3, 125-132

Vonder Mühll, D., Hoelzle, M., and Wagner, S., 1994: Permafrost in den Alpen. Die Gewissenschaften, 12, 149-153

Wang, Z., 1993; The glacier variation and influence since little ice age and future trends in northwest region, China, Scientia Geographica Sinica, 13, 97-104

Warren, H.E., and. LeDuc, S.K, 1981: Impact of climate on energy sector in economic analysis, Journal of Applied Meteorology, 20, 1431-1439.

WGMS, 1993: Glacier Mass Balance, Bulletin No. 2, W. Haeberli, E. Herren, and M. Hoelzle (eds.), World Glacier Monitoring Service, ETH Zurich, 74

Wood, F.B., 1990: Monitoring global climate change: the case of greenhouse warming, Bull. Am. Meteorol. Soc., 71, 42-52.

Woodward, F.I., Smith, T.M., and Emanuel, W.R., 1995: A global primary productivity and phytogeography model. Global Biogeochem. Cycles, 9, 471 - 490

World Food Institute. 1988. World Food Trade and U.S. Agriculture, 1960-1987. Ames: Iowa State University.

WRI, 1996 : World Resources 1996-1997. World Resources Institute. Oxford University Press

Yoshino, M., Horie, T., Seino, H., Tsujii, H., Uchijima, T., and Uchijima, Z., 1988: The effects of climatic variations on agriculture in Japan, The Impact of Climatic Variations on Agriculture. Vol 1: Assessments in Cool Temperature and Cold Regions, M.L. Parry, T.R. Carter, and N.T. Konijn, (eds), Kluwer, Dordrecht, The Netherlands, 723-868.

Yoshino M., and Jilan, S., et al., 1998 : Temperate Asia. . In : The Regional Impacts of Climate Change. Cambridge, UK, pp. 355-379.

Global Change in Respect to Tendency to Acidification of Subarctic Mountain Lakes

VLADIMIR DAUVALTER, TATYANA MOISEENKO, LUDMILA KAGAN
Institute of North Industrial Ecology Problems, Kola Science Centre, Russian Academy of Sciences, 14 Fersman St., 184200 Apatity, Murmansk region, Russia

Key words: acidification, subarctic mountain lakes, heavy metals

Abstract: The Kola Peninsula mountain lakes reflect a real situation not only of the local polluted airborne transfer but also polluted transborder emissions from Europe to Arctic. Despite of two monitoring mountain lakes (the Chuna and Chibiny lakes) are close to smelters of the Severonickel Company, local emissions very slightly affect the mountain lakes, because heavily polluted air masses do not rise high altitude. Sulphur depositions on the Chuna and Chibiny lakes catchments are 0.4 and 0.6 g/cm^2, respectively, in comparison with area at the foot of the mountain, where the deposition is 1.0-1.5 g/cm^2. The water quality of the lakes is consistent with an average value for the region: SO_4^{2-}<80 µeq/l, Ni<1 µg/l. In the Chibiny lake there is observed a moderate water buffer capacity. The analysis of sediments showed, that heavy metal concentrations exceeded in the upper of 4-5 cm layers of the Chuna lake sediments are accounted by local atmospheric emissions of smelters (as regards Ni and Cu), and general increase of Pb contamination in the atmosphere of the northern hemisphere. Diatom investigations in the sediment cores of the lakes have ascertained a tendency to acidification process. In the originally weakly acidified and sensitive Chuna lake the pH reconstructed has dropped from 6.7 to 6.2. Diatoms from the Chuna lake reflect also the toxic load- in the upper layers there occur the ugly pathological forms.

1. INTRODUCTION

The mountain lakes are the most sensitive to changes occurred in the air quality. The airborne effects upon the freshwater systems of the Kola Peninsula are the result of transboundary transfer of air masses from Europe to Arctic as well as the result of local emissions from Russian large smelters

G. Visconti et al. (eds.), Global Change and Protected Areas, 187–194.

Table. Water chemistry of the Kola mountain lakes (1993-1994)

Parameters	Units	Chuna			Chibiny		
		Average	Minimum	Maximum	Average	Minimum	Maximum
pH		6.47	6.24	6.72	7.30	7.15	7.46
Conductivity	µS/cm	16	10	25	33	32	35
Colour	° Pt	5	0	9	2	0	3
COD_{Mn}	mgO/l	1.14	0.05	2.5	0.45	0.15	0.98
Alkalinity	µeq/l	40	23	62	234	226	251
N_{tot}	µg/l	550	287	1091	154	140	167
NO_2+NO_3	µg/l	83	17	162	140	79	215
NH_4	µg/l	22	2	45	7	0	14
P_{tot}	µg/l	10	3	16	6	3	8
PO_4	µg/l	2	0	3	2.5	0	5
Si	mg/l	1.5	0.88	2.36	2.5	1.95	3.13
SO_4	mg/l	2.3	1.7	3.2	2.9	2.7	3.2
Cl	mg/l	1.3	0.6	1.5	0.85	0.8	0.9
K	mg/l	0.16	0.04	0.30	1.4	1.2	1.5
Na	mg/l	0.78	0.49	1.2	5.8	5.6	6.0
Ca	mg/l	1.3	0.85	2.0	0.47	0.33	0.68
Mg	mg/l	0.22	0.14	0.34	0.06	0.06	0.07
Sr_{tot}	µg/l	6.8	5.2	8.5	60	55	62
Sr_{lab}	µg/l	5.8	4.5	7	56	53	58
Al_{tot}	µg/l	15.3	7.0	41	60	47	77
Al_{lab}	µg/l	2.0	1.0	2.5	31	25	40
Ni_{tot}	µg/l	1.1	0.8	1.4	0.8	0.5	1.1
Ni_{lab}	µg/l	0.8	0.5	1.1	0.65	0.6	0.7
Cu_{tot}	µg/l	1.3	0.6	2.6	0.23	0.2	0.3
Cu_{lab}	µg/l	0.4	0.1	0.8	-	-	-
Fe_{tot}	µg/l	6.1	4.2	7.0	3.4	1.1	6.0
Fe_{lab}	µg/l	0.3	0.1	0.5	0.13	0.1	0.2
Zn_{tot}	µg/l	9.4	1.0	25	1.1	0.6	1.7

situated in the Kola North (Severonickel, Pechenganickel, Kandalaksha Aluminium Works etc.). The Kola Peninsula occupies the Northern part of Fennoscandia above the Polar Circle. Its central part is presented by mountain massif: Chibiny (1190 m above sea level) and Chuna (1114 m), which are separated by deep depression where the largest Imandra lake is situated. Acid effect and heavy metal pollutions on the mountain lakes are the result of the Severonickel Company activity sitting between two mountain massifs. The average annual sulphur dioxide emissions account for about 300 000 tons, Ni - 295 tons. Despite the Chuna and Chibiny lakes are close placement to the Severonickel smelters, local emissions very slightly affect the mountain lakes, because heavily polluted air masses do not rise high altitude. Sulphur deposition on the Chuna lake catchment is 0.4 gS/m^2, the Chibiny lake - 0.6 gS/m^2. In comparison with down area of this mountain (less then 200 m above the sea level), sulphur deposition is 1.0-1.5 gS/m^2. The catchments of the Kola mountain lakes are characterized by bare rocks, tundra soils and tundra vegetation- moss and, in some places, dwarf birch *Betula rana*. The Chibiny lake is smaller than the Chuna lake (area 3.5 and 12.5 hectares, respectively, water volume $0.11 \cdot 10^6$ and $1.25 \cdot 10^6$ m^3) and more quick running (resident time is 0,03 and 1 year, respectively). Multidisciplinary investigations of these lakes were carried out in 1993-1994 within the frame of Project AL:PE-2, supported by NIVA and Norwegian Ministry of Environment.

The main aims of the investigation have been: 1) ascertainment of specificity and intensity of airborne pollution impacts upon the fresh-water mountain ecosystems; 2) understanding the ecosystem of remote mountain lakes in subarctic region and its response to acid deposition and toxic heavy metals; 3) substantiation of the history trend and information criteria monitoring for airborne pollution.

2. MATERIAL AND METHODS

Water has been sampled during 1993-1994. The samples have been analyzed according to standard procedures for analysing low ionic strength waters. The analytical program includes the following components: pH, conductivity, calcium (Ca), potassium (K), magnesium (Mg), sodium (Na), chloride (Cl), sulphate (SO_4), nitrate (NO_3), fluoride (F), alkalinity, total organic carbon (TOC), total phosphorus (P_{tot}), total (tot) and labile (lab) concentrations of aluminium (Al), and heavy metals (Ni, Cu, Zn, Cr, Mn, Fe). Intercalibration of the results have been made within the frame of Project ALPE-2. The sediment cores were collected from the deepest parts (accumulation area) of the investigated lakes: 16.5 m from the Chuna lake and 6.5 m from the Chibiny lake. For sediment sampling a gravity core with automatically closed diaphragm (44 mm inner diameter) has been used [1]. The sediment cores (18 cm length from the Chuna lake and 14 cm length from the Chibiny lake) were vertically extruded and sectioned in of 1 cm layers for analyses. Values of loss of ignition (LOI) as indirect index of organic content were determined, as well as metal (Ni, Cu, Co, Zn, Cd, Pb, Sr, Mn, Fe, Ca, Mg, Na, K, Al) concentrations with atomic-absorption spectrophotometry [2]. Metal contents in the sediment subsamples from the lowest layers of the cores permit to evaluate the background values. Diatom analysis in of 10 cm upper layer of the Chibiny lake and of 18 cm upper layer of the Chuna lake sediment cores has been performed with the purpose to detect reconstruction of water pH alteration. Every layer of 1 cm has been analyzed. Subsamples of these sediment sections were prepared for diatom analyses by boiling in30% H_2O_2 adding $N_4P_2O_7$ to remove an organic matter.

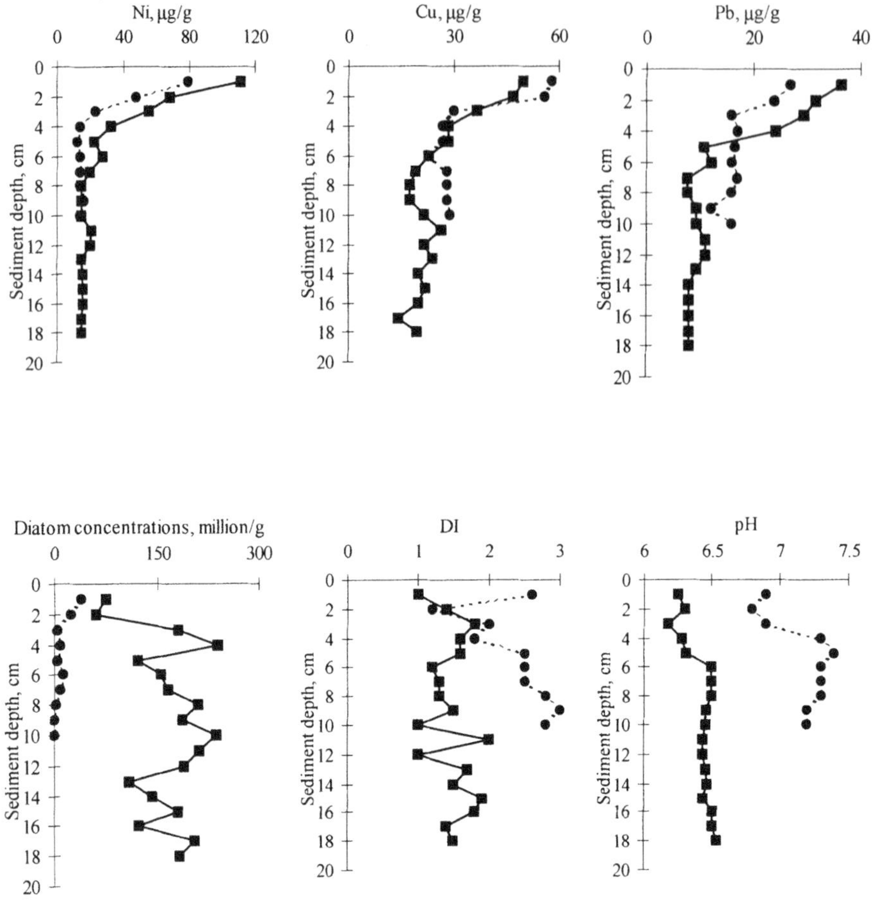

Fig. 1. Vertical distributions of concentrations of Cu, Ni, Pb (µg/g dry weight), diatom valves (million/g dry weight), diversity index (DI) and reconstructed pH values of the Chuna (■) and Chibiny (•) mountain lakes.

The valves (up to 600) in the prepared subsamples were counted to determine the relative frequencies (%) of the valves.

3. RESULTS AND DISCUSSION

The water chemistry of two lakes in the Chibiny and the Chuna mountains was different (Table).

The Chibiny lake has weak alkaline reaction - pH 7.2-7.5 and rather high buffer capacity conditioned by Na+ K cations, alkalinity 226-251 µeq/l.

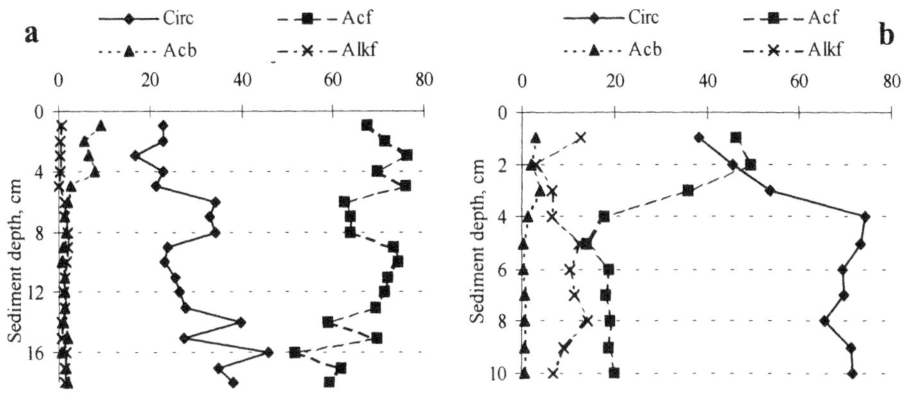

Fig. 2. The relative frequencies (%) of valves of alkaliphilous (alkf), circumneutral (circ), acidophilous (acf) and acidobiontic (acb) diatom species in sediment cores of the Chuna (a) and Chibiny (b) mountain lakes.

The Chuna lake has a weak acidic reaction - pH 6.2-6.7, a content of main cations of Ca + Mg is low. Ni and sulphate content mirrors a negligible local effect on the atmosphere pollution by smelter and despite their close placement to that of their content satisfies an average regional level ($SO_4 <$ 80 μeq/l, Ni < 1 μg/l). It is typical for the oligotrophic lakes. The content of nutrient is low: P_{tot} are compared in both lakes, the content of N_{tot} is high in the Chuna lake. The content of nutrient is increasing during winter time, and decreasing during summer time.

Local atmosphere emissions affect slightly the high mountain lakes. No heavily polluted air mass rise high altitude in the atmosphere. Sulphate content in lakes situated less than 200 m above the sea level and remote at the same distance from a smelter (< 30 km) accounts for 150 - 200 μeq/l [3].

Thus, the Chibiny and Chuna lakes reflect a real situation with airborne emissions on the whole Kola region. The values of critical loads are slightly exceeded for the Chuna lake, having more low buffer capacity in comparison with the Chibiny lake. At the same time, in the Chibiny lake there were discovered substantially high Al and Sr concentrations in a toxic labile form that is a result of negative effect of acidic precipitations. Both in the Chuna and Chibiny mountains we have discovered earlier the acidic episodes in the rivers when pH dropped up to 4.4 and 4.7, respectively [3].

Sediments of the Chuna lake are characterized by exceeded concentrations of heavy metals (Ni, Cu and Pb) in the upper of 4-5 cm core (Fig. 1). Factors of contamination (according to [4]), i.e. a quotient of concentrations from the uppermost (0-1 cm) to the lowermost (17-18 cm for the Chuna lake) layers, for Ni, Pb and Cu are 7.5, 4.6 and 2.5, respectively. Increased Ni and Cu concentrations at the upper 3 cm core layers of the Chibiny Lake were found out, those were by 4 and 2 times higher than background values, respectively (Fig. 1). Increased Pb concentrations at the upper 2 cm core layers were also observed (3 background values). The exceeding Ni, Cu and Pb concentrations in the upper 4-5 cm layers of the Chuna lake sediment core is accounted by atmospheric emissions of Severonickel Company smelters situated close to the lake (as regards Ni and Cu), a net of automobile roads and the general increase in Pb contamination in the atmosphere of the northern hemisphere [5]. The onset of increasing Ni and Cu concentration is caused by beginning of metallurgical Company activity. Increasing Pb concentrations in the Chuna lake sediments is explained, mainly, by beginning of intensive development of the Kola Peninsula.

Diatoms are classified into pH groups (acidobiontic, acidophilous, circumneutral and alkaliphilous) according to their ecological preference [6]. The percentage ratio of groups in the layer cores (10 cm in the Chibiny lake and 18 cm in the Chuna lake) is shown in Fig. 2. The diatom inferred pH values in the sediment core has been defined by the value of index B with the use of equation of linear regression [7]. Reconstruction of the lake water pH was based on defining 20 diatomic communities from the surface sediments in the similar lakes of the Kola North and values of water pH in those lakes. We have obtained the formulae for the Chibiny lake $pH=7.5-0.85\cdot \log B$, $r=0.95$, $s=\pm 0.27$ and for the Chuna lake $pH=7.13-0.6\cdot \log B$, $r=0.87$, $s=\pm 0.30$.

The neutral species predominate in the range of 3-10 cm (74%) of the Chibiny lake sediment core. Among the plankton *Cyclotella kuetzingiana* var. *kuetzingiana, C. Kuetzingiana* var. *radiosa* are the most massive. The importance of acidobiontic species is increasing, they prefer water with pH < 5.5. Among them *Eunotia exigua, E. monodon* and other (4%); acidobiontic species in the lakes of the South-West Sweden with water pH=4.5 account for 30% in the lakes of the Kola Peninsula (near Varzuga object)- up to 45%, water pH =4.8. The percent of alkaliphilous in sediments of the Chibiny lake is maximum in the layer of 7-8 cm up to 14.4%, and the lowest in the layer of 1-2 cm, up to 2.8%. Reconstruction of pH using index B gives a value of pH=7.3 (layer of 10-3 cm) and pH=6.9 (for the layer of 3-10 cm). The structure of diatomic communities changes in the layer of 4-1 cm, species diversity decreases there by 1.5 time in comparison with a layer of 10-4 cm. In the upper layers of 2 cm diatom concentration increases by 5 times (Fig. 1). Reconstruction of pH by diatoms allows to make a statement that weakly alkalized the Chibiny lake became weakly acidified.

Acidophilous species of benthic and periphytonic are predominate (up to 70%) throughout 18 cm of the Chuna lake sediment core (Fig. 2). The most massive species *Anomoeins serians* var. *brachysira* (up to 50% in the layer of 4-5 cm) and *Frustulia rhomboides* (layer of 0-1 cm, up to 18%). Species diversity of *Eunotia* (26 species) and *Pinnularia* (14 species) evidence that the lake has experienced natural acidification, according to [8]; pH calculated by index B for a layer of 18-5 cm accounts for 6.5; that may be taken a standard for the similar lakes in the Kola Peninsula. From 5 cm and higher the pH decreases to some extent up to 6.2-6.3, the same pH values are for the lake water. There is observed changing in composition of diatoms, respective by: a number of neutral species decreases by 1.5 time, alkaliphilous by 3 times via a number of acidophilous rises by 1.1 time and acidobiontic by 5%, reaching 10% in the surface layer. There are species *Eunotia exigua, E. monodon, E. robusta* var. *Diadema, Pinnularia biceps* and rare one acidophilous *Stenopterobia intermedia*. In the layer of 0-2 cm a number of diatoms sharply drops by 2.5 times and by 1.5 times there is decreasing diatom diversity. Here is a characteristics of ugly forms particularly beginning with a layer of 5 cm and higher among the species of *Eunotia -E. arcus, E. praerupta* and other. There also occur destroyed central parts of *Pinnularia viridis* var. *intermedia*. The given lake, mainly, agrees with a criteria of long-term monitoring for the mountain lakes. The historical trend of composition and diatom state detects the earlier stages of acidification and ecotoxic changes under heavy metal effect.

4. CONCLUSION

In the Chibiny lake there is observed a good water buffer capacity. A number of negative changes have been revealed in the water quality because of the acidic precipitation effect. High Al and Sr concentrations are discovered in water, particularly, their ionic toxic speciation. The apatite - nepheline syenites contain Al and Sr which are badly subject to weathering. However they more easily transit into ionic soluble speciation under the acidic precipitation effect that caused their content increase in the Chibiny mountain lake water.

The analysis of sediment chemistry in the Chuna and Chibiny lakes showed, that Ni, Cu and Pb exceeded concentrations in the upper of 4-5 cm layers of the Chuna lake sediment core is accounted by local atmospheric emissions of smelters (as regards Ni and Cu), and general increase by Pb contamination in the atmosphere of the northern hemisphere. The onset of increasing Ni and Cu concentrations is caused by the beginning of metallurgical Company activity. Increasing Pb concentrations in the Chuna lake sediments is explained, mainly, to the beginning of the Kola Peninsula intensive development.

Diatom investigations in the sediment cores of the Chuna and Chibiny lakes have ascertained a tendency to acidification process in the both lakes. Weakly alkanized the Chibiny lake by diatom composition of surface layers shows weakly acid state. Among the plankton forms there is an increasing share of acidophilous. In benthos and periphyton (the layer of 0-3 cm) there becomes the most specific acidbiontic species *Eunotia exiqua, E. monodon, E. robusta d. cliadema*. Similar reconstruction, alarming the starting point of water acidification when preserving during summer-autumn period the pH 7.0-7.4 may be provoked by acid episode effect during snowmelting period and ionic form of metals. In the originally weakly acidified and sensitive Chuna lake the pH reconstructed has dropped from 6.5 to 6.2. Beginning with the layer of 5 cm there is noticed a number of diatoms, their variety and importance of acidofills and acidobionts rises: *Eundia monodon, Pinnularia bicepts, Brachysira serians, Stenopterobia intermedia (S. signatella)*. Diatoms from the Chuna lake reflect the toxic load - in the upper layers sometimes there occur the ugly forms of species (*Eunotia*) distrusted *Pennularia diridis*.

5. REFERENCES

Dauvalter V.. Sci. Tot. Envir., 158 (1994) 51-61.

Håkanson L.. Water Research, 14 (1980) 975-1001.

Hustedt F.. Arch. Hydrobiol. Suppl. 15 (1939) 638-790.

Moiseenko T.. Ambio 27 (3) (1994) 418-424.

Norton S.A., P.J. Dillon, R.D. Evans, G. Mierle and J.S. Kahl, In: Lindberg S.E. et al. (Eds.), Sources, Deposition and Capony Interactions, Vol. III, Acidic Precipitation. Springer-Verlag, 1990, pp. 73-101.

Renberg I. and T. Hellberg. Ambio 11 (1) (1982) 30-33.

Renberg I., T. Korsman and J. Anderson. Ambio 22 (5) (1993) 264-271.

Skogheim O.K.. Rapport fra Arungenprosjektet. AS-NLN, Norway, 2, 1979.

Influence of Climate, Species Immigration, Fire, and Men on Forest Dynamics In Northern Italy, from 6000 Cal. BP To Today

THOMAS MATHIS[*], FRANZISKA KELLER[*], ADRIAN MÖHL[*], LUCIA WICK[*], HEIKE LISCHKE[**]

[*] *Institute of Geobotany, Section Paleoecology, University of Bern, Altenbergrain 21, CH-3013 Bern Switzerland,*

[**] *Swiss Federal Institute for Forest, Snow and Landscape Research, Zürcherstr. 111, CH-8903 Birmensdorf.*

Key words: forest modeling, climate change, human impact, species immigration, late Holocene pollen data, *Castanea sativa*

Abstract: Pollen records reflect integrated effects of abiotic and biotic processes such as establishment, competition, climatic change, fire history and human impact. To disentangle these processes we compared a pollen record of Lago di Annone (Northern Italy) in the time interval 6000 cal. BP till today with simulations of a forest-dynamic model (DisCForm) under different combinations of climate, species immigration, human impact, and fire scenarios. Circularity was prevented by using input data that were independent of pollen data. While species competition, climatic change, and species immigration seem to produce model outputs with little similarities with the evaluated pollen record, the simulation of fire events and human activities reflect the main patterns of the original pollen record. The scenario for human impact slightly improves the simulation output. Species composition and abundance of Insubric forests of the time investigated seems therefore to be highly determined by fire and human impact. The simulation runs show that introduced species such as *Castanea sativa* are not able to coexist with indigenous species.

1. INTRODUCTION

European forests in the late Holocene are strongly influenced by dynamic processes that result in a temporal change in the presence and abundance of

G. Visconti et al. (eds.), Global Change and Protected Areas, 195–208.

tree species. Forest development is not only influenced by autogenic processes (changes in resource availability resulting from e.g. competition) but also by external factors such as species immigration, climatic change, and human impact.

It is assumed that changing climatic conditions are the most important factors in forest dynamics [1]. Several records of climate history in Europe reveal distinct phases of high or low temperatures, whereas the reconstruction of the temporal progression of precipitation stays fuzzy [2].

However, also other factors have the potential to have influenced forest dynamics. To our knowledge important factors like fire or human impact were not included in dynamic models up to now; this fact may be explained by the difficulty to get precise data on men's activity in the late Holocene. Information about human impact on forest development remains a topic with disparate views and insufficient data. Furthermore it is difficult to separate human impact on forest dynamics from natural influences such as fire.

Pollen and macro-particle analyses deliver reliable data to understand the long-term changes in vegetation. It is for example possible to detect the point of time when a tree species migrated back from the refuges after the glaciation [3].

All these forces that add to the change of long-living ecosystems and their interplay need to be understood to assess the elasticity of forests to past and future disturbances and environmental changes. Computer simulations help to formalise relevant determinants and to figure out the importance of each of these factors [4]. The aim of our study is to try to understand these forces and to disentangle these factors.

The ability to simulate natural forest systems provides the opportunity to assess whether or not introduced tree species such as *Castanea sativa* would be outcompeted by indigenous species [5] and to which degree the current vegetation of the considered region is influenced by human impact.

2. DATA, MODEL AND METHODS

2.1 Experimental Setup

To study the influence of the various factors on forest dynamics, simulations with a dynamic forest model were run for a specific site in northern Italy. The model outputs have been compared to pollen data of this site. Assumptions on the factors were formulated as scenarios for the input variables of the model, which were supported by independent data.

The following combinations of scenarios were studied successively:

Effects of...	immigration	precipitation co	precipitation var	temperature co	temperature var	Fire	Men
Succession (1b)		x		x			
Climate (1c)			x		x		
Immigration(1d)	x	x		x			
Fire (1e)	x		x		x	x	
Men (1f)	x		x		x	x	x

To examine the influence of system intrinsic dynamics, namely succession, a simulation was run (1b) with constant inputs for climate, and without fire events, human impact and a later immigration of species. Impact of climate change was simulated by changing temperature and precipitation inputs (1c). In further simulations under constant climate species were not allowed to establish before their recorded immigration times (1d). Additionally, the impact of fires (e) and finally humans (f) are studied.

2.2 Location

Lago di Annone (226 m a.s.l.) is one of several lakes situated near the southern end of Lago di Como (Brianza) in the transitional zone between the Southern Alps and the Po plain.

Lago di Annone is a fairly big lake with a water surface of 5.5 km^2, a maximum water depth of about 14 m (in the eastern basin), and a catchment area of 22.5 km^2 that reaches altitudes of 1261 m a.s.l. in the north (Mt. Rai) and 922 m a.s.l. at Mt Barro in the north east. Its outlet flows into Lago di Como [6].

The climate of the study area today is of the insubrian type, characterised by relatively mild and dry winters and warm and humid summers, with maximum precipitation in spring and autumn [6].

The natural vegetation around the lakes in Brianza is largely destroyed by man. The forest sites on the slopes at Monte Barro can be attributed to the submediterranean vegetation complex covering the lowlands at the foothills of the Alps in northern Italy up to about 800 m a.s.l. [7].

A pollen core was taken at about 6 m water depth in the southern part of Lago di Annone. It contains a complete Holocene sequence and is a summary of different altitudes. That is the reason why simulations were performed for 197 m, 500 m and 800 m a.s.l. which were averaged in the following way: 197 m counted 70%, 500 m 20% and 800 m 10%.

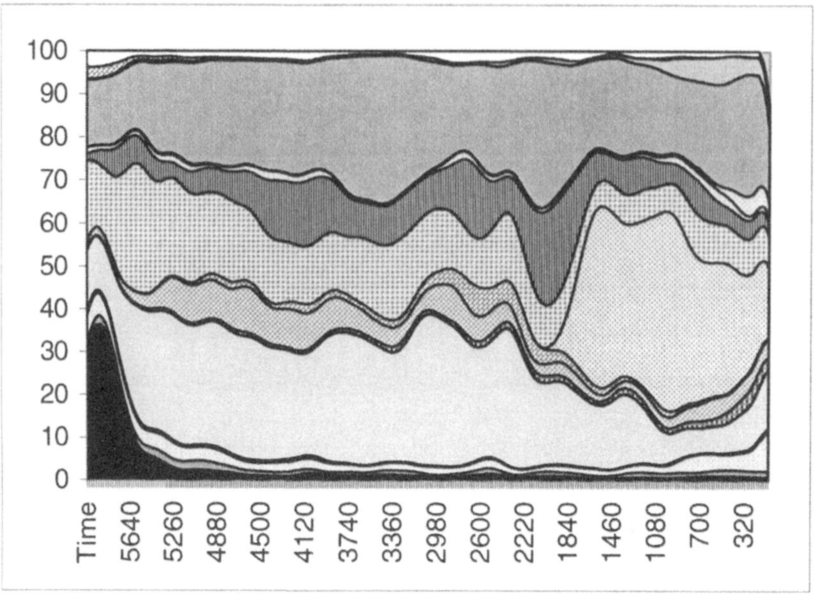
100
90
80
70
60
50
40
30
20
10
0
Time
5640
5260
4880
4500
4120
3740
3360
2980
2600
2220
1840
1460
1080
700
320

Fig 1a

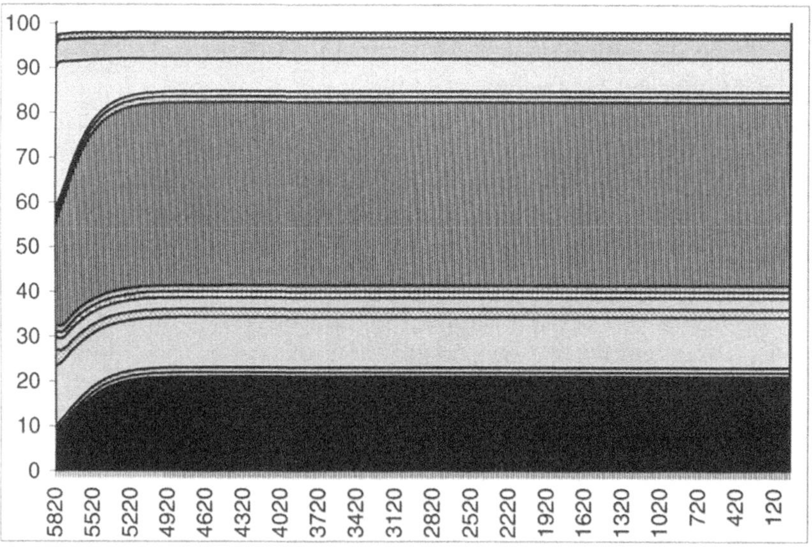
100
90
80
70
60
50
40
30
20
10
0
5820
5520
5220
4920
4620
4320
4020
3720
3420
3120
2820
2520
2220
1920
1620
1320
1020
720
420
120

Fig. 1b

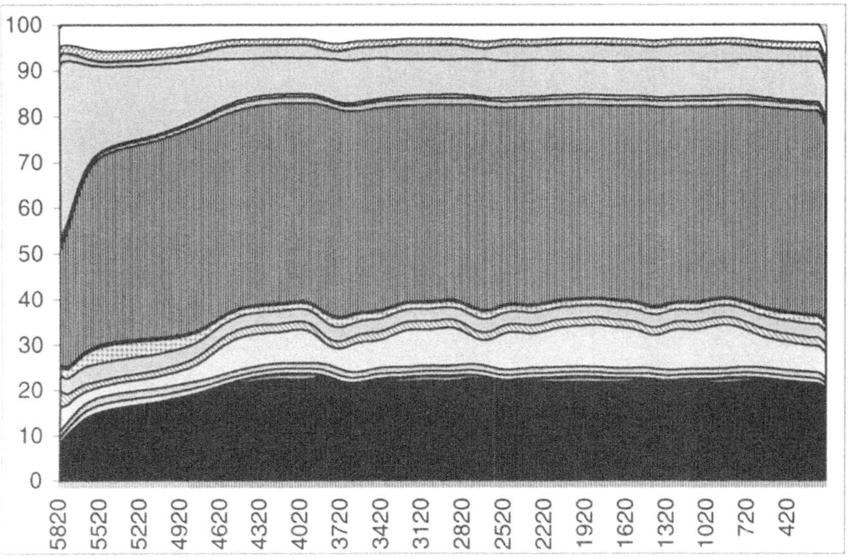
100
90
80
70
60
50
40
30
20
10
0
5820
5520
5220
4920
4620
4320
4020
3720
3420
3120
2820
2520
2220
1920
1620
1320
1020
720
420

Fig. 1c

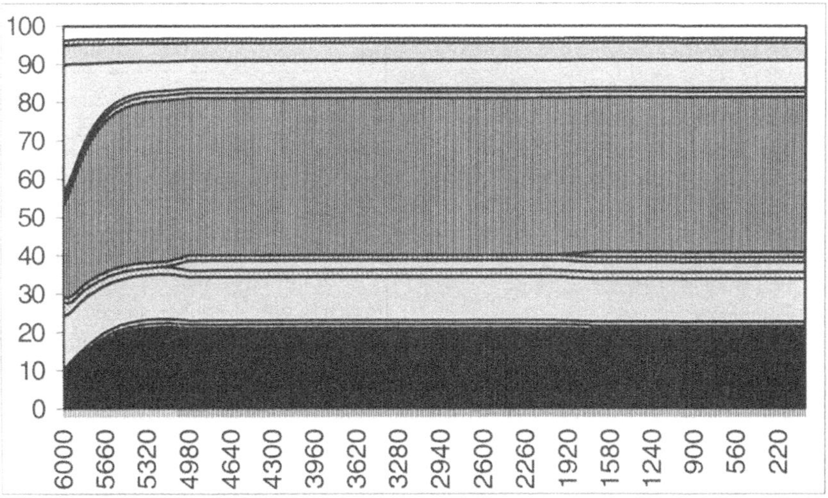
100
90
80
70
60
50
40
30
20
10
0
6000
5660
5320
4980
4640
4300
3960
3620
3280
2940
2600
2260
1920
1580
1240
900
560
220

Fig. 1d

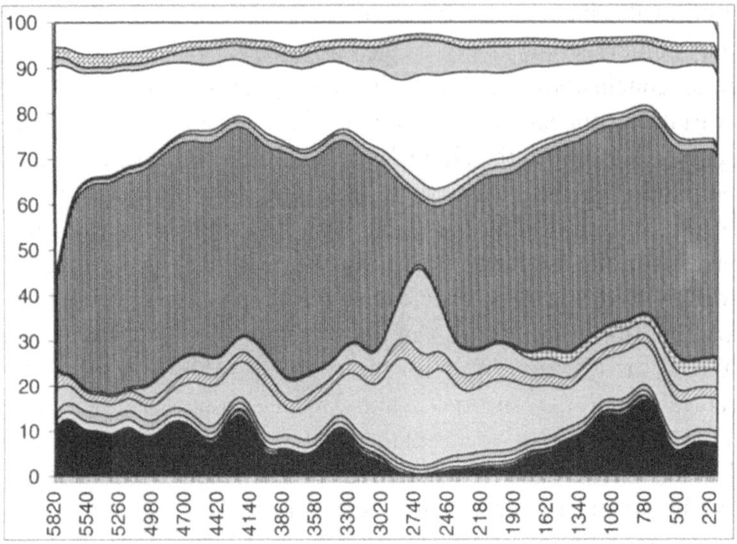
100
90
80
70
60
50
40
30
20
10
0
5820
5540
5260
4980
4700
4420
4140
3860
3580
3300
3020
2740
2460
2180
1900
1620
1340
1060
780
500
220

Fig. 1e

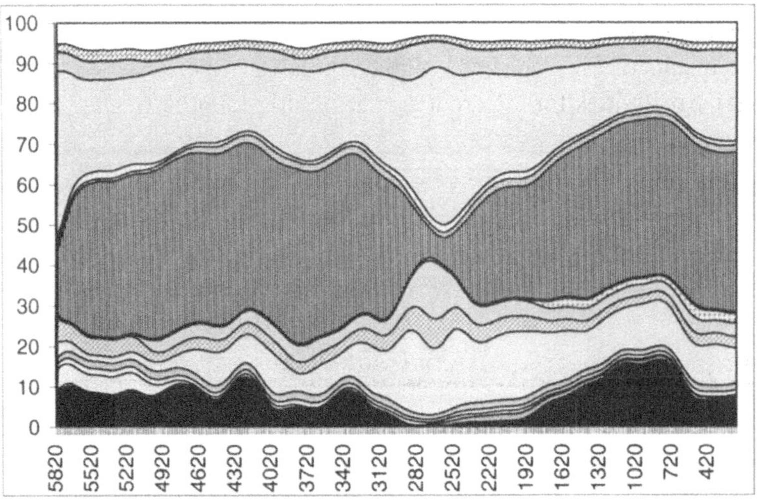
100
90
80
70
60
50
40
30
20
10
0
5820
5520
5220
4920
4620
4320
4020
3720
3420
3120
2820
2520
2220
1920
1620
1320
1020
720
420

Fig. 1f

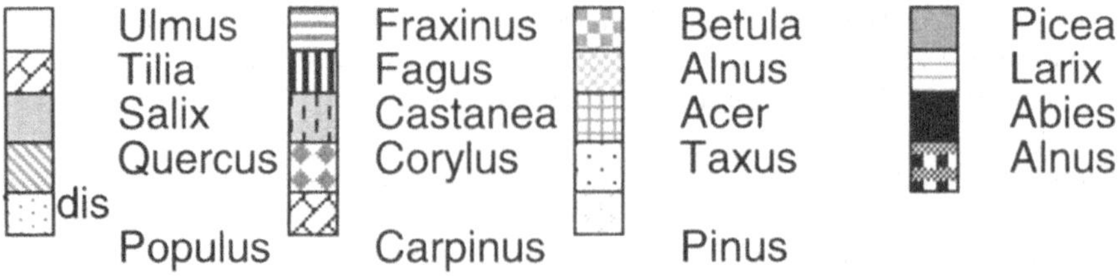
Ulmus
Tilia
Salix
Quercus
dis
Populus
Fraxinus
Fagus
Castanea
Corylus
Carpinus
Betula
Alnus
Acer
Taxus
Pinus
Picea
Larix
Abies
Alnus

Fig. 1 Comparision of pollen record (Wick 1996) with forest model simulations under input scenarios

a C_{14}-calibrated pollen record from Lago di Annone.

b – e simulations with DiscForm. Simulated pollen percentages are determined from basal area of species, corrected by factors of Iversen (Faegri and Iversen 1975)

b all species present, constant climate, average temperature and precipitation

c all species present, transient temperature and precipitation

d species immigrate, constant temperature and precipitation

e species immigrate, transient temperature and precipitation, fire history

f species immigrate, transient temperature and precipitation, fire history, human impact

2.3 Model input: immigration, climate and fire

The model inputs for immigration, climate and fire were deduced and introduced in the same way as described in Keller et al. (in prep.) [8].

To determine the immigration dates we compared the pollen diagrams of all the lakes in the region that have been investigated by pollen analysis: Lago di Ganna [9], Lago di Origlio [10], and Lago di Muzzano [11].

The establishment period was defined between the date for the first appearance of each species in the pollen profile and a point of time with local appearance (defined by exceeding a specific pollen threshold).

During the establishment period the abundance of the species increase exponentially in the way as described in Keller et al. (in prep) [8]. All dates are given as calendar years before present (BP i.e. before 1950 AD).

The actual climatic conditions at Lago di Annone were described with a meteorological record from Olginate (4 km east of the study site). The average values for the period 1973-1985 were 12.21°C and 1159.8 mm of precipitation per year (Istituto Italiano di Meteorologia, pers. com.).

The temperature scenario for the 6'000 years BP of simulation was introduced according to Gamper (1993) [2] who averaged fluctuations of the glaciers in the Swiss Alps. The precipitation scenario is based on a combined pollen- and lake-level reconstruction by Guiot et al. (1993) [12] for grid point 9°E 46°N for the time after 9930 cal. BP.

We interpolated the reconstruction linearly between 6750 cal. BP with a precipitation anomaly of –25 cm/year and today (1950 AD) with a value of 0 cm/year (calibration level).

The fire events were deduced from Tinner et al. (1999) [11] as described in Möhl & Wick (in prep.) [13].

In the model fire appearance is formulated as a process, whose probability (frequence) is proportional to the values of the charcoal data.

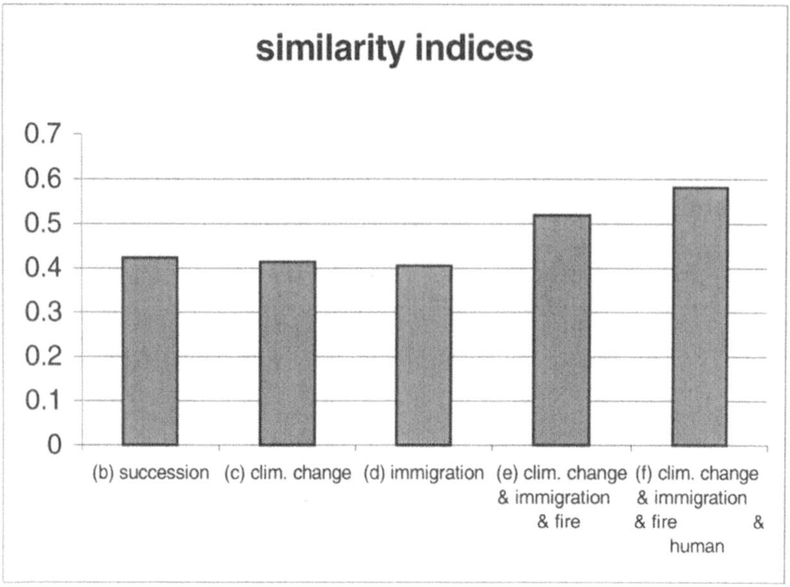

Fig. 2
Similarity of forest model simulation and pollen data over whole simulation period for the different input scenario combinations

2.4 Human impact

To imitate human impact we have to refer to special indicators that reveal human activity, such as *Cerealia* pollen for agriculture [14], but because of poor pollen dispersal cereals generally are considerably underrepresented in pollen records. Therefore the quantification of Neolithic forest clearances based on cereal pollen is not possible.

Archaeological studies [15] and interpreted charcoal data [16] assume that fire played a major role in Neolithic forest clearances.

Tinner et al. (1999) [17] show significant correlations between charcoal influx and human indicators in the pollen diagram (e.g. pollen percentages of *Plantago lanceolata* and total herbs) at Lago di Origlio (southern part of Switzerland) situated 41 km Northwest of Lago di Annone. We refer to these charcoal influx data as indicator of human activity and used them above a defined level to represent human activity in the catchment area of Lago di Annone.

Whenever the charcoal values exceed 10 mm^2 cm^{-2} yr^{-1} we have to suppose human activity. This acceptance is supported by Clark et al. (1995)

[18] who show that in natural forests charcoal values do not exceed this limit.

Fire clearances

The model takes into account that men practised fire clearances. Archaeological models assume that such cleared areas were then kept open for agriculture for about 30 years following a fire event [19].

We used the following definition for the percentage (P) of the burnt area that is affected by man:

$P = (a - 10\ mm^2\ cm^{-2}\ yr^{-1}) * a^{-1}$ where a is the charcoal influx ($mm^2\ cm^{-2}\ yr^{-1}$)

The model simulates this agricultural area by stopping seedling establishment on fire-cleared areas for 30 years.

Human browsing

Studies of goat/sheep faeces [20] revealed that the fodder consisted partially of leaves of some deciduous tree species. Akeret et al. (1999) [21] suggest that all tree species that grew around a settlement and that produce leaves digestible for goats and sheep have been used since the Neolithic or even the Mesolithic period.

In the model the browsing factor has an impact on sapling establishment. The precursor models [22], [23] already incorporated the browsing by sheep and goats, which results in a low probability for establishment [24] and diminishes the growth of both sapling and young or suppressed trees, modelled by a lower probability of sapling establishment. The assumption cannot be facilitated by quantitative data for sapling mortality because corresponding studies are not found in the literature [25].

In addition to this natural browsing factor we calculated a value that is representative for the human use of trees for goat/sheep-fodder. This impact is simulated in the same as explained above.

We assume that browsing by goats and sheep correlates with charcoal data: Whenever there are high charcoal values that indicate a general high human impact we suppose that the goat/sheep-browsing is increased as well.

For every data unit that exceeds the level of 10 $mm^2\ cm^{-2}\ yr^{-1}$ we used the following equation to calculate the values for goat/sheep-browsing:

$$a = 0.2 * (\text{charcoal} - 10)$$

The variable »a" is rounded to an integer that represents classes 1 to 9. This value indicates the importance of goat/sheep-browsing.

2.5 Modelling forest dynamics

Many kinds of models simulate forest development [26]. According to Bugmann (1994) [24] forest-gap models are a good tool to simulate forest dynamics under changing climatic conditions because a variety of phenomena can be considered, ranging from age structure and species composition to primary productivity and nutrient cycling. Several studies using forest-gap models to simulate Holocene vegetation development have already been published [27], [4], [28]. The DisCForm model employed [22], [23] a descendant of the forest-gap model ForClim. DisCForm differs from a conventional forest-gap model in that it does not consider single trees but summarises the tree-population densities in several height classes. The spatial variability expressed in gap models by stochastic simulations of numerous gaps is represented by theoretical descriptions of the spatial (Poisson) tree distribution in each height class. This results in a spatial distribution of light availability and consequently in the rates of change within each height class, thus determining the dynamics of the population density. From the tree population densities in the height classes the biomass per species in t/ha can be calculated. Because the simulation is no longer stochastic, this distribution-based approach is much faster than the traditional gap-model approach and suitable for numerous simulations over several millennia.

2.6 Validation of the model results with pollen data

The pollen data from the Lago di Annone [6] were used for comparison with the model simulations and to validate them.

To compare the model output and the pollen analysis, the simulated biomass per species has to be converted to pollen percentages as represented in the pollen record. Therefore the model output was adjusted with the help of the conversion factors of Iversen for pollen representation [29] as used in Lotter & Kienast (1990) [30] and in Lischke et al. (1999) [4].

To have a quantification of the resemblance of the simulation and the pollen analysis a similarity index can be calculated. First model outputs and pollen results are smoothed with a Gaussian Low Pass Filter resuming every 400 years. Then the similarity index [31], [32] is calculated for each simulation in the same way as in Lischke et al. (1999) [4].

3. RESULTS

The various combinations of scenarios for climate, species immigration, fire and human impact (1b - 1f) allow us to test if climatic change, successional dynamics and species immigration can be used as explanations for the appearance and abundance of taxa in the pollen record (fig. 1a), or whether or not fire events and human impact have to be taken into account. In Fig. 1b - 1f the simulation results, in Fig. 2 the quantitative similarity values are represented. Figure 1b shows the typical result with the presence of every simulated taxon and of recent precipitation and temperature data. The simulated 6000 years show the successional changes that last for about 800 years. This contrasts the dynamics in the pollen record for the whole time period of 6000 years. After this transient phase the simulation shows that *Fagus silvatica,* deciduous *Quercus* species, *Abies alba,* and *Alnus glutinosa/incana* dominate the forest. The low similarity index reveals that interspecific competition alone does not seem to be able to explain the changes in the pollen record. To assess the effect of a changing climate we ran a simulation with transient precipitation and temperature. However, Fig 1c shows no significant characteristics and the slight fluctuations do not represent any patterns according to the pollen record.

No significant improvements can be attained in Fig. 1d (immigration data, constant temperature and precipitation). At the beginning of the simulation *Fagus* is in the establishment phase and shows a fast and efficient development towards a high dominance, which is inconsistent with the pollen record. The introduced species *Castanea sativa* shows an inconsiderable portion in contrast to the pollen record. In Fig. 1e the simulated fire events result in an important improvement of the similarity index. *Abies* decreases progressively from the start of the simulation to about 2300 cal. BP but afterwards increases significantly. The share of *Quercus* increases and the portion of *Alnus* increases slowly. The characteristic share of *Castanea* in the pollen record cannot be achieved and also the quota of *Fagus* stays constantly strong in this version. The simulation of human impact (Fig. 1f) shows the effect of a higher portion of *Quercus* and a decreasing portion of *Fagus*. These two slight improvements can be reflected by a higher similarity (Fig. 2).

4. DISCUSSION

The trials to disentangle factors by computer simulations are satisfying, although the pollen record shows some properties that cannot be explained by the input parameters and data.

Although multiple data sources avoid circularities there are still some uncertainties: the climate scenario according to Gamper (1993) [2] gives information about temperature anomalies on the northern part of the Alps, and these anomalies cannot simply be transferred to the insubrian climate. It cannot be determined ultimately whether or not the charcoal values represent the fire history and the human scenario of the catchment area concerned.

The way fire, fire clearances and goat/sheep browsing affect the tree species in the model had to be based largely on assumptions. Also commonly used percent representation of pollen data and the thus required conversion of biomass to pollen percent introduce further uncertainties to the simulation-pollen study as discussed in Lischke et al. (1999) [4] and Lischke et al. 1998 [33].

In contrast to Lischke et al. (1999) [4] it is not possible to be certain that the climate scenario can explain main characteristics. A distinct improvement is shown by the simulation of fire. The suppression of *Abies*, the increased proportion of the pioneer species *Alnus,* and a slight increase of *Quercus* suggest that fire is an important factor in the forest dynamics of this area. The amplification of this characteristic by incorporating human impact, such as goat/sheep browsing and fire clearances, suggests that in this region there was an influence of man on forest dynamics. In the simulation the increasing proportion of *Abies* after 2000 cal. BP could be the consequence of the decrease of fire frequency [11].

This characteristic in simulations can be explained by a strong influence by the Romans who favoured *Castanea sativa* by reducing fire (e.g. less litter at the ground) and restraining competing species such as *Fagus* or *Abies*. The simulations reveal that at low altitudes *Abies* has a potential to gain a clear abundance in natural forests, whereas *Castanea* seems not to be able to compete against indigenous species.

This is not consistent with observations in forests that reveal significant occurrences and rejuvenation of *Castanea* on acidic soils where chestnut cultivation has been abandoned [33] . The model does not take into account some important soil properties such as pH or nutrient availability. These factors could be simulated by specific growth parameters that are adjusted to certain soil properties. A model with differentiated input parameters could check if on acidic soils chestnut forests can exist without human influence.

5. CONCLUSIONS

The simulation of forest dynamics in the region of Como does only correspond significantly with the pollen record if fire events and human activity are incorporated.

Therefore in the investigated region fire and human impact seem to be important factors that influence forest dynamics.

In contrast to other studies [4] the analysis reveals that the climate scenario and the species immigration are no determinants that explain main characteristics of the pollen record of the time window 6000 BP until today.

Simulation results affirm that the introduced *Castanea sativa* is not able to establish and coexist in natural forests with insubrian climate.

6. REFERENCES

Akeret Ö., J.N. Haas, U. Leuzinger, J. Stefanie. Veg. Hist. Archaeobot (submitted) (1999).

Akeret Ö., S. Jacomet, Veget. Hist. Archaeobot 6, (1997) 235-239.

Berglund, B.E. (Ed.). Handbook of Holocene Paleoecology and Paleohydrology, Chichester, 1986.

Bugmann H.K.M., ETH, (Ed.) Diss. No. 10638, Zürich, 1994.

Clark J. S., P.D. Royall, Quat. Res. 43 (1995) 80-89.

Conedera M., P. Stanga, B. Oester, P. Bachmann, In: Dynamics of Mediterranean Vegetation Landscape (submitted) (1999).

Cormack R.M., J. Royal Stat. Soc. 134 (1971) 321-353.

Durand R., H. Chaumeton, Paoline, (Ed.), Gli alberi, Milano, 1991.

Erny-Rodmann C., E.Gross-Klee, J.N. Haas, S. Jacomet, H. Zoller, Jahrb. Schweiz. Ges. Ur- u. Frühgesch. 80 (1997) 27-56.

Faegri K., J. Iversen, Munksgaard, (Ed.), Textbook of Pollen Analysis, Copenhagen, 1975.

Favre P., S. Jacomet, Hist. Archaeobot 7, (1998) 167-178.

Gamper M., In: B. Frenzel, (Ed.), Solifluctuation and climatic variation in the Holocene. Paläoklimaforschung 387, Stuttgart, 1993.

Guiot J., S. Harrison, I.C. Prentice, Quart. Res. 40 (1993) 139-149.

Jacomet S., C. Brombacher, M. Dick, In: Schweiz. Landesmuseum, (Ed.), Die ersten Bauern, Zürich, 1990.

Keller F., T. Mathis, A. Möhl, H. Lischke, L. Wick, B. Ammann, F. Kienast, prep. for J. Ecosyst.

Lischke H., A. Guisan, A. Fischlin, J. Williams, H. Bugman, In: P. Cebon, U. Dahinden, H. Davies, D. Imboden, C. Jaeger (Eds.), A view from the Alps: Regional perspectives on climate change, MIT Press, Boston, 1998, pp. 309 - 350.

Lischke H., A. Lotter, A. Fischlin, Ecology (submitted) (1999).

Lischke H., T.J. Loeffler & A. Fischlin, Theor. Popul. Biol. 54(3) (1998) 213-226.

Lischke H.. Nat. Res Mod. 1999 accepted.

Loeffler T.J., H. Lischke, Natural Ressource Modeling (submitted) (1999).

Lotter A., F. Kienast, Geol. Surv. Finland, Special Paper 14, (1990) 25-31.

Möhl A., L. Wick, in prep.

Näscher F.A., ETH, (Ed.) Diss. No. 6373, Zürich, 1979.

Oberdorfer E., In: Beitr. naturk. Forsch. SW-Deutschl., 1964, pp. 141-187.

Schneider R., K. Tobolski, Diss., Universität Bern, 1985.

Solomon A.M., D.C. West, J.A. Solomon, In: D.C. West, H.H. Shugart, D.B. Botkin, (Eds.), Forest succession: Concept and application. Springer, New York, USA, 1981

Tinner W., M. Conedera, Boll. Soc. Tic. d. Sc. Nat. 83 (1995) 91-106.
Tinner W., M. Conedera, E. Gobet, P. Hubschmid, M. Wehrli, B. Ammann, Holocene (submitted) (1999).
Tinner W., P. Hubschmid, M. Wehrli, B. Ammann, M. Conedera, J. Ecol. 87, (1999) in press.
Tzedakis P.C., K.D. Bennett, D. Magri, Nature 370 (1994) 513.
Wick L., Diss., Universität Bern, 1996.
Winiger J., In: Schweiz. Landesmuseum, (Ed.), Die ersten Bauern, Zürich, 1990, pp. 297-306.
Wolda H., Oecologica 50 (1981) 296-302.

Koenigia Islandica (Iceland Purslane) – A Case Study of a Potential Indicator of Climate Change in the UK.

BARRY MEATYARD
Environmental Sciences Research and Education Unit, Institute of Education, University of Warwick, Coventry,

Key words: Koenigia, climate change, Scottish vegetation, environmental indicator.

Abstract: *Koenigia islandica* (Iceland purslane) is an annual arctic-subarctic species that is found in only two locations in the UK. On the Isle of Mull, Argyll, Scotland, it grows at the southerly limit of its W.European distribution. The habitat requirements of *Koenigia* are specialised. It is predicted that an annual species at the limit of its geographical range is potentially sensitive to climatic change. The abundance of the plant has been monitored annually since 1994 and fluctuations in population levels have been compared with meteorological data over the same period. Preliminary results indicate that *Koenigia* populations are sensitive to temperature and rainfall during the growing season. Analysis of trends in the Scottish climate suggests an increased frequency of higher air temperatures in the spring and a decrease in summer precipitation. The implications of this for *Koenigia* are discussed.

1. INTRODUCTION

Iceland Purslane is a diminutive annual plant with a widespread distribution in mountain and periglacial regions in the high latitudes. Its circumpolar distribution is well documented by Hultén [1]. In Britain it is confined to two localities on islands on the west coast of Scotland [2] and it is listed as a rare plant in the UK Red Data Book [3]. However the presence of its pollen in fossil deposits indicates that its distribution during the Late Weichselian extended further east and south [4]. In its present distribution it is considered to be a relic of a flora that existed towards the end of the last glaciation in the UK having retreated from its previous wider distribution [5]

G. Visconti et al. (eds.), Global Change and Protected Areas, 209–217.

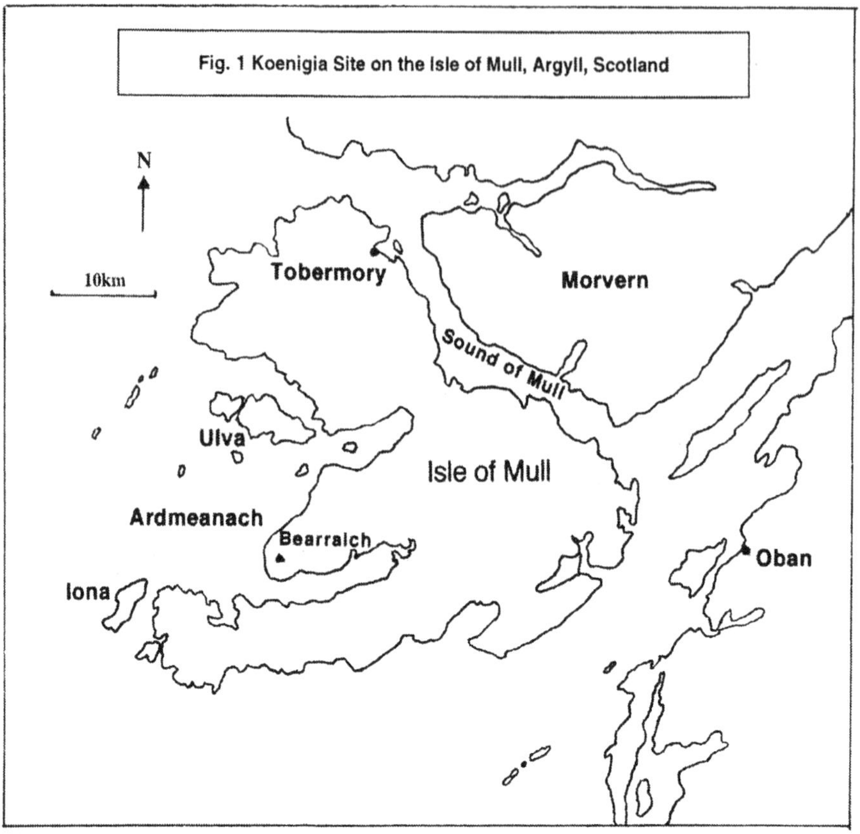

Fig. 1

The contemporary presence of *Koenigia* in the UK was not confirmed until 1950 when B.L. Burtt was examining material labelled as *Peplis* in the herbarium at Kew and discovered the incorrectly labelled *Koenigia* which had been collected on Skye in 1934 [6]. Its distribution was extended to the Isle of Mull in 1956 [5, 7].

The late addition of *Koenigia* to the UK flora is an indication of the remoteness and inaccessibility of the mountains of the Scottish islands which to some extent still exists today. On Skye it is restricted to the summit plateaux of hills around the Storr and on Mull it is found around the summit of Bearraich, and on other hills, on the Ardmeanach peninsula (Fig. 1). On Mull it is at the southern limit of its western European distribution (latitude 56° 21′N, longitude 6° 9′W).

The study site on Bearraich that forms the focus of this paper is within an area designated as a Site of Special Scientific Interest under UK legislation. The site is owned by, and is under the management and protection of, the National Trust for Scotland, a charitable, non-governmental organisation.

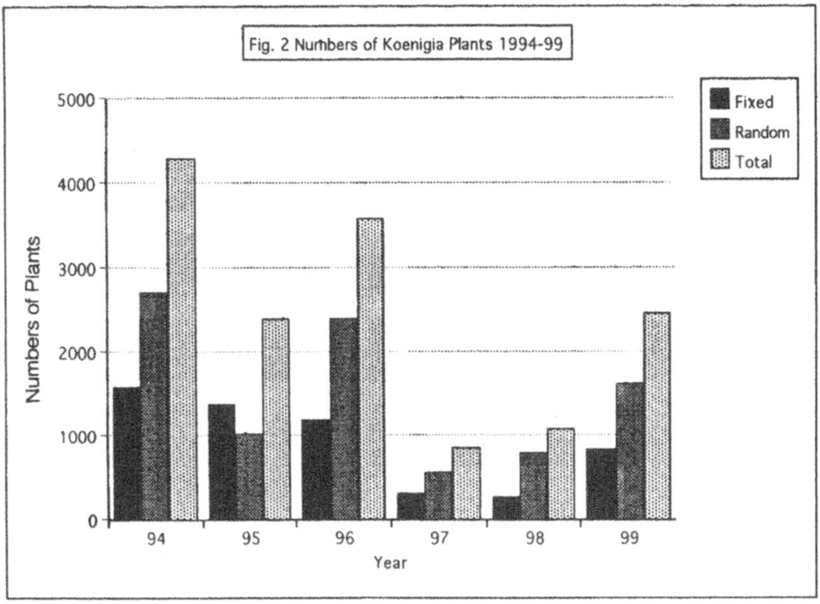

Fig. 2

The habitat of *Koenigia* on both Mull and Skye is highly specialised and comprises open basalt gravel terraces at or near the summits of hills ranging in altitude from around 400 m to 700 m, although on Mull the maxima are within the lower part of this range and the study site is at an altitude of 385m. The gravel particles range from 0.5 cm to 5 cm in diameter beneath which is a finely graded silt. This skeletal soil is derived from the Tertiary lavas which form a major feature of the geology of these islands. The terraces are very sparsely vegetated and their highly characteristic structure and origin are a matter of some debate [8]. It seems likely that climatic factors are involved and also biotic influences such as trampling by red deer (*Cervus elephus*), large populations of which are present on Ardmeanach. Comparison of the shapes of the terraces in 1998 with an aerial survey of 1946 showed many clearly recognisable features which indicate that they are relatively stable - at least over the last 50 years. The terraces on Mull are washed by rain water fed flushes, the pH of which has previously been recorded as pH 5.25 to pH 5.8 [8]. However during the period of this study the pH of the flushes has been recorded in the range pH 5.5 to pH 6.8.

The plant community of which *Koenigia* is a feature is defined in Floristic Table M34 of the UK National Vegetation Classification [9]. This community includes *Carex viridula ssp. oedocarpa, Deschampsia caespitosa* and *Juncus triglumis* as well as *Koenigia* among its dominants. M34 is derived from data obtained from Skye and on Mull *Sedum villosum* and *Sagina nodosa* are also additional common associates. Since *Koenigia* is an annual, and on Mull is at the southern limit of its European distribution, its populations there may be regarded as potentially sensitive to, and possible indicators of, climatic change. An indication of this was reported at the European Conference on Global Change in Mountain regions in Oxford in 1997 (10). In a recent study of 107 plants of the mountains of Norway, Sætersdal and Birks have predicted *Koenigia* to be sensitive to an applied climatic change model based primarily on an increase in temperature in July and January of 2˚C and 4˚C respectively and have estimated that its summer temperature optimum is 7˚C [11]. Based on the data obtained in this study precipitation also appears to play a significant role in determining its abundance from year to year. Observations during field work in 1995, 1997 and 1998, years in which *Koenigia* population levels were reduced, indicated that these years were particularly dry, with reduced flush evident in the gravel terraces. In particular the spring season of 1997 was considered to be very dry on Mull and there is anecdotal evidence of remote houses experiencing water shortages on the island during May and June. It was thus predicted that *Koenigia* was sensitive to weather conditions - particularly in the early part of the year. To test this relationship climate data was obtained from the UK Meteorological Office for the years 1994 to 1998, the data for 1999 not yet being available. There is no recording weather station on Mull, and data was therefore obtained from the neighbouring island of Tiree. The weather station on Tiree (at 12m) is of much lower altitude than the *Koenigia* site, but overall trends in precipitation and temperature are likely to be comparable since the weather on the west coast is largely determined by the major Atlantic weather systems which move eastwards over the islands.

2. THE MONITORING OF KOENIGIA ON MULL

Since 1994 annual counts have been made on sample sites around the summit of Bearraich, on the Ardmeanach peninsula on the Isle of Mull (see Fig. 1). The technique has been to determine species density by counting all *Koenigia* plants in 1m square quadrats placed either in fixed locations or randomly in selected, easily identified areas of the terraces. The fixed locations and the areas for random sampling are both identified with 100% accuracy due to the recognition of natural rock and vegetation features. The accuracy is checked each year using field assistants to identify locations from photographic records. Sixty five quadrats are counted each year, of which 21 are in fixed positions and 44 are placed randomly within eleven defined areas of the terraces.

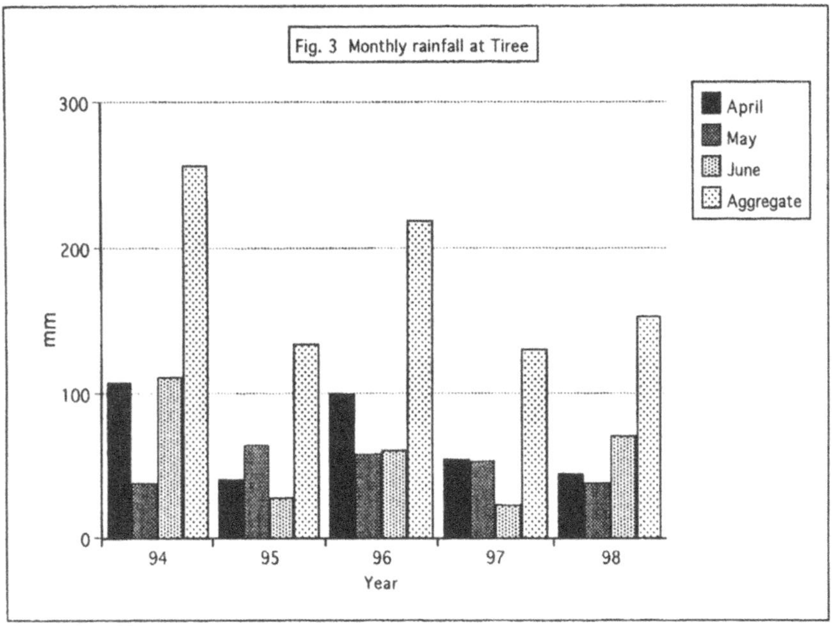

Fig. 3

The eleven areas that are randomly sampled each cover an area of approximately 30 square metres. Counts are made in the last two weeks of July each year.

The results for the years 1994 to 1999 are summarised in Fig. 2 which shows the total numbers of plants counted each year and a break down of those counted in fixed and randomly placed quadrats.

3. METEOROLOGICAL DATA

All figures are derived by abstraction from the data supplied by the UK Meteorological Office.

a) Rainfall

Fig. 3 shows the rainfall data for Tiree during the months of April, May, and June. Details of the life cycle and phenology of *Koenigia* are under investigation but it is held that seed germinates in late May and early June [12] and from observations during this current work the plant is in flower by the first week of July. Rainfall during May and June and the residue of April rain held in the peat deposits surrounding the terraces which feeds the flushes is likely to have an influence on seed germination.

The data presented in Fig. 3 indicates that rainfall levels were low during the spring season on the west coast of Scotland in 1995, 1997 and 1998.

b) Temperature

An analysis of temperature trends at the study sites has been done by applying a correction of 3.73°C to the data obtained for Tiree, based on a lapse rate on the west coast of Scotland of -1°C for every 100m increase in altitude [13].

Since the summer temperature optimum for *Koenigia* has been estimated to be 7°C [11] the graphs in Fig. 4 have been constructed by extracting maxima in excess of 10.73°C to represent days at the study sites when the temperature is in excess of 7°C. Any day during which the mean daily temperature exceeds 10.73°C at Tiree is defined in this study as a 'warm day'.

Fig. 4 shows the number of warm days, as defined above, during the seed germination season. Alternative analysis of the data using a range of baseline temperature criteria also indicates a similar pattern of increased numbers of warm days in 1995, 1997 and 1998.

4. DISCUSSION

The data presented in Fig. 2 indicates that *Koenigia* populations are dynamic with numbers fluctuating considerably from year to year. Even fixed quadrats that contain no plants one year may have several in the following year, suggesting either seed immigration in the flush or a bank of dormant seed, not all of which germinates in any given year.

However the overall pattern in both the fixed and random quadrats appears to be similar and it can be seen that there was an overall decline in abundance in 1995, 1977 and 1978, although plant numbers have recovered this year (1999) to the previous low of 1995. This recovery may be indicative of a healthy dormant seed bank in the soil.

The years in which *Koenigia* populations are recorded as decreased coincide with those in which there is a decrease in rainfall and an increase in the number of warm days during the growing season in comparison with years of relative abundance.

Detailed statistical analysis of the meteorological data and its correlation with the abundance of *Koenigia* is currently being undertaken and will be the subject of a future paper when the 1999 figures are released.

However preliminary analysis suggests that both the trends and the correlation are significant. It thus appears that the abundance of *Koenigia* on Mull is at least partly determined by the pattern of weather in the early part of the year with both temperature and rainfall being implicated.

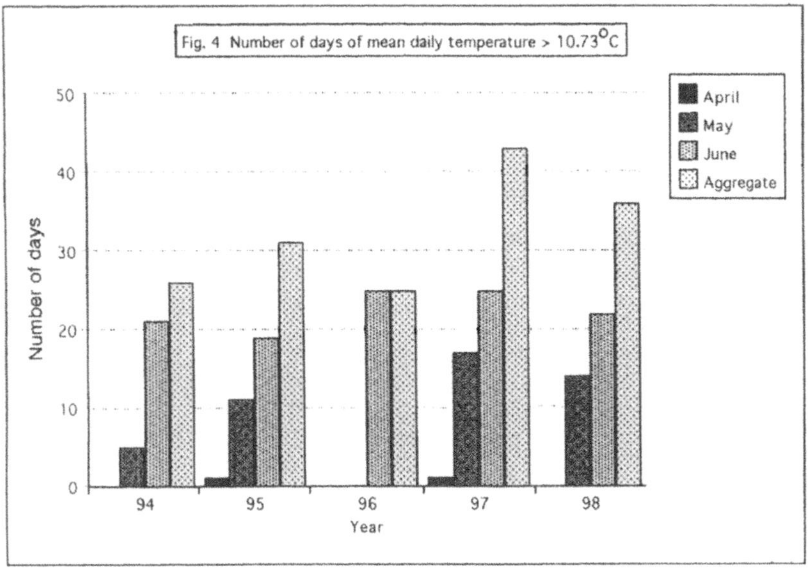

Fig. 4

5. CLIMATE TRENDS IN SCOTLAND

Cannell *et al.* have analysed trends in the Scottish climate and calculate that the dry summer of 1995 had a 1 in 80 year probability , but also that such conditions are predicted to occur three times in the period 1997-2050 [14]. In the event the springs and early summers of 1997 and 1998 were equally dry on the west coast. Also Harrison [13] has reviewed Scottish climate records from 1964 to 1993 and reports significant changes in recent years, there having been in particular an increase in winter but decrease in summer precipitation and increased air temperature in the spring. If such a trend continues the long term effects could significantly affect *Koenigia* populations in the future. Evidence suggests that, historically, most plant species have responded to climate change by migration, particularly in the post-glacial period [15]. Unlike the situation reported for mountain vegetation in the mountains of mainland Europe where there has been an upslope migration of plant communities, apparently in response to climate change [16, 17], on both Mull and Skye this option is not available since in these locations *Koenigia* already grows at the maximum altitudes of the hills concerned. The restricted availability of suitable habitat in other parts of Scotland is probably also a factor in limiting the potential for migration to other localities.

6. CONCLUSION

The results of this study indicate that the abundance of *Koenigia* on Mull is influenced by weather conditions during the growing season and *Koenigia* is therefore potentially sensitive to longer term climatic variation. Clearly more work needs to be done on *Koenigia* on Mull and this project is scheduled to continue with annual monitoring for the foreseeable future. Since a 30 year period is generally considered to be the minimum on which to base climate trends it would seem that there is still scope for establishing the relationship between the abundance of *Koenigia* and any changes in the Scottish climate.

7. ACKNOWLEDGEMENTS.

The author would like to acknowledge the support and help of the following individuals and organisations in this work:

The Climate Services Unit at the Meteorological Office
Scottish Natural Heritage
The National Trust for Scotland
James Fenton
Lynne Farrell
Clive Jermy
Phillip Lusby
Student Members of the Brathay Exploration Group Mull Expeditions 1994, 96 and 98.

The financial support of the British Ecological Society and the University of Warwick is gratefully acknowledged.

8. REFERENCES

Burtt B.L., *Koenigia islandica* in Britain, Kew Bulletin (1950) 173

Cannell M.G.R., D. Fowler, and C.E.R. Pitcairn, Climate change and pollutant impacts on Scottish vegetation, Botanical Journal of Scotland 49 (2) (1997) 301-313.

Godwin H., The History of the British Flora, Cambridge University Press, Cambridge, 1997

Grabherr G., M. Gottfried and H. Pauli, Climate effects on mountain plants, Nature 369 (1994) 448

Harrison S.L., Changes in the Scottish climate, Botanical Journal of Scotland 49 (2) (1997) 287-300.

Hultén E., The Circumpolar Plants II Dicotyledons, Almquist and Wiskell, Stockholm, 1994, p64.

Huntley B., How plants respond to climate change: migration rates, individualism and the consequences for plant communities, Annals of Botany 67 (1991) 15-22.

Jermy A.C. and J.A. Crabbe (Eds) The Island of Mull - a survey of its flora and environment, The Natural History Museum, London, 1978.
Lusby P. and J. Wright, Scottish Wild Plants, The Stationery Office, Edinburgh, 1996.
Lusby P., Scottish Rare Plant Project, Royal Botanic Garden Edinburgh, pers comm.
Meatyard B.T., In M. Price (Ed) Global Change in the Mountains, Parthenon, New York and London, 1999.
Pauli H., M. Gottfried and G. Grabherr, Effects of climate change on mountain ecosystems - upward shifting of alpine plants. World Resource Review 8 (1996) 382-390.
Ratcliffe D.. *Koenigia islandica* in Mull, Trans. Proc. bot. soc. Edinb. 39 (1960) 115-116
Rodwell J.S. (Ed) British Plant Communities, Vol. 2, Mires and Heaths, Cambridge University Press, Cambridge, 1991, pp 329-330.
Sætersdal M. and H.J.B. Birks, A comparative ecological study of Norwegian mountain plants in relation to possible future climatic change, Journal of Biogeography 24 (1997) 127-152.
Stace C.A., New Flora of the British Isles, Cambridge University Press, Cambridge, 1997
Wigginton M.J., British Red Data Book I, Vascular Plants, JNCC, Peterborough, 1999.

Semi-Objective Sampling Strategies as One Basis for a Vegetation Survey

KARL REITER, KARL HÜLBER AND GEORG GRABHERR
Dep. for Vegetation Ecology and Conservation Biology, Inst. of Plant Physiology, University of Vienna Althanstr. 14,1090 Vienna

Key words: sampling design, modelling, GIS, remote sensing, plant communities

Abstract: Satellite images or aerial photos are often the most appropriate method of gaining a first general view of large areas. In order to detect different spatial patterns (e.g. vegetation patterns, soil patterns), aerial or satellite images can assist ecological studies. The revealed patterns could provide the basis for the selection of sampling points. Until recently, there has been much debate on how to optimally design sampling programs at large spatial scales. The problem of locating the sampling points in the area of interest, and how many sampling points should be taken, remains largely unresolved. Random sampling is too laborious for comprehensive surveys, whereas traditional subjective sampling, which has been commonly used in e.g. vegetation ecology, violates basic scientific principles for quantitative assessments (e.g. reproducibility, comparability, statistical analysis). One compromise is stratified random sampling. This process of stratification divides the space into subunits (strata) based on factors like precipitation, land management, or habitat constants, e.g. habitat type or elevation. Modern tools for computer-assisted data handling, especially Geographical Information System (GIS) and programs for image processing, have greatly simplified the selection of strata. We present a case study which aims to describe meadow vegetation in a 150 Km2 area of the Prealpine region of Lower Austria based on stratified random sampling to minimize field work, and to maximize the reliability of the result, i.e. the description of the variability, the character of the vegetation in the whole research area based on a vegetation distribution model. Main emphasis was to establish monitoring system for detection of land use change effects in this unique meadow vegetation in the mountain belt. The procedure of stratification was based on satellite data (LandsatTM with a pixel size of 28.5 m x 28.5m) and the use of a digital elevation model (DEM) with a grid size of 250 m x 250 m. The classification of all the input data-sets resulted in a total 20 strata. Each stratum consists of several disconnected sub areas so called sampling regions. Within each Stratum five sampling regions were randomly selected for further analysis. Within these sampling regions, several relevs

G. Visconti et al. (eds.), Global Change and Protected Areas, 219–228.

were subjectively selected resulting in 30 plant communities. These group were found to closely resemble community types revealed from a previous vegetation survey of the meadow-land of that area. The applied satellite images, as well as the coarse grained DEM-data (commercially available data sets), provide sufficient information to delineate sampling regions but not to fix relev locations in the regions. Whether the application of satellite images with a pixel size of less than 10 meters will lead a new approach, which allow a point centered sampling design in contrast to the sampling regions concept, will be shown in the near future. Based on the relevs and the derived plant community types with high degree of representation, a monitoring system which covers the different ecological situations in a particular area, was established.

1. INTRODUCTION

Classical research work in vegetation ecology - the Braun–Blanquet approach[1] in particular – is characterized both by the description of the floristic composition and by the description of the abiotic factors occurring at a particular site. Recording of elevation, exposition, inclination, soiltype , etc. are standard methods for the description of the so called relevés[1]. Knowledge derived from the survey of the relationship between plant composition and abiotic site-factors may provide the basis for predicting current vegetation cover. This simple principle is the basis of the up-scaling approach, whereby, as in this study, site–factors are derived from the analysis of satellite images and digital elevation models. Availability of site-factors and knowledge of vegetation/site relationship allows modelling of current vegetation cover, and possibly to predict future vegetation patterns, as well as to reconstruct the past [2][3].

Research work based on this approach is now made possible by high powered computers in combination with geographical information systems (GIS) and remote sensing methods .

This paper is focused on the design of semi-objective sampling strategies via the application of spatial-analytical methods combined with multivariate analysis. Until recently, there has been much debate on how to optimally design sampling programs for a particular region [4][5]. The problem of where to arrange the sampling points (e.g. relevés) in the area segment of interest, and how many sampling points should be taken, is still to be exemplified by case studies[6].

Random sampling is often too laborious for comprehensive surveys, whereas traditional subjective sampling[1], as it is used frequently in vegetation ecology, violates basic scientific requirements, e.g. reproducibility, comparability.

Table 1: Plant communities shown in column a were identified by a sample based on satellite images, DEM and a transect (current study). Plant communities in column b for the same study area (termed as second study) were identified based on methods used by Reiter and Grabherr [10]

	a	B
Angelico-Cirsietum oleracei	+	+
Astrantio-Trisetetum	+	+
Carlino acaulis-Brometum	+	+
Crepido-Cynosuretum	+	+
Euphorbio verrucosae-Caricetum montanae	+	+
Festuco commutatae-Cynosuretum	+	+
Filipendulo vulgaris-Arrhenatheretum	+	+
Onobrychido viciifoliae-Brometum	+	+
Poo-Trisetetum	+	+
Ranunculo bulbosi-Arrhenatheretum	+	+
Campanulo rapunculoidis - Brachypodietum sylvatici	+	
Caricetum davallianae	+	
Homogyno alpinae-Nardetum	+	
Gymnadenio-Nardetum		+
Juncetum sylvatici		+
Lolio-Cynosuretum		+
Pastinaco-Arrhenatheretum		+

However, an element of random selection of items must be maintained if valid statistical calculations are to be made, e.g. methods that can effectively predict vegetation cover over the whole area [7]. One compromise might be a stratified random sampling design [8]. The process of stratification divides the space into subassemblies based on factors like precipitation, seasonal management or habitat constants like habitat type or elevation. The present study compares two sampling strategies. The methodological description is focused on the method of interpretation of satellite images together with the analysis of digital elevation models, comparing results by study within the same area but based on different methods (see below). The overall aim of both studies was to describe the different vegetation types in the Steirische / Niederösterreichische Kalkalpen. Main emphasis was to establish a monitoring system for detection of land use change effects in a scenario of decreasing human influences in the study area. The aim of this paper is to compare results of the two studies that address the same question yet used different methods.

2. STUDY AREA

The area considered for this study is one of the most extreme examples of Alpine agriculture. The Steirische / Niederösterreichische Kalklalpen belong to the northeastern most part of the Alps where steep limestone mountains of a true alpine character (Hochgebirge) are clustered. Natural to seminatural landscapes consisting of mountain forests and the whole spectrum of alpine elements (e.g. rocks, screes, mountain rivers) dominate. Agriculture is restricted to valley bottoms and is exclusively of a dairy farming type. Summer pasturing (Almwirtschaft) had been established in medival times and is still maintained. Biodiversity is very high in terms of plat species richness and landscape diversity. The agricultural area contribute a comparable small but distinct part to the overall diversity. From a biodiversity point of view the area is mainly to be considered for it's exceptional rich meadows whose extent has dramatically decreased during the last decades. The whole area belongs to the Alpine Biogeographic Region of "Fauna – Flora – Habitat Directive of the European Union" and some parts are proposed as potential NATURA2000 protected areas. The research area proper covers nearly 120 km².

3. MATERIAL AND METHODS

Based on the pioneer work of Orloci and Stanek [9], the authors of the present paper have provided a number of relevant case studies [10].

The factors actually used for stratification (design of strata) depend on the study topic, on scale, as well as on the availability of digital data sets, suitable for analysis via computer managed information systems (preferably GIS). Satellite images and digital elevation models (DEM) might be appropriate data sources for analysis with geographical information systems (GIS) focusing on sampling design. The term sample is used by many authors in a different meaning and a source of confusion.

In this paper the word sample is used in its common statistical meaning, i.e. a collection of sampling units [11]. The sampling units are termed in this paper as sampling regions or relevés [1] (a representative vegetation sampling unit of a plant community).

3.1 Material

Within the context of this paper, `current study` refers to a sample inside a arbitrary transect, which runs from the NE to SW in the study area, based on the analysis of satellite images and digital elevation models.

3.2 Digital elevation models

Digital elevation models (DEM) are the digital representations of the shape of the earth's surface. Elevation data are helpful to analyse, to model and to identify phenomena which are associated with the earth's topography. Based upon the capabilities of ARC/Info's (the leading commercially available software for GIS) surface modelling tools, it is possible to derive information about surface topography and hence to calculate elevation, aspect or inclination and present these as polygons.

The accuracy of an analysis based upon spatial site factors depends on the scaling of these factors and on the resolution of the DEM. In the current study we used a coarse grained DEM with a resolution of 250 m (distance between two meshpoints of a regular grid, where elevation is measured). The scaling of the range-classes was based on Austria's range zones [12], exposition was divided into four classes (north, east, south, west) and the inclination was scaled into seven classes (<1.5°, <3°, <6°, <12°, <24°, <45°, <90°).

3.3 Satellite images

We used LandsatTM satellite images with a pixel resolution of a ground size of 30 m x 30 m, which have the advantage of being readily distributed, available, and of temporal continuity.

The latter is of central importance in vegetation ecology for quantifying temporal changes in vegetation patterns. Furthermore, only this sensor provides the spectral channels of the middle and the thermal infrared which are most relevant for ecological interpretation.

For the current study, the infrared channels 4 (Reflective-Infrared), 5 (Mid-Infrared) and 7 (Mid-Infrared) were chosen for the classification (image classes = ie., vegetation cover) of the pixels and for the process of segmentation.

3.4 Steps from spatial data to the sample.

The presented design of a sample is subdivided into six steps

- Segmentation of satellite images

- Classification of satellite images
- Analysis of digital elevation model
- Intersection of all derived spatial layers -> strata of first order level
- Reclassification of the strata -> strata of second order level
- Random selection of sampling units

Initially, a segmentation procedure was applied to define regions based on distinct naturally occurring ground objects (e.g. vegetation types). According to our experience best results of segmentation can be achieved with the Woodcock – Harward algorithm . This algorithm is based on the nested-hierarchical model that explicitly recognizes all the different scaled objects, which appear in a typical fine grained landscape of Middle Europe [13]. To identify existing clusters in the data set, the next step was an unsupervised classification carried out by the Isoclustering procedure, which is part of Erdas - IMAGINE. All calculations are only performed on cell values in a multivariate space e.g. constructed by the three spatial layers, which represent the bands 4, 5, 7. The calculated variance and covariance is derived form variation within and between the different layers. The estimated clusters (= image classes representing e.g. vegetation types) we evaluated using scattergrams and ellipses. The next step of spatial stratification was the intersection of all the layers (image classes, elevation, aspect, inclination) to determine different sectors of space (strata of first order level). The whole area of a stratum has not to be connected and is independent of other strata (spatial separation). We subsequently reclassified strata of first order level based on remaining variance between them in e.g. spatial characteristics. The aim of this further classification was to group strata, but strata of unique content remained as individual objects (will not be grouped). The outcome of these steps were 20 different strata types, which can be described for the research area for subsequent analysis. After stratum selection a simple random sample of size n sampling units (=sampling regions) at each stratum has to be defined in this way. The selection without replacement is preferable, so that repeats are not allowed. An entity of random selection is a region, which represents an independent part of a stratum. Within regions the sampling points are chosen subjectively so as to describe the different vegetation types within the chosen region. In the current study we selected a fix amount of five regions per stratum.

The step of random selection was concentrated on a diagonal transect (from NE – SW) trough the research area. This transect covers all the determined strata of the whole area and each stratum was represented by a minimum of five regions. The use of such a transect aims to combine two objective approaches – stratified random sampling and the systematic approach represented by the arbitrary transect, resulting in 100 sampling regions (5 regions per 20 strata).

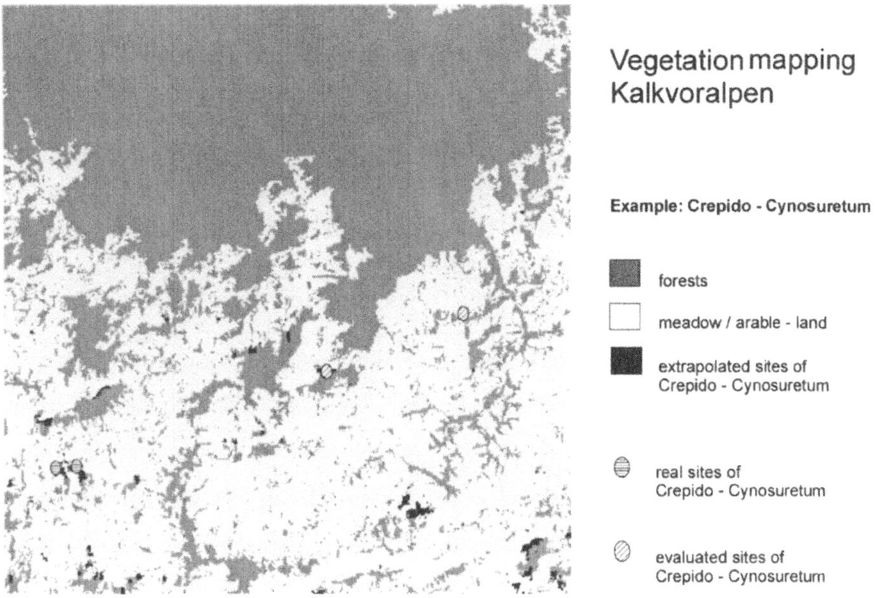

Fig. 1 Potential distribution of the Crepido-Cynosuretum in high – montane/subalpine belt in the research area enumerated by a linear model based on the semi-objective sampling design. Evaluation of the distribution by the results of a second study.

The methodological background of the investigation with which the current study is compared will not be described in detail in this paper [10] and will termed in the following as second study. It's aim was the same (description of meadowland in the same study area; fieldwork by A. Hofstätter, M. Pühringer) but the data and analysis were based on commercially available maps and DEM without the use of transects and satellite images. The data base of that study was restricted to 33 strata with only the selection of one region per stratum.

4. RESULTS AND DISCUSSION

For the sampling – regions of the current study 108 relevés (only meadow land) were recorded during fieldwork in period of two weeks carried out by one person. These were classified using TWINSPAN [14], resulting in the identification and description of 13 plant communities (syntaxa) of meadowland [15].

Table 1 shows plant communities identified by the current study (column a) and of the second study (column b). Of a total of 16 plant communities were identified with three and four additional plant communities identified by one or the other of the two methodological approaches used.

The spatially most extensive plant communities of the research area (Ranunculo bulbosi-Arrhenatheretum, Astrantio-Trisetum and Festuco communatae-Cynosuretum) were identified by both methods. The plant communities which were identified by only one of either approaches, are not typical for meadow land but are typical for the surroundings (e.g. Juncetum sylvatici, Caricetum davallianae).

The rare occurrence of Lolio-Cynosuretum for either of the two methodological approaches will not agree with descriptions [16][17] for comparable meadowland environments. This is shown by the fact that this plant community is verified only by three relevés in the second study and none in the current study.

An important point shown by the comparison is the absence of Pastinaco-Arrhenateretum based on the transect method used in the current study. This plant community is typical for parts of the research area and was verified by ten relevés by the second study. The spatial analysis of this plant community shows that its actual distribution extends beyond the study transect. We conclude that the combination of a systematic approach using arbitrary transects and stratified random sampling will decrease the quality of the results with respect to identification of plant communities.

With exceptions the consistent occurrence of some distinct syntaxons within different strata offers the opportunity to predict possible sites of a plant community. A simple linear relationship between the occurrence of a syntaxon and individual strata and its abiotic characteristics is not always clearly obvious. But the appearance of a syntaxon inside a stratum shows the potential opportunity that this syntaxon could be verified in one or all regions of a stratum. For example fig. 1 shows the potential distribution of the Crepido-Cynosuretum in high – montane /subalpine belt in the research area enumerated by a linear model based on the semi-objective sampling design. The model is based on an environmental data set that reflects the environmental gradients in the study area as expressed by species distribution patterns.

However, the predicted occurrence of Crepido-Cynosuretum in fig. 1 is only potential and in reality the plant community actually occurs on only a few sites. But it is shown, that some of potential sites contain this plant community as shown by the second study. This clearly provides no proof for the quality of the model but it shows that the semi-objective sampling approach based on spatial factors, will, with further development (e.g. increasing the quality of satellite images and DEM) , enhance the predictive capability of the model.

The semi-objective approach of sampling design presented in this paper could be characterized by the terms "objective by determine the sampling region – subjective by investigations inside the region". This study shows a useful compromise between objectivity (e.g. for modelling) and time saving fieldwork. Modern tools for computer-assisted data handling, especially Geographical Information Systems (GIS) and programs for image processing, have greatly simplified selections of strata as well as the selection of sites for relevés.

5. SUMMARY

The study purpose was to describe the types of mountain meadows in an 120 km^2 large area in the Prealps of Lower Austria. We have shown, that the use of GIS and remote sensing methods provide a sampling design with a high degree of objectivity and representativeness. With the data set of 108 so selected relevés, together with habitat data derived from the GIS, a predictive model of the meadow vegetation down to particular community types of this area could be developed.

In conclusion, the new methods of spatial data analysis provide semi-objective sampling strategies and may help to save time for reconnaissance of the area of interest. In this context, the classification of plant communities, syntaxonomy in particular, should enhance acceptance of this approach.

6. REFERENCES

Braun-Blanquet J., Pflanzensoziologie Grundzüge der Vegetationskunde, 3. Aufl, Springer, Wien, New York, 1964.

Gottfried M., H. Pauli and G. Grabherr, G., Prediction of vegetation patterns at the limits of plant life: A new view of the alpine-nival ecotone. Arctic and Alpine Research, 30 (1998) 207-221.

Grabherr G. and A. Polatschek, Lebensräume und Flora Voralrbergs, Vorarlberger Verlag-sanstalt, Dornbirn, 1986.

Grabherr G., L. Mucina and T. Ellmauer, Die Pflanzengesellschaften Österreichs. Teil I: Anthropogene Vegetation, Fischer, Jena, 1993.

Grabherr G., G. Koch, H. Kirchmeir and K. Reiter, Hemerobie österreichischer Waldökosysteme, MAB- Berichte Band 18, Universitätsverlag Wagner, Innsbruck, 1998.

Green R.H., Sampling design and statistical methods for environmental Biologists, John Wiley and Sons, New York Chichester Brisbane Toronto, 1979.

Greig-Smith P., Quantitative plant ecology; 3rd Edition, Blackwell Scientific Publications, Oxford,London, Edinburgh, 1983.

Guisan A., J.P. Theurillat and F. Kienast, Predicting the potential distribution of plant species in an alpine environment, Journal of Vegetation Science 9 (1998) 65-74.

Hill M.O., TWINSPAN, A FORTRAN program for arranging multivariate data in an ordered two-way table by classification of the individuals and attributes, Cornell University, Ithaca, New York, 1979.

Jongman R.H.G., C.J.F. ter Braak, O.F.R. van Tongeren, Data analysis in community and landscape ecology, Cambridge university press, 1995.

Klapp E., Grünlandvegetation und Standort, Verlag Paul Parey, Berlin, 1965.

M. Mühlenberg, Freilandökologie, 2. Auflage UTB 595, Quelle und Meyer, 1989.

Manly B.F.J., The design and analysis of research studies, Cambridge University Press, 1992.

Orloci L. and W. Stanek ,Vegetation survey of the Alaska Highway, Yukon Territory: types and gradients, Vegetatio 41 (1980) 1–56.

Reiter K. and G. Grabherr, Digitale Höhenmodelle als Grundlage der Stichprobenwahl bei Vegetationsanalysen, Verh. Zool.-Bot. Ges. Österreich 134 (1997) 389-412.

Wildi O., Analyse vegetationskundlicher Daten, Theorie und Einsatz statistischer Methoden, Veröff. ETH Zürich 90. Heft, 1996.

Woodcock C.,and V.J. Harward, Nested-hierarchical scene models and images segmentation, Int. J. remote sensing, Vol. 13. No.16 (1992) 3167-3187

Simulating the Impact of Climate Change on Drought in Swiss Forest Stands

BARBEL ZIERL
Swiss Federal Institute for Forest, Snow and Landscape Research, Zurcherstr. 111, 8903 Birmensdorf, Switzerland

Key words: climate change, drought, modelling, Swiss forest stands, water balance, WAWAHAMO.

Abstract: The main objective of this study is to investigate and assess the possible effects of a changing climate on hydrological processes in Swiss forest stands. For this, sensitivity studies on temperature and precipitation as well as model runs with eight different climate scenarios were carried out using the hydrological model WAWAHAMO (from Waldwasserhaushaltsmodell, German). WAWAHAMO is designed to simulate the water balance for the entire forested area of Switzerland on a daily basis. It is a simple bucket model, which predicts soil moisture content, evapotranspiration, interception, snow cover and drainage, assuming a fixed bucket size defined by the reservoir capacity of the soil. Currently, the model runs at a spatial resolution of 1km.Emphasis is put on the occurence of drought as a limiting factor for forest condition. For this, an annual drought index is defined by the ratio of yearly totals of actual and potential evapotranspiration. Simulations of the annual drought index for the years 1969 to 1998show a clear pattern of current drought over Switzerland. In dry forest stands as in the main valley of Valais and near the northern border the sensitivity studies and climate scenarios indicate a high sensitivity of the annual drought index to the possible climate change.

1. INTRODUCTION

In nature many processes are closely linked to the water cycle. For forest stands the available water amount is one of the most limiting factors influencing forest condition. Especially drought can be a serious cause of stress for forest trees and affect forest health and growth [1,2].

G. Visconti et al. (eds.), Global Change and Protected Areas, 229–243.

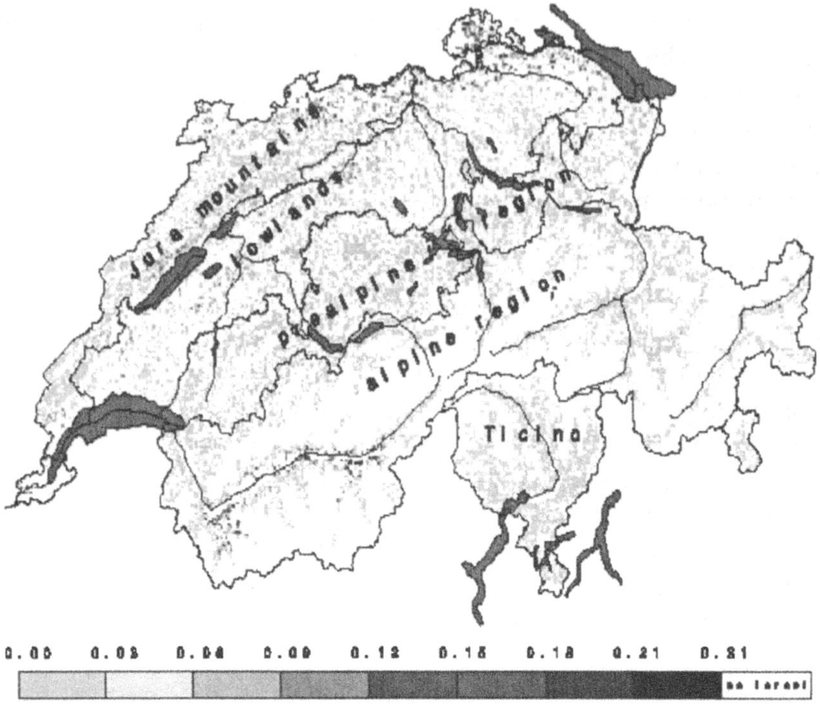

Figure 1: Spatial distribution of the 30-years-mean value of the annual drought index for the time period 1969-1998 as simulated by the hydrological model WAWAHAMO.

Apart from precipitation, the most significant component of the hydrological budget of a forest stand is evapotranspiration. Since evapotranspiration is mainly determined by the current climate condition, a future climate change, as predicted by several climate scenarios, might strongly influence future water availability. The extent of this effect has been subject of many research papers [3,4,5,6]. But, so far most hydrological research studies have concentrated on a catchment scale. Because there is not enough data and computing times to run the catchment scale models on a regional scale, the hydrological model WAWAHAMO (from Waldwasserhaushaltsmodell, German) was developed to simulate the water balance of the entire forested area of Switzerland. In order to assess possible impacts of climate change on the water balance of Swiss forest stands sensitivity studies on temperature and precipitation were carried out. Furthermore, model runs with eight different climate scenarios were performed. The current paper shows the results of these studies for selected sites and for specific regions.

Table 1: Description of the selected forest stands. z: geographical height in m; AWC: available water capacity in mm; LAI: leaf area index.

number of forest stand	region	z	mean tree species	AWC	LAI	hydrological situation
(1)	prealpine region	1605	spruce	184	9.6	wet
(2)	Ticino	395	chestnut	141	9.5	dry
(3)	Valais (western part of alpine region)	1373	pine	108	5.7	very dry

Interpreting the results, focus was set on the occurrence of strong drought events and their temporal and spatial distribution.

2. INPUT DATA

The research area includes the entire forested area of Switzerland. It is characterised by strong variations in topography, soil properties, climatic conditions, forest structure and tree species composition. Careful attention is payed to these spatial variations, since all these data are needed to model hydrological processes. At the moment, model application is performed at a spatial resolution of 1 km, since this is the resolution of the currently available forest data. All other input data have to be interpolated to this model grid. For a detailed description of the methods see Zierl [7].

2.1 Topographical data

To represent the complex alpine landscape, information on elevation, slope, exposition, relief shading, day length (maximum sunshine duration), and height of the horizon have to be considered in detail. A digital terrain model (DTM) provided elevation data on a 25 m grid. All other parameters were derived from this grid using the Geographical Information System ArcInfo.

2.2 Climate data

WAWAHAMO requires daily observations of climatic variables. Within Switzerland there are approximately 100 weather stations run by the Swiss

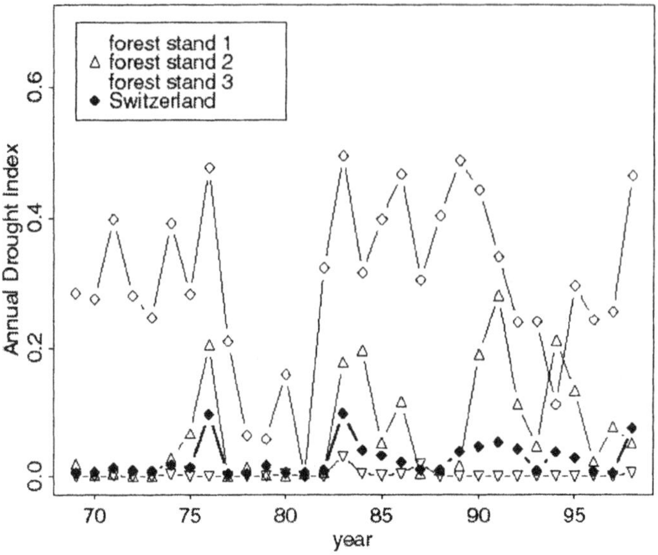

Figure 2: Temporal variation of the annual drought index for the time period 1969-1998. Thick line: Swiss average; thin lines: three selected forest stands as described in table 1.

Meteorological Institute providing data on air temperature, air humidity, wind speed, cloud cover, absolute sunshine duration and pressure. Furthermore, daily precipitation totals are available from around 400 stations

The daily observations were interpolated to the model grid for a time period of 30 years (1969-1998). Here, emphasis was put on the correct representation of altitude dependencies, especially the representation of temperature inversions within the winter months.

2.3 Soil data

To simulate soil water content and soil matric potential, data on available water capacity and permeability of the soil are required. Values of available water capacity were estimated on the model grid by external drift kriging [8] using information from a soil map [9], a digital terrain model, and 725 soil profiles. Values of permeability were directly taken from the soil map [9].

2.4 Forest data

Stand information such as ratio conifer to deciduous trees, vegetation cover, tree species, tree height and breast height diameter were available on a 1 km grid from the Swiss National Forest Inventory [10]. Values for an average leaf area index were derived from these data by allometric equations [11].

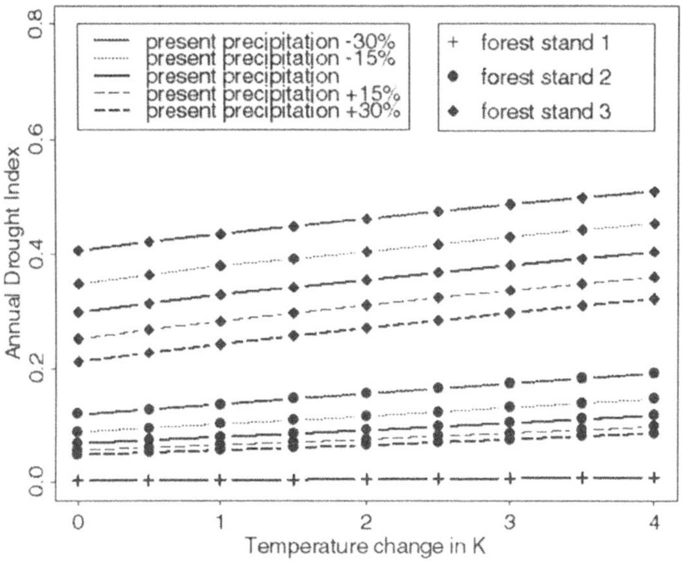

Figure 3: Evolution of the 30-years-mean value of the annual drought index (ADI) dependent on temperature(x-axes)and precipitation change (different lines)for three selected forest stands (table 1).Points: simulated values. Lines: linear interpolation.

3. THE WAWAHAMO MODEL

The water balance model WAWAHAMO is documented in detail by Zierl [7]. The water balance of a forest can be calculated on a daily basis according to:

$$\Delta S = P + SM - SA - D - SR - ETA - I \qquad (1)$$

(Symbols as explained in list of symbols, section 8). The change in soil water content ΔS is determined by precipitation P, snow melt SM, snow accumulation SA, drainage flux D, surface run-off SR, actual evapotranspiration ETA and interception evaporation I.

Because of lack resp. inaccuracies of required soil physical data the application of a detailed hydrological model is not very useful. For this reason a simple bucket model is chosen, which predicts soil moisture content, evapotranspiration, interception, snow cover and drainage, assuming a fixed bucket size defined by the reservoir capacity of the soil.

The model is implemented at a spatial resolution of 1km resembling the resolution of the currently available forest data. In the following the main components of the water balance are briefly explained.

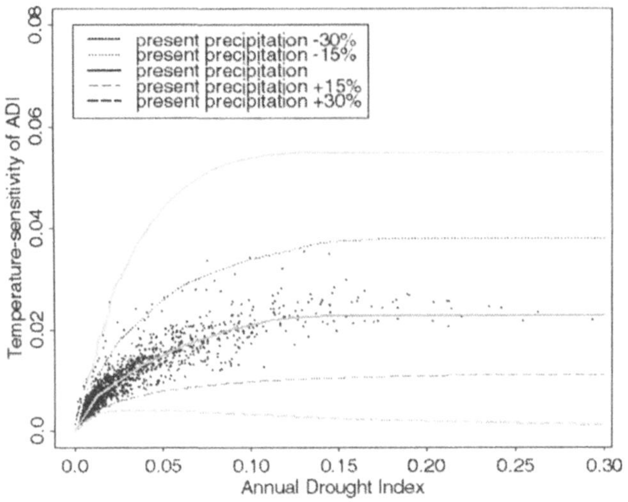

Figure 4: Temperature sensitivity of the 30-years-mean value of the annual drought index (ΔADI/ΔT)dependent on present ADI (x-axes) and precipitation change (different lines). Points: simulated values for present precipitation rates. Lines: fitted curves to the simulated values for the five different precipitation changes.

3.1 Precipitation and snow processes

Precipitation *P* is either intercepted by the trees, infiltrated into the soil or stored in the snow cover. When temperature falls below a defined threshold, precipitation is assumed to fall as snow and accumulate on the land surface (*SA*). The melting of the snow cover (*SM*) is simulated by a snow model following a description by Gurtz et al. [4].

3.2 Drainage and surface run-off

It is assumed, that when the soil bucket is filled, the surplus of water will percolate to deeper soil layers. In the model this water excess is called the drainage flux *D*. In most hydrological models surface run-off *SR* plays an important role. In forests, however, it is often limited due to high porosity and high infiltration rates. For this reason surface run-off is neglected.

3.3 Evapotranspiration

Apart from precipitation, the most significant component of the hydrological budget of a forest stand is evapotranspiration. It is calculated by the Penman-Monteith equation [12,13].

The equation requires meteorological and plant physiological data, which

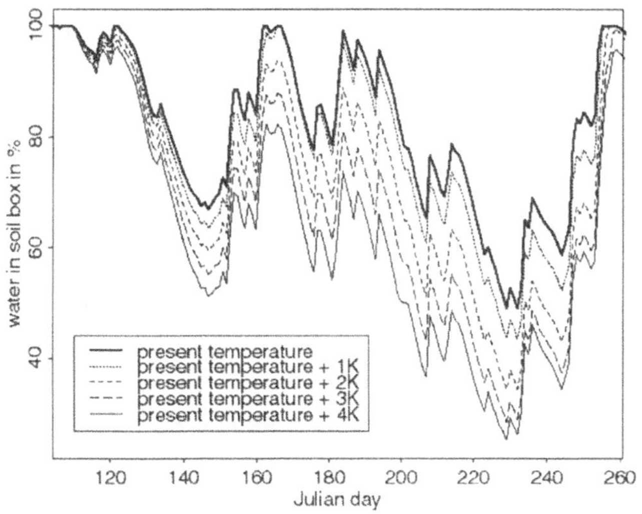

Figure 5: The relative water content of the soil box of a forest stand in the valley of Valais from May until September assuming different temperature conditions.

$$\lambda \cdot ET = \frac{\in R + \dfrac{\rho c_p \delta_e}{r_a}}{\in + \gamma \left(1 + \dfrac{r_c}{r_a}\right)} \cdot \Delta t \tag{2}$$

either can be directly taken from the input data set or have to be simulated by the model WAWAHAMO.

3.3.1 Radiation

Net radiation *R* is the most important parameter calculating evapotranspiration *ET* (equation (2)). Besides meteorological and sun physical parameters net radiation is strongly controlled by the topographical parameters described in section 2.1. For this reason a model was developed which estimates the radiation budget taking the topographical situation into account. It is based on models from Kienast [14] and Marks and Dozier [15].

3.3.2 Canopy resistance

The most common way of parameterizing the response of stomata to environmental factors at canopy scale are the Jarvis-type models [16,17]. They describe the stomatal resistance r_{st} in the form of a minimal resistance $r_{st,min}$ multiplied by the product of independent stress functions F_i without synergy

$$r_{st} = r_{st,\min} F_1(R_s) \cdot F_2(T) \cdot F_3(\delta e) \cdot F_4(\Psi) \quad (3)$$

Each function F_i varies from unity to infinity. For a detailed description see Lhomme et al. [17]. In order to obtain the canopy resistance r_c needed in equation (2), the resistance to soil evaporation r_{so} dependent only on soil water potential Ψ is connected in parallel to the stomatal resistance r_{st} with a weighting factor depending on the leaf area index *LAI* of the forest stand [18].

$$\frac{1}{r_c} = \frac{1 - f^{LAI}}{r_{st}} + \frac{f^{LAI}}{r_{so}} \quad (4)$$

3.3.3 Soil water potential

The soil water potential Ψ strongly controls the stomatal resistance r_{st} of the forest trees (equation (3)). To derive soil water potential from the water content of the soil bucket the Brooks & Corey equation [19] is used. For this, the hydrological model distinguishes between 3 permeability categories: sand, silt and loam.

3.3.4 Phenology

In order to predict the seasonal development of the leaf area index *LAI* needed to determine the canopy resistance r_c a sequential model of chilling and forcing, as described by Kramer [20] is implemented in the hydrological model. It calculates the date of bud burst dependent on the tree species as a function of chilling temperatures in winter and forcing temperatures in spring. Leaf loss in autumn starts when the average temperature of the previous four days drops below 5°C.

3.3.5 Potential and actual evapotranspiration

In the following, potential evapotranspiration (ETP) is defined by the evapotranspiration rate without water stress.

This is calculated with equation by adjusting the stomatal resistance r_{st} only to radiation R and temperature T, i.e. multiplying the minimum stomatal resistance only with the stress functions F_1 and F_2 in equation (3). To determine the actual evapotranspiration (ETA), the stomatal resistance r_{st} is additionally adjusted to the soil matric potential Ψ and the specific humidity deficit δe. Thus, ETA represents the evapotranspiration rate taking the current water stress situation into account.

3.4 Interception and evaporation

A proportion of the daily rainfall dependent on the ratio of conifer and deciduous trees and the precipitation rate are intercepted by the forest canopy (*I*). Hereby, the maximum interception storage is a function of leaf area index and vegetation cover.

The rate of evaporation of intercepted water is calculated by the Penman Monteith equation (2) with the canopy resistance set equal to zero [21].

4. DROUGHT

Drought stress occurs in situations where ETA is less than ETP. Therefore, a ratio of the yearly totals of ETA and ETP can be used as indicator for drought stress. An annual drought index (ADI) is defined according to

$$ADI = 1 - \frac{\sum_{I=I}^{365} ETA_{UI}}{\sum_{I=1}^{365} ETP_I} \tag{5}$$

4.1 Spatial variation

The temporal averaged ADI for the period 1969 to 1998 shows a clear spatial pattern within Switzerland (figure \ref{fig1}).

High values of ADI are simulated in the main valley of Valais (in the western part of the alpine region), the Jura mountains, the lowlands and the southern part of Ticino.

The highest values up to 0.53 are reached in the main valley of Valais and at the northern border, which was expected because of low precipitation and high potential evapotranspiration rates in these regions.

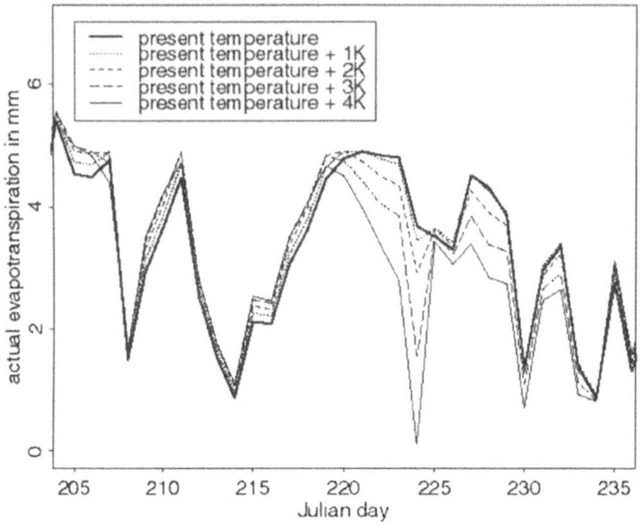

Figure 6: The actual evapotranspiration for August 1998 of a forest stand in the valley of Valais assuming different temperature conditions.

4.2 Temporal variation

The temporal variation of ADI averaged over Switzerland and for three selected forest stands are shown in figure fig. 2. A short description of the forest stands is given in table 1 forest. The interannual variation differs substantially between dry and wet forest stands. Whereas there is almost no variation at the wet site (forest 1), the dry sites (forest 2, 3) show a rather large variation in ADI. In general the temporal variation of ADI is highly correlated to its mean value.

5. CLIMATE CHANGE

CO_2 concentration is increasing, temperature is likely to rise and precipitation patterns might change [22]. These potential climatic shifts are expected to strongly influence the water budget and thus the water supply of forest stands. In order to investigate the hydrological response of forest stands to climate change sensitivity and scenario studies were carried out.

5.1 Sensitivity studies

To study the sensitivity of ADI towards temperature and precipitation changes, the hydrological model was run with the 30-years-data set (1969-1998) with increased temperatures and increased or decreased precipitation rates.

The results for three selected forest stands are given in figure 3. Both, the impact of changing temperature and precipitation strongly depends on the present hydrological situation. Whereas forest 1, a wet forest stand in the prealpine region, shows nearly no response at all (all lines in figure fig3 lie one upon another), the dry forest stands (2 and 3) experience a strong increase of ADI.

Generally, the temperature sensitivity, defined as $\Delta ADI/\Delta T$, does strongly depend on present ADI and predicted precipitation rates (figure 4). The higher the present ADI the stronger is the impact of increased temperatures. Furthermore, in forest stands with limited soil water supply the future precipitation plays an essential role. If precipitation is reduced by 30 %, the temperature sensitivity might double from 0.023 K^{-1} to 0.055 K^{-1}.

The reason for this result is a substantial change in the seasonal distribution of the ratio of ETA and ETP. Due to the increased temperatures the stomatal resistance r_{st} is reduced resulting in an increase of ETP all over Switzerland. In contrast to this, ETA changes in a more complex manner.

As long as the soil moisture content does not limit ETA, the increase of ETA will roughly resemble the increase of ETP, resulting in an unchanged ratio of ETA and ETP, i.e. an unchanged ADI. This is the case in forest stands with large precipitation excess, that can compensate for the intensified consumption.

However, in dry forest stands the intensified water consumption can not be compensated by precipitation and, thus, leads to a strong reduction of the soil water content during the summer months (figure 5). As a result, the reduced soil water content limits ETA much stonger. This effect might even lead to an ETA decrease despite of increased temperatures (figure 6). In this case, ADI is substantially increased.

5.2 Climate scenarios

Regional scenarios for the future Alpine climate provide information on the future evolution of temperature and precipitation. For the current study eight climate scenarios (either empirical, semiempirical or model based) as described by Gyalistras et al. [23] are taken into consideration.

Table 2 lists the absolute temperature and the relative precipitation changes predicted by these scenarios.

Corresponding driving variables for the hydrological model were obtained by adjusting temperature and precipitation of the 30-years-data set (1969-1998) according to the predicted variable changes given in table 2.

Considering the different climate scenarios (figure 7), the 30-years-mean

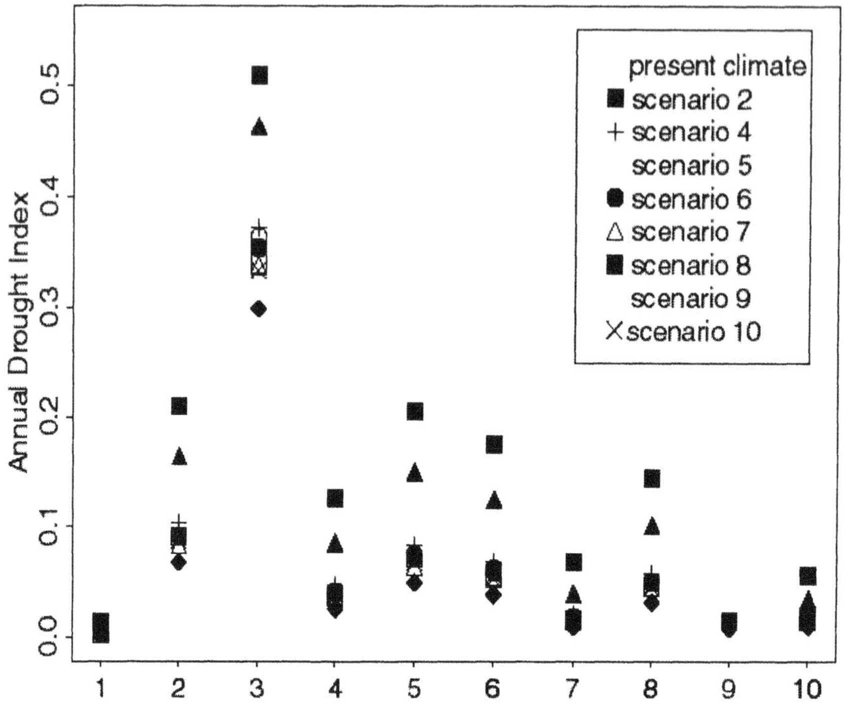

Figure 7: The annual drought index at present climate conditions and as predicted by the climate scenarios. x-axes: 1-forest 1, 2-forest 2, 3-forest3, 4-Switzerland, 5-Jura mountains, 6-lowlands, 7-prealpine region, 8-western art of the alpine region, 9-eastern part of the alpine region, 10-Ticino.

value of ADI rises a least slightly for almost the entire forested area of Switzerland when compared with the present climate.

As expected, the strongest effects have scenarios 8 and 9, predicting very high temperatures and strongly reduced precipitation rates during the growing season. The increased winter precipitation in both scenarios results mainly in larger drainage fluxes during snow melt in spring and is not able to compensate for the reduced summer precipitation.

The smallest impacts on ADI are predicted by scenarios 6, 7, and 10. Due to increased precipitation rates during the growing season the effect of raised temperatures on evapotranspiration can be compensated to a considerable extent. As for the sensitivity studies, the change in ADI is highly correlated to the present values of ADI, i.e. the drier the present forest stand the more intense is the impact of the climate change on the soil water supply. The strongest effects are simulated for forest stands in the main valley of Valais and near the northern border. Again, the reason is the limited ETA due to reduced soil water content during the summer months.

6. CONCLUSION

The used sensitivity studies and scenarios are not forecasts but application-oriented descriptions of conceivable future climate changes geared for the examination of possible impacts [23]. Therefore, one may not definitely conclude that the predicted climate change will increase the exposure of Swiss forest stands to drought. The considered sensitivity studies and scenarios, however, show a serious prolongation and strengthening of drought periods and, thus, a substantial increase of ADI in currently dry forest stands as in the main valley of Valais and near the northern border. Forest stands in these areas might experience substantial reductions in water availability leading to fundamental stress situations.

However, the extent to which the soil water supply in Swiss forest stands might be reduced during the growing season strongly depends on the seasonal distribution of precipitation. The scenarios under investigation show a large variety of resulting ADIs mainly dependent on the predicted rainfall distribution. To get more reliable statements about the impact of climate change on the water availability in forest stands, further research has to be performed on this point.

7. REFERENCES

Anonymous, Bodeneignungskarte der Schweiz. Grundlagen fur die Raumplanung. Bundesamt fur Raumplanung, Bern, 1980.

Brooks R.H., and A.T. Corey, Hydraulic properties of porous media, Hydrology Paper No. 3, Colorado State University, Fort Collins, Colorado, 1964.

Brutsaert W. (Ed.), Evaporation into the atmosphere, Kluwer Academic Publishers, Dordrecht, 1982.

Bugmann H.K.M., On the Ecology of Mountainous Forests in a Changing Climate: A Simulation study, Diss ETH Zurich No. 10638, 1994.

Gurtz J., A. Baltensweiler, H. Lang, L. Menzel, J. Schulla (Eds.), Auswirkungen von klimatischen Variationen auf Wasserhaushalt und Abfluss im Flussgebiet des Rheins, vdf, Hochschulverlag ETH Zurich, 1998.

Gyalistras D., C. Schaer, H.C. Davis, H.Wanner, Future Alpine Climate, in Views from the Alps, P.Cebon, U. Dahinden, H. Davies, D.M. Imboden, C.C.Jaeger (Eds.), The MIT Press Cambridge, (1999) 171-223.

Herbst M., G. Hormann, Climatic Change 40 (1998) 683-698.

Houghton J.T., L.G. Meira Filho, B.A. Callander, N. Harris, A. Kattenberg, and K. Maskel (Eds.), IPCC: Climate Change 1995: The Science of Climate Change, Cambridge University Press, Cambridge U.K., 1996.

Innes J. L., (Ed.), Forest Health, Its Assessment and Status. CAB International, Oxon (UK), 1993.

Jarvis P.G., Phil. Trans. R. Soc. Lond. 273 (1976) 593-610.

Kienast F., internal report, Landschaftsdynamic und -magament, Federal Institute of Forest, Snow and Landscape Research, Birmensdorf, Switzerland, 1998.

Kramer K., Phenology and growth of European trees in relation to climate change, Thesis Landbouw Universiteit Wageningen, 1996.

Lhomme J.-P., E.Elguero, A. Chehbouni, G.Boulet, Water Resoures Research, Vol.34, No.9 (1998) 2301-2308.

Marks D., J. Dozier, Arch. Met. Geoph. Biokl. Ser. B, 27 (1979) 159-187.

Monteith J.L. (Ed.), Vegetation and Atmosphere, Vol. 1: Principles. Academic Press, 1975.

Muller-Edzards C., W. de Vries, J.W. Erisman (Eds.), Ten Years of Monitoring Forest Condition in Europe, European Commission, compiled by: Federal Research Center for Forestry and Forest Products, 1997.

Rutter A.J., In: T.T. Kozlowski(Ed.), Water deficits and plant growth, Vol.2, Plant water consumption and response, Academic Press, 1968, pp.23-84.

Schulla J. (Ed.), Hydrologische Modellierung von Flussgebieten zur Abschatzung der Folgen von Klimaanderungen, Diss ETH Zurich No. 12018, 1997.

Schweizerisches Landesforstinventar (LFI), 1983-85: Datenbankauszug vom 25.6.1997. Eidg. Forschungsanstalt fur Wald, Schnee und Landschaft (WSL), Birmensdorf.

Thompson N., I.A. Barrie, M. Ayles (Eds.), The Meteorological Office Rainfall and evaporation calculatoin system: MORECS. Hydrological Memorandum NO.45, 1981.

Tiktak A., H.J.M. van Grinsven, Ecol. Model. 83 (1995) 35-53.

Zierl B., Schatzung der nutzbaren Feldkapazitat flachendeckend fur die Waldbestande in der Schweiz, Seminar fur Statistik, ETH Zurich, Switzerland, 1999.

Zierl B., The hydrological model WAWAHAMO, internal report, Landesinventuren (LI), Federal Institute of Forest, Snow and Landscape Research, Birmensdorf, Switzerland, 1999.

8. LIST OF SYMBOLS

ADI	annual drought index	[]
D	drainage	[mm]
ET	evapotranspiration	[mm]
ETP	potential evapotranspiration	[mm]
ETA	actual evapotranspiration	[mm]
F	stress functions	[]
I	interception	[mm]
LAI	leaf area index	[m^2m^{-2}]
P	precipitation	[mm]
R	radiation budget	[Wm^{-2}]
R_S,	shortwave radiation budget	[Wm^{-2}]
S	soil water storage	[mm]

SA	snow accumulation	[mm]
SM	snow melt	[mm]
SR	surface run-off	[mm]
T	temperature	[K]
X_i	value of X at t[me step I	[]
c_P	specific heat capacity	*[Jkg^-1 K^-1]* → $[Jkg^{-1}K^{-1}]$
r_a	aerodynamic resistance	$[sm^{-1}]$
r_c	canopy resistance	$[sm^{-1}]$
r_{so}	soil resistance	$[sm^{-1}]$
r_{st}	stomatal resistance	$[sm^{-1}]$
$r_{st,\ min}$	minimal stomatal resistance	$[sm^{-1}]$
δe	specific humidity deficit	$[hPa]$
Ψ	soil matric potential	$[MPa]$
λ	latent heat	$[kJkg^{-1}]$
ρ	air density	$[kgm^{-3}]$
ϵ	change of saturation vapour pressure with temperature	$[hPaK^{-1}]$
γ	psychrometrie constant	$[hPaK^{-1}]$
ΔADI	change in ADI	[]
ΔT	change in Temperature	$[K]$
Δt	time step	$[s]$

Forecasted Stability of Mediterranean Evergreen Species Considering Global Changes

LORETTA GRATANI, ANTONIO BOMBELLI
Dipartimento di Biologia Vegetale, Università degli Studi di Roma "La Sapienza", P.le A. Moro, 5, 00185, Roma, Italy

Key words: *Q. ilex*, *P. latifolia*, *C. incanus*, leaf life span, leaf inclination, leaf morphology, leaf anatomy

Abstract: Plant communities of the Mediterranean climate Regions are exposed to high temperatures, high radiation and water stress during summer; they are dominated by evergreen sclerophyllous species and drought semi deciduous species. To define the adaptive strategies, anatomical and morphological leaf traits of *Quercus ilex* L., *Phillyrea latifolia* L. (typical evergreen sclerophyllous species) and *Cistus incanus* L., (a drought semi deciduous species), growing in the Mediterranean maquis along Rome's coast line (Italy) were analysed. The typical evergreen sclerophyllous species have long leaf life span (from 1 to 4 years), steeper leaf inclination (average 56°), higher specific leaf mass (average 21.3 mg cm^{-2}) and the highest leaf thickness (average 324 μm). The semi deciduous species have a lower leaf life span (from 4 to 8 months), a lower leaf inclination (44°±13°), a lower specific leaf mass (14.7±1.5 mg cm^{-2}) and a lower leaf thickness (244±15 μm). The more xeromorphyc species (*Q. ilex* and *P. latifolia*) may be at a competitive advantage considering the forecasted air temperature increase in the Mediterranean basin. Increasing drought stress may in fact determine a shortening of leaf life span that may prove to be critical for *C. incanus*. Knowledge of plant response to stress factors is important in the perspective of climatic changes.

1. INTRODUCTION

Global change is considered to affect various ecosystems on earth [1] [2] and monitoring plant response to climatic change has been identified as a

G. Visconti et al. (eds.), Global Change and Protected Areas, 245–252.

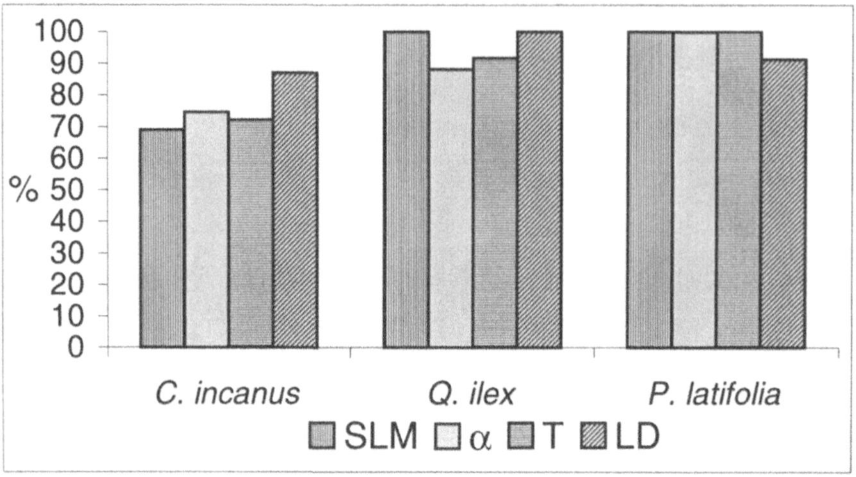

Figure 1. Specific leaf mass (SLM), leaf inclination (α), total leaf thickness (T) and leaf density (LD) measured at full leaf expansion for *C. incanus*, *P. latifolia* and *Q,. ilex*. Each leaf trait has been calculated as percentage of the maximum value among the examined species.

Valori massimi assoluti di SLW, Spessore e inclinazione fogliare per le 3 specie

		C. incanus	*Q. ilex*	*P. latifolia*
valori reali	SLM	14.7	21.3	21.3
	α	44	52	59
	T	244	310	338
	LD	0.60	0.69	0.63
valori percentuali	SLM	69	100	100
	α	75	88	100
	T	72	92	100
	LD	87	100	91

crucial component of global research programs [3] particularly in those areas where changes are expected to occur [4]. The increase in CO_2 concentration is predicted to produce an increase in average global temperature of about 3°C by the year 2050 [5]. A temperature increase may drastically modify water stress and eventually lead to changes in vegetation [6]. Not all plants are able to withstand the same types or the same intensities of environmental stress [6], consequently there is great international concern about the possible extinction of certain species [7].

Mediterranean type ecosystems comprise of a vegetation form which is of world wide importance and distribution, therefore the impact of global change in Mediterranean–type ecosystems is of great global significance. Moreover the shrublands' relatively complex structure makes them excellent models to determine and understand the impact of rising CO_2, air temperature and drought on ecosystems functioning [8]. Vegetation in Mediterranean type ecosystems include a large variety of plant species; it is dominated by evergreen sclerophyllous species and drought semi deciduous species which show differences in anatomical, morphological and physiological leaf traits [9, 10, 11]. During the Mediterranean summer, dehydration stress coincides with high light and high air temperatures; under these conditions leaves are the most exposed and therefore the most vulnerable plant organs [6]. High light intensity, if accompanied by high infrared, such as sunlight, may give rise to secondary stress injury: 1) sunheat is a secondary radiation induced heat stress injury; 2) drought injury is a secondary radiation induced stress due to the evaporative effect of absorbed energy under high insolation conditions [12]. Since leaves of most evergreen maquis species cannot effectively utilise the full irradiance, leaf inclination in all probability, provides the most powerful means of regulating light interception [13]. In order to forecast the future presence of the most representative evergreen species of the maquis, a community which dominates large areas in the Mediterranean climate we analysed adaptive strategies of *Q. ilex, P. latifolia* (typical evergreen sclerophyllous species) and *C. incanus* (a drought semi deciduous species). Since the evolution of leaf structure has occurred in concert with the evolution of leaf inclination properties [14] our general approach was to determine the relationship between leaf inclination and other structural leaf traits (specific leaf mass, leaf thickness, leaf density and leaf life span).

2. MATERIALS AND METHODS

2.1 Study area

The study was conducted in the Mediterranean maquis developing along the coast near Rome, in the Castelporziano Estate (41°45' N; 12°26' E). The area's climate is a Mediterranean type and most of its annual rainfall (727 mm) was distributed in autumn and in winter.

The average minimum air temperature of the coldest month (February) was 4.1°C and the average maximum air temperature of the hottest month (August) was 30.8°C (data by the Castelporziano Meteorological Station for the period 1987-1998).

All field measurements were made in 1998 on randomly chosen *Quercus ilex* L., *Phillyrea latifolia* L. and *Cistus incanus* L. plants (three per species),

Table 1: Morphological and anatomical leaf traits at full leaf expansion of *C. incanus*, *P. latifolia* and *Q. ilex*. Standard error is shown.

	C. incanus	C. incanus	*P. latifolia*	*Q. ilex*
Leaf inclination (°)	-37±12	44±13	59±7	52±11
Reduction of PAR (%)	20	32	47	37
Specific leaf mass (mg cm^{-2})	14.0±1.3	14.7±1.5	21.3±2.2	21.3±1.9
Leaf density (g cm^{-3})	0.64±0.03	0.60±0.02	0.63±0.03	0.69±0.05
Total leaf thickness (μm)	220±14	244±15	338±18	310±21
Abaxial Cuticle (μm)	7.2±0.7	7.2±0.7	16.8±1.7	9.7±1.8
Abaxial Epidermis (μm)	22.3±3.5	22.3±3.5	19.9±3.1	14.1±1.4
Palisade Parenchyma (μm)	107.2±9.1	107.2±9.1	141.2±15.7	163.2±16.0
Spongy Parenchyma (μm)	93.1±12.1	93.1±12.1	138.0±16.6	106.4±15.5
Adaxial Epidermis (μm)	10.8±1.0	10.8±1.0	15.8±1.8	9.3±0.8
Adaxial Cuticle (μm)	3.3±1	3.3±1	6.7±0.7	7.2±0.9

representative of the population, during the coldest (02/02/98 and 06/03/98) and the hottest (06/07/98 and 03/08/98) months of the year.

2.2 Photosynthetic active radiation and leaf inclination

Photosynthetic active radiation (PAR) was measured using a Li-COR 190SB Quantum Sensor. The reduction of PAR incident on a sloping leaf surface (RI) was calculated by: RI=100 (1 - I_s/ I_o); where I_o was the fraction of PAR intercepted by a horizontal surface and I_s the fraction of PAR incident on a sloping leaf surface [15].

Leaf inclination was obtained by measuring the angle from the horizontal of the dorsal leaf surface [16] on forty leaves for each species. Angle measurements were made by a hand held clinometer (Suunto Co. Model PM-5/360PC). The repeatability of measurements was ±5°.

2.3 Leaf morphology and Leaf anatomy

Forty one year old leaf samples at full expansion were collected and were dried at 90°C to constant weight. Leaf area (excluding petiole) was measured using the Image Analysis System (Delta-T Devices, LTD). Specific leaf mass (SLM) was calculated as the ratio of leaf dry mass to unifacial leaf area [17]. Leaf density (LD) was calculated by the ratio of leaf dry mass and leaf volume (g cm^{-3}), according to [18]. Leaf life span was analysed *in situ* monitoring the number of nodes and leaf cohorts, according to [17], since flushing patterns were known [10]. *C. incanus* leaves were characteristically folding along the midrib. The degree of concavity was measured by a goniometer on prints of ink dipped leaf cross sections (n=120). Leaf sections were hand cut from fresh leaves, dehydrated in 90% ethanol and analysed by light microscopy [19]: lamina thickness, palisade and spongy layer thickness, dorsal and ventral epidermis thickness, dorsal and ventral cuticle thickness were measured.

2.4 Statistics

All statistical tests were performed using a statistical software package (Statistica, Stats off USA). The differences in leaf inclination during the day and between winter and summer were tested by one way ANOVA and Tukey test for multiple comparison. Leaf traits differences were determined by t-test, by analysis of variance (ANOVA) and by Tukey test for multiple comparison.

3. RESULTS

3.1 Leaf orientation and incident PAR

P. latifolia and *Q.ilex* showed the steepest leaf inclination (59°±7 and 52°±11 respectively) in summer ($P<0.01$) effecting the highest RI (47% and 37% respectively). *C. incanus* showed a seasonal trend of leaf inclination: in winter α was negative (-37°±12°) while in summer it was positive (44°±13°) ($P<0.01$). *C. incanus* showed, moreover, a particular trend of leaf margin: leaves in winter had marginal leaf folding along the midrib axis (29°±9°), higher in summer (76°±11°) ($P<0.01$).

3.2 Leaf morphology and leaf anatomy

Morphological and anatomical leaf traits of the analysed species are shown in table 1. *P. latifolia* was characterised by the smallest leaf surface area (3.6±0.6 cm^2), three times lower than *Q. ilex*. SLM was significantly higher in *Q. ilex* and *P. latifolia* (average 21.3 mg cm^{-2}) than in *C. incanus* ($P<0.01$).

The microscopical analysis of sun leaves revealed variations among the species.

There were significant differences in cuticle and epidermis thickness between the dorsal and the ventral surfaces ($P<0.01$). *P. latifolia* showed a particularly thick dorsal cuticle (16.8±1.7μm) while *C. incanus* the thinnest (7.2±0.7 μm). *P. latifolia* the highest total leaf thickness (338±18 μm) ($P<0.05$). *Q. ilex* the highest LD (0.69±0.05 g m^{-3}) and *C. incanus* the lowest (0.60±0.02). The *P. latifolia* foliar sclereids occurred diffusely within the mesophyll and were vertically oriented, occupying 5% of the total mesophyll volume.

4. DISCUSSION

Leaf longevity is the essential feature which allows it to assume its function [20] and leaf phenology is assumed to change with climatic change [21]. Leaf longevity of various plants has reviewed [17] obtaining relationships between leaf longevity and leaf traits. Our results on the whole confirm this trend: the typical evergreen sclerophyllous species have long leaf life span (from 1 to 4 years), steeper leaf inclination and higher leaf density, that is opposite in the semi deciduous species which has a low leaf life span from 4 to 8 months. Evergreenness in *C. incanus* is maintained by sequentially formed leaf cohorts rather than by extended leaf longevity. The higher leaf inclination in *Q. ilex* and *P. latifolia*, reducing the amount of solar radiation [22, 23], may be a preventive means to photoinhibition of water stressed leaves [24] and it is particularly important for the evergreen sclerophyllous species that maintain their leaves for more than one year. The ability of a plant to grow in a particular habitat depends on both environmental and biotic factors; any factor which reduces photosynthetic carbon fixation may reduce productivity, and therefore the ability to compete in its habitat. If photoinhibition causes a significant limitation of net carbon fixation, it may affect the distribution of a species [25]. The anatomy and orientation of the foliar sclereids of *P. latifolia* suggest a light guiding function [26] that allows a better distribution of the reduced incident radiation (by the steeper leaf inclination) in the spongy layer. By the seasonal leaf dimorphism and leaf folding, leaf inclination is less important in *C. incanus* and the adjustment of leaf inclination from –37° in winter to +44° in summer leaves increases RI during drought. Moreover the increase in leaf folding, the high leaf pubescence and the reduced leaf surface area in summer substitute for the lower α and the lower SLM. In all three species SLM significantly correlated with total leaf thickness ($P<0.01$), LD ($P<0.01$) and leaf inclination ($P<0.05$) (Table 2). In a large number of Mediterranean woody species, leaf shedding occurs at the time of intense spring shoot growth [27, 28, 29] and leaf senescence and shedding in some drought semi-

Table 2: Summary of significant (**= P<0.01; *= P<0.05) regression relationships between total leaf thickness (μm), specific leaf mass (mg cm^{-2}), leaf density index (g cm^{-3}), leaf inclination (°). n=14.

Leaf traits (x - y)	Relationship	r
(total leaf thickness - specific leaf mass)**	y= 0.0623x	0.90
(specific leaf mass - leaf density index)**	y= 0.2115Ln(x) + 0.0102	0.78
(specific leaf mass - leaf inclination)*	y= 1.2507x + 18.821	0.60

deciduous species are accelerated by water deficits and high air temperatures [6]. *Quercus ilex* and *Phillyrea latifolia*, the more xeromorphic species, by the higher SLM, total leaf thickness, leaf density and leaf inclination (Fig. 1) may be at a competitive advantage, considering the forecasted air temperatures and drought increases in the Mediterranean Basin [30]. Increased drought stress may be a discriminant of species persistence in the distribution area, determining a shorter leaf life span but, probably without any important morphological and physiological implications in *P. latifolia* and *Q. ilex.* On the contrary, a shortening of leaf life span in *C. incanus,* the species with the shortest leaf life span, the lower SLM, total leaf thickness, leaf density and leaf inclination, may prove to be critical. Drought and low nutrient availability are common to Mediterranean type ecosystems, and the two interact strongly [6]. In particular it appears that the leaf life form plays a much larger role in leaf nutrient content than geographical distribution and soil nutrient regime [31], and that nutrient reabsorption from senescing leaves contribute largely to the high nutrient use efficiency [32]. Nutrient necessity becomes increasingly dependent on reabsorption, particularly when the soil supply is low during drought periods; under this condition a faster leaf fall in *C. incanus* probably may compromise its future presence. The extended leaf longevity in the evergreen sclerophyllous species may be interpreted as a nutrient conservation mechanism that enhances resource use efficiency.

5. REFERENCES

Bolhàr-Nordenkampf H.R., G. Draxler, In: D.O. Hall, J.M.O. Scurlock, H.R. Bolhàr-Nordenkampf, R.C. Leegood, S.P. Lang (Eds.), Photosynthesis and Production in a Changing Environment, Chapman and Hall, London, 1993, pp. 91-112.

Bottner P., M.M. Coûteaux, V.R. Vallejo, In: J.M. Moreno, W.C. Oechel (Eds.), Global Change and Mediterranean-Type Ecosystems, Ecological Studies 117, Springer-Verlag, New York, Berlin, Heidelberg. 1995, pp. 306-325.

Christodoulakis N.S., K.A. Mitrakos, In: J.D. Tenhunen, F.M. Catarino, O.L. Lange, W.C. Oechel (Eds.), Plant Response to Stress, NATO ASI Series G 15, Springer-Verlag, Berlin, Heidelberg, 1987, pp. 547-551.

Comstock J.P., B.E. Mahall, Oecologia, 65 (1985) 531-535.

Ehleringer J.R., In: R.W. Pearcy, J.Ehleringer, H.A. Mooney, P.W. Rundel (Eds.), Plant Physiological Ecology, Field Methods and Instrumentation, Chapman and Hall, London, 1989, pp. 117-135.

Field C.B., F.S. Chapin III, P.A. Matson, H.A. Mooney. Ann. Rew. Ecol. Systematics 23 (1992) 201-235.

Grabherr G., M. Gottfried, H. Pauli, Nature 369 (1994) 448.

Gratani L., Acta Oecol. 17 (1996) 17-27.

Gratani L., Ecol. Medit. XX (1994) 61-71.

Gratani L., M.F. Crescente, Ecol. Medit. 23 (1997) 11-19.

Gratani L., Photosynthetica 31 (1995) 335-343.

Groom Q.J., N.B. Baker, P.L. Steve, Physiol. Plant. 85 (1991) 585-590.

Hope A.S., In: J.M. Moreno, W.C. Oechel (Eds.), Global Change and Mediterranean-Type Ecosystems, Ecological Studies 117. Springer-Verlag, New York, Berlin, Heidelberg, 1995, pp239-253.

Jonasson S., Oikos 56 (1989) 121-131.

Kao W.Y., I.N. Forseth, Plant Cell Environ. 14 (1991) 287-293.

Kao W.Y., I.N. Forseth, Plant Cell Environ. 15 (1992) 703-710.

Kao W.Y., T.T. Tsai, Plant Cell Environ. 21 (1998) 1055-1062.

Karabourniotis G., J. Exp. Bot. 49 (1998) 739-746.

Kikuzawa K., Vegetatio 121 (1995) 89-100.

Kutbay H.G., M. Kilinç, Vegetatio 113 (1994) 93-97.

Levitt J., Responses of Plants to Environmental Stresses, Academic Press, New York, 1972.

Merino O., R. Villar, A. Martín, D. García, J. Merino, In: J.M. Moreno, W.C. Oechel (Eds.), Global Change and Mediterranean-Type Ecosystems, Ecological Studies 117, Springer-Verlag, New York, Berlin, Heidelberg, 1995, pp. 225-238.

Mooney H.A., F.S. Chapin III, Tree 9 (1994) 371-372.

Moore P., Nature 585 (1980) 535.

Moreno J.M., W.C. Oechel (Eds.), Global Change and Mediterranean-Type Ecosystems, Ecological Studies 117, Springer-Verlag, New York, Berlin, Heidelberg, 1995.

Pereira J.S., G. Beyshlag, O. Lange, W. Beyshlag, J.D. Tenhunen, In: J.D. Tenhunen, F.M. Catarino, O.L. Lange, W.C. Oechel (Eds.), Plant Response to Stress. NATO ASI Series G 15, Springer-Verlag, Berlin, Heidelberg, 1987, pp. 547-551.

Pereira J.S., M.M. Chaves, In: J.M. Moreno, W.C. Oechel (Eds.), Global Change and Mediterranean-Type Ecosystems, Ecological Studies 117, Springer-Verlag, New York, Berlin, Heidelberg, 1995, pp. 140-160.

Prichard J.M., I.N. Forseth, Am. J. Bot. 75 (1988) 1201-1211.

Reich P.B., M.B. Walters, D.S. Ellsworth, Ecol. Monogr. 62 (1992) 365-392.

Schlesinger W.H., B.F. Chabot, Bot. Gaz. 138 (1977) 490-497.

Smith W.K., D.T. Bell, K.A. Shepherd, Am. J. Bot. 85 (1998) 56-63.

Watson R.T., H. Rodhe, H. Oeschger, U. Siegenthaler, In: J.T. Houghton, G.J. Jenkins, J.J. Ephraums, Climate Change, The IPCC Scientific Assessment, Cambridge University Press, Cambridge UK, 1990.

Birds as Bio-Indicators of Long-Transported Lead in the Alpine Environment

MARIÁN JANIGA
Tatra National Park Research Centre, 059 60 Tatranská Lomnica, Slovak Republic

Key words: altitudinal transect, birds, Tatra Mountains, lead long-range atmospheric transport

Abstract: Numerous investigations on mosses and epiphytic lichens have demonstrated that there is a high level of pollution by lead in alpine regions, i.e. the ecosystems which substantially form the national parks and reserves throughout the Europe. Mosses, lichens are very suitable organisms for biological monitoring of air pollution due to their specific physiological features which easily enable to measure the heavy metal deposition. However, in order to establish the temporal and spatial scale of a potential problem, connected with lead, it is necessary to consider its total residence time in all compartments of the environment. As for an example, the accumulation of heavy metals in wildlife tissues in relation to long-range atmospheric transport has not been extensively studied. Furthermore, the studies have not been done by surveying a well orientated transect along an altitudinal gradient. This study's aim was to find bioindicators of lead pollution using birds. It was performed in the five different habitats of the Tatra Mountains, Slovakia. Along altitudinal gradients 112 bodies of 41 species of dead birds were collected. The transects were mainly on south-eastern slopes. Lead concentration in the bones of the most applicable bioindicator - Alpine Accentor *(Prunella collaris)* reflected that lead was deposited in the alpine areas of the Tatra Mountains as long-range air pollutants. For granivorous and frugivorous birds which live in forest and rural areas, the high individual variation of lead in the bones probably reflected the local sources of lead polution. Lead was found at higher level in granivorous and frugivorous birds than in birds predominantly eating evertebrates. Concentrations of lead in bones were significantly lower in birds of prey and owls.

G. Visconti et al. (eds.), Global Change and Protected Areas, 253–259.

1. INTRODUCTION

There is evidence of increased deposition of the metals over large mountain areas of Europe. The major source regions are found in central and eastern Europe [1, 2]. The investigations have demonstrated that there is a high level of pollution by lead in alpine regions, i.e. the type of ecosystems which substantially form the national parks and reserves throughout the Europe [3, 4, 5]. The main aim of this study was the examination of heavy metal accumulation in the bird tissues in relation to altitude. The use of certain bird species as biomonitors of regional or long-range atmospheric transport of lead was also explored.

2. MATERIAL AND METHODS

All birds were found dead in the field, and sent or brought to the TANAP Research Center Laboratory. Most cadavers arrived intact, some were collected from a taxidermist, and all were obtained in the period from 1991 to 1996. Only those cadavers were used in the study which were collected from transects along altitudinal gradients on slopes mainly within the south-eastern High Tatra Mountains, Slovakia. Since the toxic responses in birds being exposed to chronic input of lead via long-range atmospheric transport are not likely to occur in terms of acute toxicity, we investigated the bone tissues of the birds. Skeletal lead in birds may represent 90 per-cent of the total body lead content [6]. Therefore, lead in the avian bones provides a good index of chronic exposure [7]. Laboratory preparation of samples and analytic instrumentation are described in our previous study [8]. The sterna of 112 birds were examined in total.

The lead concentrations were reported as mean and SE in mg per kg dry weight for the comparison to numerous other studies, but the lead levels showed a highly skewed distribution in most of the sample groups. Therefore, a non-parametric approach to the analysis of the data was necessary [9]. The significance of differences between groups of birds was tested in the Mann-Whitney rank sum test. When $p < 0.05$, the data were considered as significantly different.

The feeding habits (granivorous, omnivorous, etc.) of each bird species was defined according to the basic and general ornithological literature. Since the feeding habits of species determine their risk to lead exposure [10], we created the clusters of species according to their type of the diet.

The lead levels in the bones of birds were compared among the clusters, and consequently among the following habitats: rural areas and agricultural fields - buffer zone of the Tatra National Park; ecotones or transition zones between forest and farm land; forest zone of the Tatra NP; villages and settlements in the forest; and alpine rocks and meadows.

3. RESULTS

Lead concentrations in bones of birds of the same feeding habit (especially granivorous) were higher in alpine areas than in forest, agricultural, or rural habitats (Fig.1). When the lead levels in the sterna of birds of different taxa of birds were compared, the lead concentrations were low in bones of birds predominantly eating animal diet (Fig.2). Mean and median lead levels were consistently high in the Alpine Accentors *(Prunella collaris).*

4. DISCUSSION

Lead belongs to the group of pollutants being transported as far as a few thousand kilometres, depending on meteorological conditions. Until present times, it has been uncertain, to what degree the concentration levels of heavy metals in the remote regions, such as alpine ecosystems, were harmful to the environment [11]. Some previous studies have indicated that wet deposition of lead is highly correlated with total deposition of lead in natural ecosystems [12, 13]. Lead is absorbed only in a limited extent by roots [14]. In the study on lead contamination of herbivorous animals, Kalas and Lierhagen [15] showed that hare *Lepus timidus*, black grouse *Tetrao tetrix*, and willow grouse *Lagopus lagopus* had similar gradients with a tenfold increase in liver lead concentrations from Northern to Southern Norway. The same pattern of lead increase was found in the bones of chamois in the transect from the East to Western Tatras [5]. Lead concentration levels in bird bones indicate that partially or fully granivorous birds have a higher degree of exposure to lead in the environment than other birds species. Scheuhammer [10] has also described greater susceptibility of granivores to lead exposure. In our study, of interest was the relatively high degree of lead exposure that Crows, Finches, Thrushes, and especially Alpine Accentors have in the environment. Many factors have a significant influence on lead concentrations in the bones of birds, making it difficult to decide whether lead originates from long-range air pollution or local sources. One of the useful and simple methods to differentiate local and transboundary sources is, according to our opinion, comparison of variances. Although Song Thrushes or Carrion Crows displayed relatively high averages of lead in the sterna, the standard errors of the data were very high,

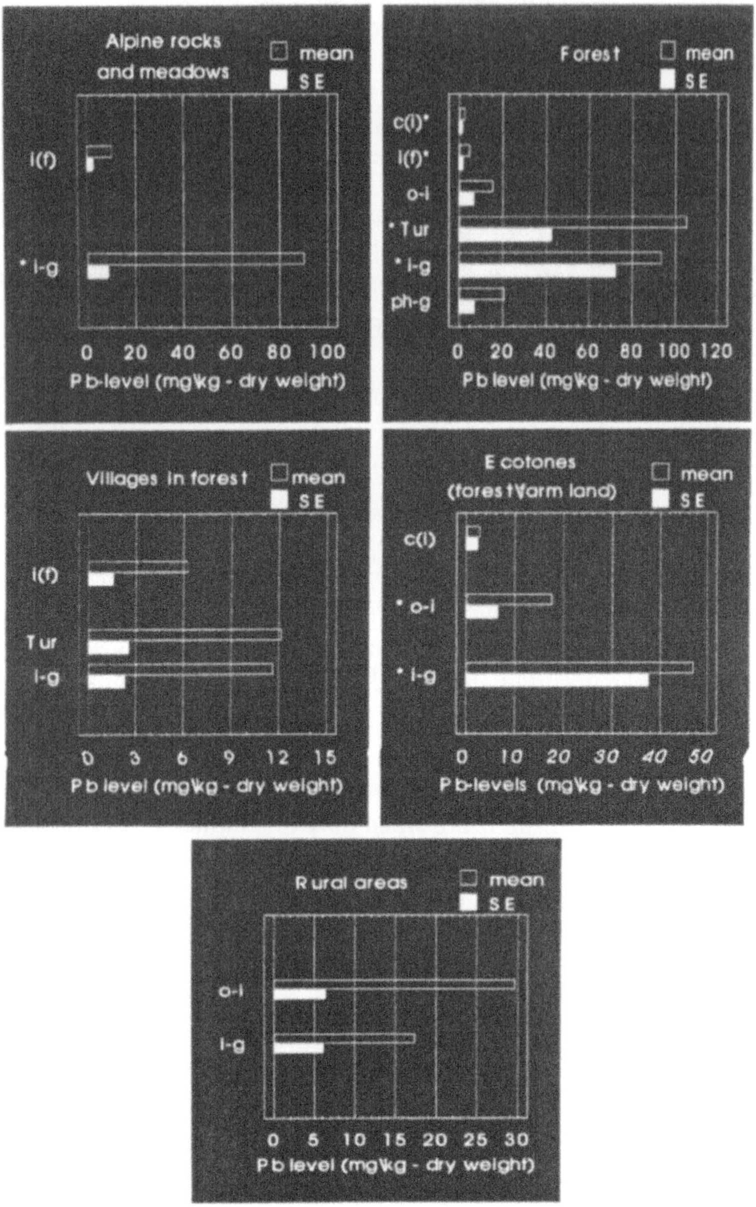

Fig. 1. Sterna lead concentrations in birds from five different habitats of the Tatra mountains, Slovakia. c - carnivores, i - insectivores (eating insect but often other evertebrates), f - frugivores, g - granivores, o - omnivores, ph - phytophagous birds, Tur - birds of family Turdidae. Alpine: i-g (n=5) and i(f)(n=3) significantly differ; Forest: c(n=3), i (4), o-i (4), Tur (8), i-g (12), ph-g (6) - within left or right column, groups not sharing a common symbol are significantly different; Villages: i (5), Tur (7), i-g (13) – groups do not differ; Ecotones: c (8), o-i (4), i-g (9) - group not sharing an asterisk in column is significantly different; Rural: o-i (18), i-g (3) – groups do not differ significantly.

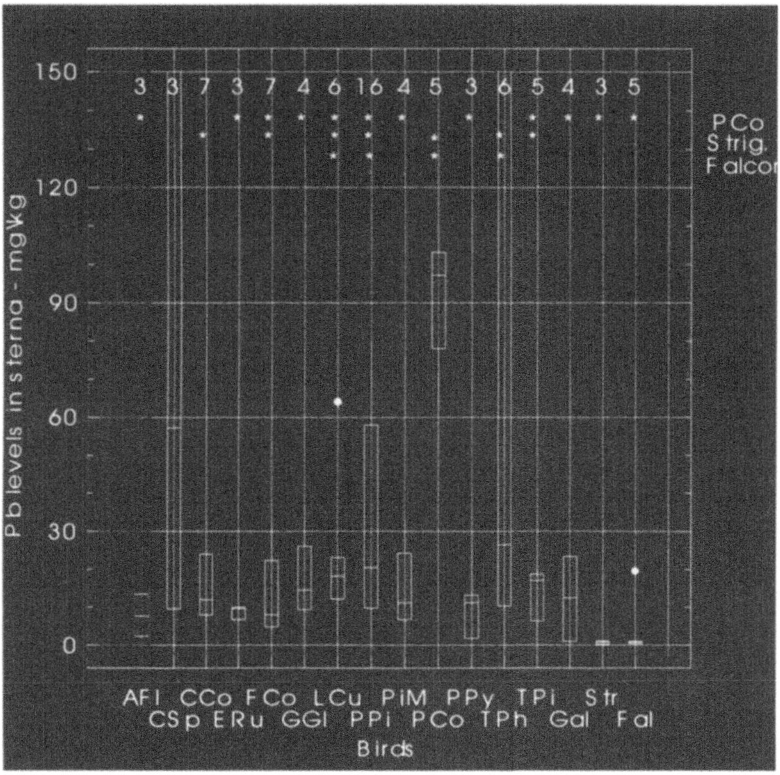

Fig. 2. Lead concentrations in sterna of different taxonomic groups of birds. Central line = median, box = difference between first and third quartiles, vertical lines denote minima and maxima, points represent extreme outliers. Numbers denote the sample sizes. Asterisks indicate significant difference (Mann-Whitney rank sum tests, $p < 0.05$) between *Prunella collaris, Strigiformes, Falconiformes* (right column) and the other groups of birds (x-axis). Taxa abbreviations: *AFl – Acanthis flammea, CSp – Carduelis spinus, CCo – C.. coccothraustes, ERu – Erithacus rubecula, FCo – Fringilla coelebs, GGl – Garrulus glandarius, LCu – Loxia curvirostra, PPi – P. pica, PiM – Picoides major, PCo – Prunella collaris, PPy – P. pyrrhula, TPh – Turdus philomelos, TPi – Turdus pilaris, Gal – Galliformes* (Game Birds), *Str – Strigiformes, Fal – Falconiformes* (Raptors).

indicating contributions of local origin, i.e. droplets, pellets found by some birds. These individuals probably possessed high levels of lead while the other crows or thrushes had low lead concentrations in the bones. On the other hand, Alpine Accentors also had very high levels of lead in the bones, but standard errors (Fig.1) of the data were low indicating stable long-range lead pollution of alpine areas. All measured accentors had high lead levels in their bones. In the alpine environment, the Alpine Accentor is considered to be suitable for studies of uptake of lead deposited by long-range air polution also because a part of the population is resident in the area throughout the year. Lead in the mountain ecosystems is probably transmitted by water;

therefore, the experiments with lead in the drinking water of birds may be helpful in understanding of the effects of levels of lead in the sterna of Tatranian birds. Ringed turtle doves that received a 100 mg\ml lead in their drinking water for 11 weeks had higher bone lead concentrations than controls, adult males averaged 62 mg\g, dry weight, bone lead, compared with 4 mg\g, dry weight, in controls. Bone lead in females was substantially higher (796 mg\g, dry weight) in lead-treated birds compared with controls - 8 mg\g, dry weight [16]. Columbid birds usually need 30 to 50 g of water per day, so treated turtle doves received, at least, app. 230,000 mg of lead in their water in total. Progeny of lead-treated parents had higher lead concentrations in bones, and testes weights were lower in lead treated males where spermatozoan numbers tended to be lower. From this point of view, we may assume that lead consumption can pose a hazard to reproduction processes in Alpine Accentors. The mean life expectancy of accentors is app. 3 - 4 years (Janiga, unpublished data), and the lead levels in their bones ranged from 63 to 110 mg\g, dry weight, i.e. more than average lead concentration in the males of turtle doves. In pheasants *Phasianus colchicus* living in the Silesia District, Poland, the mean values of lead in the bones ranged from 42 to 91 mg\g, dry weight [17]. That again is less than the lead level in the sterna of Tatranian accentors. The Upper Silesia Industrial Region has been declared an ecological disaster zone due to extremely high emissions of gasses and dust. It is one of the most important sources of long range air pollutants in the Central Europe, and it is not far from the chain of the Tatra Mountains. The experience from our study on alpine birds justifies the impression that similar studies could be adequate in projects aimed at assessing environmental hazards of long-range lead pollution in the alpine areas. Moreover, our results show that species-specific differences have to be carefully considered when planning schemes for monitoring of air pollution.

5. REFERENCES

Baars A.J., Arch. Environ. Contam. Toxicol. 28 (1995) 471-486.

Berg T., O. Royset and E. Steinnes, Atmos. Environ. 29 (1995) 353-369.

Esselink H., F.M. van der Geld, L.P. Jager, G.A. Posthuma-Trumpie, P.E.F. Zoun and

Janiga M. and M. Žemberyová, Arch. Environ. Contam. Toxicol. 35 (1998) 70-74.

Janiga M., M., B. Chovancová, M. Žemberyová and I. Farkašovská, Proc. 2nd World Conf. Mt. Ungulates (1998) 145-150.

Janiga M., Proc. Eur. Conf. Env. Soc. Change in Mountain Regions – 1997 (1999) 97-99.

Jenkins C., C.R.Acad.Sci.Paris, 281 (1975) 1187-1189.

Kabata-Pendias A. and H. Pendias, Trace elements in soils and plants, CRS, Press Inc., Florida, 1984.

Kalas J.A., J.A. and S. Lierhagen, Terrestrial monitoring of ecosystems. Metal concentrations in the livers of hares, black grouse, and willow ptarmigan in Norway. Norwegian Institution for Nature Research, Report no. 137, Trondheim, 1992.

Kendall R.J. and P..F. Scanlon, Environm. Pollution (Ser. A) 26 (1981) 203-213.

Kendall R.J. and P.F. Scanlon, Arch. Environ. Contam. Toxicol. 11 (1982) 265-268.

Nybo S, P.E. Fjeld, K. Jerstad and A. Nissen, Environ. Pollution, 94 (1996) 31-38.

Rühling A., L. Rasmussen, K. Pilegaard, A. Mäkinen and E. Steinnes, NORD 1987 (1987) 1 - 44.

Scheuhammer A.M., A.M (1991). Environ.Pollut. 71 (1991) 329-375.

Šoltés R., Oecologia Montana 1 (1992) 31 - 36.

Swiergosz R., R. (1993), In: I.D. Thompson (Ed.), Forests and Wildlife Towards the 21st Century,. Proceedings of the IUGB XXI Congress, Halifax, Vol.2., 1993, pp. 14-19.

UN ECE - United Nations Economic Commission for Europe, Task force on heavy metals emmissions. State-of-the-Art Report, 2nd Edition. Prague, 1995.

Zechmeister H.G., Environ. Pollut. 89 (1995), 73-80.

Annual Estimations of Ecophysiological Parameters and Biogenic Volatile Compounds (BVOCs) Emissions in Citrus Sinensis (L.) Osbeck

[1]MANES FAUSTO, [1]DONATO EUGENIO, [1]SILLI VALERIO AND [2]VITALE MARCELLO

[1]University of Rome "La Sapienza", Department of Plant Biology, P.le A. Moro, 5 I-00185 Rome

[2]University of Molise, Faculty of Mathematical, Physical and Natural Sciences, Via Mazzini, 8 I-86170 Isernia

Key words: Photosynthesis, Transpiration, Biogenic Volatile Organic Compounds (BVOCs), Emissions, modelling, *Citrus sinensis* (L.) Osbeck.

Abstract: Chemical and physical interactions among trace gases (CH_4, N_2O, VOCs) contribute towards greenhouse effects. This paper by using a original simulation model analyses the annual trends of net photosynthesis, transpiration, and water use efficiency, the total primary production and finally the emission of monoterpenes from orange groves located in the Burriana plain (Spain). The net photosynthetic rate showed a bell-shaped trend with maximum average values of 3.5 $\mu molCO_2$ m^{-2} s^{-1} calculated for the months of June and July. The canopy transpiration rate reached maximum values of approximately 2.5 $mmolH_2O$ m^{-2} s^{-1}. Monoterpene emissions displayed a relevant increase during spring time reaching the highest values of approximately 70 ng monoterpene · m^{-2} s^{-1} during the summer period and a progressive decrease during the successive months. The loss of carbon by monoterpene emissions for the different plant organs is evaluated at approximately 0.404% of the total carbon fixed by photosynthesis (559.77 gC m^{-2} y^{-1}) On a fractionated basis the contribution of the leaves is 0.083%, those of the fruits is 0.243% (for a period lasting from July to December), and only for the May period 0.079% for the flowers.

1. INTRODUCTION

At present, many research projects are being carried out to define and to

G. Visconti et al. (eds.), Global Change and Protected Areas, 261–270.

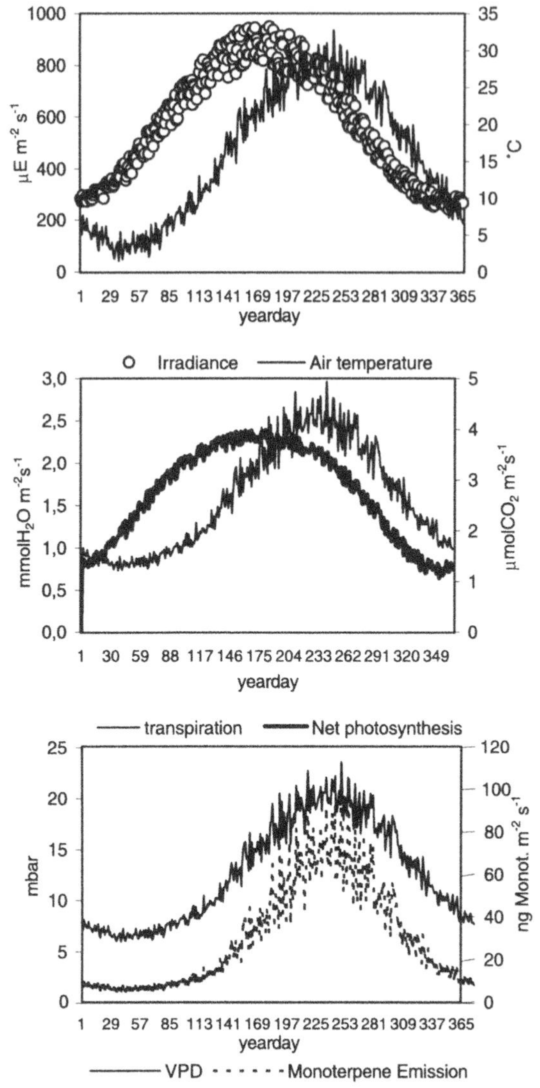

Figure 1 Top: Daily average valuers of irradiance and air temperature; middle: Net photosynthetic and Transpiration rates calculated for the Citrus sinensis canopy; Vapour Pressure Differences and Monoterpene emission rates calculated for each day of the year.

characterise the phenomena that are involved in greenhouse effects [1], [2], such as the increase of the carbon dioxide concentration, of the temperature in the atmosphere [3], and the chemical and physical interactions among trace gases (CH_4, N_2O, VOCs) on atmospheric compounds [4].

Moreover, an important outcome of these investigations was the idea that emissions of trace amounts of BVOCs might contribute towards ozone formation in the atmosphere [5]. Plant communities are formidable transformation systems of solar energy, and consequently many investigations are focused on the energy-mass exchange processes (CO_2, trace gases and water) at different spatial and temporal levels [6].

New methodologies and portable instruments have allowed plant communities to be studied at increasingly complex levels, and the great quantity of experimental data has led to the creation of a data base used for different types of "future predictions", from the net photosynthesis estimation [7] [8] to water use efficiency by individual plants or plant community for different regions of the Earth [9]; [10]. In this context, three generations of mathematical models were succeeding; from the first, mainly based on energy transfer in the atmosphere [11], to those of the second generation that considered plants as elements engaged in energy and mass processes in the atmosphere and soil [12]; [13], up to those of the third generation in which functional processes such as photosynthesis and stomatal conductance are integrated in the plant coverage of a given area by remote sensing devices [14]; [15]. Mathematical models are also used to predict the emission of biogenic volatile compounds (BVOCs) at different scales. Recently, accurate ecophysiological studies have been carried out at different scales to measure the emission rates of BVOCs from leaves [16]; [17] to plant canopies and communities [18]; [19]. The quantification and fate of biogenic fluxes from vegetation have been the main objective of many European (Biogenic Emissions in the Mediterranean Area, B.E.M.A. Project 1992-1997; ECOVOC, 1997) and North American (Southern Oxidant Study in the U.S., S.O.S.; BOREAS experiment in Canada) projects, concerning also the study of the relationships between chemical gas traces and the atmosphere.

The principal aim of this paper is to simulate, by a plant simulation model previously published by us [8], the annual trends of net photosynthesis, transpiration, and water use efficiency and to estimate the total primary production. Moreover, it will estimate the emission of biogenic volatile compounds (monoterpenes) coming from orange groves located in the Burriana plain (Castellon, Spain).

2. STRUCTURE MODEL

The model is of deterministic type, organised in sub-routines for calculating variables input data. Further information about its structure can be found in [8]. In this context, we summarise the following model inputs:

a) daily average solar radiation (μmol photons · m^{-2} s^{-1}), photoperiod (h), dark period (h),average daily temperature (°C),and meteoric precipitation

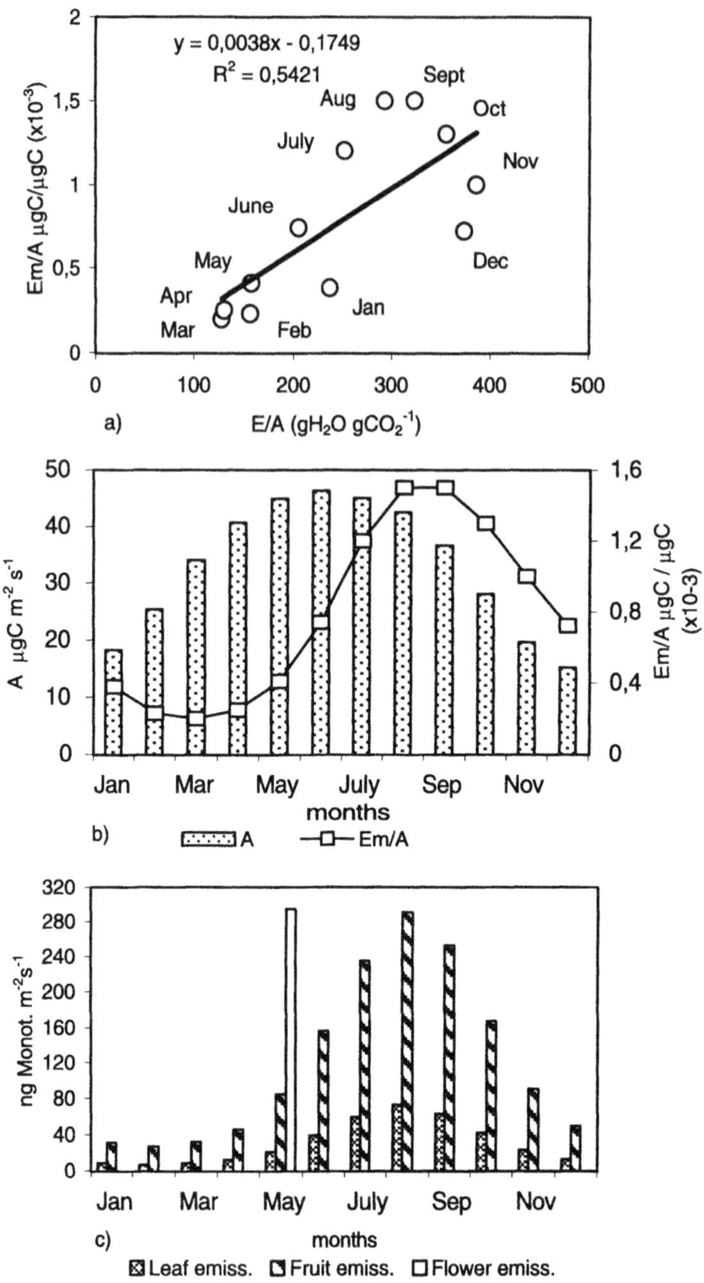

Figure 2 - a) Relationship between emission rates (Em) normalised by net assimilation (A) and the transpiration (E) and net assimilation (A) ratio; b) Annual trends of both net assimilation (A) and emissions normalised (Em/A) calculated for each month of the year; c) Monoterpenes emitted by different organs of the plant, calculated for monthly averages during the year.

(mm) measured for a period of 30 years (1960-1990);

b) leaf area index (LAI, $m^2 \cdot$ leaf m^{-2} ground), light extinction (k), photorespiration value (μmol $CO_2 \cdot m^{-2} s^{-1}$), quantum yield of photosynthesis ((μmol $CO_2 \cdot$ μmol photons $^{-1}$), maximum photosynthetic rate at saturation light (μmol $CO_2 \cdot m^{-2} s^{-1}$), canopy stomatal conductance (mol $H_2O \cdot m^{-2} s^{-1}$), light intensity under the canopy (Qi, μmol photons $m^{-2} s^{-1}$), dark respiration (μmol $CO_2 \cdot m^{-2} s^{-1}$), air-leaf water vapour pressure (mbar)and leaf temperature (°C);

Model output: daily means of net assimilation (μmol $CO_2 \cdot m^{-2} s^{-1}$), canopy transpiration (mmol H_2O $m^{-2} s^{-1}$), annual primary production (T $CH_2O \cdot ha^{-1}$ $year^{-1}$), annual canopy transpiration (Kg H2O $\cdot ha^{-1}$ $year^{-1}$); Monoterpene emission (ng Monoterpene $\cdot m^{-2} s^{-1}$).

The model considers the forest canopy as a homogeneous three-dimensional leaf of depth proportional to the LAI (total leaf area index not projected). Recent workshops on global ecological issues have identified leaf area index (LAI) as the most important single variable for measuring vegetation structure over large areas, and relating it to energy and mass exchange [20]; [21].

Consequently, this model is designed to be particularly sensitive to LAI, and LAI is used as the principal independent variable for calculating canopy interception, transpiration, respiration, photosynthesis, carbon allocation and biogenic emissions.

Most soil processes, such as root water uptake, are only inferred from their control of canopy processes and states, because only aboveground conditions are amenable to routine validation by remote sensing devices.

3. TEST-SITE AND LEAF PHYSIOLOGICAL MEASUREMENTS

The test site (Burriana, Castellon – Spain) was chosen for the carrying out of the B.E.M.A. Campaigns (1995-1997) to evaluate the biogenic fluxes from the orange grove stands extending for 2074 hectares.

Moreover, long-term climatic records were available, from 1960 to the 1990's.

Measurements of environmental and physiological parameters were carried out in two orange grove test-sites named A (near to the seashore) and B (2.5 Km inside the Burriana plain), on current leaves of *Citrus sinensis* (L.) Osbeck trees, by using a portable instrument (AutoCiras I, PP Systems – Hutchin, UK).

Further details can be found in the B.E.M.A. Reports [22]; [23].

Table 1 Parameter values for model simulations, using the Burriana plain as the test-site.

Variable	Value	Unit	Description
k	4.3	Dimensionless	Canopy light extinction coefficient
LAI	2.95	m^2 leaf · m^{-2} ground	Leaf Area Index
Kc	-0.5	$\mu molCO_2 \cdot m^{-2}s^{-1}$	Dark respiration
Q.Y.	0.0352	$\mu mol\ CO_2 \cdot \mu mol$ photons	Quantum Yield
As	5	$\mu molCO_2 \cdot m^{-2}s^{-1}$	Max. photosynthetic rate at light saturation
r	-0.3	$\mu molCO_2 \cdot m^{-2}s^{-1}$	Day respiration
Cp	1010	$J \cdot (Kg\ °C)^{-1}$	Specific heat at constant pressure
ρ_a	1.22	$Kg \cdot (m^3)^{-1}$	Density of dry air
λ	2470E3	$J \cdot (Kg)^{-1}$	Latent heat of vaporisation of water
γ	0.655	mbar · s^{-1}	Psychrometer constant
gs	44E-3	mol $H_2O\ m^{-2}s^{-1}$	Average stomatal conductance
Em_s	25	ng Monoterpene $m^{-2}s^{-1}$	Emission rate at standard condition (30°C, 1000 $\mu E\ m^{-2}s^{-1}$ of PAR)
β	0.09	$(°C)^{-1}$	Empirical factor monoterpene emissions

4. CALCULATION OF MONOTERPENE EMISSIONS

Biogenic emissions were calculated by using Tingey's equation [24] $Em = Em_s \cdot e^{\beta(t - ts)}$ and reviewed by Guenther [25] in relation to the determination of an empirical factor of emission, β (Table 1).

5. RESULTS

The average daily values of the principal environmental and physiological parameters calculated by the model are shown in Figure 1a-c. The maximum daily average values of the irradiance and temperature occur in the summer months, with values between 800-1000 $\mu E\ m^{-2}\ s^{-1}$ and 27-30 ° C respectively (Fig. 1a).

The net photosynthetic rate showed a bell-shaped trend with maximum average values of 3.5 $\mu molCO_2\ m^{-2}\ s^{-1}$ calculated for June-July (Fig. 1b). The canopy transpiration rate reached maximum values of approximately 2.5 $mmolH_2O\ m^{-2}\ s^{-1}$ (Fig. 1b). Vapour pressure difference (Fig. 1c) showed higher values (about 22 mbar) corresponding to higher ambient air temperature. Monoterpene emissions displayed a relevant increase during spring time reaching the highest values of approximately 70 ng monoterpene · $m^{-2}\ s^{-1}$ (Fig. 1c) during the summer period and a progressive decrease during the successive months.

Table 2: Synoptic table showing the average values regarding all measurements carried out on current leaves of *Citrus sinensis* plants located in two test sites, during the three B.E.M.A. campaigns.

Site	PPFD µE m-2s-1	RH %	T air °C	Photo. µmolCO2 m-2s-1	Transp. mmolH2O m-2s-1	Cond. mmolH2O m-2s-1	VPD mbar	n	WUE µmolCO2 mmolH2O-1	E/A mmolH2O µmolCO2-1
B June '97	1827	45.1	28.4	1.8	0.68	35.3	25.4	225	2.65	377.4
C June '97	1587	51.3	27.5	2.8	0.65	42.5	20.8	311	4.31	232.0
B April '96	1504	37.2	23.1	2.6	0.63	35.2	20.2	181	4.13	242.1
C April '96	1485	35.5	22.5	3.7	0.67	37.8	19.9	153	5.52	181.2
B July '95	1865	58.8	30.5	4.8	1.26	75.0	21.2	163	3.81	262.5
C July '95	1818	50.2	30.9	2.7	1.00	45.3	25.9	201	2.7	370.4

Figure 2 a shows the emission rates normalised by net assimilation (Em/A, µgC / µgC) vs. leaf transpiration and net photosynthesis ratio, E/A (gH_2O / gCO_2) estimated data by the model. It is possible to observe the seasonal relationship between the two parameters (Fig. 2a) and the uncoupling of annual trends of both Em/A ratio and net assimilation A (Fig. 2b), thus confirming the temporal shift between the production and biogenic monoterpene emissions. It is important to note that the emission rates were quite different among different plant organs; in fact, as one can observe in the Fig. 2c, in May monoterpene rates emitted by orange flowers were approximately 3.7 times higher than those coming from leaves. In general, the monoterpene emission rates of the fruits were 4 times higher than emission rates of the leaves (Fig. 2c) [22].

6. DISCUSSION AND CONCLUSIONS

The climatic conditions of Burriana plain come under the classical Mediterranean-type, which is characterised by a summer drought period and a peak of precipitation in autumn. These climatic conditions play an important role in the BVOC emission process and, consequently, on the formation of photochemical oxidants [26] and particle formation [27].

In particular, it is interesting to highlight the coupling between the annual trends of photosynthesis and irradiance and between the annual trends of canopy transpiration rate and air temperature.

These findings can be explained if one considers that the analysed agro-ecosystems are submitted to intensive irrigation during the summer drought periods to avoid water-limiting conditions.

Consequently, the non-limiting water conditions justify the next coupling response, in the time, observed for the annual trends of transpiration vs. temperature, whereas the response of photosynthesis is linked to the light availability.

Moreover, the physiological values estimated by the model both for photosynthesis and transpiration are in agreement with the data obtained by Manes et al. [28] regarding the ecophysiological characterisation of *Citrus sinensis* in Burriana during the B.E.M.A. campaigns (Table 2).

It is noted that notwithstanding the non-limiting water conditions the *Citrus* plants exhibited low levels of CO_2 assimilation, part of which may be used for monoterpene emission.

The estimated values of monoterpene emissions are in agreement with data obtained by Seufert et al. [22]; [23] by branch cuvette measurements [29] for orange plants.

As pointed out by Manes et al. [30], with regard to other Mediterranean species, the emission rates normalised by net assimilation (Em/A, µgC / µgC) were related to a functional index as E/A (leaf transpiration and net photosynthesis ratio, [6]), and they highlighted the different values for plants with internal reservoirs (i.e. *Citrus sinensis*, and *Pinus pinea*) with respect to plants that do not have mesophyll reservoirs (i.e. *Quercus ilex*).

The uncoupling between Em/A and A data is analogous to that observed in *Pinus pinea* plants, indicating a temporal separation between Monoterpene production and their emission.

Finally, the loss of carbon by monoterpene emissions by different plant organs is approximately 0.404% of the total carbon assimilated by photosynthesis (559.77 gC m^{-2} y^{-1}) and considering that the contribution of the leaves is 0.083%, those of fruits is 0.243% (for a period lasting from July to December), and of flowers 0.079% for the May period only.

These data are comparable with those derived by calculations of the terrestrial emission of Monoterpene values reported by Fall [26], and by annual primary productivity of terrestrial ecosystems reported by Whittaker [31]; in particular, the percentage value of the monoterpenes emitted with respect to the total carbon fixed at global level was approximately 0.62 %. These data help to estimate the amount of biogenic volatile compounds emitted by orange groves into the atmosphere during the year, thus permitting the contribution of these compounds to atmospheric processes such as ozone formation in the Mediterranean area to be evaluated. Moreover the model can help to evaluate how the BVOCs emission contribute to global change processes and reciprocally as biogenic emission are expected to change with altered composition and climate.

7. REFERENCES

A.A.V.V., In: Report on 3rd B.E.M.A. measuring campaign Burriana, Valencia – Spain, July 12-23 1995. Environment Institute, Join Research Centre, European Commission; EUR 17305 EN, 1996.

A.A.V.V., In: Report on 4th B.E.M.A. measuring campaign Burriana, Valencia – Spain, April-May 1996. Environment Institute, Join Research Centre, European Commission; EUR 17336 EN, 1997.

Bertin N., M. Staudt, U. Hansen, G. Seufert, P. Ciccioli, P. Foster, J.-L. Fugit and L. Torres, Atmos. Env. 31 SI (1997) 135-144.

Botkin D.B., In: Remote sensing of the biosphere. Report of the Committee on Planetary Biology. National Research Council, National Academy of Sciences, New York, 1986.

Chaves M.M. and J.S. Pereira J. Exp. Bot. 43 (1992) 1131-1139.

Dickinson R.E., Geophys. Monogr. Am. Geophys. Union. 29 (1984) 58.

Ehleringer J.R and C.B. Field, In: Scaling physiological processes: Leaf to Globe, Academic Press, San Diego, 1993.

Fall R., In: Reactive hydrocarbons in the atmosphere, N.C. Hewitt (Ed.), Academic Press, San Diego, 1999, pp. 41-96.

Guenther A., P.R. Zimmermann, P.C. Harley, J. Geophys. Res. 98 (1993) 12609-12617.

Hall F.G. and P.J Sellers J. Geophys. Res. 100 (1995) 25383.

Jarvis P.G., Plant, Cell and Environment 18 (1995) 1079-1089.

Lenz R., T. Selige and G. Seufert, Atmos. Env. 31 SI (1997) 239-250.

Loreto F., P. Ciccioli, A. Cecinato, E. Brancaleoni, M. Frattoni and D. Tricoli, Plant Physiol. 110 (1996) 267-275.

Manes F., G. Seufert, M. Vitale, E. Donato, O. Csiki and V. Silli, Phys., Chem. of the Earth, 1999, in press.

Manes F., M. Vitale, E. Feoli, M. Scimone and E. Canfora, Fresenius Envir. Bull. 7 (1998) 71-78.

Manes F., M. Vitale, P. Ciccioli, G. Seufert, In: *The role of vegetation emissions in tropospheric chemistry*. European Geophysical Society, XXII General Assembly, Wien 21-25 April, 1997.

Meng Z., D. Dabdub and J.H. Seinfeld, Science, 277 (1997) 116-119.

Monson R.K., P.C. Harley, M.E. Litvak, M. Wildermuth, A. Guenther, P.R. Zimmermann and R. Fall, Oecologia 99 (1994) 260-270.

Randall D.A. et al. J. Clim. 9 (1996) 738.

Sellers P.J, Y. Mintz, Y.C. Sud, A. Dalcher, J. Atmos. Sci. 43 (1986) 305.

Sharkey T.D., F. Loreto and C.F. Delwiche In: Trace gas emission from plants, T.D. Sharkey, E.A. Holland, H.A. Mooney (Eds.), Academic Press, San Diego, 1991, pp. 153-184.

Shugart H., In: A.M. Solomon and H. Shugart, (Eds), Vegetation dynamics and global change, Chapman and Hall, London, 1993, pp. 3-21.

Solomon A.M. and W. Cramer, In: A.M. Solomon and H. Shugart, (Eds), Vegetation dynamics and global change, Chapman and Hall, London, 1993, pp. 25-52.

Tingey D.T., R. Evans and M. Gumpertz, Planta 152 (1981) 565-570.

Trenberth K. (Ed.), Climate systems modelling, Cambridge University Press, Cambridge, 1992.

Valentini R., S. Greco, G. Seufert, N. Bertin, P. Ciccioli, A. Cecinato, E. Brancaleoni and M. Frattoni, Atmos. Env. 31 SI (1997) 229-238.

Versino B., Atmos. Env. 31 SI (1997) 1-3.

Webb W.L., W.K. Lauenroth, S.R. Szarek and R.S. Kinerson, Ecology, 64 (1983) 134-151.

Whittaker R.H., In: Communities and Ecosystems, London: Macmillan, 1970.

Williams M., Y. Malhi, A.D. Nobre, E.B. Rastetter, J.Grace and M.J.P. Pereira, Plant, Cell and Env. 21 (1998) 953-968.

Wittwer S., In: Land related global habitability science issues. NASA Tech. Mem. 85841, 1983.

A Multiscale Study to Analyse the Response of Vegetation to Climatic Conditions

[1]MANES FAUSTO, [1]BLASI CARLO, [2]ANSELMI SILVIA, [1]GIANNINI MONICA

[1]University of Rome "La Sapienza", Department of Plant Biology, P.le A. Moro 5, 00185 Italy Rome. Telephone/Fax: 0039 06 49912448; E-mail: manes@axcasp.caspur.it

[2]University of Molise, Via Mazzini 8, 86170 Isernia Italy

Key words: Plant Communities, Drought-Stress, Multiscale Analysis, Gas Exchanges, Remote Sensing, Leaf Area Index (LAI) Map

Abstract: The climatic changes predicted for the next century need to better understand the effects of water stress on the ecophysiological response of the plants in the Mediterranean areas. Gas exchanges, water potential and Leaf Area Index measurements were carried out during three years summer campaigns. Moreover remotely sensed analysis were performed to analyse vegetation indices. Measurements performed at leaf level on the four species have shown in *Fraxinus ornus* L. a reduction of more than 50% in net CO_2 assimilation in the hottest hours of the day. The two species belonging to Quercus genus showed a more costant assimilation rate, where the evergreen Mediterranean Stone Pine shows a reduction in net photosynthesis during midday hours. Leaf area index average data from field measurements have enabled to discriminate broadleaf deciduous communities (LAI=4.21±0.53) from coniferous evergreen communities (LAI=3.2±0.28). Remote sensing vegetation index results conducted on a multitemporal basis are in agreement with the response obtained from the field functional and structural data. The multiscale integrated experimental approach applied in this study seems a useful tool to define more clearly the ecological response of the main woody species of the Circeo National Forest and their sensitivity to global changes.

1. INTRODUCTION

Among the abiotic factors which affect the geographic distribution and

G. Visconti et al. (eds.), Global Change and Protected Areas, 271–279.

ecophysiological performance of plants, climate is probably the most relevant (1, 2). While predictions for average annual temperature change in the Mediterranean Basin agree on a 2-3 °C increase by 2050 the predictions of precipitation change remain more uncertain (3, 4). The climatic changes predicted for the next century may determine a decoupling between distribution of ecosystems and climatic conditions. Since a rapid climatic variation has been hypothesized, rather than a migration or a shifting of entire plant communities, the effect may be a new functional and structural reorganization of the actual plant communities (5, 6, 7). It is thus important to contribute to the study of "plant and ecosystem functional types" as suggested by many authors to describe and predict responses to environmental change (8, 9).

As a consequence of climatic changes the intensity of environmental stress and its effects on plants may increase (5, 10). This phenomenon is expected to become more pronounced, especially in areas with a climate of Mediterranean type (11), where drought will very likely become an increasingly large stress factor. Consequently, it is important to understand more clearly the effects of water deficit on the ecophysiological response of the plants in the Mediterranean areas where the limited availability of water in the soil would result in a reduction of plant growth (12).

Since photosynthesis is the main process determining primary production, it may be relevant to study gas exchange responses during the dry season for the main plant species which constitute the physiognomy and biomass bulk of major plant communities. Recently Solomon and Kirlenko (1997) (13) in a model exercise under delayed immigration processes show that the terrestrial biosphere provides a source rather than a sink for atmosphere carbon. To analyse the development of plant communities and to compare communities in different environmental conditions the Leaf Area Index is commonly used. Lavorel et al., 1998 (7) underline that LAI measurements will improve models simulating plant and soil water status and fluxes and plant simulation model to predict gas exchange processes and primary production (14).

Moreover multitemporal remote sensing studies are a useful tool to produce land use and land cover maps, and recently also to analyse the functional conditions of the vegetation by vegetation index analysis (NDVI, TM5/TM4) (15).

The aim of the present study is to analyse, using a multiscale integrated approach (16), the response at the leaf and canopy level to summer climatic conditions of a forest located in a Mediterranean area which is part of the Circeo National Park an Italian MAB (Man and Biosphere) Reserve. Moreover it was developed a correlation between NDVI and LAI to produce Leaf Area Index maps.

2. MATERIALS AND METHODS

Study site - The Circeo National Park is located along the coast of Lazio (Italy) and is characterised by a climate with seasonal photoperiodism, concentrated precipitation in the autumn months and a dry summer period. The Circeo forest is a remnant of the nearly 11,000 hectares which made up the ancient forest of Terracina. The forest (about 3,190 hectares) is located on a series of continental dunes of reddish-yellow sand. Mixed deciduous oak woods constitute the most extensive type of vegetation and are dominated by *Quecus cerris* L., *Quercus frainetto* Ten., *Quercus robur* L., *Fraxinus ornus* L., *Fraxinus oxycarpa* Bieb. and *Carpinus betulus* L.. Facies with hygrophilous or xerophilous plant communities can be found in relation with the microclimate in mixed deciduous oak woods.

Gas exchanges and water potential measurements - In 1997, 1998 and 1999 measurements of daily photosynthetic activity and traspiration rate were carried out on *Quercus cerris* L., *Quercus frainetto* Ten., *Fraxinus ornus* L., and *Pinus pinea* L. using a portable IRGA instrument CIRAS-AUTO (PP Systems, UK) according to Manes et al., 1997 (17). Physiological data are referred to at least 15 measurements made on leaves exposed to a photon flux density higher than 1000 $\mu Em^{-2}s^{-1}$. Throughout the measurement campaigns environmental parameters were recorded (air temperature, relative humidity and photon flux density of photosynthetic active radiation, PPFD). Pre-dawn and daily leaf xylem water potential measurements were performed using a portable pressure chamber (PMS Instrument Co, USA) accordingly to Sholander and Hammel, 1964 (18).

Field Leaf Area Index measurements (LAI) - By using a portable instrument (LICOR LAI-2000), on July-August 1996, 1998 and 1999 the different vegetation types, previously identified with satellite data processing and field survey, were analysed for LAI evaluation according to Welles and Norman, 1991 (19). The measurements were performed in mixed deciduous oak woods and areas afforested with Stone Pine. Transects were established, along which measurements were carried out every 10 meters. For each test site, a physiognomical-structural survey was made (20)

Remotely sensed analysis - The satellite data used in this study are Landsat TM images taken on 17 August 1994, 19 July 1995, 6 August 1996, 9 August 1997 and 27 July 1998 (Frame 19031) (provided by Telespazio s.p.a.), geocoded by the nearest neighbour method that yielded a square mean error of less than 1 pixel. All processing was performed by ILWIS 2.2 (Integrated Land Water and Information System).

Remote sensing satellite data were processed to obtain multitemporal vegetation indices [normalised difference vegetation index (NDVI) and [TM5 / TM4] which provided information on the photosynthetically active

Table 1. Main vegetation types recognised for the Circeo National Park Forest. Abbreviation for every class is also reported.

Classes	Coverage Subclass (%)	Vegetation types
MDOF	1 (>90%)	Mixed Deciduous Oak Forest dominated by Quercus cerris L., Q. frainetto Ten.,
MDOF	2 (60-90%)	Mixed Deciduous Oak Forest dominated by Quercus cerris L., Q. frainetto Ten.,
MDOF	3 (<60%)	Mixed Deciduous Oak Forest dominated by Quercus cerris L., Q. frainetto Ten.,
RPp	1 (>80%)	Reforestation with Pinus pinea L.
RPp	2 (<60%)	Reforestation with Pinus pinea L.

biomass and on water stress conditions of various types of vegetation (20).

The main vegetation types detected in the field and recognised in the images are listed in Table 1, which also illustrates the level of coverage for different subclasses. (The abbreviations explained in Table 1 have been used for all Figures).

3. RESULTS

The study area is located in a Mediterranean climatic region with 900 mm of mean annual precipitation, and winter and summer mean annual temperatures of 10 °C and 25 °C respectively. A preliminary analysis of climatic data for two twenty one year periods shows a slow reduction of winter rainfall and a more intense drought condition in the summer season (Fig. 1).

Gas exchange measurements performed at leaf level on the four species have shown a different response to summer environmental climatic conditions. In particular, among the deciduous species (Figs. 2a, b) *Fraxinus ornus* shows a reduction of more than 50% in net CO_2 assimilation during the hottest hours with a maximum value of about 4 µmol early in the morning (Fig. 2a); the two species belonging to Quercus genus showed a relative costant assimilation rate during the day with a maximum value for *Q. cerris* and *Q. frainetto* of about 11 $\mu molCO_2/m^2s$. The evergreen Mediterranean Stone Pine also shows a reduction in net photosynthesis during midday hours and a maximum value in the early hours of the morning of about 7 µmol (Fig. 2a).

The lower transpiration rate can be observed in *F. ornus* while more than three times these values were found in Quercus species with a slight increase during the hottest hours, more regular in *Q. frainetto.* A noticeable decrease

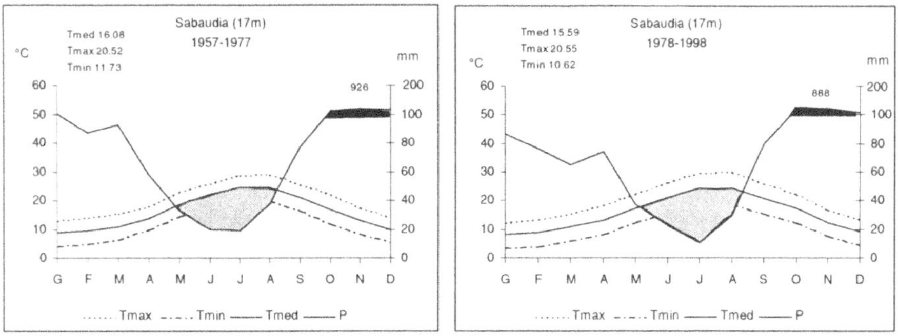

Fig. 1. Bagnouls-Gaussen diagrams Sabaudia station, Circeo National Park.

for *P. pinea* was recorded during the day (Fig. 2b).

Leaf water potential measurements carried out during the hottest hours showed lower values than -3.5 MPa for *F. ornus* and less negative values for the two Quercus and the Pinus species (Fig. 2d).

Leaf Area Index average data from field measurements of broadleaf deciduous (LAI=4.21±0.53) and coniferous evergreen communities (LAI=3.2±0.28) are statistically different (P=0.0000).

Remote sensing vegetation index results (Figs. 3a, b) performed on a multitemporal basis are in agreement with the response obtained from the field ecophysiological and structural data. In particular, NDVI, as shown in LAI measurements, has higher values for mixed deciduous forest. The water stress index (TM5/TM4) reaches higher values in broadleaf deciduous communities than evergreen communities. Moreover the multitemporal analysis performed for a 5 year period highlights the optimum growth conditions during 1996 and 1998.

By correlating NDVI and LAI field data ($y=11.776x^{1.9291}$; $R^2=0.822$) we have obtained a Leaf Area Index maps for the entire area which allows us to estimate the LAI values (measured vs. estimated: $y=0.8396x+0.574$; $R^2=0.815$) for the different vegetation types on a multitemporal basis (Fig. 4).

4. DISCUSSION AND CONCLUSIONS

A different response strategy to water stress is found among the four species. *P. pinea* and to a smaller extent *F. ornus* are characterized by a tolerance to summer drought conditions. The incapacity to recover in a short time is more evident in *F. ornus* for which pre-dawn water potential remains at very negative values without night - time recovery.

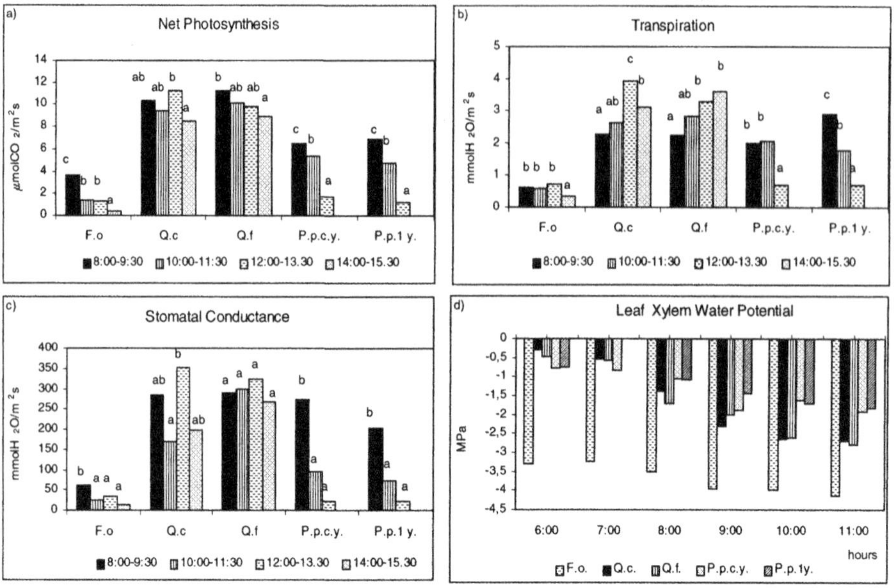

Figs. 2a, b, c and d. Diurnal course of net photosynthesis activity (a),transpiration rate (b), stomatal conductance (c) and leaf xylem water potential (d) measured on the analyzed species in the Circeo National Park Forest.
Different letters refer to significant differences (significance level=0.05).
F.o. = *Fraxinus ornus* L.; Q.c. = *Quercus cerris* L.; Q.f. = *Quercus frainetto* Ten.; P.p.c.y. = *Pinus pinea* L. current year needls; P.p.1.y. = *Pinus pinea* L. 1 year needls.

However when environmental conditions are favourable higher values of photosynthesis for this last species were measured (11 $\mu molCO_2/m^2s$ - spring 1998). A different behaviour may be observed for Quercus species for which the daily trend of gas exchanges seems more related to the diurnal trend of microclimatic conditions. As observed during these field surveies *Q. frainetto* appears better adapted to water shortage and high temperature than *Q. cerris* in relation to stomatal control capacity (Fig. 2c). These findings are also supported by a previous laboratory experiment (21). When it was imposed serious water stress the two species showed high sensitivity although at different degree (21).

NDVI multitemporal analysis which exhibits a similar trend between deciduous and evergreen vegetation types.

The increase of the index observed in broadleaf communities from 1994 to 1996 is related to good climatic conditions and to a gypsy moth infestation occurred in 1993 (Fig. 3a). It is also interesting to note the sharp reduction of the values in both communities during 1997 attributable to a more pronounced and prolonged summer drought period supported by climatic data. A confirm to this stress condition which has determined less leaf biomass production derives from lower LAI estimation values (Fig. 3c).

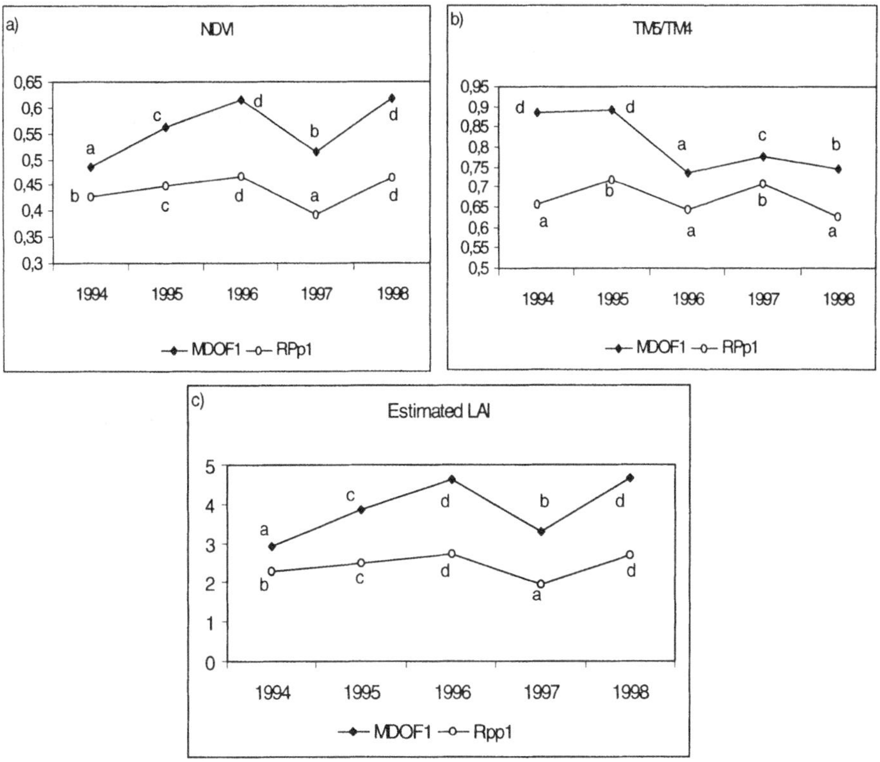

Figs.3a, b and c. Multitemporal trends of Normalised Difference Vegetation Index (NDVI) (a), Water Stress Index (TM5/TM4) (b) and estimated LAI (c) for Mixed Deciduous Oak Forest (MDOF1) and for reforestation with *Pinus pinea* L. (RPp1) calculated by remote sensing satellite data for the Circeo National Park Forest

Also TM5/TM4 index values, as expected, increase in 1997 for both communities analysed.

The results presented in this paper highlight as the relictual forest of the Circeo National Park is subjected to natural biotic and abiotic stress factors and to environmental changes due to anthropic impact both of local and global origin. The multiscale integrated structural and functional experimental approach applied in this study seems a useful tool to define more clearly the ecological response of the main woody species and the sensitivity of deciduous stands to global changes. Moreover the LAI maps derived applying NDVI/LAI correlation allow to better evaluate, to model and to monitor vegetation response to stress factors in Mediterranean areas. In particular, the synoptic view of the response of the different vegetation types to the annual variation of climatic conditions is well highlighted by the annual trend of LAI data and from the LAI maps and by ecophysiological measures from which it is possible to hypothise a higher sensitivity for deciduous communities.

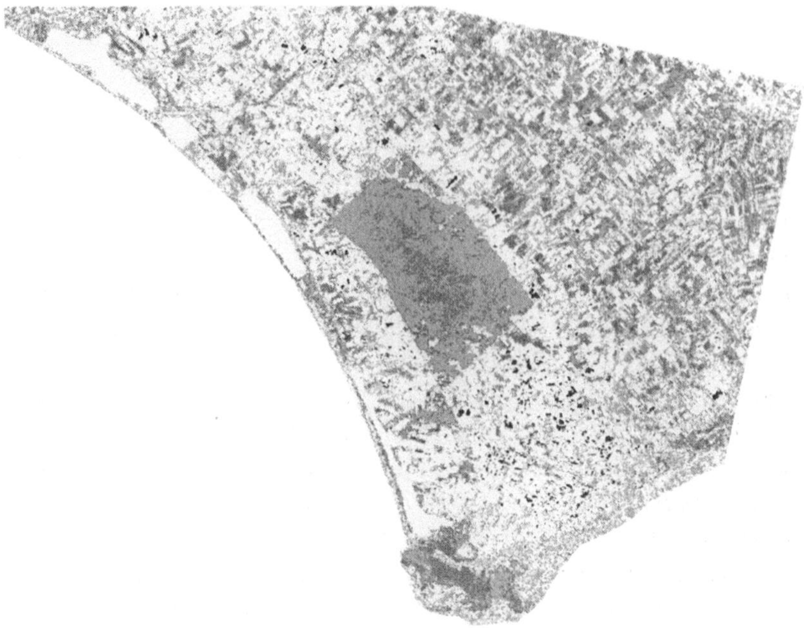

Fig 4. Estimated LAI map (Scale 1:270000) for 1998 using Landsat TM image (27/07/1998) of the Circeo National Park and surrounding areas.

It is however necessary to better analyse the continuous anthropic impact on the territory and more in general the global change effects which may determine landscape modifications (e.g. ecosystems fragmentation) that need continuous analysis of the functional and structural conditions of natural ecosystems. The main target is to find better management solutions and to develop monitoring actions especially in protected areas.

5. REFERENCES

Cubash U., H. von Storch, J. Waskewitz & E. Zorita, Climate Res. 7 (1996) 129-149

Donato E., M. Giannini, A. Tinelli, M. Vitale, F. Manes, Atti del VII Congresso Nazionale della Società Italiana di Ecologia (S.It.E. - Atti 17) (1996) 149-152

Hobbs R.J., In: T.M. Smith, H.H. Shugart and F.I. Woodward, (Eds.), Plant Functional Types their relevance to ecosystem properties and global change, Cambridge University Press, 1997, pp. 66- 90

Jarvis P. G., Plant, Cell Environ. 18 (1995) 1079-1089

Jeftic L., J.D. Millian and G. Sestini, (Eds.), Climatic Change and the Mediterranean, Edward Arnold, London, New York, Melbourne, Auckland, 1992

Lavorel S., Canadell J., Rambal S. and Terradas J., Global Ecology and Biogeography Letters 7 (1998) 157-166

Levitt J., Responses of Plants to Environmental Stresses. Water, Radiation, Salt and Other Stresses, Vol. II, 2nd Edition, Academic Press, New York, 1980
Manes F. and C. Blasi, Fresenius Environmental Bulletin 4 n°3 (1995) 183-188
Manes F., G. Seufert and M. Vitale, Atmos. Environ. 31 SI (1997) 51-60
Manes F., In: Global Change, International Geosphere - Biosphere Program. Ricerche Italiane, Workshop CNR - Roma - 25-26 marzo 1996, pp. 97-100
Manes F., M. Vitale, E. Feoli, M. Scimone and E. Canfora, Fresenius Environmental Bulletin 7 (1998) 71-78
Manes F., S. Anselmi, E. Canfora and M. Giannini, Proceedings EUROPTO Series 3222 (1998) 246-252
Moreno J.M. and W.C. Oechel, (Eds.), Global Change and mediterranean type ecosystems, Springer-Verlag, New York, 1995
Pereira J.S. and M.M. Chaves, In: J.M. Moreno and W.C. Oechel, (Eds.), Global Change and mediterranean type ecosystems, Springer-Verlag, New York, 1995, pp140-160
Sholander P.F. and H.T. Hammel, Proc. Natl. Acad. Sci USA 52 (1964) 119-125
Smith T.M., H.H. Shugart and F.I. Woodward, (Eds.), Plant Functional Types their relevance to ecosystem properties and global change, Cambridge University Press, 1997
Solomon A.M. and A.P. Kirilenko, Global Ecology and Biogeography Letters 6 (1997) 139-148
Stow D.A., In: J.M. Moreno and W.C. Oechel, (Eds.), Global Change and mediterranean type ecosystems, Springer-Verlag, New York, 1995, pp.254-286
Walter H., Vegetation of the earth and ecological systems of the geobiosphere, Springer-Verlag, Berlin, 1984.
Welles J.M. and J.M. Norman, Agronomy J. 83 5 (1991) 818-825
Woodward F.I. and D.J. Beerling, Global Ecology and Biogeography Letters 6 (1997) 413-418

Phytotoxic Ozone Effect on Selected Plant Species in a Standardized Experimental Design

MANES FAUSTO, CAPOGNA FRANCESCA, GIANNINI MARIA ANTONIETTA, SILLI VALERIO
University of Rome "La Sapienza", Department of Plant Biology, P.le A. Moro 5, 00185 Rome, Italy

Key words: tropospheric ozone, Trifolium repens (clover clone), biomonitor, leaf injury, biomarkers, ICP-Crops Programme.

Abstract: The most phytotoxic atmospheric pollutant is ozone which represents a stress factor for forest species and for the growth and the production of many plant species used in agriculture, resulting in great economic loss. Experimental protocols for passive and active biomonitoring studies to assess plant injury induced by ozone have been adopted by the United Nations/Economic Commission for Europe (UN/ECE) International Cooperative Programme (ICP-Crops), which aims to quantify crop responses to ozone, contributing to the determination of critical levels for this pollutant. Since 1988, our research group has participated in the ICP-Crops programme, conducting studies in the area of Rome with the aim of characterising the ecophysiological response to tropospheric ozone of species of agrarian interest. In this context, in 1997-1998 we have conducted studies on clones (an ozone-sensitive, NC-S clone, and an ozone resistant, NC-R clone) of *Trifolium repens* L. cv. Regal at the Botanical Garden of the "La Sapienza" University of Rome, in accordance with the ICP-Crops standard protocol. During the studies, environmental parameters and ozone concentrations were measured. In clover clone experiments, comparison of the two clones in 1997-98 experimental periods revealed both morphologic and biochemical differences associated with the response to tropospheric ozone. The results relative to visible injury, peroxidase activity and chlorophyll content confirmed the greater sensitivity of the NC-S clone to tropospheric ozone.

1. INTRODUCTION

Over the last three decades the pollutants which characterise the chemical

G. Visconti et al. (eds.), Global Change and Protected Areas, 281–288.

composition of the atmosphere over the Europe have generally changed. The photochemical oxidants, ozone in particular, have become major atmospheric pollutants. In fact North America and Europe, because of their geographical position, their industrial development, and the presence of evergreen vegetation, represent areas where photochemical smog often occurs (1, 2). The most phytotoxic atmospheric pollutant is ozone which represents a stress factor for forest species and for the growth and production of many plant species used in agriculture (3), resulting in great economic loss. In 1988, the Economic Commission for Europe of the United Nations (UN/ECE) created the international research programme "International Cooperative Programme on Crops" ("ICP-Crops") with the aim of studying the effects of atmospheric pollutants and other stress factors on cultivated species such as grain, tobacco, bean, and white clover (4). Since certain species show high levels of sensitivity to specific pollutants, environmental quality can be evaluated with bioindication and biomonitoring studies, and decreases in agrarian crop-yields caused by pollutants can be determined. These studies contribute to improving appropriate methods for evaluating the effects on plants of atmospheric pollutants, integrating biological information with information on pollutant concentrations, as determined by physico-chemical monitoring. The present paper presents the morphological and biochemical results of studies conducted on clones of *Trifolium repens* L. cv. Regal under field conditions.

2. MATERIALS AND METHODS

In 1997 and 1998 the field studies were conducted at the Botanical Garden of the "La Sapienza" University of Rome. During the studies, environmental parameters were measured (i.e., maximum and minimum temperature, relative humidity), as were environmental ozone concentrations using a UV photometric analyser (model 1108, Dasibi Environmental Corp., U.S.A.). The two clones of *Trifolium repens* cv. Regal were used, one sensitive to ozone (NC-S) and the other resistant (NC-R) (5), in accordance with the ICP-Crops standard protocol (6, 7).

After 28 days of growth in climatic chambers (Labco, model CT 15) in conditions of controlled temperature and humidity, 20 plants of each clone were transferred outdoors. In the period June-October of the years 1997 and 1998, five harvests were conducted for each year.

Approximately every 28 days, prior to harvest, both morphological parameters (above-ground biomass production and leaf injury) and physiological parameters (chlorophyll content and peroxidase activity) were measured in healthy and

Table 1. Leaf injury, biomass production and physiological data during experimental period 1997-1998.

		1997				1998			
Harv	Clone	Leaf injury %	Dry weight g	POD ΔO.D./g.min	Chl a μg/ml	Injury class more present **	Dry weight g	Chl a μg/ml	POD ΔO.D./g.min
1°	NC-R	0	12,29	1436	6,8	C0	2,89	7,71	595
	NC-S	12,35	7,91*	2722*	5,98	C1	2,83	6,89*	1302*
2°	NC-R	0,75	32,95	1329	6,65	C1	6,42	7,26	1470
	NC-S	61,75	20,43*	5606*	4,87*	C4	9,03*	6,4	4014*
3°	NC-R	2,5	26,17	1250	6,06	C0	8,02	7,32	1565
	NC-S	41,25	8*	2190*	4,69*	C3	8,04	6,38*	3560*
4°	NC-R	0	16,52	1525	6,18	C0	8,84	6,25	1985
	NC-S	18	4,23*	2295*	5,17*	C1	8,7	5,49*	3710*
5°	NC-R	0	12,41	1700	5,88	C0	4,25	5,41	1900
	NC-S	0,75	4,89*	1780	5,28*	C0	4,01	4,68*	2060

*Differences between clones are statistically significant with $p \leq 0{,}05$

**C0= no injury, C1= very slight injury, C2= slight injury, C3= moderate injury, C4= heavy injury, C5= total injury

fully expanded leaves. The chlorophyll content was measured using a spectrophotometric method (Jasco UV-visible, Japan), following an experimental protocol in accordance with Singha & Townsend (8) and Wellburn (9). Peroxidase activity was measured according to Manes et al. (10).

3. STATISTICAL ANALYSES

Data were analysed with the SPSS statistical software (SPSS Corp., USA) using the Student-Newman-Keuls test at $p \leq 0.05$. Different letters in figures indicate significant differences between clones.

4. RESULTS AND DISCUSSION

Comparison of the two clones in the two experimental periods revealed both morphologic and biochemical data (Table 1) as a response to tropospheric ozone.

The data relative to the environmental parameters are shown in Table 2.

Table 2. Environmental Parameters during experimental periods 1997-1998.

HARVEST 1997	HARV. DATE	T mean (°C)	RH (%) (9 GMT)	O_3 mean (ppb, 7hrs)	AOT40 (ppb.h)
1°	01/07/97	25,1	56,8	54	2962
2°	31/07/97	26,9	53,4	61	4159
3°	01/09/97	26,4	57,4	60	3985
4°	02/10/97	25,5	54,9	56	3559
5°	13/11/97	19,3	61,3	31	925
1998					
1°	08/07/98	26,5	56,1	60*	5649*
2°	06/08/98	28,6	54,9	72	6504
3°	02/09/98	29,3	62,4	70	5660
4°	13/10/98	23,4	67,5	42	1865
5°	13/11/98	17,8	62,6	26	61

* Ozone values obtained from measurements in 1998 and from the comparison of 1997-1998 data.

4.1 Leaf injury

In 1997, the NC-S plants had typical ozone-induced leaf injury (Fig. 1), as observed by Heagle et al. (11), whereas the NC-R plants showed no important ozone-induced injury (Fig. 2).

During the period of the second harvest, when the average environmental ozone concentration was at its highest (average 61 ppb and 4159 ppb.h of AOT40) (Table 2), the maximum average percentage of injured leaves per pot was observed in NC-S plants (62% of total leaves) (Fig. 3a). In the fifth harvest (the last), when the average concentration of tropospheric ozone was lowest (average 31 ppb and 925 ppb.h of AOT40), only three pots of NC-S plants showed ozone-injured leaves (5% of total leaves).

These data are in agreement with studies performed by Benton et al. (12) which report that leaf injury in *T. repens* occurs when average ozone doses exceed 35 ppb for 7 hours. Also in 1998, the NC-S plants showed greater leaf injury with respect to NC-R plants as described by 5 damage classes (Fig. 3b).

In the second harvest, which corresponded to the highest ozone levels, injury classes were highest in both NC-R and NC-S plants, compared to the other harvests (Fig. 3b). In fact, class 5 (C5) injury was identified in 15.4% of the NC-S plants and the remaining 84.6% of plants showed injury of class 2, 3, or 4. In the successive harvests, progressively lower classes of injury were assigned to both clones.

Fig. 1

Fig.2

4.2 Peroxidase activity

In 1997, the peroxidase activity of NC-R clones remaind basically constant for all harvests. The NC-S clones showed significantly higher values of peroxidase activity compared to the NC-R clones in the first, second, third and fourth harvest (Tab.1). In particular, in the second harvest, when the ozone concentration was highest, the greatest difference between the two clones was observed; in the fifth harvest, corresponding to the lowest ozone levels, no significant differences were observed (Tab.1). The peroxidase activity observed in 1998 (Tab.1) confirmed the 1997 data, showing a high correlation ($r^2 = 1.0$, $p = 0.000$) between peroxidase activity and AOT40 levels. In light of these results, peroxidase activity can be considered as a useful biomarker of biochemical-physiological alterations induced by pollutants (13, 14).

4.3 Chlorophyll *a* content

Both in 1997 and 1998, in all harvests the chlorophyll *a* content was lower in the NC-S clone, with respect to the NC-R clone (Tab.1).

This seems to confirm the degrading action exerted by ozone on photosynthetic pigments, observed by other studies (15).

4.4 Biomass production

In 1997, the relevant difference in the response to ozone between the two clones was observed in biomass production, which should, in the absence of

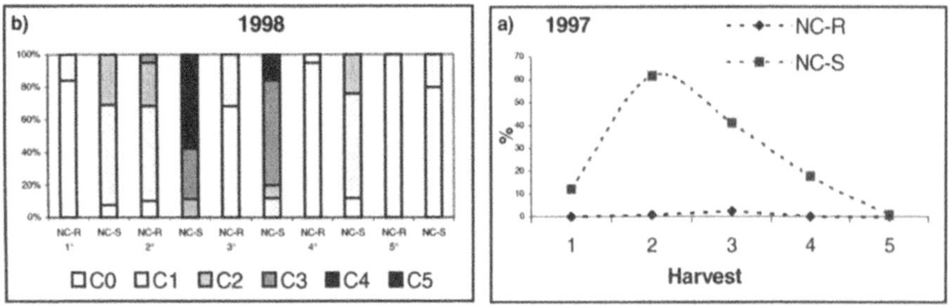

Fig. 3a, b Leaf injury in NC-R and NC-S in 1997-1998

this pollutant, remain similar between sensitive and resistant plants (5); in this case, significant differences were observed between the NC-S and the NC-R clones in all harvests, with the greatest difference in the third harvest (Fig. 4a). By contrast, in 1998, no significant differences in biomass production were observed between the two clones (Fig. 4b), except in the second harvest, during which the biomass production was significantly higher in the NC-S clone.

This result may be related to the lower growth of the NC-R clone, also existing since the first development phase in climatic chambers.

Thus, in both 1997 and 1998, the results relative to visible injury, peroxidase activity and chlorophyll content confirmed the greater sensitivity of the NC-S clone to tropospheric ozone. When evaluating the biomass data, it's necessary to underline the state of growth of the plants considering this from the initial phases following the planting of the cuttings. In fact, the growth state of the clones in the period preceding the successive transplanting could influence the future growth of the plants, irrespective of the environmental ozone concentrations.

5. CONCLUSIONS

Environmental ozone concentrations, measured in the study site in diverse years, demonstrate that the area of Rome has one of the highest levels of this atmospheric pollutant when compared to the other European sites participating in the ICP-Crops programme. Moreover, the effects observed on the bioindicator species show how the levels of this pollutant are phytotoxic and capable of determining a reduction in agrarian production.

The *T. Repens* clones represent an efficient system for the biomonitoring of tropospheric ozone that does not require antiozonants for obtaining a control system.

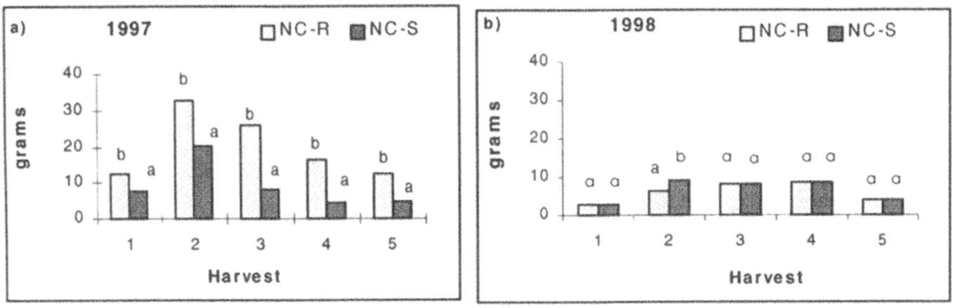

Fig. 4a, b Dry weight measured in NC-R and NC-S in 1997-1998

In particular, the results relative to biomass production in 1997 demonstrate the possibility of estimating decreases in the average harvest using this biomonitoring system.

The results of this research contribute to characterising the response of agrarian species to tropospheric ozone, to defining standard protocols for bioindication studies in the field, and to develop the knowledges in selecting sensitive plant species in experiment on air quality control. During the ICP-Crops programme, experimental protocols that included plants with wider ranges of adaptation to climatic conditions were developed; these plants were also characterised by a greater specificity of their response to ozone. In the complex frame of global change it is important to consider the relationships existing among the increase of temperature, that could influence the plant response to environmental factors, photochemical ozone formation and VOCs (Volatile Organic Compounds) emission.

An useful tool is the development of suitable bioindication kit; we are testing in field experiments a kit based on clover clones (NC-R, NC-S) to develop a monitoring approach of the most important gaseous pollutants (O_3, NO_x, SO_2) in rural and urban areas.

6. REFERENCES

Allegrini I., M. Cortiello, F. Manes, P. Tripodo, The Science of the Total Environment 141(1994) 75-85.

Ashmore M.R., J.N.B. Bell, A.I. Rutten, Ambio 14 (1985) 81-87.

Benton J., J. Fuhrer, B.S. Gimeno, L. Skärby and G. Sanders, Water, Air and Soil Pollution 85 (1995) 1473-1478.

Heagle A.S., J.E. Miller and D.E., Sherrill and J.O. Rawlings, New Phytol. 123 (1993) 751-762.

Heagle A.S., J.E. Miller and D.E.Sherrill, J. Environ. Qual. 23 (1994) 613-621.

Kangasjärvi J., J. Talvinen, M. Utriainen & R. Karjalainen, Plant, Cell and Environment 17 (1994) 783-794.

Manes F., R. Federico, F. Bruno, Phytopath. Medit. 25 (1986) 76-79.

Manes F., R. Federico, M. Cortiello and R. Angelini, Phytopath. Medit. 29 (1990) 101-106.
Millan M., R. Salvador, E. Mantilla, B. Artinano, Atmospheric Environment 30 (1996) 1909-1924.
Sarkar R.K., A. Baneryee and S. Mukherji, Environ. Pollut. (Ser. A), 42 (1986) 289-295.
Singha S. and E.C. Townsend, HortScience 24 (6) (1989) 1034.
UN/ECE ICP-Crops, Experimental Protocol for the ICP-Crops, 1998 Season. Institute of Terrestrial Ecology. Centre fo Ecology & Hydrology 1998.
UN/ECE ICP-Crops, ICP-Crops Experimental Protocol for 1997. The Nottingham Trent University, UK. 1997.
UN/ECE Workshop Report, University of Kuopio, Department of Ecology and Environmental Science. L. Kärenlampi and L. Skärby (Editors)1996.
Wellburn A.R., J. Plant Physiol 144 (1994) 307-313.

Plant Invasions in Central European Middle-Mountains: A Result of Global Change ?

LENKA SOUKUPOVÁ
Institute of Botany, Academy of Science of Czech Republic (GACR 502/94/0768 and 206/99/1411) CZ-252 43 Průhonice, Czech Republic

Key words: : Hercynian mountains, altitudinal zonation, A-O systems, air pollution, forest decline, spreading of plants

Abstract: The success of alien species for migration beyond their primary area of distribution, i.e. plant invasions, is affected by multiple factors which should be readily distinguished into two distinctive categories: physical/abiotic and biotic factors. While the prevailingly deterministic physical events can be broadly stated for the whole Earth, the highly stochastic biotic events are less predictable for a large region, mainly due to evolutionary changes in the genome of individual organisms/species and in the constellation of genomes within particular communities/ecosystems. In mountain areas, a better understanding of plant invasions requires both a detailed knowledge of the physical environment, and comprehensive data with regard to the participant flora, vegetation and landscape ecology. Similar prerequisites are also needed for the analysis of relationships between plant invasions and climate change. In Central Europe, favourable conditions exist for these studies in the eastern Hercynian mountains, where meteorological, geological and botanical data have been gathered for several centuries. These central European middle-mountains are exposed to two inseparable components of global change: climate change (temperature, sunshine duration) and regional change (pollution, precipitation). More than one-century-long measurements on Sniezska Peak confirm a continual climate warming (annual means some 0.6 C higher) and increasingly erratic precipitations. The final quarter of the 20th century has seen a continual increase of SO_2, NO_x and harmful pollutants which has caused an imbalance in plant and animal populations/communities, of which the most obvious manifestation is the catastrophic decline of coniferous forests. Alien plant populations are spreading in an explosive way on the summits of the middle-mountains, their successive penetration into native plant associations threatening the biogeographical status of national parks.

G. Visconti et al. (eds.), Global Change and Protected Areas, 289–299.

1. INTRODUCTION

The success of alien species in migrating beyond their primary area of distribution, called invasion, is affected by many factors, of which two distinctive categories should be distinguished: physical/abiotic and biotic. In plants, the key abiotic factors predetermining the establishment of their populations/communities are climate (temperature, precipitation, radiation and wind), nutrient availability (soil, bedrock but also air composition), and water status (moisture and water surface). The setting of these factors can be broadly valid for the whole Earth, their various combinations being reflected in spatial/regional variations and their changes being suitable for deterministic models [1].

Biotic factors, in contrast, differ even among individuals of the species. Genetic diversity, phenotypic plasticity, growth performance, life-cycles, strategy, energy efficiency and allocation, ecophysiological adaptability, interspecific interactions, disease resistance, defence mechanisms and migratory/dispersal potential, all predetermine the manifestation of a species population within a community. In a stable environment, with unchanging abiotic factors, the species assemblage of a community/ecosystem would reflect the successful combination of the individual genomes available within a given region. With time, the genomes are subjected to evolutionary changes, the process inherent to all biota [2]. This implies that even in an unchanging environment the community assemblage will change. Over centuries and millennia, the regional environments of the Earth undergo everlasting changes, and the species assemblages in regional communities/ecosystems/landscapes are adapting to new conditions by continual shifts in composition - including the invasions of better adapted specimens that have arrived successfully into a particular region.

A series of analyses about recent invasions have been carried out [3, 4, 5, 6]. Two kinds of approaches used in studying biological invasions were distinguished and compared [7]. (i) Deterministic ones, coming from the Lotka-Volterra formulations of population growth, that implicitly or explicitly assume a constant environment, and thus consider invasions as a result of genetic variation [8] or occasions when community resistance is overcome [9]. (ii) Individualistic ones, based on stochastic non-equilibrium models [10], that assume that species remain the same and their dynamic response follows changes of the environment. For a shorter period of a few decades, the latter approach with its less–untested assumptions seems to be more realistic, though the predictability of such a stochastic process for a larger region is very uncertain.

Mountains represent highly variable environments in both space and time. The altitudinal differentiation from harsh summits down to mild valleys, features the close accumulation of contrasting ecotopes. In close proximity, the communities are established composed of stress-tolerant biota, on the one hand, and of highly competitive species on the other, the frequent hazards enabling the occurrence of opportunistic species with high reproductive potential. A great variety of species exposed to dynamic environmental variability thus cradles microevolutionary processes and genetic diversification.

2. CENTRAL EUROPEAN MIDDLE-MOUNTAINS AND THEIR ECOSYSTEMS

South of the Fenno-Sarmatian Platform, an old system of vast uniform plains which covers much of northern and eastern Europe, the mountain ranges are situated of the Hercynian Platform, the large geological unit which forms western and central Europe's middle-mountains and river basins. The easternmost extension of this Hercynian platform is the Bohemian Massif. Spread over a length of about 1500 km, the Hercynian middle-mountains are represented by an archipelago of separate ranges, including (i) the easternmost Sudetes with Jeseníky, Snieznik/Kralický Sněžník and the topmost Giant Mts. (Krkonoše/Karkonosze/Riesengebirge), the Bohemian Forest with the neighbouring Bavarian Forest, Ore Mountains (Erzgebirge/Krušné hory), Czech-Moravian Highlands, Fichtelgebirge and Black Forest, (ii) the northern branch with Thüringer Wald to Harz and Ardennes, and (iii) the western branch with the Vogesen, Morvan and Central Massif [11, 12]. One of the major characteristics of this geological/geographical unit is that this slightly undulating relief of rounded summits and wide valleys, a result of Mesozoic and Tertiary denudation, contains comparatively few limestones and dolomites. The dominant crystalline schists and granites only seldom weather to form base-rich soils and provide a fertile medium required for abundant plant life. The topmost summits are formed by particularly hard rocks, such as gneiss and mica schists. Intensive remodelling of the landscape, uplifted to its present height in the late Tertiary, proceeded during the Pleistocene Ice Ages. The last glaciation was of the Alpine type, with independent glaciers in north- and east-facing valleys; the spectacular glacial cirques, cirque lakes, moraines and waterfalls are witness to this ancient era of glaciation. Periglacial cryogenic processes have created polygonal soils, cryoplanation terraces, frost cliffs, block-fields and tors, scattered over summit areas. Plio-Pleistocene retrograde erosion has formed non-graded river valleys with alternations of deep-cut segments and the accumulations of natural leveés and floodplains, suitable for the establishment of valley raised-bogs.

The prevailing climate in the Hercynian mountains is submaritime. Basic meteorological parameters are monitored by several summit stations, e.g. Feldberg [13], Szrenica [14], Sniezka [15] and Praděd [16]. Being exposed to rather strong western winds ([17], [18] , the solitary ranges in their topmost zone receive a high level of precipitation. However, the mountains lack perennial snow and permafrost. In winter, a major proportion of precipitation falls as snow and hoarfrost; snow lies on the ground from mid–October till mid–May. Strong winds differentiated by the dissected relief cause a very uneven deposition and secondary translocation of snow; for example, the depth of snow in February ranges between 0.2 m on wind-exposed summits, to 8-10 m on the sheltered slopes of glacial cirques and nivation hollows, and where avalanches can regularly occur.

The highest summit (Sněžka/Sniezka Peak: 1602 m) bears reference to the size of these middle-mountains, although they include a remarkable range of altitudinal vegetation belts or zones. These zones, under the current climatic conditions, range from mixed broadleaved woodland (*Quercus petraea, Q.robur, Carpinus betulus, Acer spp.*) at about 500 m a.s.l. and below, through mountain beech forests (*Fagus sylvatica, Abies alba, Acer pseudoplatanus*) up to 900-1000 m, through Norway spruce forests (*Picea abies, Sorbus aucuparia*) between 1000 and 1250 m, to krummholz stands (*Pinus mugo*) and alpine tundra established at higher than 1450 m. Alpine tundra includes numerous alpine and arctic phenomena, such as an alpine treeline, alpine grasslands (*Juncus trifidus*), vegetated polygonal structures (*Nardus stricta, Carex bigelowii, Festuca supina*), patterned mires (*Trichophorum cespitosum*), avalanches, snowpatches, landslides, screes, rockfalls and solifluction.

This general altitudinal zonation, however, is markedly altered by local relief, and soil and mesoclimatic factors. This is most apparent in the varying position of the timberline. A causal explanation for such variation has been suggested in the theory of anemo-orographic systems [11]. The theory explains the multiple and combined effects of mountain landforms and atmospheric factors on the composition, distribution and development of ecosystems, including the microevolution of biota. In the Hercynian ranges, the long valleys running west-to-east (windward funnel valleys) gather and streamline the flow of air currents from the foothills up to the accelerating summit plateau, thus creating a "nozzle" between two parallel ridges; to the east, southeast or northeast of each of these plateaux, several leeward vortex cirques occur. Air-borne particles, such as snow, water droplets, dust, seeds, and even pollutants are thus unevenly drifted and deposited across the surface of such mountains. In winter, high snow drifts regularly accumulate in the lee of the mountains, and become a source of frequent avalanches. Throughout the Postglacial period, nivation and avalanches have regularly disturbed the spontaneous succession towards forest. Nivation and avalanches are responsible for the diverse features of soils, and for the continual exposure of small outcrops of nutrient-rich rocks that have served as habitats for rare nutrient-demanding, and even calciphilous, plant species. The deposition of fertile dust and humus has contributed, moreover, to the development of species-rich tall-herb communities. All the species-rich communities are situated in the leeward vortex zones of anemo–orographic systems. For example, in Velká Kotlina Corrie, the largest glacial cirque of Jeseníky covering 1 km^2, about 350 vascular plants and 200 bryophytes have been recorded, including three endemic species [19]. In the Giant Mts., 37 endemic vascular plants occur in an area of about 16 km^2 [20], including the endemic woody species *Sorbus sudetica* [21]. On the whole, the summits and upland plateaux, exposed to the cooling effects of accelerated air currents, maintain a subarctic pattern, which includes exposed cliffs, cryogenic soils and string-and-flark mires, with a number of glacial relics of both cryptogamic and vascular plants, e.g., *Rubus chamaemorus, Salix lapponum, Pedicularis sudetica* ssp. *sudetica*. Being highly stabilised over many thousands of years, A-O systems have served as a favourable refuge, and even evolutionary cradle for some of the most remarkable species of the central European flora and fauna.

3. ENVIRONMENTAL CHANGES

During the Postglacial period, secular changes of forest composition had proceeded slowly across the altitudinal zonation. The highest treeline was reached in the Atlantic Period, when mixed fir-beech forests prevailed in the Hercynian mountains [22]. These shifts corresponded both to changes in the biogeographical distribution of plant species, and to variations in climate. Palaeoecological data suggest that in central Europe the establishment of a changed vegetation type has coincided with differences in mean air temperature of about $2^{o}C$ [23]. Similar temperature difference was derived from the manifestation of a different climax species in the altitudinal zonation of vegetation in the Alps [12]. In the Giant Mts., the altitudinal gradient of temperature is steeper (about $0.60^{o}C$ per 100 m); the altitudinal ecological range for the manifestation of climax species is therefore expected to be narrower (about $1.5^{o}C$). From this assumption we may expect major shifts in the composition of plant communities following a comparable temperature change.

More than one-century-long temperature measurements on Śnieżka Peak (the topmost summit of the Hercynian mountains) suggest continual warming. The geographical situation and altitudinal exposure of the Giant

Mts. are responsible for a number of climatic peculiarities of these mountains. In the summit area, the mean annual temperature is about 0.5 °C, and over the large upland plateaux situated above 1300 m temperatures below 0°C last about 5 months. An evaluation of measurements from the abovementioned peak showed that, since 1901, the mean annual air temperature has increased by about 0.6°C [24]. This fits with observations from the Carpathians, where an increase of 1°C has been detected since 1850 [25]. A decrease of sunshine duration is characteristic for Hercynian summits since 1940 [26], the highest decrease being observed in summer (Śnieżka by 2.0 hrs/year, Feldberg by 2.5 hrs/year), whereas in winter an increase of sunshine duration was apparent above 800 m. This decrease in sunshine is associated with an increase of cloudiness in central Europe, to some extent enhanced by the rapid increase of atmospheric pollution since the mid–seventies, and suggests a decrease of continentality in central European mountains.

The prevailingly strong westerly winds bring humid oceanic air and annual precipitation reaches 1600 mm. A deficit in precipitation, reaching to about 140 mm per year in the Sudetes, has been detected since 1970-1972 [27], and its distribution coincides with the increased concentration of SO_2. The effect of atmospheric pollution on the higher occurrence of so–called secondary types of circulation has been described as possibly resulting in erratic precipitation [28].

An analysis on almost 100 avalanche tracks occurring in the Giant Mts. [29] showed that between 1962/63 and 1998/99 the area affected by avalanches has increased by 20 %. However, the distribution of plant communities indicates that, within the observed 36 years, the largest avalanches ever observed have not occurred on the period of observation. A relationship between avalanche activity and pollution-induced deforestation was not confirmed.

4. AIRBORNE POLLUTION AND FOREST DECLINE

Atmospheric pollution markedly affects many of the Hercynian mountains, in particular, eastwards of the Black Triangle, an area of large power stations fuelled by sulphur-rich coal and extensive chemical industry situated in the region of northwestern Bohemia, southeast Saxony and southwestern Silesia where annual production of SO_2 of the power stations reaches 900 000 t. Monitoring of the chemical composition of the air, rainfall, snow and rime for nine years during the eighties showed that the average daily concentration of SO_2 ranged between 13 and 20 μg/m^3 [30].

The second-most abundant pollutant in the Sudetes are nitrates, their deposition in the Giant Mts. even reaching 470 μg/m^2/day [31].

The concentration of SO_2, if acting for a long period, becomes a dangerous source of stress for Norway spruce, the dominant tree species of the upper montane forest. Physiological changes in needles appear when SO_2 concentrations in the air reach 15 μg/m^3 [32].

The coaction of other pollutants [31] and environmental factors should also be taken into account in the damage and dying-off of trees. For example, heavy deposits of rime, which developed within 24 hours at 1000 m altitude, showed these features: pH 2.89, conductance 540 mS/l, SO_2 79.4 mg/l , NO_3 42 mg/l, Ca 6.6 mg/l, Cd 7.8 μg/l, Pb 1180 μg/l and Cu 87.7 μg/l. Combed out of the air by needles and twigs of coniferous trees, the horizontal precipitation of a fairly drastic chemical composition can, upon melting, flow down to the bottom of trees, intoxicating soil, and damaging roots and soil organisms.

At the present time, more than a half of the coniferous forests in the eastern Hercynian mountains are seriously stressed and/or disturbed by air deposition of pollutants [33, 34]. The obvious reduction in needles and foliage is accompanied by reduced radial increment, fructification and viability of seeds. The decline of ectomycorrhizal basidiomycetes is apparent, and infestations by insect and fungi pests have increased. Between 1965 and 1971, and 1976 and 1981, in stands of Norway Spruce, the larch bud moth *Zeiraephera diniana* expanded over an area of about 20,000 ha in the Ore and Giant Mts., respectively; bark beetle has been spreading in the Bohemian Forest since 1993; krummholz *Pinus mugo* has suffered from the caterpillars of *Neodiprion sertifer* and from the fungus *Scleroderis lagerbergii*.

The retreat of coniferous forests, accompanied by the disappearance of bryophytes and acidophilous forbs, has been followed by the explosive spread of perennial grasses (*Calamagrostis villosa, Deschampsia flexuosa*). This, together with the decrease of soil acidity, has inhibited the successful establishment of woody species [35]. Also the above-mentioned long-term "creative" function of the Sudetic A-O systems has dramatically changed because of air pollution. The increased flow rates of polluted air in the upper parts of windward valleys have affected the vitality of montane coniferous forests, damaged solitary trees at the timberline, and disturbed krummholz. In the leeward cirques where ash and aerosol particles are deposited, the decline of forests hs proceeded to a much larger extent, even at lower altitudes. Large-scale deforestation on steep slopes results in am imbalanced water discharge [36], a deforested landscape losing its cooling capability [37], as well as high floods possibly occurring (such as in the case of the Morava River in July 1997, the extent of flooding being clearly the largest in the past 1,000 years). For successful reafforestation substitute tree species are used, sometimes even of alien origin such as *Picea pungens*. At higher elevations, growth conditions for trees are severe, substitute plantations of krummholz *Pinus mugo* have been established below the original timberline.

5. RECENT PLANT INVASIONS

Physiognomically less striking, but more dangerous from the evolutionary viewpoint, appear to be the changes in herb invasions. Epidemiological models applied to explain species manifestation in patchy and dynamic environments [7, 38], distinguish between stratified and neighbourhood diffusion [39]. Stratified diffusion describes the occasional arrival of an alien species to a new area, while neighbourhood diffusion describes the adual spread outwards of these newly–occupied outposts. Both kinds of spreading are documented for the Hercynian Mts.

A recent example of a stratified diffusion is the invasive establishment of *Rumex longifolius* DC. in the Giant Mts. [40]. This species, native to Scandinavia and adapted to cold environments by its allocation of assimilates being spread over the whole season, was documented in the Hercynian mountains first in 1961 from the Giant Mts. and from the Bohemian Forest. In spite of the Alps, where a domestic origin for this weed has been suggested [41], invasion of the summit areas of both Hercynian ranges has proceeded rapidly. In 1986, the Ore Mts. first became invaded, and since 1991 spreading of the species below 700 m a.s.l. has been observed. Initially the species only occupied disturbed road and paths margins, from where it gradually penetrated into partly–disturbed mountain grasslands and pastures (i.e. neighbourhood invasion).

While the spreadof *Rumex longifolius* has progressed from colder sites downwards, most of the plant invasions observed have advanced from the foothills upwards. Mountain tracks and their margins (touristic paths and roads) perform an important network for the spread of invasive species to the very summits of mountains [42,43,44]. In some species altitudinal maxima are temporary (e.g. *Agropyron repens, Bellis perennis, Descurainia sophia*), whereas other invaders become successful on disturbed sites permanently (*Rumex alpinus, Carduus personata, Urtica dioica, Taraxacum officinale,* or exotics like *Heracleum mantegazzianum* [45], *Telekia speciosa*) and/or even penetrate the communities of alpine grasslands (*Senecio nemorensis, Cirsium arvense, Carduus heterophyla*), enter the forest understory (*Impatiens parviflora, Calamagrostis epigeios*) or invade wetlands (*Epilobium ciliatum*). Most of these invaders are specific to nitrogen-rich sites [46].

Their successful establishment is surprising, especially in the tundra zone where the low availability of nitrogen is supposed [47].

Tundra communities also undergo important changes due to the expansion of grasses, namely *Deschampsia flexuosa* and *Anthoxanthum alpinum* [48,49]. The spread and long-term performance of these species in tundra grasslands was proved by experimental fertilisation by nitre [50]. An evaluation of climatic parameters on the growth of grasses has been controversial, the development of *Deschampsia flexuosa* being correlated to precipitation while shoot formation of *A. alpinum* was found to be controlled by temperature [51]. This implies that the spreading of a species at higher elevations of the Sudetes is most probably related to the modified availability of various forms of nitrogen. However, it is difficult to separate the effect of a high deposition of airborne nitrates from that of denitrification being accelerated due to increasing temperatures.

The invasions of alien species bring about potential threads to enable evolutionary processes. The emergence of hybrids between native and introduced species and introgressants has been observed in the Giant mountains between *Viola lutea* subsp. *sudetica*, a native species of mountain grasslands, and *V.tricolor*, a species introduced in the mid 1970s [52]: the hybrids have invaded plant communities where their parents have been previously unsuccessful. In some cases, hybridization may result in a decrease of diversity caused by genetic assimilation of a rare population by a numerically larger one, the loss of locally–adapted populations, or due to outbreeding depression [53]. Such processes constitute serious conservation threats for the maintenance of the unique biological diversity of protected areas in the Central European middle-mountains.

6. REFERENCES

Adler W., Verh. zool.-Bot. Ges. Österreich 129 (1992) 153-158.

Carter R.N. and S.D. Prince, Nature 293 (1981) 644-645.

Chapin III F. S., Nature 377 (1995) 199-200.

Cliff A.D., P.Haggett, J.K. Ord and G.R.Versey, Spatial diffusion. Cambridge,1981.

Di Castri F. et al., Biological invasions in Europe and the Mediterranean Basin. Kluwer Academic Publ., Dordrecht, 1990

Dostálek J., Příroda 10 (1997) 159-182

Drake J.A. et al., Biological invasions: A global perspective. John Wiley and Sons, Chichester, 1989.

Drukman I., K. Migala and M.Sobik, In K. Migala and J.Pereyma (eds.): Climatological aspects of environmental protection in mountain areas. Práce Instytutu geograficznego, ser.C meteorologia i klimatologia, 4 (1997) 67-73.

Dubicka M., In K. Migala and J.Pereyma (eds.): Climatological aspects of environmental protection in mountain areas. Práce Instytutu geograficznego, ser.C meteorologia i klimatologia, 4 (1997) 31-41.

Dubicki A. & K. Tomczyński, In J. Sarosiek and J.Štursa (eds.): Geoekologiczne problemy Karkonoszy, tom I, Poznań, 1998, pp. 125-132.
Dubicki A., Ecological Engineering 3 (1994) 291-298.
Ellenberg H. et al., Zeigerwerte von Pflanzen in Mitteleuropa, Scripta geobotanica, Göttingen, 1992.
Fisher R.A., Ann. Eugen. 7 (1937): 355-369.
Glovicki B., In J. Sarosiek and J.Štursa (eds.): Geoekologiczne problemy Karkonoszy, tom I, Poznań, 1998, pp. 117-123.
Havlik D., Landesanhalt für Umweltschutz Baden-Württemberg, Karlruhe (1982): 148-212.
Hengenveldt R., J.Biogeogr. 15 (1988) 819-828
Herben T., F.Krahulec, V.Hadincová and S.Pecháčková, Functional Ecology 9 (1995) 767-773.
Hošek J. and R.Kaufman, Celková atmosférická depozice ekologicky významných látek do lesních ekosystémů Krkonošského národního parku. AGNOS Hořovice, 1995.
Houghton J., Global warming: the complete briefing. Lion Publishing, Oxford, England, 1995.
Husáková J., Silva Gabreta 1 (1996) 115-121
Jeník J., Alpinská vegetace Krkonoš, Králického Sněžníku a Hrubého Jeseníku 1961. Nakladatelství Československé akademie věd, Praha, 1961.
Jeník J., In K. Migala and J.Pereyma (eds.): Climatological aspects of environmental protection in mountain areas. Práce Instytutu geograficznego, ser.C meteorologia i klimatologia, 4 (1997) 9-21.
Jeník J., L. Bureš & Z.Burešová, Folia Geobot. Phytotax. 15 (1980) 1-28.
Klimeš L. and J. Klimešová, Preslia 63 (1991) 245-268.
Kovanda M., Čs. ochrana prírody 2 (1965) 47-62.
Krahulcová A., F.Krahulec and J.Kirschner, Folia Geobot.Phytotax. 31 (1996) 219-244.
Kubíková J., Vegetatio 93 (1989) 101-108.
Kubínová D. and F. Krahulec, Opera concortica 34 (1997) 79-89.
Kwiatkowski J., and T. Holdys, In Jahn A. (ed.) Karkonosze polskie, Ossolineum, Wroclaw/Warszawa, 1985, pp.85-116.
Lorenc H., Meteorologia 19 (1994)1-50.
Ložek V., Příroda ve čtvrtohorách. Academia Praha, 1973.
Málková J., J.Malinová and H.Oslejšková, Opera concortica 34 (1997) 105-132
Materna J. and V.Mejstřík, Zemědělství a lesní hospodářství v oblastech se znečištěným ovzduším. SZN, Praha, 1987.
Mayr E., In J. Huxley, A.C.Hardy and E.B.Forf (eds.) Evolution as a process, London, 1954, pp.157-180.
Migala K. et al., In Fischer Z. (ed.) Problemy ekologiczne wysokogórskiej czesci Karkonoszy, Intytut ekologii PAN, Warszawa,1995, pp.51-78.
Obrebska-Starklowa B. et al., In K. Migala and J.Pereyma (eds.): Climatological aspects of environmental protection in mountain areas. Práce Instytutu geograficznego, ser.C meteorologia i klimatologia, 4 (1997) 23-26.
Ozenda P., Die Vegetation der Alpen. Gustav Fischer Verlag, Stuttgart/New York, 1988.
Pyšek P. et al. (eds.) Plant invasions: general aspects and special problems. SPB Academic Publishing, Amsterdam, 1995.
Pyšek P., M.Kopecký, V.Jarošík and P.Kotková, Diversity and Distributions, 4(1998) 9–16.
Rejmánek M. and D.M. Richardson, Ecology 77 (1996) 1655-1661.
Rieseberg L.R., In D.A.Falk and K.E.Holsinger [ed.]: Genetics and conservation of rare plants, Oxford University Press, Oxford, 1991.

Ripl W., J. Pokorný, M. Eiseltová and S. Ridgill, In M. Eiseltová (ed.): Restoration of lake ecosystems a holistic approach, Slimbridge, 1994, pp. 16-35.

Rybníčková E.and K. Rybníček, In: Berglund B.E.et al. (eds.): Palaeoecological events during the last 15 000 years, Wiley, Chichester, 1996, pp.473–505.

Schulze E.-D., O.L.Lange . & R.Oren (eds.): Forest Decline and Air Pollution. Ecological Studies 77, Springer Verlag, Berlin/Heidelberg, 1989.

Soukupová L., M. Vosátka and O.Rauch, In J. Sarosiek and J.Štursa (eds.): Geoekologiczne problemy Karkonoszy, tom I, Poznań, 1998, pp. 307-312.

Šourek J., Květena Krkonoš. Academia Praha,1969.

Spusta V. and M.Kociánová, In J. Sarosiek and J.Štursa (eds.): Geoekologiczne problemy Karkonoszy, tom I, Poznań, 1998, pp. 75-79.

Streseman E.and E.Nowak, J.Ornitothol. 99 (1958) 243-296.

Štursa J., Annales silesiae 18 (1988) 224-232.

Štursová H. and M. Kociánová, Opera concortica 32 (1995) 69-72.

Štursová H., Opera concortica 22 (1985) 79-120.

Tejnská S. and J. Tejnský, Meteorol.zprávy 11 (1958) 62-66.

Whittaker R.H., Biol.Rev.42 (1967) 207-264.

Can Testate Amoebae (Protozoa) and Other Micro-Organisms Help to Overcome Biogeographic Bias in Large Scale Global Change Research?

EDWARD A.D. MITCHELL, DANIEL GILBERT, ALEXANDRE BUTLER, PHILIPPE GROSVERNIER, CHRISTER ALBINSSON, HÀKAN RYDIN, MONIQUE M.P.D. HEIJMANS, MARCEL R. HOOSBEEK, ALISSON GREENUP, JONATHAN FOOT, TIMO SAARINEN, HARRI VASANDER, JEAN-MICHEL GOBAT

Department of Plant Ecology, Institute of Botany, University of Neuchâtel, Neuchâtel, Switzerland: Testate amoebae analyses for all sites, vegetation analyses for the Swiss site and data analyses.

Laboratoire de Biologie Comparée des Protistes. URA CNRS 1944, Université Blaise Pascal Clermont-Ferrand II, France: Analyses of micro-organisms.

Department of Environmental Sciences, Wageningen Agricultural University, The Netherlands: Water chemistry analyses and vegetation analyses for the Dutch site.

Department of Animal and Plant Sciences, University of Sheffield, UK: vegetation analyses for the British site.

Department of Natural Sciences, Kalmar University, and Department of Plant Ecology, Uppsala University, Sweden: vegetation analyses for the Swedish site.

Department of Ecology and Systematics and Department of Forest Ecology, University oj Helsinki, Finland: vegetation analyses for the Finnish site.

Key words: testate amoebae, micro-organisms, vegetation, biogeography, gradient, elevated CO_2.

Abstract: To monitor global change, large scale long term studies are needed. Such studies often focus on vegetation, but most plant species have limited distribution areas. Micro-organisms by contrast are mostly cosmopolitan in their distributions. To study the relationships between organisation groups, we analysed the testate amoebae (Protozoa), vegetation, and water chemistry of five Sphagnum peatlands across Europe. Inter-site differences were more pronounced for the vegetation than for testate amoebae species assemblage. Testate amoebae represent a useful tool in multi-site studies and environmental monitoring of peatlands because: 1) the number of species is much higher than for plants, 2) most species are cosmopolitan and are therefore less affected by biogeographical distribution patterns than plants; thus differences in testate

G. Visconti et al. (eds.), Global Change and Protected Areas, 301–310.

amoebae assemblages can be interpreted primarily in terms of ecology, 3) testate amoebae can be used to analyse and monitor small scale (cm) gradients that play a major role in the functioning of peatland ecosystems. We further studied the effect of elevated $C0_2$ on microbial communities in the same peatlands. Elevated $C0_2$ increased the biomass of heterotrophic bacteria and decreased the biomass of medium size protozoa (mostly small testate amoebae). These effects suggest changes in community functioning that may have feedback effects on other components of the ecosystem.

1. INTRODUCTION

To monitor global change, large scale long term studies are needed. Such studies often focus on vegetation dynamics, but even within Europe, most plant species have limited distribution areas. Micro-organisms by contrast are in most cases cosmopolitan in their distributions and may be useful biomonitors. In conservation biology there is a growing interest in relationships in species diversity of high- order taxa, but the results are highly variable [1, 21. Studies of species composition among such taxa could reveal to which degree they are dependent on one another and to the same or different environmental variables, and help to explain the variability in species diversity. Only few studies of community composition include species from different high-order taxa. Furthermore, although most global change research topics are related to belowground processes such as plant growth, litter decomposition or methane emissions from peatlands, soil organisms still represent only a marginal part of the research effort.

Peatlands contain 20-30 % of the world's soil organic carbon (SOC) [3] and are nearly continuous carbon sinks. Therefore any modifications of their functioning could have important feedback effects on global carbon cycle and on global warming. Hydrodynamics and hydrochemistry are determinant factors for peatland development [4]. The relationships to environmental gradients are well known for the vegetation [5], and especially for Sphagnum [6, 7, 8], and for the animal communities, especially invertebrates [9]. These different groups of organisms respond to - and can be indicative for - these parameters. However, less is known about the degree to which the relationship between these structuring groups are consistent across major climatic gradients and to which degree they react to environmental changes. Sphagnum peatlands are extreme environments for the soil fauna, being acidic, nutrient-poor and mostly wet [10]. Thus many groups of soil organisms are absent, litter decomposition is slow and is due mostly to the action of micro-organisms. The rising atmospheric concentration of CO_2

may influence both plant production and litter decomposition [11] and therefore the rate of peat accumulation.

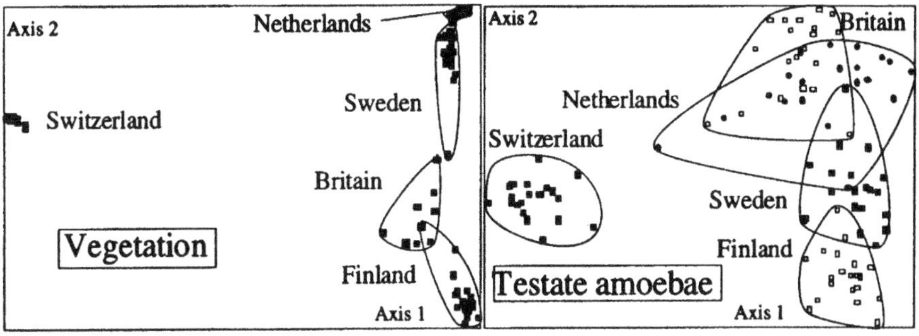

Fig.1: DCA scatter diagram (axes 1 and 2) of vegetation and testate amoebae communities in five Sphagnum peatlands in Europe.

Plant production is also affected by the physico-chemical characteristics of the soil and the availability of nutrients both of which are in part influenced by micro-organisms. Soil micro-organisms and the processes for which they are responsible are constrained by energy (carbon) input in the soil which is likely to increase as CO_2 concentrations in the atmosphere rise. Elevated CO_2 may influence micro-organisms indirectly either through increased and/or change in the quality of above and below-ground production and exudation.

To illustrate the importance of biogeographical bias and to show how micro-organisms may be useful tool for global change research, we present:

1. a study on the vegetation, testate amoebae and water chemistry in five Sphagnum peatlands across Europe,
2. a study on the effect of elevated CO_2 on microbial communities in the same five peatlands.

2. METHODS

Study sites

Five Sphagnum peatlands were chosen to be representative lawn communities in ombrotrophic or near-ombrotrophic peatlands in the major climates where Sphagnum peatlands occur in Europe.

1. Boreonemoral (Southern Sweden) (57.5°N, 14.3°E). The Swedish site, (Kopparás) is weakly minerotrophic, but smaller spots are ombrotrophic. The water supply are both soligenous and, to a lesser extent, topogenous. The mire surface slopes faintly (I: 164) towards SSW. The lawn

vegetation at the studied part of the mire is dominated by Eriophorum angustifolium, Calluna vulgaris, Andromeda polifolia, Narthecium ossifragum or Scirpus caespitosus. Predominant peat mosses are Sphagnum magellanicum, S. papillosum and S. rubellum.

2. Continental-boreal (Ilomantsi, eastern Finland) (62.47'N, 30.56'E). The Finnish site (Salmisuo) is weakly minerotrohic, but small ombrotrophic spots exist. The water supply is both soligenous and, to a lesser extent, topogenous. The vegetation is dominated by graminoids, such as Eriophorum vaginatum and Carex pauciflora and, to a lesser extent, by dwarf shrubs, such as Andromeda polifolia, Empetrum nigrum, Chamaedaphne calyculata and Vaccinium oxycoccos. On lawn communities, the dominant peat mosses are Sphagnum balticum and S. papillosum.
3. Nemoral, strongly Atlantic (North West England) (54°N, 1°W). The British site (Roudsea) has been drained but a secondary vegetation is well established an is dominated by Eriophorum vaginatum, Scirpus caespitosus, Erica tetralix and Sphagnum papillosum. The site. Most of the surface is covered by lawn communities with isolated hollows.
4. Nemoral, mild Atlantic (the Netherlands) (52.50'N, 6.40'E). The Dutch site (Dwingelo) is a small (150×50m^2) peat area in the State Forestry Dwingeloo, province of Drenthe from which peat sods were extracted for the use of the BERI experiments in Wageningen. Up to 1955 the site was used by local fanners for peat-cutting. The vegetation nowadays consists of a mosaic of secondary successional stages including pools, carpets and hummocks. Common species are Erica tetralix, Vaccinium oxycoccos and Sphagnum magellanicum.
5. Subalpine (Jura, Switzerland) (47°N, 7°E). The Swiss site (La Chaux-des-Breuleux) lies at an altitude of 1000 m a.s.l. The mire was drained and peat was mined until the end of World War II.

A bog vegetation has re-established through a natural regeneration process [12] and a mosaic of lawn, hummocks and depressions is now well developed with Eriophorum vaginatum, Carex nigra and Vaccinium oxycoccos. The dominant mosses are Sphagnum fallax and Polytrichum strictum.

Comparative study of vegetation, testate amoebae and water chemistry.

In each site 20 plots (1 -m diameter circles) were randomly selected in a homogeneous patterned part of the mire (10 in Britain). The vegetation was analysed at the beginning of the 1996 growing season using the point-intercept method [13]. We chose a homogeneous subplot (35×22.5 cm, grid intervals 2.5 cm, adding up to 150 points).

Specific frequencies of all species were calculated. Authorities for plant species follow Corley et al. [14] for mosses, Grolle [15] for hepatics, and Tutin et al. [16] for vascular plants.

Testate amoebae shells were extracted from Sphagnum mosses (3-5 cm depth) sampled at the beginning of the 1996 growing season from all plots of the five sites. The extraction and identification of testate amoebae shells follows Warner [17].

Water samples were collected from all plots at the five sites using millipore soil moisture samplers (Rhizon, Eijkelkamp B.V., NL).

The samplers were inserted in the moss carpet as close as possible to the water table, and connected to pre-evacuated glass bottles. The water samples were taken at the beginning of the 1996 growing season and analysed for DOC (dissolved organic carbon), pH, and major cations and anions.

The data were analysed in two ways:

Vegetation and testate amoebae data were first submitted to separate detrended correspondence analyses (DCA, Program CANOCO, [18]) to show the relationships between the sites and samples. Both analyses were carried out using the same option in the program: detrending by 26 segments, no data transformation.

Mantel permutation tests were performed between pairs of similarity matrices to test the overall relationships between all combinations of the following data sets:

1. deep-rooted vascular plants (Carex spp, Eriophorum spp., Erica tetralix, Calluna vulgaris, Empetrum nigrum, Scirpus caespitosus, Narthecium ossifragum and Rhychospora alba, 11 species),
2. mosses, lichens and liverworts (hereafter cryptogams, 12 species),
3. testate amoebae,
4. water chemistry.

We used the Steinhaus similarity index for the amoeba and vegetation matrices (transformed in a distance matrix for comparison) and the Euclidean distance for the water chemistry matrix [19]. All computations were done using the R-package for data analysis [20]. The permutations were restricted within five blocks representing the sites. Testate amoebae and plant species occurring only in one sample were deleted.

CO_2 enrichment experiment.

We studied the effect of elevated CO_2 on microbial communities in the same five Sphagnum peatlands. Mini-FACE (small Free Air CO_2 enrichment systems, [21]) were used to raise atmospheric CO_2 concentrations to 560 ppm in 1-m-diamter plots starting in spring 1996. At the end of the experiment (end of summer 1998) 10-20 Sphagnum samples 5 cm in length were taken in all plots of the five sites. The samples were stored in glutaraldehyde solutions (2% final concentration) kept in the dark at 4°C and analysed for heterotrophic bacteria, cyanobacteria, micro-algae, heterotrophic flagellates, ciliates, testate amoebae, nematoda, rotifera and fungi. Biovolumes of all observed organisms were estimated assuming geometrical shapes and converted to carbon biomass using conversion factors from the literature following Gilbert et al. [22].

3. RESULTS

Comparative study of vegetation, testate amoebae and water chemistry.

A total of 44 plant and 54 testate amoebae species were recorded. The different sites clearly differ in vegetation composition. Most plant species were present in only one or two sites. Each site appeared to have either several characteristic testate amoebae species, or different frequencies of some common species, as compared with other sites. However, inter-site differences were more pronounced for the vegetation than for testate amoebae species assemblage (Figure 1). Mantel tests with limited permutations revealed significant relationships between the water chemistry and deep-rooted vascular plants ($r = 0.25$, $P = 0.04$) and between the water chemistry and testate amoebae ($r = 0.\ 18$, $P = 0.02$). Furthermore, the relationship between the cryptogams and testate amoebae data sets was marginally significant ($r = 0.67$, $P = 0.07$). For the other three pairs of data sets, the relationships were clearly not significant.

CO_2 enrichment experiment.

In terms of biomass, heterotrophic bacteria, fungi, testate amoeba and micro-algae were the dominant groups of micro-organisms in the control plots (ambient CO_2). Rotifera, nematoda, cyanobacteria, ciliates and flagellates accounted for only a low proportion of the microbial biomass. Under elevated CO_2 the biomass of five groups increased: bacteria (average CO_2 effect over the five sites: + 32 %), cyanobacteria (nano cyanobacteria: + 96 %, micro-cynaobacteria: + 18), algae (micro-algae: + 10%; nano-algae: + 7 %), small protozoa (heterotrophic flagellates and small ciliates; + 44 %) and nematoda (+ 58 %). The biomass of five groups was lower in the elevated CO_2 plots medium-size protozoa (large ciliates and small testate amoebae; -33 %), large protozoa (large testate amoebae; - *15* %), fungi spores (-29 %) fungi mycelium (- 10 %), rotifera (- 29 %). Bacterial density and biomass was on average higher in the C02 enriched plots in all five sites. Other groups reacted differently depending on the sites. The treatment effect was significant for two groups (ANOVA with additional model: treatment, site, water table depth, site x treatment): bacteria and medium size protozoa (i.e. the large ciliates and small testate amoebae), and marginally significant for nematoda.

4. DISCUSSION

Comparative study of vegetation, testate amoebae and water chemistry.

Important floristic differences exist between the five sites, but these are in part due to plant distribution patterns. Many of the species occurring only in one or two sites are absent or rare in the other countries. By contrast differences in testate amoebae assemblages are less pronounced but reflect better the ecological differences between the sites. To some extent the different organism groups behave similarly: the Swiss site differs from the others both in amoebae composition and in vegetation. The reason might be its successional status. On the other hand, in accordance with suggestion from other studies [1, 2] we find that organism groups are not well correlated, the inter- site differences being more marked for the vegetation than for the testate amoebae communities, and the relative positions of the sites on the first two axes of the DCA being different. These differences may be due to biogeography (large scale plant distribution patterns are much more marked than for amoebae), phenology (the amoebae are able to encyst during unfavourable periods such as summer drought whereas the plants can not) and/or to the existence of vertical gradients of abiotic variables within the mosses and underlying young peat. These observations support the use of testate amoebae in experimental mires studies which cover a wide geographical area. Of particular interest is the greater diversity of testate amoebae as compared to cryptogams: 54 species versus 12 species in our data set and a higher diversity at all sites.

The Mantel tests reveal interesting relationships between the different data subsets. Deep-rooted (i.e. down to about 30 cm) vascular plants were significantly related to the water chemistry, but cryptogams were not. Testate amoebae were related to the water chemistry and also to the mosses in which they live, but not to plants rooted as much as 40 cm below the surface. Thus, testate amoebae reflected the chemistry of the water and to a lesser extent the botanical composition of the moss carpet in which they lived. Deep-rooted plants and mosses may well respond to different factors and can indeed be considered as two different functional groups. Such differences were more marked in relatively dry situations. This was particularly obvious in the Swiss site where the deep-rooted plants (i.e. Eriophorum vaginatum, E. angustifolium, Carex nigra and C. rostrata) had their roots at the interface between the original peat partly mineralised when the peatland was drained and the younger overlying peat. Soil chemical properties at this depth differed from those closer to the surface. These results bring to light an aspect of peatland ecosystems that is often overlooked: the existence of vertical as well as horizontal ecological small scale gradients below the surface. This relates to the crucial distinction between ombrotrophic (rain water fed) and minerotrophic (influenced by surrounding mineral soil) conditions of mires: Deeply rooted species may be exposed to minerotrophic conditions while the bryophytes at the surface may undergo ombrotrophic conditions. For the plants this is a excellent case of life form niche separation [23], enabling resource acquisition from different sources. At or near the surface, the importance of small scale ecological gradients for the growth of mosses has recently been established in Swiss peatlands [12, 24, 25], but what happens below the surface is certainly also important for the functioning of peatland ecosystems.

$C0_2$ enrichment experiment.

Elevated CO_2, may have caused a increase in sugar exudation by plants from which heterotrophic bacteria and their predators benefited. The reaction of predator organisms to elevated CO_2, reflected their feeding habits. There was a trend for an increase in the biomass of nematoda and the smaller size class of protozoa (<500 μm^3) under elevated CO_2 suggesting these organisms could benefit from the higher bacterial biomass available. By contrast the biomass of larger predators decreased and their relative importance in the microbial community was lower under elevated CO_2, suggesting these organisms could not benefit from the increased biomass of bacteria and smaller predators. Testate amoebae feed on a wide range of organisms including fungi, pigmented organisms and small metazoa but bacteria are probably not an important part of the diet except for the smaller species [26]. The observed changes suggest changes in community functioning that may have feedback effects on other component of the ecosystem. However, further studies are needed to get insight on the processes involved (e.g. abundance and activity of micro-organisms are not necessarily related, trophic relationships are still largely unknown or at best approximate [27]).

In a global change perspective, a good understanding of below-ground processes is crucial for predictions of the response of peatlands to environmental change [28]. The modelling of peatland- atmosphere exchanges is crucial for global change scenarios. Recognising vertical gradients in peatlands may help to better model their functioning. Given the diversity of soil organisms and the general species- or group-specific response of organisms to elevated $C0_2$ and other environmental perturbations, we need to identify which species or group react to changes and what consequences these modifications might have on belowground processes. It is generally recognised that population and community level phenomena exert a strong influence on ecosystem structure and function, but we know little about how ecological processes operate in belowground systems or how they influence relationships at higher scales [29]. The observed change in microbial community structure suggests CO_2 enrichment might have affected the functioning of microbial communities. The reaction of one group (bacteria) fits into our understanding of soil functioning but the reaction of another groups (medium size protozoa) can not be fully explained. This illustrates some of the challenges soil microbiologists are presently facing. To get further insight into soil processes, more attention is needed on microbial communities. Indeed in many aspects, soil represents a kind of last frontier in ecology [27]. Clearly, if we want to establish reliable predictive models of ecosystem functioning in a high CO_2environment, soil processes and organisms can no longer be considered as a "black box".

So, can testate amoebae (Protozoa) and other micro-organisms help to overcome biogeographic bias in large scale global change research? Distribution patterns suggest testate amoebae indeed represent a useful tool in multi-site studies but more work is needed to confirm the usefulness of micro-organisms in global change research although these first results are encouraging.

5. ACKNOWLEDGEMENTS

This work has been done as part of the BERI project (Bog Ecosystem Research Initiative, project NR. ENV4-CT95-0028), funded by the European Community. The Swiss part of the BERI project was funded by the OFES (Swiss Federal Office for Education and Science, Project NR. 95.0415). BERI is part of TERI (Terrestrial Ecosystem Research Initiative) and was officially recognised to contribute to the Core Research Programme of GCTE (Global Change & Terrestrial Ecosystems), a core project of IGBP (International Geosphere - Biosphere Programme). We thank Daniel Borcard for statistical advice.

6. REFERENCES

Andrus R.E., Can J. Bot. 64 (1986) 416-426.
Borcard D., C. Vaucher-von Ballmoos, Ecoscience 4 (1997) 470-479.
Bridgham S.D., J. Pastor, J.A. Janssens, C. Chapin, T. Malterer, Wetlands 16 (1996) 45-65.
Buttier, Ph. Grosvemier, Y. Matthey, J. Appl. Ecol. 35 (1998) 800-810.
Buttler A., Vegetatio 103 (1992) 113-124.
Clymo R.A., J. Ecol. 61 (1973) 849-869.
Corley M.F.V., A.C. Crundwell, R. Dúll, 0. Hill, A.J.E. Smith, J. Bryol. 11 (1981) 609- 689.
Curtis P.S., E.G. O'Neill, J.A. Teeri, D.R. Zak, K.S. Pregitzer, Plant Soil. 165 (1994) 1-6.
Gilbert D., C. Ambiard, G. Bourdier, A.-J. Francez, Microb. Ecol. 35 (1998) 83-93.
Gilbert D., C. Amblard, G. Bourdier, A.-J. Francez, E.A.D. Mitchell, Année Biol. in press.
Glaser P.H., J.A. Janssens, D.l. Siegel, J. Ecol. 78 (1990) 1021-1048.

Gobat J.-M., M. Aragno, W. Matthey. Le Sol Vivant. Presses Polytechniques et Universitaires Romandes, Lausanne, Switzerland, 1998.
Gorham E., Ecol. Aplic. 1 (1991) 182 - 195.
Grolle R., J. Bryol. 12 (1983) 403-459.
Grosvemier Ph., Y. Matthey, A. Buttler, In: B.Wheeler, S.Shaw, W.Fojt, Robertson A, (Eds.) The restoration of temperate wetlands. J Wiley & Sons, Chichester UK, 1995, pp. 437-450.
Grosvemier Ph., Y. Matthey, A. Buttler, J. Appl. Ecol. 34 (1997) 471-483.
Grubb P.J., Biol. Rev. 52 (1977) 107-145.
Lee J.A., J. Ecol. 86 (1998) 1-12.
Legendre P., A. Vaudor, The "R" package - Multidimensional analysis, spatial analysis. Département de sciences biologiques, Université de Montréal. 1991.
Legendre P., L. Legendre, Numerical ecology. Second English Edition. Elsevier, Amsterdam, 1998.
Miglietta F., M.R. Hoosbeek, J. Foot, F. Gigon, M. Heijmans, A. Peressotti, T. Saarinen, N. van Breemen, B. Wallen, Environmental Monitoring and Assessment (in press).
Norby R.J., M.F. Cotrufo, Nature 396 (1998) 17-18.
Prendergast J.R., B.C. Eversham, Ecography 20 (1997) 210-216.
Prendergast J.R., R.M. Quinn, J.H. Lawton, B.C. Eversham, D.W. Gibbons, Nature 365 (1993) 335-337.
Rydin H., Adv. Bryol. 5 (1993) 153-185.
ter Braak C.J.F., CANOCO - a FORTRAN program for canonical community ordination (version 2. 1). Microcomputer power, Ithaca, New York. 1988-1992.
Tutin T.G., V.H. Heywood, N.A. Burges, D.H. Valentine, S.M. Walters, D.A. Webb, Flora europaea. 5 vol. Cambridge University Press, Cambridge. 1964-1980.
van Breemen N., Trends Ecol. Evol. 10 (1995) 270-275.
Wamer B.G., In: Wamer B.G., (Ed.) Methods in quatemary ecology. Geoscience Canada, reprint series S. Geological association of Canada, St Johns, 1990, pp 65-74.

Effects of Elevated Atmospheric CO_2 and Mineral Nitrogen Deposition on Litter Quality, Bioleaching and Decomposition in A Sphagnum Peat Bog.

A. SIEGENTHALER[1], E.A.D. MITCHELL[1], E. VAN DER HEIJDEN[2], A. BUTTLER[1], PH. GROSVERNIER[1] AND J. M. GOBAT[1]
[1]*Laboratory of Plant Ecology, Institute of Botany, University of Neuchâtel, Switzerland.*
[2]*Department of Plant Biology, University of Groningen, Haren, The Netherlands.*

Key words: decomposition, litter quality, elevated CO_2, nitrogen deposition, bioleaching, peat bogs.

Abstract: A brief overview of an attempt to link the effect of elevated CO_2 and nitrogen deposition on litter quality and decomposition in a *Sphagnum* peat bog is given. Litter of three common species (*Eriophorum vaginatum*, *Polytrichum strictum* and *Sphagnum fallax*) was collected from field plots after two years of pre-treatment in two parallel experiments: a) Elevated atmospheric CO_2 experiment, b) mineral nitrogen fertilisation experiment. The litters were put into litterbags, leached and inserted into field plots for 3 months, where they decomposed under specific treatment. Distinction between effects of initial litter quality and decomposition on mass loss in the bioleaching and/or in field decomposition process could be tested using a particular set-up in which cross-effects of pre-treatment and treatment were considered.

1. INTRODUCTION

Twenty times more CO_2-C is cycled annually along the terrestrial photosynthesis-decomposition pathway than the annual net addition to the atmosphere. Small changes in net primary productivity or in decomposition of soil organic carbon (SOC) could significantly influence the net increase of atmospheric CO_2. Because elevated CO_2 may strongly influence (1) the net primary productivity and specific composition of natural vegetation, and (2) the chemical composition of plant material, and therefore the decomposability of plant litter, strong feedback into the SOC pools are to be

G. Visconti et al. (eds.), Global Change and Protected Areas, 311–321.

expected [1]. Peatlands contain 20-30 % of the world's soil organic carbon (SOC) [2]. If growing, they constitute an almost continuous carbon sink and are therefore a key element of the carbon cycle. Any modification to their functioning could have significant feedback effects on global warming.

Under growth limiting conditions (nutrients, water), and according to the carbon-nutrient balance hypothesis, elevated atmospheric CO_2 would either directly or indirectly increase carbon based secondary or structural compounds (CBSSC) as well as total non-structural carbohydrates (TNC) of growing plants [3].

The opposite effect could be expected for higher mineral nitrogen deposition. These direct or indirect hypotheses were extended and referred to by [4] as a 'carbon supply model of secondary plant metabolism' or as the 'amino acid diversion model of secondary plant metabolism', respectively. There are few answers to how these changes in living plants affect the litter quality and the ensuing decomposition process [5]. We therefore set up an experiment to try to link litter quality and decomposition aspects in a Sphagnum peat bog in a perspective of climate change.

Litter of three common species (*Eriophorum vaginatum*, *Polytrichum strictum* and *Sphagnum fallax*) was collected from field plots after two years of pre-treatment in two parallel experiments: a) Elevated atmospheric CO_2 (560 ppm) / ambient CO_2 (360 ppm), b) mineral nitrogen fertilisation experiment: 30 [kg/ha/a] NH_4NO_3 / 0 (control). The litters were put into litterbags, leached under artificial rain and inserted into field plots for 3 months, where they decomposed under a specific treatment.

This research was part of the EU BERI project (Bog Ecosystem Research Initiative).

2. METHODS

2.1 Study site

The field experiment has taken place in an ombrotrophic peat bog in La Chaux-des-Breuleux, in the Swiss Jura, (47°15'N, 6°55'E, alt: 1000 m) from 27.3.1998 to 1.9.1998. The mire was drained and the peat was exploited until the end of Word war II. The mean daily temperature in the warmest month is 15°C and -5°C in the coldest month. Annual precipitation is 1390 mm and snow covers the site 80 to 120 days a year. Nitrogen deposition is 10-30 [kg/ha/a]. The vegetation is dominated by *Eriophorum vaginatum, Carex nigra, Vaccinium oxycoccos.* The dominant mosses are *Sphagnum fallax, Polytrichum strictum, Aulacomnium palustre* [6].

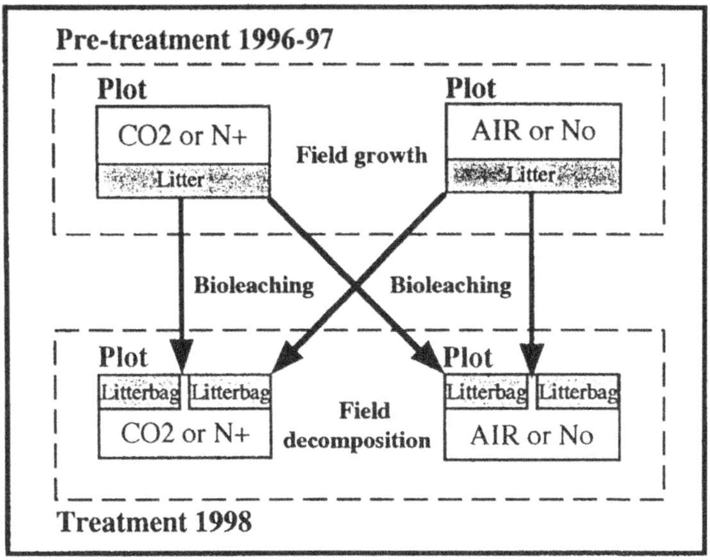

Figure 1: experimental set-up showing a) cross-effects between pre-treatments and treatments, b) the bioleaching and field decomposition processes. One unit is composed of a (pre-)treatment and its control. (Pre-)treatment were assigned randomly to a plot.

2.2 Experimental set-up

Litter from *Eriophorum vaginatum*, *Polytrichum strictum* and*Sphagnum fallax* plants produced during the previous two seasons of pre-treatment (1996 and 1997) were collected from the plots. About 0.3 g of litter material were put into 300 polyester litterbags (25 μm mesh) [7, 8, 9], briefly autoclaved at 110 °C and then exposed to seven cycles of: 1) 12 hours artificial rain (fine spray of distilled water, 6000 mm/12h), 2) 12 hours thawing at –20°C and 3) 10 hours drying at 65°C in a pulsed-air oven. The bags were weighed after each cycle until their dry weight stabilised. The litterbags were leached before being reinserted in the peat surface for the following purposes: 1) To standardise the litter before a short-term field experiment [7]. 2) Together with pre-treatment and/or treatment, to gain a better idea of the qualitative or quantitative contributions of either bioleaching or active decomposition processes on the overall litter degradation. Finally, they were inserted into the field plots for decomposition under particular treatment for a period of three months. At the end of the season, the bags were retrieved, cleaned carefully and dried to constant weight at 65°C. By considering cross-effects between pre-treatments and treatments we could distinguish between the effects of initial litter quality and decomposition on mass loss during the field and the overall decomposition processes. It was also possible to determine the effect of initial litter quality on mass loss during the bioleaching process. Every combination of pre-treatment and treatment effects was replicated: a) three times for *Eriophorum* in the CO_2 enrichment bloc and one time in the nitrogen fertilisation bloc, b) three times for *Polytrichum*, c) two times for *Sphagnum* (Fig. 1).

2.3 Pre-treatments and treatments

The site was divided into two experimental blocs, with no cross-effects between them. The first bloc is equipped with a MiniFace (Free Air Carbon dioxide Enrichment) system [10]. On the second bloc, NH_4NO_3 was finely sprayed every third week on five one metre-squared plots (six applications in total). Each dose was given in two litres of distilled water per plot. Control plots received two litres of distilled water. The treatments were randomly assigned to the plots and there were five replicates per treatment (4 treatments x 5 replicates = 20 plots).a) Atmospheric CO_2 enrichment experiment: CO_2= 560 ppm CO_2 / AIR = ambient (360 ppm CO_2) b) Mineral nitrogen fertilisation experiment: N+=30 [kg/ha/yr] NH_4NO_3 / N_0 = 0 [kg/ha/yr] (control).

2.4 Mass loss

The bags were dried to constant weight at 65°C and put in a desiccator while cooling [11].

The data was expressed and compared using three different ratios. (ML: mass loss; A: litter dry weight with bag before exposure /g; B: litter dry weight with bag after exposure /g; T: emptied litter bag dry weight after exposure /g; C_i: compound dry weight /g)

$$\%\,Total\ ML = \frac{(A-T)-(B-T)}{(A-T)} * 100 = \sum_{1}^{i} \frac{(C_{iA} - C_{iB})}{(A-T)} * 100$$

$$\%\,ML\,ratio = \frac{\%\,C_i\ ML}{\%\,Total\ ML}$$

$$C_i\ contribution\,to\ ML = \frac{C_{iA} - C_{iB}}{(A-T)-(B-T)}$$

2.5 Chemical analyses

Eriophorum vaginatum litter was analysed initially (t_0, 12 samples), after bioleaching (t_1, 20 samples) and after field experiment (t_2, 40 samples).

The 65°C oven-dried material was pulverised through a 0.2 mm mesh grid and homogenised. All analyses were colorimetric and duplicated except for the lignin and total solubles. Mean light extinction values were taken to calculate the concentrations on a percentage dry weight basis. The following analyses were done: a) quantification of lignin in Monocotyledons according to Morrison [12], b) determination of total solubles, soluble sugars and starch according to Fales [13, 14], c) determination of soluble phenols according to Singleton [15], d) determination of total C and total N using CarloErba Elemental Analysor EA1108 CHN, e) determination of bacterial ATP activity using bioluminescence (1997) [16], f) PH determination by titration on gravity water from the first 10 cm of the moss carpet.

2.6 Statistical analyses

Distribution and normality of variables were first checked using box plots and histograms. Those containing too many and/or too distant out layers were analysed using ranked values. Variables were analysed separately using an additional three-way (pre-treatment, treatment, unit, pre-treatment x treatment) ANOVA model. The analyses were done separately for bioleaching, field and overall processes. To perform these ANOVAs we used the S-PLUS (1995) program package. The significance level was set at 5%, but probabilities between 5% and 10 % were considered worth mentioning in the text as being 'almost significant'.

3. RESULTS

3.1 Species specific decomposition

Elevated CO_2 pre-treatment significantly increased % total ML of *Eriophorum vaginatum*, (9 %, $P<0.001$) significantly decreased that of *Polytrichum strictum* (-7 %, $P<0.05$) and had no significant effect on *Sphagnum fallax* in the bioleaching process. However, in the field decomposition process, the pre-treatment did not significantly affect *Eriophorum vaginatum* and *Sphagnum fallax* but had a strong negative effect on *Polytrichum strictum* (-67 %, $P<0.0001$) % total ML. As a result, the pre-treatment had a significant positive effect on the overall % total ML of

Eriophorum vaginatum (10 %, $P<0.05$) and a significant negative effect on *Polytrichum strictum*. (-15 %, $P<0.0001$).

Elevated CO_2 treatment significantly increased % total ML in the field decomposition and overall processes for *Eriophorum vaginatum* (45 %, $P<0.05$, 12, $P<0.05$, respectively) but had no effect on the other two species (Fig. 2). Nitrogen pre-treatment had no effect on % total ML in the bioleaching process. However, in the field decomposition process, there was a negative trend which was almost significant for *Eriophorum vaginatum* (-41 %, $0.05=P<0.1$). All treated litters lost less mass in the field and the overall processes in spite of not being significant. This was not the case in the bioleaching process for *Eriophorum vaginatum* and *Sphagnum fallax*, which lost more mass (n.s.). In the cases of *sphagnum fallax* and *Polytrichum strictum* the treatment had an almost significant negative effect on % total ML in the field decomposition process (-60 %, -63 %, $0.05=P<0.1$, respectively).

3.2 Litter quality

Remark: soluble sugar concentrations were on average <1 % dry weight and were therefore neglected. Expressing the concentrations and losses on a structural dry weight basis did not alter the results.

Elevated CO_2 pre-treatment did not significantly affect the initial litter quality of *Eriophorum vaginatum*. The higher almost significant C-to-N and Lignin-to-N ratios in pre-treated plots were mainly due to a 13 % lower almost significant N concentration rather than to an increase in C. Secondary compounds like lignin and soluble phenols were almost significantly lower (7 % and 9 % respectively) in pre-treated litter. Starch concentration was almost significantly enhanced by 8 %. The total solubles were 6 % lower in elevated CO_2 plots (n.s.).

Due to a limited amount of litter material in some plots of the nitrogen fertilisation bloc, the initial litter of the five replicates for pre-treatment and control had to be pooled. The differences in initial concentrations could not be tested. However, the results remained very informative. Lignin and soluble phenols content were enhanced by the pre-treatment by 7 % and 87 % respectively. However, the C-to-N and L-to-N ratios were slightly lower because of lower C and higher N concentrations in pre-treated litter.

3.3 Qualitative bioleaching and decomposition

The higher (9 %, $P<0.001$) % total ML for CO_2 pre-treated litter in the bioleaching process could only be explained (linear regression, $R^2=0.492$, $P=0.0238$) by a higher almost significant C loss, yet neither the starch loss nor total solubles loss could account for it. However, in the field decomposition process, there was a 766 % higher but not significant starch loss which could over-balance the significant 32 % decrease in total solubles loss, and together with an enhanced N loss (42 %, n.s.) match the positive not significant % total ML.

Lignin, together with the phenol loss (both enhanced by the treatment), accounted for more than half of the enhanced litter loss in the field process. This wasn't the case for N, which lost (121 %, n.s.) less in the treated plots. As in the pre-treatment, the total solubles loss was significantly reduced by treatment (-74%, n.s.).

The bioleaching process was not affected by the nitrogen fertilisation pre-treatment in terms of % total ML. In the field decomposition process, there was an almost significant decrease (-41%) of % total ML coupled with a general negative effect on all compounds loss except for C and soluble phenols. One part of this reduced % total ML could be attributed to a net immobilisation of nitrogen (95 %) and total solubles (107 %) which were greater in the pre-treated plots. The pre-treatment tripled the soluble phenols loss significantly ($P<0.001$) which represent 46 % of the initial soluble phenols dry weight and contributed to 17 % of % total ML (only 4 % in the control).

The treatment decreased the % total ML in the field decomposition and overall processes (-70 %, n.s., -9 %, n.s.). However, the carbon loss was almost significantly decreased (-12 %) and all the other compounds followed the same trend. An enhanced (43 %) starch loss was the only exception found, although was not significant.

3.4 Bacterial ATP activity and PH

The 1997 experiment showed that nitrogen treatment significantly decreased the bacterial ATP activity in the litterbags and slightly decrease the gravity water PH (-25%, $P<0.05$, -2%, n.s., respectively).

4. DISCUSSION

4.1 Species-specific decomposition

Eriophorum vaginatum and *Polytrichum strictum* reacted in the opposite way to CO_2 pre-treatment and treatment during both processes. Pre-treated *Sphagnum fallax* only reacted similarly to *Polytrichum strictum* during the field decomposition process.

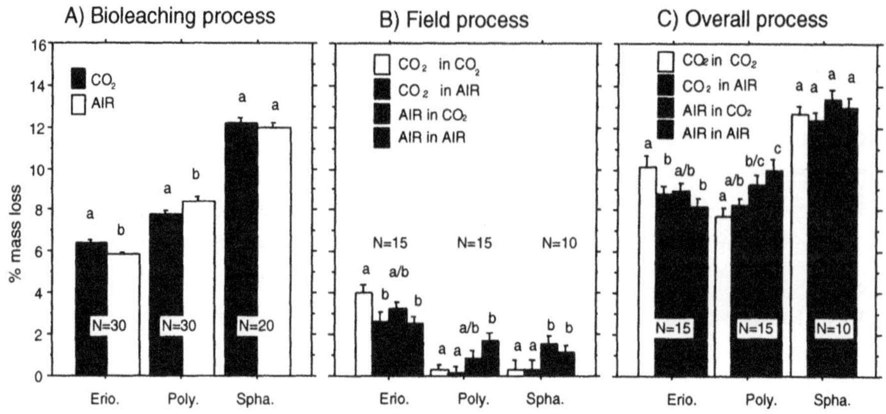

Figure 2. Species-specific effects of pre-treatment and treatment on litter mass loss (values followed by different letters are significantly different (P<0.05). Error bars indicate SEM. N represent the number of bags used for each nominal variable).

This may emphasise disparities in the physiology between vascular plants and mosses. While lignin and soluble phenols concentrations were not enhanced in elevated CO_2 pre-treated *Eriophorum vaginatum* litter, starch concentration was almost significantly increased. This accumulation of starch could be due to an 'energy overflow' from enhanced rate of photosynthesis [4]. Hence, the higher % total ML for pre-treated *Eriophorum vaginatum* litter could be explained by a high contribution of starch and N rich compounds (amino-acids or proteins) to the litter ML in the field decomposition process. It seems that N was not limiting growth enough to increase either directly or indirectly CBSSC. However, mosses (*Sphagnum* and *Polytrichum)* which have no root system, rely more on nitrogen deposition for their N nutrition than do vascular plants which can extend their root system for greater uptake [17, 18]. Therefore, nitrogen could become a strong limiting factor for these plants under elevated CO_2 conditions. Under limited growth and in accordance with the carbon-nutrient balance hypothesis [3, 4, 19], more secondary compounds could be produced. If these compounds happen to be easily leachable secondary compounds such as soluble phenols, they may inhibit microorganisms' activity and active decomposition in the field process. This could explain the negative effect of CO2 treatment on *Polytrichum strictum* ML in the field decomposition. These remain assumptions, and require further investigation.

4.2 Lignin and decomposition

The fact that lignin contribution to the litter ML was on average 60 % higher in the field than in the bioleaching process could show that even in the relatively short period of field decomposition lignin had already been selectively degraded by microorganisms or soil enzymes. It implies that early colonisers may as well be lignolytic fungi or bacteria. Fungi and bacteria were indeed shown to be the dominant groups of microorganisms in our experimental site (34 % and 50% of total microbial biomass respectively) [20]. No taxonomical and enzymatical studies have been undertaken for these two groups. According to [21] principally responsible for lignocellulose degradation are aerobic filamentous fungi, and the most rapid degraders in this group are Basidiomycetes. Furthermore, the lignin contribution to the *Eriophorum vaginatum* litter ML during field decomposition was lower in nitrogen pre-treated and treated plots as compared to control albeit not significant. This may as well reflect a inhibition of microbial activity.

4.3 Inhibition of decomposition

Despite the fact that these results can not be tested, the fact that nitrogen pre-treated litter contained 6 % more lignin and 86 % more soluble phenols should retain our attention. This could reflect a nutrient limited growth and could match with the carbon balance hypothesis. Since the TNC were not enhanced together with lignin and soluble phenols, the 'amino-acid diversion model of secondary plant metabolism' which doesn't consider increased levels of TNC as the major trigger, could be favoured.

This hypothesis rather states that increased accumulation of phenolics stems from a decrease use of a common precursor (phenylalanine or tyrosine) for protein synthesis [4]. Higher nitrogen deposition, even using NH_4NO_3, might decrease PH [22]. This acidification could in turn decrease exchangeable cations (Ca^{2+}, Mg^{2+}), increase NH_4^+ and NO_3^- leaching, and decrease decomposition [23, 24]. All these effects would lead to a nutrient limited growth. Our experiment has even shown that the pre-treatment significantly ($P<0.0001$) increased the soluble phenols loss in the overall process (46 % of their own weight has been lost). Results from a previous experiment in 1997, made with litter that did not undergo bioleaching, have indicated that the bags' bacterial ATP activity decreased with the same amount of NH_4NO_3 used.

4.4 Quantitative versus qualitative mass loss

In the case of *Eriophorum vaginatum*, the bioleaching ML and the field decomposition ML were both compared to the overall ML (Fig. 3).

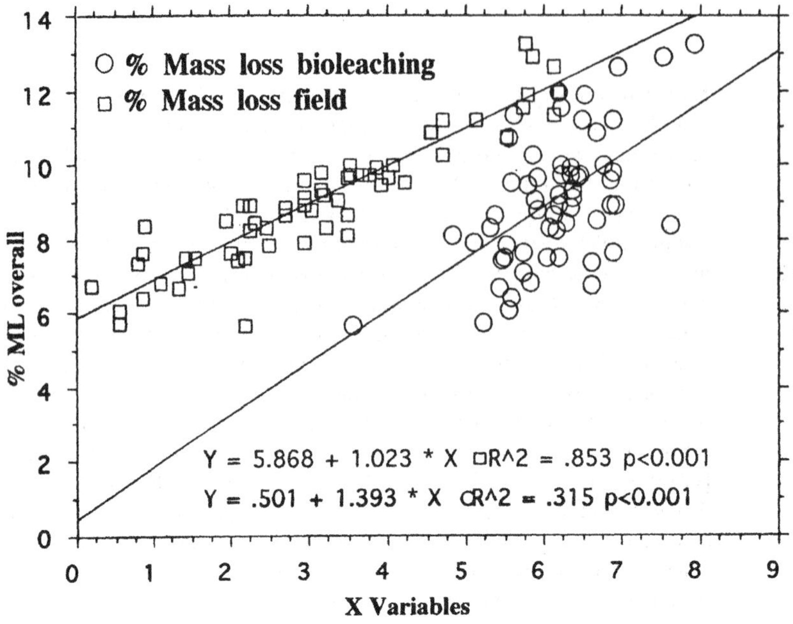

Figure 3: Linear regression for the comparison between a) overall % ML and Bioleaching (n=60)% ML, b) Overall % and filed (n=60)% ML. Example taken for Eriophorum vaginatum litter of the CO2 enrichment bloc.

The comparisons show that: a) bioleaching is quantitatively more important (higher regression coefficient), b) field decomposition is establishing differences between (pre-) treatments and therefore qualitatively more important (higher R^2) for the pattern of the overall ML.

5. REFERENCES

Belyea L.R., OIKOS 77:3 (1996) 529-539.

Bryant J.P., F.S. Chapin, P.B. Reinhardt, T.P. P.B. Clausen, Oecologia 72 (1987) 510-514.

Coulson J.C., J. Butterfield, Journal of Ecology 66 (1978) 631-650.

Cresser M., L. Yesmin, S. Gammack, A.K. Dawod, M. Billett, L. Sanger, In: J.H. Tallis, R. Meade, P.D. Hulme, BIANKET MIRE DEGRADATION. Causes, Consequences and Challenges, The Macaulay Land Use Research Institute, Aberdeen, 1997, pp. 153-159.

Fales F.W., J. Biol. Chem. 193 (1951) 113-124.

Gorham E., Ecol. Aplic. 1 (1991) 182-195.

Grosvernier P., Y. Matthey, A. Buttler, Journal of Applied Ecology 34 (1997) 471-483.

Hammel K.E, In: G. Cadisch, K.E. Giller, (Eds.), Driven by Nature, CAB INTERNATIONAL, 1997, pp. 33-45.

Lambers H., Vegetatio 104/105 (1993) 263-271.

Lee J.A., J.Tallis, S.J. Woodin, Ecological Change in the Uplands, Blackwell Scientific Publications, London, 1988, pp. 151-162.

Lumac bv, LUMAC, A step ahead in rapid microbial testing systems, P.O. Box 31101, 6370 AC Landgraaf, The Netherlands.

Miglietta F., M.R. Hoosbeek, J.Foot, F.Gigon, M. Heijmans, A. Peressotti, T. Saarinen, N. van Breemen, B. Wallen, Environmental Monitoring and Assessment (in press).

Mitchell E., D. Gilbert, C. Amblard, A. Buttler, Ph. Grosvernier, J.-M. Gobat, The microbial communities at the surface of five Sphagnum-dominated peatlands in Europe: structure and effects of elevated CO_2, (In prep.).

Morrison I.M., J. Sci. Fd Agric. 23 (1972) 455-463.

Norby R.J., M.F. Cotrufo, Nature 396 (1998) 17-18.

O.W. Heal, J.M. Anderson, M.J. Swift, In: G. Cadisch, K.E. Giller, (Eds.), Driven by Nature, CAB INTERNATIONAL, 1997, pp.3-30.

O'Neill E.J., R.J. Norby, In: G.W. Koch, H.A. Mooney, (Eds.), arbon Dioxide and Terrestrial Ecosystems, Academic Press, San Diego, 996, pp. 87-103.

Palm C.A., A.P. Rowland, In: G. Cadisch, K.E. Giller, (Eds.), Driven by Nature, CAB INTERNATIONAL, 1997, pp. 56-70.

Penuelas J., M. Estiarte, TREE 13 (1998) 20-24.

Proctor M., In: J.H. Tallis, R. Meade, P.D. Hulme, BlANKET MIRE DEGRADATION. Causes, Consequences and Challenges, The Macaulay Land Use Research Institute, Aberdeen, 1997, pp. 153-159.

Schinner F., (Ed.), Methods in soil Biology, Springer-Verlag, Berlin/Heidelberg, 1996, pp. 110-121.

Singleton V.L., In: H.F. Linken, J.F. Jackson, (Eds.), Modern methods of plant analysis, Springer-Verlag, Berlin, 1988, pp. 200-207.

Steinnes E., J.E. Hanssen, J.P. Rambaek, Water, Air, and Soil Pollut. 74 (1994) 121.

Analysis of the Environmental Impact Caused by Introduced Animals in the Clarion Island, Archipelago of Revillagigedo, Colima, Mexico

PEDRO MENDEZ-GUARDADO
Departamento de Geografia y Ordenación Territorial, C.U.C.S.H., Universidad de Guadaiajara. Av. De los Maestros y Av. Mariano Barcena, C.P. 44260, Guadaiajara, Jalisco, México

Key words: Endemic, impact, biosphere reserve, Mexico, erosion depletion.

Abstract: The Clarion Island is located to the west of Manzanillo, Colima, Mexico, to about 1,000 km. It belongs to the Archipelago of Reviliagigedo and represents the furthest point of Mexico's possession. Because of the importance of the archipelago's biota, the Mexican government declared it as a biosphere reserve in 1994. Unfortunately, the introduction of exotic animals (pigs, rabbits, and sheep's) have altered the natural ecosystem of the island, and endangered the presence of endemic species. Nowadays, there are several areas were the vegetation has been depleted causing the runoff of the soil, creating a desertic landscape, especially on the east of the island. In 1997 we set four plots of 9 m^2 to prevent the animals to grass in these places, and assess the natural recovery of the vegetation.

1. INTRODUCTION

The Archipelago of Reviliagigedo is composed of four volcanic islands; Socorro, Clarion, San Benedicto and Roca Partida. They are truly oceanic islands whose natural and singular ecosystems had not been well studied by the Mexican scientist. Clarion is the second one in size and biodiversity (after Socorro) with about 24 kM2. It is located among the Coordinates of 18022' North and 114"44' West (see figure No. 1) and is part of an aggregate of elevations originated in the ocean floor at about 4,000 meters of depth.

As it is well known the evolutionary processes of this type of islands are totally different from those in islands near the continent or that were connected at some time to it. In the first case it is reasonable to see the

G. Visconti et al. (eds.), Global Change and Protected Areas, 323–329.

influence that exerts the species of the continent in the colonization and propagation of the organisms, because the proximity facilitates its passage to island. In the last ones, at the time of the separation they already had a defined flora and fauna which explains the abundance of species commonly found in the mainland coasts. Its a relatively low island since its higher altitude is a hill named "Pico de la Tienda" located on the East with only 308 meters of height. Mostly the entire area of the island is over 150 meters of altitude, and only the south has slow slopes ending in a small bay. The rest of the coastline has cliffs of up to 200 meters high. Clarion has been named "La Isla Verde" (The green island) because of the presence of the Tribulus cistoides L. a small herb whose foliage still green even during the dry season. Such a plant it's well distributed through the whole area and from the distance the island looks green and covered with vegetation.

In spite of its humid appearance, there is not fresh water on the island, so the animals must have modified their natural requirements of this liquid. There are nevertheless, two small depressions on the south, which collect water during the rainy season, although the water is very salty, and last only for a few months. Mostly all the vegetation is herbaceous and only a few shrubs with a height of no more than 3 meters compose the flora. The Karwinskia humboldtiana (the taller specie) is found in the plateau of the island at approximately 150 meters of altitude and because of the strong wind form the North, all the individuals are twisted towards the South. The hills were the slope exceeds 20% are covered by an Euphorbia whose latex produces irritation in the human skin (Euphorbia californica).

The "Gannets" (Sula sula) use these plants to nests and to hatch their young. The importance that this Archipelago represents for Mexico can be described as follows:

1. Economic, since with its possession the area of the Mexican patrimonial sea is increased enormously and with it the possibility of operating its resources conscientiously. The marine species of economic importance that can be captured in the surroundings of the island are numerous; among them it is possible to enumerate the Tuna, the Lobster, the Gilded one, etc.
2. Strategic, being the last border of the Mexican territory to the West, it represents a conflicting point for its monitoring and control, and
3. Naturalist, by its richness in species as much animals as vegetals, as well as of mineral resources.

The Archipelago in general has being considered by some authors as the "Galapagos of Mexico", this is mostly due to the percentage of endemic species found. In the case of Clarion it has been considered that around 20% of the vegetal species are endemic to the island or to the archipelago as a whole.

2. OBJECTIVE OF THE PROJECT

The fundamental objective of this project is the knowledge of the environmental conditions that prevail at the moment in the island and the form in which the degradation of its resources can be stopped or at least lessen. A second objective is to compare the current situation of the island with those of Socorro Island and probably Cleofas island (islas Marias). Covering with it net oceanic islands as well as a continental one and analyse the affinities and differences of species, its origin, dispersion and distribution.

3. METHODS

This investigation leans of the cartography processed in the project on cartography of the islands of the Archipelago of Revillagigedo done in the Department of Geography of the University of Guadaiajara. The scale that is used for fieldwork is 1:5,000 and for cabinet can be 1:2,000. First a reconstruction of the landscape of the island on the basis of the information compiled has been made. So a map of original vegetation has been elaborated. In agreement with it, we have found that some species that once were the dominant ones, nowadays have almost disappear (typical case of the Opuntia engelmanii Salm Dyck). With this map and the current one, the changes on the vegetation composition and its possible affectation will be determinated to assess the dynamics of the ecosystems of the island. Two excursions are made per year (April and November) with duration of three weeks each one. The trips are planned at this time to collect the plants growing after the rainy season as well as the ones growing in the dry period. In addition, it is not possible to have fieldtrips during the rainy season (June-October) due to the presence of hurricanes. Two examples of every vegetational specimen is collected, one of them is send to the Botanical institute of the University of Guadaiajara for its identification, and the other one stays in the Department of Geography. In the base map a grid of square kilometres covering the total area of the island has been drawn. The samplings are being collected in each one of the squares to determine which are the most affected areas. Then they will be categorized grouping them according to the current conditions in which they are. In the first explorations the routes were made in transects covering the total area for its better recognition. In order to obtain a greater precision in the boundary of the areas a portable Geographic Positioner is used (GPS).

The first vegetational map is being corroborated on the field.

Figure 1: Localization of the Clarion Island in the Archipelago of Revillagigedo.

With the help of the topographic map we are analysing the drainage and soil behaviour to combine this information with the type of vegetation, length and degree of the slop and the physical characteristics of the soils. Overlapping the soil, vegetation and erosion maps will be possible to analyse the dynamics of the elements of the environment and to determine the current state of the landscape, as well as its possible solutions. Ali the information obtained in field is introduced in a database using the dbase, and EXCEL programs. And IDRISI, ILWIS and CorelDraw are used to make the maps and have the final presentation.

4. SOME IMPORTANT DATES FOR THE ARCHIPIELAGO

The first reports of the Archipelago date from 1533, during the fever of the discovery of the Seas of the South when the Ship San Lazaro under the command of Captain Hernando de Grijaiva arrived at Socorro Island.

Nine years later, in 1542 Ruy Lopez de Villalobos in his attempt to open a route towards the Australian Continent discovered San Benedicto and Clarion.

Although the islands were discovered from the beginnings of the Mexican Colony, the first botanical reports appear many years later. In 1839 the English Corvettes Sulphur and Sterling to the control of the captain Edward Belcher visited the archipelago obtaining samples of flora and fauna. It is from this trip that George Barclay and Bentham publish their first scientific study of the Archipelago.

In 1861 the Government of President Don Benito Juárez, grants the possession of the Archipelago of Revillagigedo to the State of Colima with the condition of build a prison of Maximum security eliminating therefore the death pain.

In 1925 the Academy of Sciences of California makes one more of its trips collecting samples and presenting to the scientific community its findings.

In April of 1979 the President of Mexico Jose Lopez Portillo hoists for the first time the National flag in Clarion island and from that day on there is a permanent presence of personnel of Mexico's Navy preserving the sovereignty of the Country on that territory.

In 1980 a couple of pigs were brought to the isle and after breeding them for a while, they were set free.

In 1983 a couple of rabbits were also brought to the island and were released to colonize it.

In June of 1994 the Mexican Government declared the Archipelago of Revillagigedo as a Biosphere Reserve

5. CLARION PROBLEMS

For some reason, three species of mammals (Pigs, Rabbits and Lambs) have been introduced to the island, so the natural ecosystem of it has been altered enormously. We are trying to assess the impact caused by different factors: no biotic (hurricanes, fires, drought, etc.), biotic (with the introduction of species such as: pigs, lambs and rabbits) and anthropogenic.

In 1980 a pair of pigs was taken to the isle. For some time they stayed in controlled, but somehow months latter they were left free so their population grew without control.

In 1983 it was the turn for a pair of rabbits to be released propagating in an incredible way. As for the Lambs no one knows when were they released in Clarion, but we think they were brought from Socorro island some time between 1978 and 1979. As the time pass by and with the good adaptation shown from these species, their population increased enormously so the vegetal species of the island began to undergo the damage of this (we could consider) new plague.

The Pigs remain mostly in the South and Middle East of the island although can be observed in the West in the surroundings of Mount Gallegos. The Rabbits have distributed through all the surface of the island. The Lambs have restricted their area to the East of the island

6. RESULTS

Apparently Clarion is the oldest island from the Revillagigedo archipelago. The soils are very deep volcanic, and there is not a historical registry of any volcanic activity.

Forty-three vegetal species have been collected (table no. 1) from which a good percentage corresponds to individuals of the Poaceae, Gramineae and Fabaceae families.

With respect to the fauna, it is interesting the presence of two endemic species; "La culebra roja de Clarion" (Masticophis anthonyi) and the "Lagartija azul" (Uta clarionensis). Individual that have called the attention of the scientists. In addition to these species the "Saltapared de Clarion" (Troglodytes tanneri) and the "Paloma de Clarion" (Zenaida macroura clarionensis) are some of the most common endemic species

The introduction of mammalian is affecting drastically the flora and fauna's population. The diminution on the frequency of some species previously frequently reported has been observed. So is the case of Opuntia sp. that previous to 1979 was well distributed on the South and North of the island, and nowadays, it is restricted to rocky zones of the SouthWest where the slope exceeds 35%. The same case happens to the species of the Fabaceae family that has diminished considerably their number.

The alterations found in the ecosystems can be disastrous if we take into account that a good part of the flora and fauna of the island is endemic to the Archipelago.

With the introduction of these organisms the nutritional chain of the species has been modified. Now the Crows (endemic species Corvux corax clarionensis) are feeding on the viscera's and eyes of the rabbits, the pigs also consume them and finally the rabbits are depleting the original vegetation.

The loss of vegetal cover can be observed in several zones of the island. At the same time, the competition for food and space among the original and the introduced species is having its effect on the first ones which are losing the battle since they do not have the capacity to resist the effects. The beaches of the island (on the South) are visited by marine turtles that deposit their eggs in the sand, unfortunately the pigs are in alert to excavate and eat them.

7. CONCLUSIONS

Due to the negative effects that have appeared with the introduction of the species of mammals (pigs, rabbits and lambs) in the island, the original ecosystem has been altered. Having as a consequence the degradation of the natural resources.

The diminution in the population of some species and the erosion of the soil in the island, are some of the worries of the Mexican scientists. It is observed that the ignorance of the adverse effects that can be caused with the introduction of exotic species in protected zones causes serious damages to the original ones. In the case of the Archipelago of Revillagigedo the problem is even greater because a good part of the species that inhabit it are endemic, thus any alteration in its ecosystem can produce considerable damages.

8. BIBLIOGRAPHY

Brattstrom, Bayard H. The cactus of the Revillagigedo Islands, Mexico. 1953. Cact. Succ. Journal 25(6): 181-182.

Brattstrom, Bayard H. Y T.R., Howell. 1956. The Birds of the Revillagigeda islands, Mexico. Condor. 58:107-120

Evertt, W.T.. 1988. Notes from Clarion Island. Condor. 90:512-513.

Gorman, M. L. 1991. Ecoiogía insular. Ediciones Vedra. Barcelona, Espana. 101 pp.

Howell, S.N. G. and S. Webb. 1990. The Seabirds of the Revillagigedo Islands, Mexico. Willson Bulletin. 102(1): 140-146.

Howell, S.N. G. and S. Webb. 1995. A guide to the Birds of Mexico and Northern Central America. Oxford University Press. 851 pp.

Johnston, M.I. 1931. The Flora of the Revillagigedo. Il Proc. Calif. Acad. Sc. 41. Ser. 20 (2): 9-104

Medina, G. M., 1978. Memoria de la expedici6n científica a las Islas Revillagigedo, abrildel954. Universidad de Guadalajara. 335pp.

Reyes Vayssade, M. (Coord). 1992. Cartografía Hist6rica de las Islas Mexicanas. Secretaría de Gobernaci6n. México, D.F.

Rzedowski, J. 1978. La Vegetaci6n de México. Edit. Limusa. 2da. Edici6n, 432 pp.

Secretaria del Medio Ambiente Recursos Naturales Y Pesca, et al. Reservas de la Bi6sfera y otras areas naturales protegidas de México. 1995. México, D.F.

Vivo' Escoto, Jorge, A. 1979. Clari6n: la verde isla mexicana, más aiejada dei Pacifico. Secretaría de Marína, Armada de México.

High Mountain Environment as Indicator of Global Change.

GEORG GRABHERR, MICHAEL GOTTFRIED AND HARALD PAULI
Department of Vegetation Ecology and Conservation Ecology, Institute of Ecology and Conservation Biology, University of Vienna

Key words: Mountain Environment, Global Warming, Plant Physiology.

Abstract: The recent debate on global change effects on the living world is still characterized by lack of confidency as depending on rather hypothetical assumptions and predictions. This is a consequence of missing global ecological observation networks comparable to those which have been established for climate, glacier movements, atmospheric composition etc. Alpine biota such as grasslands, dwarf shrub heaths, alpine steppes, the tropical paramo, vegetation fragments at the high elevational limits of plant life, snow beds offer a broad spectrum for permanent long term ecological monitoring as it could be demonstrated by the few already available examples such as biodiversity changes in the Central Alps during this century. Most of these biota are simply structured, show clear habitat dependences, thus indicating change of one dominating environmental factor, and are in many high mountain regions still in a natural or at least seminatural state which means that direct anthropogenic effects could be excluded, or clearly described. As indicators for short term effects (<10 years) we recommend 1. flowering phenology of vascular plants, 2. composition and abundance of moss species; for medium term effects (<50 years) 1. structural parameters of plant communities such as rank/frequency relations, or horizontal individual distribution pattern, 2. species composition of plant communities, appearance of exotics and pathogens, 3. abundance, appearance and disappearance of selected animals (soil and above ground). A simple functional classification based on CO2- requirement, thermal and water relations, mobility and trophic characters is provided for selecting such species. For long term observations (>50 years) also landscape patterning might be useful especially close to tree line. Efforts for developing a global network of alpine monitoring sites should be started immediately. The next generations will that gratefully acknowledge.

G. Visconti et al. (eds.), Global Change and Protected Areas, 331–345.

1. INTRODUCTION

The increased awareness of the greenhouse effect, of atmospheric nitrogen deposition, of acid rain, and land use change, has lead to a broad discussion how these globally recognized impacts may alter biotic systems in all parts of the world, climate change effects in particular 1,2,3,4,5,6. Most of the predictions are either derived from probabilisitic model approaches, or proxy-analysis using paleorecords, or from simple experiments, still being nothing else than hypothesis with a more or less realistic background. Only a few studies are based on direct long term observations of natural or seminatural ecosystems 7,8,9 (fig.1). They provide the only hard data for the global change debate which is still rather controversal.

There are several reasons for this lack of confidency how the living world may and do change under changed conditions. First of all there does not exist a consistent ecological observation system such as the worldwide meteorological or glaciological networks. A second problem arise from the fact that ecological systems are exposed to a multitude of environmental factors each having a different influence on the system. This holds true in particular for seminatural or anthropogenic systems where the human influence adds to the natural determinants. Thirdly, biodiversity of most ecological systems is high, and species may react often individualistically. Species composition might change in an unpredictable direction which not necessarily means that ecosystem functioning will also change. In simple terms: A meadow might maintain as a meadow with its typical primary productivity, energy flow or nutrient turnover but species composition will be different.

As a consequence ecological systems which could be considered as reliable indicators for environmental change effects should be easy to monitor for a long time, should depend clearly on that environmental factor which is of interest, and, finally, should be comparatively simple. Exactly these requirements are optimally realized by alpine ecosystems, those at the climate dependent limits of life in particular.

2. HIGH MOUNTAIN ENVIRONMENTS – AN OVERVIEW

High mountains which means mountains with an alpine zone occur from the tropics up to the Arctic. Alpine zones in ecological terms are per definition free of trees and large shrubs, therefore being easily recognisable in high mountain landscapes as the area above a tree line 8.

Tab. 1 Functional classification of plants and animals according to CO2, water, thermal regulation, mobility, and trophic characters.

Funktional type	CO_2-requirement	Water balance	Thermal reg.	Mobility
Vascular plants	Autotrophic	Homoiohydric	Poikilothermic	Immobile (rooting)
Mosses/lichens	Autotrophic	Poikilohydric	Poikilothermic	Immobil (long dispersal)
Fungi	Heterotrophic	Poikilohydric	Poikilothermic	Immobil (long dispersal)
Mammals, birds	Heterotrophic	Homoiohydric	Homothermous	Mobil (carnivor/herbivor)
Reptiles, amph. Many arthropodes	Heterotrophic	Homoiohydric	Poikilothermic	Mobil (carnivor /erbivor)
Molluscs, some arthropods	Heterotrophic	Poikilohydric	Poikilothermic	Immobile (herbivor)

Exceptions are high mountains of large deserts or those of the high Arctic. Typical for alpine climates is that they are low temperature environments in general, that frost can occur throughout the year, that soil temperatures are very low, and that extreme temperature gradients act at one and the same time on one and the same organism (e.g. roots in frozen soil, leaves warming up at the same time in the sun up to +30°C). Such extreme temperature conditions may also change rapidly in time. For a detailed account consult the recently published textbook of C. Körner 10)

In many high mountains the treeless zone can be subdivided in the alpine zone sensu strictu and the nival zone. The alpine zone is composed of a variety of habitat types most of them inhabited by closed vegetation types, and covered by well developed soil profiles. Typical habitats are dwarf shrub heathlands, just above or intermingeled with the treeline, grasslands or alpine steppes, windswept ridges, snow beds, fens and sources, rock faces which host characteristic plant and animal communities, screes and moraines. Stable alpine sites in the tropics which are not exposed to recurrent disturbance, or which are not to wet or to rocky, are the habitat of the unique paramo of which giant rosette plants are a characteristic element 11,12,13,14,15,16 .

The alpine zone in a strict sense expand over an altitudinal distance of about 1000m then often abruptly disintegrating into that what is generally termed nival zone 13. This zone – thought to be coincident with the climatic snow line - is frequently effected by cryoturbation (i.e. freezing of soil during the night, thawing during the day) which causes a mechanical stress on rooting plants. Only species poor plant assemblages can be found under these extreme conditions, however, species composition still indicates differences in snow cover or in the chemical composition of the substrate 17.

Biotic interference between plant species is low but high between plants and animals. Many animal species, most of them being small insects, rely for example on the many cushion plants typical for nival environments. Either they feed on them or they are using the relatively warm microenvironment in them or do both.

The ultimate limits for vascular plant life can be set at 1000m above the closed alpine vegetation belt, however, this is very variable from mountain system to mountain system. Cryptogams may exceed these limits, and bacteria as well as soil algae have been found up the highest Himalaja – peaks.

Man has influenced high mountain ecosystems in many parts of the world 10, 18 especially where mountain pastures could be used for summer farming (e.g. European mountains, Caucasus, mountains of High Asia, Himalaja, Hindukush, mountains of the Sahara). Others have left as true wilderness such as many boreal mountains, the Japanese Alps, the paramo of some tropical mountains, the Andes of Patagonia). Hunters have shot out large herbivores to extinction or decreased their numbers significantly, and in so doing, changed the grazing pressure on alpine grasslands and steppes which, however, has never been as selective as in the lowlands 19. Extinction of predators has changed the overall consumer systems.

Modern trends are: 1. Increased grazing by the life stock in areas still keeping traditional farming but experiencing population growth (e.g. Himalaja, Hindukush, Andes), 2. Decrease of agricultural use or change to modern agroindustrial techniques, and/or increase of "hard" tourism in developed countries (e.g. Alps, Pyrenees), 3. Logging up to timberline, mining (Rocky Mountains, Tien Shan, Altai), 4. Exploitation of wild stock by hunting, increase of recreation activities and tourism in former wilderness areas (Rocky mountains, tropical mountains, mountains of High Asia). Though in many mountains the anthropogenic influence must be considered as important, many alpine areas, nature reserves and national parks in particular, and even more, nival areas, are still hosting ecosystems in a natural or close to natural state. It is this opportunity, i.e. to find refugia of naturalness, which makes high mountains so interesting for monitoring environmental change effects on biotic systems.

3. WHAT TO OBSERVE ?

The decision what characters of alpine biota should be taken as indicators, and for what, depends on their sensitivity to a particular environmental factor and the time perspective. It's also a matter of the spatial scale. Species composition of alpine grasslands for example may have changed long before effects become obvious on the landscape scale. The most important characters for indicating environmental changes might be: 1. Flowering phenology of vascular plants as indicator for short term (1 – 5 years) fluctuations , or changes in fluctuations in a long term, 2. Ecosystem functions such as primary productivity, nutrient uptake as integrative measure (soil and atmosphere) for medium term effects (5-10 years), 3. Community structures (vertical and horizontal patterns, eveness) which integrate soil and atmospheric effects, and puffer interannual fluctuations, for medium to long term effects (>10 years), 4. Plant species composition and abundance as integrative indicators (soil and atmosphere) for medium to long term trends, 5. Landscape change for displaying very long term trends (>50 years), 6. Appearance or disappearance of functional types (including animals) for medium term effects, 7. Soil parameters and selected soil organisms to display medium to long term trends, 8. Appearance of exotics for medium term effects and display of anthropogenic effects. In the following these characters are checked for their indicator potential, where they have to be observed and in what habitat, what their specific advantages are, or what has to be considered as a disadvantage respectively.

3.1 Ad 1) Flowering phenology of vascular plants

Start of flowering reacts quite sensitive to the start ,and temperature conditions, of the early summer season which, interestingly, does not have an influence on the time of fruit setting and ripening 20 .

However, early dates of flowering might be a good indicator in seasonal climates (i.e. everywhere except the tropics) for a warming. During a warming experiment with Open Top Chambers e.g. flowering of the arctic/alpine dwarf shrubs Dryas octopetala could be enhanced significantly (fig.2). It's also the best character to study interannual variation but has to be observed for a long time to document long term effects. But there is a warning to be stated: One should be aware that some plants depend phenologically on day length, others might be opportunistic 10 .

3.2 Ad 2) Primary productivity, nutrient uptake

Primary productivity is a widely recommended measure in ecological monitoring. In alpine ecosystems it is generally low but is certainly correlated with climatic factors such as temperature and precipitation 10 . It integrates temperature effects on the belowground organs as well as on the aboveground parts. Productivity can even be stimulated by grazing. For the interpretation of productivity data one has to consider further facts such as: change in precipitation may influence productivity in arid regions.

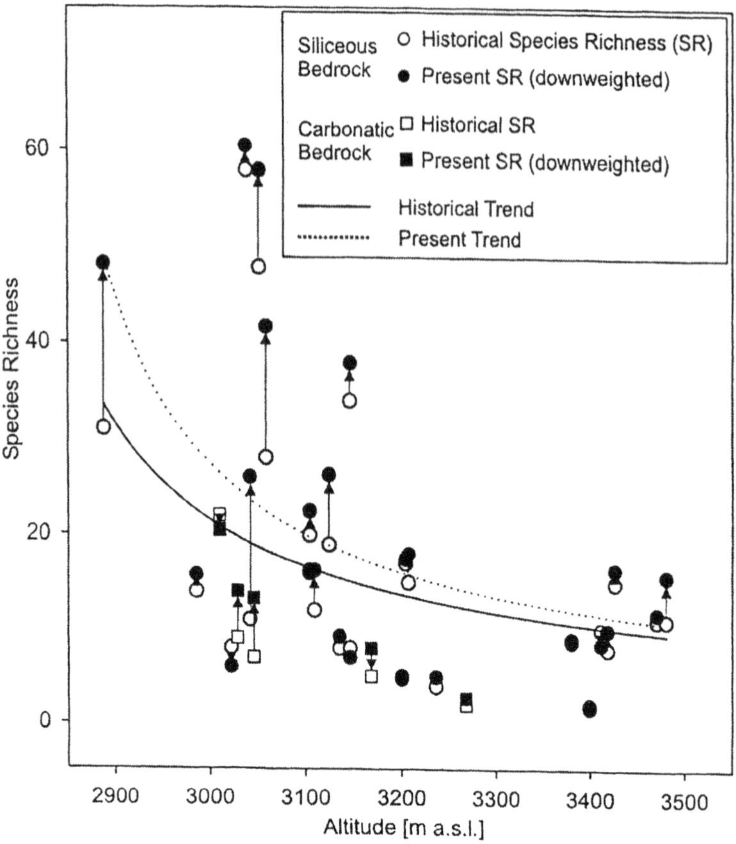

Fig. 1 Comparison of historical (open symbols) and recent (black symbols) species richness of vascular plants at 30 mountains from the Central Alps exceeding an elevation of 3000m. Values for species richness are downweighted for rare species considering that species found in low numbers today, and missing in the historical record, meight had been overlooked the historical author. Thus the given figures are „at least – values". Quadrats are for rocky limestone mountains, circles for siliceous bedrocks. Arrows indicate increase of species richness. Fig. based on 6 but modified and with new records for limestone mountains.

Increased nutrient levels stimulate productivity significantly most pronounced in areas with alpine ruderals such as morain fields or screes. This has to be considered especially in areas exposed to atmospheric nitrogen deposition.

As plant species may react very specifically, productivity of selected species (e.g. classified into ruderals, competitors and stress tolerators 21..) might be the better character to observe than ecosystem productivity, the latter also being very destructive.

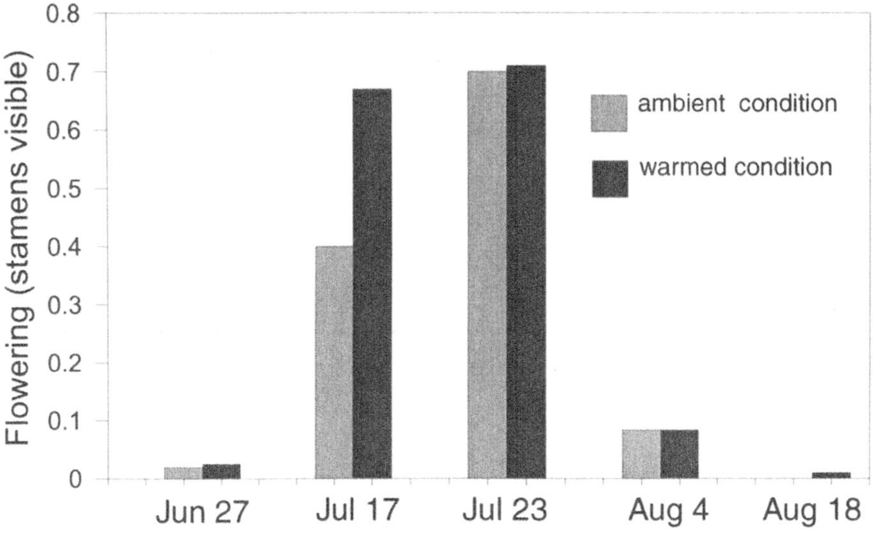

Fig.2 : Phenological patterns of the arctic-alpine plant Dryas octopetala in situ at Svalbard (Spitsbergen) under natural and warmed conditions in Open Top Chambers. The difference on July 17 is significant at $P<0.001$. 20.

As plant species may react very specifically, productivity of selected species (e.g. classified into ruderals, competitors and stress tolerators 21..) might be the better character to observe than ecosystem productivity, the latter also being very destructive. Furthermore standard deviation exceeds as a rule 20% of the mean even in the most homogeneous communities 22 which means that only large changes in ecosystem productivity can be detected.

Thus primary production of whole ecosystems are not recommended as a good indicator. If it is used it should not be measured in time intervals closer than 5 – 10 years.

3.3 Ad 3) Communities structure

Alpine vegetation types are vertically simply structured, often composed of a low moss or lichen layer and a layer of dwarf shrubs, rushes, sedges or grasses. The height of the plant layer seldom exceeds more than 0,5m, except in the tropical paramo where it might be a few metres. As changes in canopy height may only become obvious if e.g. a low community has changed to a tall one which is also indicated by species composition change this parameter is certainly not very interesting. If considering abundance or frequency structure late successional vegetation types are usually characterized by an uneven distribution of species. One or a few species occur frequently, a group of intermediate frequency is associated, rare species occur at low numbers (fig.3). Species distribution in azonal habitats like screes, moraines, or in the nival zone, is probably more even. Eveness seems to decrease with altitude (fig.3). Regarding horizontal patterns individuals of dominating species such as rushes or sedges often show a regular distribution whereas associated species are clustered or even randomly dispersed. These types of patterns could be used as indicators for structural changes regardless of the specific species composition.

Changes in spatial or frequency patterns can be caused by climate induced changes of the competitive balance between the species given in a particular community. That includes effects on the vegetative systems as well as on the generative ones, and, therefore, can be taken as an integrative indicator which might be of medium to long term relevance. Especially rank/frequency analysis (fig.3) is simply to be applied and is not destructive. Relevant data can be collected even under expedition conditions.

3.4 Ad 4) Plant species composition

All alpine and nival plant community types are composed of several to many vascular plant species as well as cryptogams 16,23 . Late successional types are generally richer. Frequently more than 50 species may occur in such communities. Early successional types or types under recurrent disturbance (e.g. screes), rock communities, snow beds, fens, and the communities of the nival zone are less rich, but often host rare endemic plants. All these community types can be classified according to their floristic composition most of them not only by the overall species composition but also by character species which are restricted in their distribution to a particular community type as it has been so frequently documented by the Central European school of phytosociology 23. These character species are excellent ecological indicators and may show – if they appear or disappear – whether a site has become wetter or dryer, cooler or warmer, less or more nutrient rich, or exposed to increased disturbance. Thus species composition is probably the most effective way to follow, and document, environmental changes. In addition it should be considered that alpine environments offer a high variety of habitats and communities in close vicinity to each other being indicators of different environmental situations. Changes can be found along short distances, therefore, even small

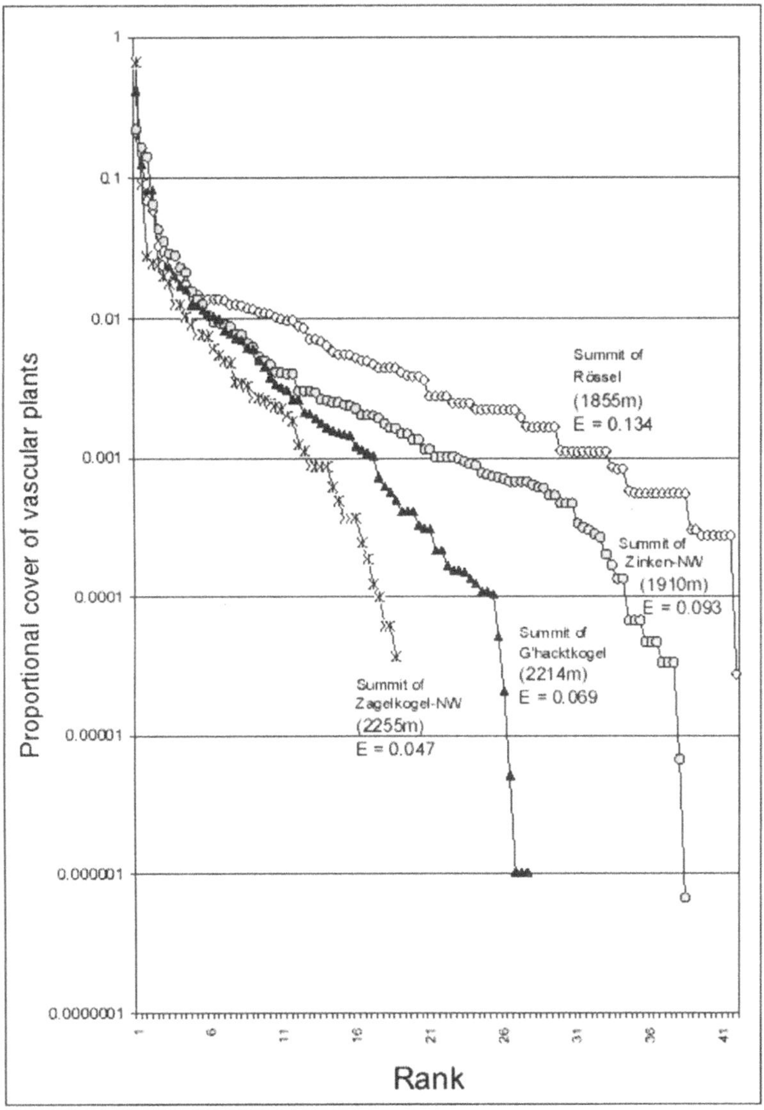

Fig. 3 : Rank/frequency diagramm of vegetation at four summits of different elevation from the Northeasternmost Alps. Data from 16 quadrats (1m2) at each summit. Elevation of the summit is given in metres, E = eveness calculated according to Simpson's Index standardized by the maximum value.

alpine study areas provide an unique experimental field where the result of the natural experiments has only to be described and to be interpreted. As alpine and nival species are long living perennials with low generative production and only short distance dispersal, interannual fluctuations will be puffered but long trends will become clearly displayed as it was shown by a comparative study of alpine summit floras (fig.1).

For monitoring practice it is relevant that data on plant species composition can be taken without distruction of the habitat, and in a comparatively short time. Time demand depends on the parameter chosen to express quantitative characters such as cover or abundance. Very sophisticated quantitative measurements has to be applied if short term effects should be detected like for the ITEX – protocolls 20 . For long term studies presence/absence might be sufficient. The latter means that monitoring sites could be established simply by a list of species in a site of appropriate size whose precise position is measured e.g. by a GPS. Even in remote areas observation sites could be established and actual states of species composition documented for future generations. Derived parameters such as species richness or diversity indices can be supplied if sites from different biogeographical regions should be compared, e.g. for a global trend analysis.

3.5 Ad 5) Landscape physiognomy

The plant cover in alpine environments is of less importance for the physiognomic appearance as well as for the ecological funtioning of landscapes than it is below timber line. This is even more pronounced in the nival zone. The dominance of long living perennials puffer short term effects. Thus observations of whole landscapes is only relevant in a very long term perspective. However, comparison of old photographs from areas close to or at tree line has shown that especially at this ecotone landscape studies might be of relevance 24 . Similarly erosion and solifluction can be dedected easily by photographic documents, deflation on steep slopes or increased rock slides in particular. The latter has become of interest as warming may lead to thawing of permafrost layers. Monitoring of snow melt patterns or movements of glaciers is also relevant for the landscape scale and may be taken as a complementary tool to the observation of alpine or nival biota.

For monitoring photographs from the ground or air photographs with high resolution should be applied. Documentation of the photographs must be supplied with details on the position of the photographer and at least a rough description of the main vegetation types.

3.6 Ad 6) Funtional types – abundance and composition

The definition of funtional types as an ecological concept, and defining ecosystems according to the composition of functional types as an alternative to floristic or faunistic classifications, has become very popular in the global change debate 25. However, clear links between the environmental factors and functional types either defined as life forms, or by ecophysiological constitution, or combinations of both, are only to be found at very broad levels such as a global comparison 26 but not on a local scale, and when going into details. Alpine plant communities for example may be composed of all the classical life forms except phanerogames (i.e. trees and shrubs). Even succulents may occur in snow beds, i.e. in a habitat where water shortage might be only exceptionally be a problem for plants. The appearance of a, "snowbed succulent" such as Sedum alpestre in an alpine grassland of the Alps might therefore not being an indication for a dryer climate but exactly for the opposite, i.e. a change to a climate with more snow.

However, when applying the functional type concept in the following way, it will have exploratory power (tab.1): Vascular plants e.g. can be taken as CO2 autotrophic, homoiohydric (i.e. regulation of the internal water potential), poikilothermic (i.e. changing body temperature with sorrounding temperatur), immobile organisms (i.e. by setting roots). The roots in the soil live under very different conditions than the aboveground parts and, therefore, vasular plants integrate soil and air conditions. Mosses and lichens are like vascular plants, i.e. CO_2-autotrophic, poikilothermic, but poikilohydric, immobile but long distance dispersing (by spores), and soil independed. They indicate primarily changes in atmospheric but not in soil conditions, species appear faster (i.e. by long distance spore transport through the air), short term effects can be detected at the compositional level. Lichens in particular are excellent indicators for air pollution.

From an ecological perspective fungi are like mosses but are not CO_2-autotrophic. They depend often on host plants or are partners of a symbiosis with a vascular plant, and are soil dependent. They can only be identified by their fruiting bodies which only appear occassionally. Therefore fungi are not recommended for monitoring though mycologists would stay with a different opinion.

Among animals of which all are CO_2-heterotrophic, and most of them homoiohydric the following types can be distinguished: 1. Homoiothermic, mobile animals and these subdivided into carnivors and herbivores (mammals, birds), 2. Poikilothermic, mobile animals and these subdivided into carnivores and herbivores (reptiles, amphibians, many arthropodes), 3. Poikilothermic, immobile animals (almost herbivores; molluscs, some arthropode groups). This list shows that considering a wide spectrum of organisms one may find groups which indicate different aspects of a given environment. Synchronic monitoring of these groups will improve the interpretation of the observed changes tremendously. As an example, the increase of species richness at Alpine summits during this century might have been caused not only by the observed warming but also by the evident increase of atmospheric CO_2 6. If data on CO_2-independend animals would had been available the potential CO_2-effect could be excluded, or proved.

Monitoring animals is not easy, and even simple approaches are difficult to design. However, at least a few selected species according to their indicator potential should be included in monitoring programmes at least the detection of appearing or disappearing species.

3.7 Ad 7) Soil indicators

As far as closed vegetation types occur well developed soil profiles are their substrate though often built on subfossil layers from historical periods of different climate, or before human influence has appeared. In the nival zone soils maintain in early successional stages as a result of frequent solifluction, low decomposition rates and low rates of weathering of the parent material.

From an ecological point of view soils are not only the substrate for rooting plants but also habitat for a large variety of soil organisms 27. These organisms could be used as indicators for changes of the soil environment, the temperature regime in particular. According to the low mobility of soil organisms soils are often refugia for animals formerly widely distributed under a different climate. Disappearance of a particular group, therefore, might be a clear indication for a change, however, interpretation might not always be easy. This observation of particular organisms should be combined with integrating measurements such as decomposition rates or enzym activities.

For practice such groups should be checked which are taxonomically well known, and which are always present throughout the growing season, and easy to be sampled such as collembolas or nematodes. As lumbricides are absent from most alpine soils they are not recommended for monitoring programmes.

3.8 Ad 8) Appearance of exotic plants or animals, or pathogens

High alpine pastures have – where they occur – always experienced a steady input of plants and animals from other regions. Land use change might therefore also effects species composition above timber line. Species adapted to grazing might disappear or be decreased in their quantities. These grazing indicators are generally known in well studied mountain areas such as the Alps, and can be treated accordingly. Much more difficult to handle are effects created by introduced herbivores or the extinction of native once by hunting. In the Alps for example the Ibex was shot to extinction about 200 years ago except a small hunting refugium in the Italian Alps.

As Ibex also graze up to the limits of plant life the absence of this large animal for about 200 years in most parts of the Alps must have had consequences for alpine vegetation.

Compared to the lowlands appearance of exotic plants and animals in recent times has been still a rare event in most alpine areas. Locally, plant species from different biogeographical regions escaped from alpine gardens or were planted on purpose to enrich local floras with ornamental species. Revegetation of ski runs or mining areas have used sometimes exotic plant species from industrial seed poducers, most of them having failed to establish. In a long term, however, introduction of alien plants may increase and could be considered for both, indication of human impacts and indication for a more favorable climate.

A group of organisms whose indicative value should not be neglected are pathogens, those on plants in particular. More favorable conditions may cause epidemies of such organism at higher altitude as it has been already observed for tropical diseases at lower altitudes 28.

4. CONCLUSIONS

Alpine and nival biota which are close to ,or reach up to, the low temperature limits of plant and animal life are excellent indicators for ecological relevant climate change effects, warming or cooling in particular. The monitoring cannot replace direct measurements of the relevant environmental factors but – as integrating their effects over time – may be the best measures to detect trends, and to display the ecological relevance of climatic (including change of atmospheric chemistry) or land use changes. This can be approached even by simple measurements applied to alpine/nival biota in long term intervals such as 5 to 10 years or even longer which reduces the efforts for data collection to a minimum. Relevant data sets can be sampled in all parts of the world far apart from permanent scientific institutions. Of course there is no reason to do not more than the minimum. Master stations managed by well equipped research institutions might help not only to observe as much as possible but – in so doing – will help to understand alpine ecosystems, and ecological systems as such - much better than we do it today.

A problem which has to be solved of course will be the documentation and data banking of the data sets sampled throughout the world. For a must be established where such data could be published, preferably in international journals. Officially established data bank systems such as that of the European Environmental Agency 29 should include observation data on ecosystems once having been standardized to a certain degree. Therefore our working group (see also the contribution of Pauli, Gottfried, Reiter) has started with support from the Austrian Ministry of Science an initiative which tries to establish an international network to achieve a global movement for establishing monitoring sites in alpine areas, to discuss standardized procedures, and to organize an appropriate data management.

5. REFERENCES

Archibold O.W., Ecology of World Vegetation. Chapman and Hall, London, 1995.

Billings W.D., In: D.A. Johnson (Ed.), Special Management Needs of Alpine Ecosystems, Society for Range Management Eocsytsmes, Demver, 1979.

Bolin B., B.Döös, J.Jäger and A. Warrik (Eds.), The greenhouse effects , Climate Change, and Ecosystems. SCOPE 29.

Box E.O., Macroclimate and plant forms: An introduction to predictive modelling in phytogeography. Dr.W. Junk, The. Hague, 1981.

Dudley and S. Stolton, Some like it hot.WWF-International, Gland, 1993.

Ellenberg H., Vegetation Ecology of Central Europe, Cambridge University Press, Cambridge, 1988.

Emanuel W.R., H.H. Shugart and M.P. Stevenson, Climate Change 7 (1985) 29-43.

Epstein P.R., H.F. Diaz, S.Elias, G. Grabherr, N.E. Graham, W.J.M. Martens, E. Mosley-Thompson and J. Susskind, Bull.Am.Met.Soc. 79 (1998), 409-417.

European Environmental Agency, Environmental assessment report No 2, EEA, Kopenhagen, 1999.

Fowbert J.A. and I. Lewis Smith. Arctic and Alpine Res. 26 (1994) 290-296.

Franz H., Ökologie der Hochgebirge. Ulmer, Stuttgart, 1979.

Grabherr G., E. Mähr and H. Reisigl, Oecologia Plantarum, 13 (1978) 227-251.

Grabherr G., Farbatlas Ökosysteme der Erde, Ulmer, Stuttgart, 1997.

Grabherr G., Guide des Ecosystemes de la Terre, Ulmer, Stuttgart, 1999.

Grabherr G., H. Pauli and M. Gottfried. Climate effects on mountain plants. Nature 369 (1994) 448.

Grabherr G., M. Gottfried, A.Gruber and H. Pauli, In: T. Chapin and C. Körner (Eds.), Arctic and Alpine Biodiversity, Springer, Berlin Heidelberg, 1995.

Grime J.P., Plant Strategies and Vegetation Processes. Wiley, Chichester, 1979.

Henry G.H.R. (Ed.), The International Tundra Experiment (ITEX). Short term responses of Tundra Plants to Experimental Warming. Global Change Biology 3, Supplement (1997),

Houghton J.T., L-G. Meira Filho, B.A. Callander, N. Harris, A. Kattenberg and K. Maskell (Eds.) Climate Change 1995. The Science of Climate Change. Cambridge University Press, Cambridge, 1996.

Klötzli F. and G.-R. Walther (Eds.). Conference on Resent Shifts in Vegetation Boundaries of Deciduous Forests, Especially Due to General Global Warming. Birkhäuser, Basel, 1999.

Körner C., Alpine Plant Life. Springer, Berlin, 1999.

Messerli B. and J. Ives, Mountains of the world – a global priority, Parthenon, Cranforth, 1997.

Parmesan C.. Nature, 382 (1996),765.

Pauli H., M. Gottfried and G. Grabherr, Phytocoenologia 29 (1999) (in press).

Shiyatov S.G., In: Akademiya nauk SSSR, Uralskiy nauchniy zentr (Ed.) Floristicheskie i na Urale. Akademiya nauk SSSR, Jekaterinenburg, 1983.

Smith T.M., H.H. Shugart and F.I. Woodward (Eds.), Plant Functional Types, Cambridge University Press, Cambridge, 1997.

Solomon A.M. and H. Shugart, (Eds.), Vegetation Dynamics and Global Change. Chapman & Hall, New York, London, 1993.

Walter H., Vegetation of the Earth and Ecological Systems of the Geo-biosphere. Springer, Berlin, 1985.

Wielgolaski F. (Ed.), Polar and Alpine Tundra, Elsevier, Amsterdam, 1997.

Effects of Elevated CO_2 and Nitrogen Deposition on Natural Regeneration Processes of Cut-Over Ombrotrophic Peat Bogs in the Swiss Jura Mountains.

PHILIPPE R. GROSVERNIER[1], EDWARD A.D. MITCHELL[2], ALEXANDRE BUTTLER[2], JEAN-MICHEL GOBAT[2]
[1]*Natura, Applied biology and ecological engineering, CH-2722 Les Reussilles, Switzerland*
[2]*Department of Plant Ecology, Institute of Botany, University of Neuchâtel, Neuchâtel, Switzerland*

Key words: *Sphagnum*, mosses, competition, environmental change, elevated CO_2, N deposition.

Abstract: In the Swiss Jura mountains most of the remaining ombrotrophic peat bogs have been exploited to some extent for peat. In these sites, natural regeneration processes are taking place. The dominant process is paludification, where a cut over drained surface is colonised by key species, usually either *Polytrichum strictum or Eriophortum vaginatum.* These early colonisers of bare peat surfaces create microclimatic conditions that enable the re-colonisation of *Sphagnum* mosses, usually S. *fallax*. In later stages of the succession *S. fallax* grows to form a continuous carpet and the key species gradually suffer from competition for light availability. We studied the effect of elevated CO_2 (560 ppm) and nitrogen deposition (30 kg ha $^{-1}$ year $^{-1}$) on the competition between *Sphagnum fallax and Polytrichum strictum* year in a three years field experiment (EU project BERI - Bog Ecosystem Research Initiative) using miniFACE systems (small size Free Air Carbon dioxide Enrichment). The cover and growth in length of the two species was monitored. The height difference between the emerging *Polytrichum* and the top of the *Sphagnum* mosses was also recorded at regular intervals. Effect Of CO_2: *Sphagnum* cover increased in the first year but this trend was not confirmed subsequently, whereas *Polytrichum* cover was not affected by elevated CO_2 Both *Sphagnum and Polytrichum* had a reduced growth in length under elevated CO_2. However, the growth of *Sphagnum* was less reduced than that of *Polytrichum* and therefore the height difference between *Sphagnum* and *Polytrichum* decreased. Effect of N: *Sphagnum* cover declined and *Polytrichum* cover doubled over the three years period in the high N plots. *Sphagnum* growth in length was not significant affected by N, but *Polytrichum* grew more in the

G. Visconti et al. (eds.), Global Change and Protected Areas, 347–356.

high N plots. As a results the height difference between *Sphagnum and Polytrichum* increased. These results suggest that elevated CO_2 and nitrogen deposition may have contrasting effects on bog regeneration. The positive effect of elevated CO_2 on *Sphagnum* mosses may be counterbalanced by higher N deposition levels.

1. INTRODUCTION

Sphagnum mosses are the main component of bogs and, as such, they are a major contributor to the accumulated peat characterizing these ecosystems. Covering approximately 3% of the Earth's land, northern peatlands contain 20-30% of the world's soil organic carbon [1], and, when growing, they constitute nearly continuous carbon sinks. In this respect, the observed human-induced increase of atmospheric CO_2, over the past 200 years [2] may have consequences on the growth, production and population dynamics of *Sphagnum* mosses, with possible feedbacks on peat accumulation and/or decomposition rates, and thus on the global carbon cycle.

A similar assumption can be made regarding the alteration of the global nitrogen cycle, considering that annual N fixation has been more than doubled by human activities such as use of fertilizers, fossil fuel combustion and leguminous crops [2]. There is now strong concern that such additional N inputs may affect peatland habitats, and especially raised bogs which are fed solely by atmospheric nutrient deposition [3].

The photosynthesis of *Sphagnum* mosses is known to be stimulated by higher atmospheric CO_2 concentrations [4], but this does not necessarily result in an increased biomass production. The response is indeed species dependent and related, at least, with depth to the water table [5,6]. It may also well be that, as was observed for other terrestrial ecosystems [8], the very low nutrient budget of the oligotrophic bog habitat is limiting the response of the plants to increased atmospheric CO_2 concentrations. Converging results from different sources though show that *Sphagnum* mosses respond to increased atmospheric CO_2 concentrations by a significantly reduced growth in length and an increased shoot density. On the other hand, additional atmospheric N inputs of 10-30 kgN ha^{-1} y^{-1} to the bog ecosystem result in an increased biomass production of *Sphagnum* in N-limited sites [9] but have no effect or even reduce *Sphagnum* growth and biomass production respectively in sites that are not N-Limited or at higher N deposition levels [9,5].

Combined effects of increased atmospheric CO_2 concentrations and additional N inputs were shown to affect only the relative length increment by accelerating the growth in length [5] of the mosses.

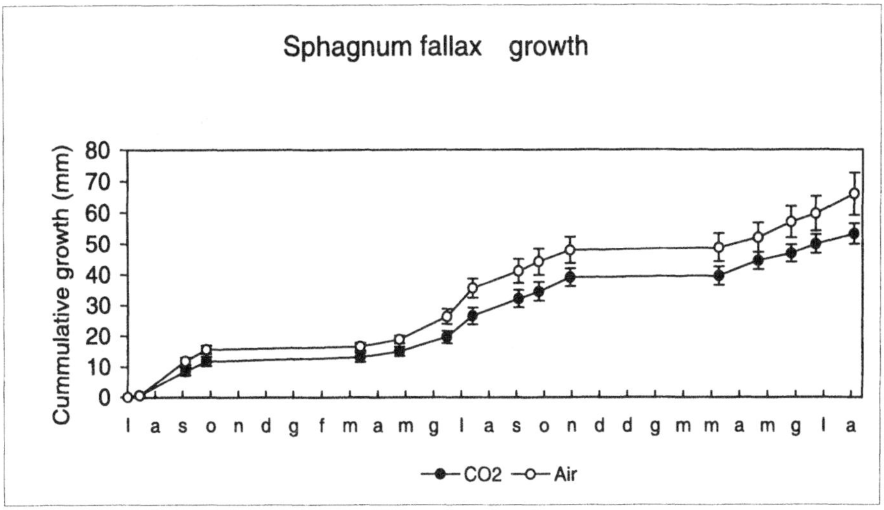

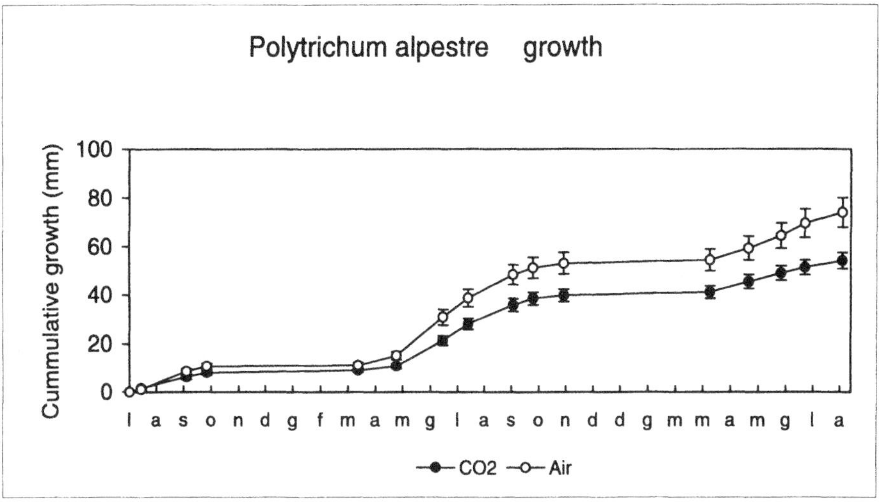

Fig. 1: Cummulative monthly growth in length (mm) of Spagnum fallax (a) and Polytrichum strictum (b) under ambient and elevated CO_2 concentration (error bars are SE).

These results suggest that: i) changes in the global C and N cycles affect the structure of the moss carpet in terms of annual length increment length increment rate and shoot density; ii) such structural changes at the community level may influence the competitive ability of *Sphagnum* mosses; iii) changes are expected to occur priority in more minerotrophic, thus less nutrient limited, sites. It is further hypothesized, that, due to the strong homeostatic properties of many ecosystems including bogs, no significant feedback on the global C and N cycles may be observed if considering only the "P-factor" (the extent to which an increase in the concentration of CO_2 increases carbon storage in the ecosystem) of individual plants or plant species [2]. But, at the community level, changes in the species composition as a consequence of competitive exclusion due to e.g. mutual shading, as was suggested for the typical bog vascular plant *Eriophorum vaginatum* [10], may result in far more importance effects [2, 11, 12].

Competition among bryophytes, and among *Sphagnum* spp. in particular, is supposed to occur extremely slowly, if at all [13]. But competitive interactions seem to take place for example between *Sphagnum fallax and S.magellanicum* under increased N deposition [14] or in successional dynamics between *Sfallax and Polytrichum strietum* [15, 16]. Successional processes are of utmost importance in the restoration of cut-over bogs, in that the outcome of the succession determines if the restored bog turns back to a C accumulating Sphagnum dominated system or into a C source system, dominated by so called "brown" mosses and vascular plants which are not peat accumulators.

This should draw the attention of policy makers and of industrial peat producers who aim at developping a sustainable peat economy. In this respect, we conducted a field experiment in the Swiss Jura mountains, in the broader framework of BERI (Bog Ecosystem Research Initiative), a project gathering partners from 5 countries that was part of the IV'" Environment Program of the European Community. In this paper we report on effects of increased atmospheric C02 or N depositions on the growth dynamics of *Sphagnum fallax and Polytrichum strictum,* growing intermixed in a previously cut-over bog.

2. METHODS

Study site

The study site, La Chaux-des-Breuleux, lies in the subalpine Jura mountains, Switzerland (47°N, 7°E) at an altitude of l'000 m a.s.l.

The mire was drained and peat was mined until the end of World war 11. A bog vegetation has re-established through a natural regeneration process [17] and a mosaic of lawn, hummocks and depressions is now well developed with *Eriophorum vaginatum, Carex nigra* and *Vaccinium oxycoccos* building the sward of the herbaceous layer.

Sphagnum fallax and Polytrichum strietum are the dominant species forming a two-layers carpet, the matrix of wich is almost exclusively made of *Sphagnum,* while *Polytrichum* shoots sprout out of it and form a sparse fur a few centimetres high over the *Sphagnum* carpet.

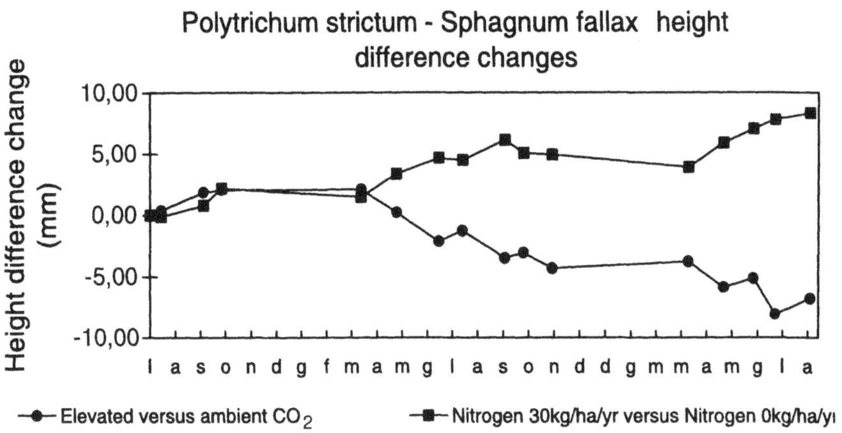

Figure 2: Monthly change in height difference between the carpet of Spagnum fallax and the top of the shoot of Polytrichum strictum under ambient and elevated CO_2 concentration.

This particularly simple two-species moss system seemed very well adapted to study interspecific competition interactions in relation to environmental changes such as increased atmospheric CO_2 concentration or N deposition. Selection of experimental plots 20 plots (l-m circles) were randomly selected in a homogeneous patterned part of the mire. 10 of them would be used for the CO_2 enrichment experiment, and 10 others for the N enrichment experiment. Each set of ten plots was subdivided in 5 replicates pairs of one treated and one control plot.

CO_2 enrichment

Mini-FACE (small Free Air CO_2 Enrichment systems) were used to raise atmospheric CO_2 concentrations to 560 ppm in the 1-m-diameter plots starting in June 1996.

The size of the plots was very well adapted to the size of the microhabitats of the typical hummock-hollow bog vegetation structure. By using a Mini-FACE, we furthermore ensured to become free of artifacts bound to comparable chamber or even open-top chamber systems and to keep as close as possible to the natural conditions of the bog [18]. Control plots were also set with the same Mini-FACE structures, ambient air being blown through the vents without additional CO_2.

Nitrogen enrichment

The natural annual N deposition in the Jura mountain is estimated to be ca. 15 kg ha^{-1} $year^{-1}$.

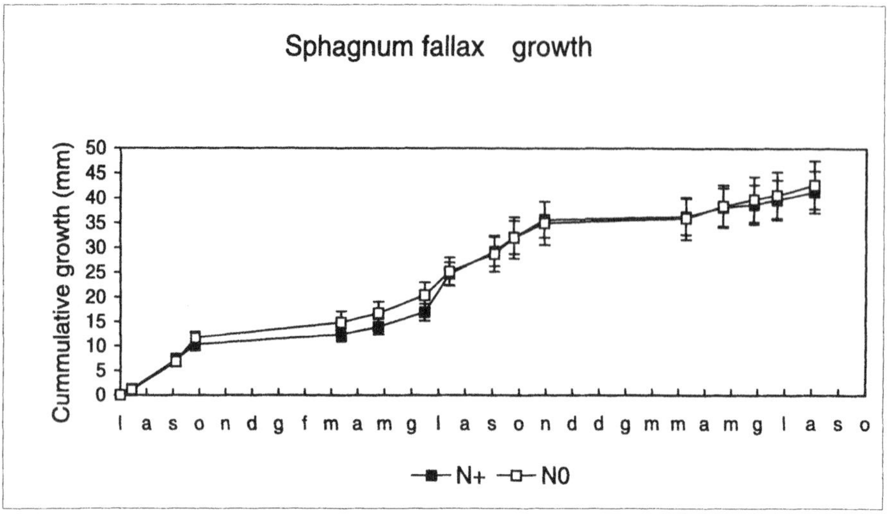

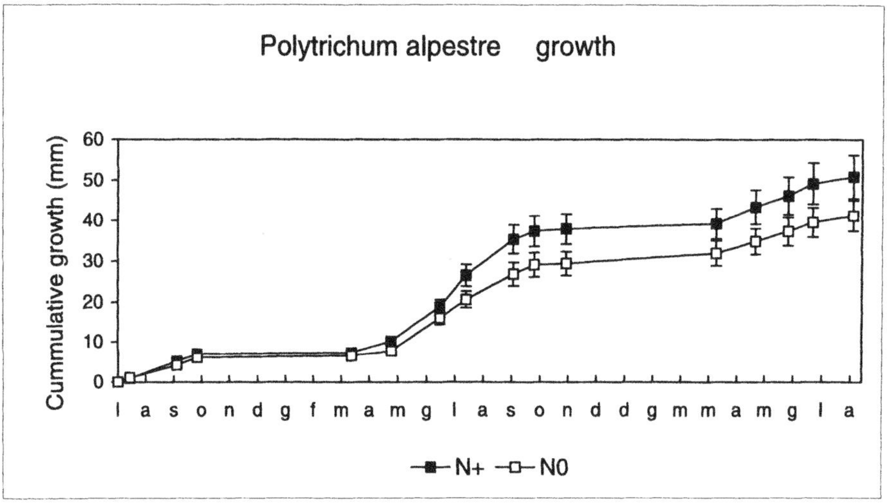

Figure 3: Cummulative monthly growth in length (mm) of Spagnum fallax (a) and Polytrichum strictum (b) under ambient and elevated nitrogen deposition (error bars are SE).

An additional increase of the annual N deposition of 30 kg ha $^{-1}$ year $^{-1}$ was simulated by watering the plots at a fortnight interval in six applications during the vegetation period.

The control plots received only deionized water at the same time and in the same amount as the treated plots.

Comparative study of *Sphagnum and Polytrichum cover*

The vegetation was analysed from July 1996 to August 1998 at a regular interval (April - July - October) using a point-intercept method [19]. We chose a homogeneous subplot (35 x 22.5 cm, grid interval 2.5 cm) to monitor plant distribution across 150 points. Specific frequencies of all species, including vascular plants in July, were calculated. Authorities for plant species follow Corley *et al.* [20] for mosses and Tutin *et al.* [21] for vascular plants.

Plant growth

The growth in length of the mosses was measured at a monthly interval during the vegetation period (roughly from April to October). To avoid destructive harvesting or disturbing repeated measurements in the small study plots, we used the "cranked wire" method [22]. 5 such wires were set into the moss carpet of each plot at the very beginning of the experiment and all were measured each time to obtain an average measurement per plot. The height of both the *Sphagnum* and the *Polytrichum* mosses were measured, and the height difference between the two species calculated.

3. RESULTS

Effects of elevated atmospheric CO_2 concentration

Under elevated $C0_2$ concentration, the shoot density of *Sphagnum fallax* increased by 25% in the first 3 months of the experiment, but then fluctuated between 10 and 20% more than at the beginning. The overall trend did not result in a significant difference between July 1996 and October 1998. The control Plots, with ambient air blown by the Mini-FACE, also underwent a similar, though less important trend towards an increasing density of the *Sphagnum* shoots, the increase being only between 5 and 10%.

No change in shoot density could be assessed for *Polytrichum strictum* after the 3 years of experiment. Seasonal fluctuations though show that the shoot density of *Sphagnum fallax is* at its maximum during the summer time, and at its minimum in early spring after snow melt. *Polytrichum strictum* on the other hand displays the opposite pattern, with stable, similar shoot densities in early spring, but lower densities.

The growth in length of *Sphagnum fallax* and that of *Polytrichum strictum were* significantly reduced under elevated atmospheric CO_2 compared to the ambient air treatment. The average difference in growth length between treatment for both moss species was already apparent by the end of the first season, after only 3 months time, and it increased throughout the duration of the experiment (fig. la and lb).

However, the growth of *Sphagnum fallax* was less reduced than that of *Polytrichum strictum* and therefore the height difference between the carpet of *Sphagnum fallax* and the top of the shoots of *Polytrichum strictum* eventually decreased (fig. 2, lower curve).

Effects of elevated atmospheric N deposition

After a slight increase of nearly 10% by the end of the first growing season, the soot density of *Sphagnum fallax* shrank significant and rapidly, under increased atmospheric deposition of nitrogen, and stabilized for the two subsequent years at an average 20% less cover than at the beginning of the experiment. Moreover, one could hypothesize that *Sphagnum fallax* would probably have totally disappeared, had the experiment lasted for one more year. Most of the plants were indeed close to death by the end of 1998, showing clear signs of chlorosis and strongly reduced vitality.

No change in percentage cover was observed in the control plots, but for a non significant overall 5% increase. On the opposite, the shoot density of *Polytrichum strictum* significant doubled during the three years experiment under increased nitrogen deposition, while no significant change in cover could be observed in the control plots. The growth in length of *Sphagnum fallax* was not significantly affected by increased nitrogen deposition (fig. 3a), while *Polytrichum stríctum* significantly grew more in the nitrogen treated plots (fig. 3b). As a results the height difference between *Sphagnum fallax and Polytrichum strictum* increased from the beginning of the second year to the end of the experiment (fig. 2, upper curve).

4. DISCUSSION

These results suggest that elevated CO_2 and increased nitrogen depositions may have contrasting effects on the regeneration of cut-over bogs. On the one hand, if bogs are supposed to act as effective carbon sinks, the observed increase in shoot density of *S. fallax,* combined with a decreasing height difference between the *Sphagnum* moss carpet and the overtopping *P. strictum,* seems to confer a competitive advantage to *S. fallax.*

The response of *S. fallax* to elevated atmospheric CO_2 in a secondary, disturbed but regenerating bog, is a structural one: the moss has been adapting the fine structure of its shoots to increase the density of the carpet it forms. Crowing denser, even if less in length, means a better water flow regulation inside the *Sphagnum* carpet and hence a better resistance to the strong fluctuating water table level of the hydrologically disturbed bog. What the plant lost in elongation was thus gained in terms of structural stability and stress tolerance. Furthermore, as the other component of the simple, specific-poor moss layer, P. *strictum,* also underwent an even more reduced growth in length, one may hypothesize that, on a longer term basis than just three years, the *Polytrichum* mosses would eventually drown into the *Sphagnum* carpet and, if not disappear totally, at least form no more than a minor companion species.

This is suggested by the observed seasonal pattern of the evolution of the shoot density and the growth in length of *P. strictum.* In early spring, while the soft carpet of *S. fallax* has been crushed down by the snow during the winter and the shoot density of *S. fallax* is at its lowest in the year, the more robust *P. strictum* benefits from favorable, wet conditions to thrive and increase its shoot density. Later on, by the time when *S. fallax* resumes growth and its shoots stand erect again, *P. strictum* has to keep pace with the growth of the invading Sphagnum carpet in order to compete for light availability. The reduced growth in length under elevated atmospheric CO_2 during this crucial period may hamper *P. strictum* in its attempt to keep its tip over the faster growing *Sphagnum* carpet. By enhancing the population structural stability of *S. fallax,* an elevated atmospheric CO_2 concentration may favour the regeneration of secondari cut- over bogs. *S. fallax* may indeed outcompete the pioneer non- *Sphagnum moss* P. *strictum,* which had been the first dweller of the bare peat fields left behind by the peat extraction [17]. Now, it is the *Sphagnum* mosses that determine the accumulation process of peat by controlling the ecological conditions of the bog environment. Moreover, peat accumulation is progressing at a faster rate in early stages of the bog's formation, while it inexorably tends towards a steady-state after a certain time [1]. On the other hand, the positive effect of elevated C02 on *Sphagnum* mosses may be counterbalanced by higher N deposition levels. In such an event, the disappearance of *S. flax,* and its concomitant replacement by *P. strictum,* means that there would be almost no more peat accumulation any more. Consequently, the hydrological properties of the bog would be disturbed, the peat would eventually dry out and mineralization processes would take place, leading the bog to become a net C source.

This emphasizes the importance that should be awarded by public policies regarding efforts to restore damaged bog ecosystems not only for biodiversity conservation aims, but also as possibly effective carbon sinks.

5. REFERENCES

Aerts R., B. Wallén & N. Malmer (1992) J. Ecol., 80:131-140

Bazzaz F.A. (1990) Annu. Rev. Ecol. Systematics, 21:167-196.

Buttler, Ph.R. Crosvernier & Y.Matthey (1998) J. Appl. Ecol. 35:800-810.

Buttler, Vegetatio 103 (1992) 113-124

Carbon dioxide and terrestrial ecosystems. Academic Press, San Diego. 1996, pp. 163-176.

Clymo R. S. (1970) J. Ecol., 58:13-49.

Corley M.F.V., A.C. Crundwell, R. Dúll, 0. Hill, A.J.E. Smith, J. Bryol. 11 (1981) 609-689.

Crosvernier Ph. R., Y. Matthey, A. Buttler, In: B.Wheeler, S.Shaw, W.Fojt, Robertson A, (eds) The restoration of temperate wetlands. J Wiley & Sons, Chichester UK, 1995, pp. 437-450.

Crosvernier Ph.R., Y.Matthey & A.Buttler (1997) J. Appl. Ecol. 34:471-483.
D.T. Tissue & W.C. Oechel (1987) Ecology, 68:401-410.
Gorham E. (1991) Ecol. Aplic. 1:182 - 195.
Heijden E. v.d., J. Jauhiainen, J. Matero, M. Eekhof & E. Mitchell (1998) In: L.J. de Kok & I. Stulen (eds), Responses of plant metabolism to air pollution and global change. Backhuys Publishers, Leiden, NL. 1998, pp. 475-478.
Jauhiainen J., H. Vasander & J. Silvola (1994) J. Bryol., 18:83-95.
Korner C. (1993) In: A.M. Soloman & H.H. Shugart (eds) Vegetation dynamics and global change. Champan & Hall, New York, pp. 53-70.
Miglietta F., M.R. Hoosbeek, j. Foot, F. Gigon, M. Heijmans, A. Peressotti, T. Saarinen, N. van Breemen, B. Wallen, Environmental Monitoring and Assessment (in press).
Oechel W.C. & C.L.Vourlitis(1996) In: C,.W. Koch & H.A. Mooney (eds),
Rydin H. (1997) Advances in Bryology, 6:135-168.
Silvola J. (1985) Holarctic Ecology, 13:224-228.
Tutin T.C., V.H.Heywood, N.A. Burges, D.H. Valentine, S.M. Walters, D.A. Webb, Flora europaea. 5 vol. Cambridge Universtity Press, Cambridge. 1964- 1980.
Twenhoven F.L. (1992) J. Bryol., 17:71-80.
Tybirk K., J. Bak & L.H.Henriksen (1995) TemaNord 1995:610.
Vitousek P.M. (1994) Ecology, 75:1861-1876.
Wallén B., U. Falkengren-Crerup & N. Malmer (1988) Holarctic Ecology, 11:70- 76.

Section 3

Socio – Economic Implications

Economic Evaluation of Italian Parks and Natural Areas

SANDRA NOTARO[1] AND GIOVANNI SIGNORELLO[2]
Istituto Agrario di S.Michele a/Adige via Mach, 1 38010 S. Michele a/Adige
Dipartimento di Scienze Economico-Agrarie ed Estimative, Università di Catania via Valdisavoia, Catania.

Key words: Italian parks and natural areas benefits, Environmental values, Non market valuation techniques, Travel Cost Method, Contingent Valuation Method.

Abstract: The estimate of the Economic Value of environmental goods and services is a fundamental aspect for environmental management decisions. However, the failure of the market in valuing this type of goods imposes the use of non market valuation techniques, as the Contingent Valuation, the Travel Cost and the Hedonic Price Methods. These methods have gained increasing trustworthiness in the international academic world. A systematic search of the available literature in Italy was conducted in an effort to review as many empirical studies as possible from the beginnings of the Italian experience in environmental evaluation to 1999 and to produce a summary measure of the value of Italian parks and natural areas. Using an L estimator (linear order statistics estimator), and four M estimators (maximum likelihood type estimators) a per capita and per hectare net economic value for Italian environmental areas have been estimated. This information could be used to estimate the benefits deriving from the creation of a new park (or protected area) in Italy, to compare with costs this creation implicates.

1. INTRODUCTION

The estimate of the Economic Value of environmental goods and services is a fundamental aspect for environmental management decisions.

G. Visconti et al. (eds.), Global Change and Protected Areas, 359–372.

Nevertheless, even if economists have long measured the value of goods that are routinely bought and sold in the marketplace, ordinary markets do not exist for environmental (public) goods. This is the case of parks and protected areas. The failure of the market in valuing this type of goods imposes the use of non market valuation techniques, as the Contingent Valuation, the Travel Cost and the Hedonic Price Methods, methods that have gained increasing trustworthiness in the international academic world. There are few years that these techniques are applied even in Italy.

This paper is the result of a systematic research of the available literature in Italy, conducted in an effort to review as many empirical studies as possible from the beginnings of the Italian experience in environmental evaluation to 1999 and to produce a summary measure of Italian Parks and natural areas values. Included were studies in journals, chapters in books, unpublished research reports, theses and conference papers. The overall effect of the selection process was to provide sufficient studies to identify interesting trends and get a broad flavor of the findings from both published and unpublished studies.

The outline of the paper is as follows. Section 2 presents the Travel Cost (TCM) and the Contingent Valuation Methods (CVM), both applied by researches here reported. The first one analyze revealed preferences with the consumption of complementarity private good; the second one examines stated preferences in a virtual market built for the good object of estimate. In section 3 a review of Italian empirical studies concerning the economic evaluation of Parks and Natural Areas will be presented. Section 4 reports the results of estimates. A pro-capita and per hectare net economic value for Italian national and natural parks and areas will be presented. This information must be compared with costs that a park (or a protected area) implicates. Section 5 contains a brief discussion of the implications of our results.

2. ESTIMATION METHODS

The Travel Cost Method estimates the WTP referring to travel expenses sustained by consumers for in situ fruition of services offered by the environmental resource. The application of the method is possible if visitors cross different distances. Then, the consequent difference in the relative costs of transport is employed to estimate the demand curve, whose theoretical justification is found within the Household Production Theory.

Many versions of the TCM exist, in relationship to the nature of data demanded, to the objective and the conditions of the estimate, to the temporal horizon considered.

The *zonal version*, the classical one, estimates the recreational curve of demand dividing the basin of use in homogeneous zones, considering the cost of trip and other variables that can condition the number of per-capita visits.

Note: Zones have to be individualized in such a way that inside them costs per trip sustained by resident visitors are approximately equal.

The application is articulated in three phases.

In the first phase it is estimated the demand curve for the entire recreational experience, $V_i/N_i = f(TC_i)$, where TC_i is the average cost per visit, V_i is the total number of visits, and N_i is the population of the ith zone.

In the second phase the demand curve for the recreational site is estimated. Hypothesizing increments in the cost per visit, the progressive variation in the number of visits to the site is calculated.

Note: On the assumptions that for equal sustained cost the rate of fruition doesn't change, and that the consumers' reaction to increases in costs is independent to the cause that has produced it.

The algorithm of execution is the following:

a) a virtual ticket is established, T, such that the cost of access is $TC_i + T$;
b) the rate of fruition $V_i^*(TC_i + T)/N_i$ is estimated, relative to the new cost for visit;
c) the rate of fruition is multiplied by the resident population:
   $\left[V_i^*(TC_i + T)/N_i\right]N_i$;
d) the number of visits estimated is added for every zone, in such a way that
   $V^*(P) = \sum\left[V_i^*(T)/N_i^*\right]N_i$

is one of new points of the second recreational demand curve.

The iteration stops when $V^*(TC_i + T) = 0$ or the curve becomes asymptotic to the x axle.

In the third phase, total consumers welfare is measured, in terms of Marshallian consumer surplus, calculating the area below the recreational demand function:

$$CS = \int_0^T V^*(T)dT \tag{1}$$

The *individual version* instead estimates the recreational demand function taking into account the number of visits performed by every single consumer as dependent variable and the cost of the trip as independent variable.

The Contingent Valuation Method estimates economic values on the base of interviews, assuming that monetary preferences possessed for an environmental good can be expressed through a market simulation process. Its theoretical foundations are those of the modern welfare economy.

Among all estimation methods it is the most versatile and the only one able to measure non use and option values.

Note: Total economic value is the sum of use, option and non use values. Whilst use value correspond only to the direct use of the natural resource, option value refers to its future use. In the particularly case of recreation in natural parks it correspond to the willingness to pay (WTP) to preserve the resource, considering that in the future the respondent could visit it. Non use value includes bequest value (WTP to preserve the natural resource so that future generations could visit it) and existence values (WTP to maintain natural resource because nature has an independent value to direct, indirect, actual and future human use).

Nevertheless, its application is not so easy as it could be seem. To have a valid and reliable expression of value, the simulated scenario has to be draw in such a way to establish satisfactory transactions and must be incentive compatible. Particularly it is recommended:

1. to describe substantially, formally and comprehensibly the environmental good variation and possible economic effects;
2. to remind other substitute goods;
3. to remember income constraints;
4. to describe substantially, formally and comprehensibly the method of payment;
5. to furnish a detailed, comprehensible and realistic description of the hypothetical market (private or political) and of the rules that govern the payment and the offer of the good;
6. the welfare measure has always to be expressed in terms of WTP, even when the theoretical correct measure is WTA;
7. information to perform validity tests must be collected;
8. qualitative researchers, as focus group, one on one interviews, pre-tests, etc., are essential to produce a valid and reliable CV study;
9. the investigation (direct interviews are recommended) has to be statistically representative;
10. data elaboration has to be conducted with rigor and analytical and interpretative ability.

Nevertheless, it is not always possible to respect all these indications. An objective and transparent examination of contents and results of every application allows the attribution of quality judgments. It is correct to consider them if results of a CV research are used by the public decision-maker or in Courts. Preferences toward the good in question come directly answering a valuation question, which can have different formats. With an open ended question it is asked the maximum willingness to pay not giving up the good. The open question is theoretically the correct format but it is often not easy for respondents to answer without some form of "assistance".

Note: Sometimes the open ended format can be used without difficulty. When respondents have familiarity with the good and with the concept to pay for it, it is supposed that they have some defined preferences and therefore no problem to give a precise economic evaluation.

With the bidding game format the respondent has to answer affirmatively or not to a particular amount. Depending on his/her declaration of assent or refusal the following amount presented will be bigger or smaller. The game will continue until the exact valuation of the good is found.

Note: The biggest problem with this format is the starting point bias.

With the payment card format a series of values is presented. The respondent will choose the nearest amount from his/her real valuation of the good. To reduce the task of answer elaboration the closed-ended (dichotomous choice) format has been introduced. With the take it or leave it approach (1-2) respondents must give a positive or negative answer to a predetermined amount. Note: This method also has defects. Between these, the tendency of respondents to give a positive answer to please the interviewer or the sponsor (yea saying) and the bias due to the choice of prices presented. But perhaps the biggest limit concerns the assumptions around the parametric specification of the evaluation function and the indirect utility function. The referendum format is similar, with the only difference that the question simulates the vote in a referendum. The dichotomous choice scheme it is not very efficient from the statistic point of view. Such awareness has brought to one recent elaboration of it, denominated double-bounded dichotomous choice. It differs from the single-bounded scheme since a second valuation question is presented. The amount of the second question proposed is conditioned by the answer interviews give to the first question. If the first answer is positive, the second one will be of a greater amount. If it is negative, the second one will be of a smaller amount. With the multiple bounded format (3), different bids are presented. The respondent has to answer for every bid if he/she is willing to pay it and his/her level of certainty: definitely yes, probably yes, not sure, probably no, definitely no. With the one and a half bounded format (OOH) (4), before the valuation question it is said that the cost of the good will fall between two defined amounts. After that, at random one of the two amounts is presented. A second valuation question will be presented only if to the smaller amount the respondent says yes and if to the bigger amount she/he says no. Otherwise the second bid will not be offered. Another recent variation of the dichotomous choice scheme is the contingent ranking. The respondent is called to order a whole defined set of alternative offers of the environmental resource, according to a staircase of decreasing preference.

Tab. 1 - Economic Evaluation of Italian Parks and Natural Areas

The good	Author(s)	type of good	type of environment	valuation method	use value	n° trip for year	per capita value	capital value at 2% rate	capital value at 1% rate	surface (ha)	capital value per ha (2%)	n° trip for year ha
De la Mandria Park (Torino), 1995	Bravi M., Curto R. (1996)	regional park		CV	5.432.000.000	776.000	7.000	271.600.000.000	543.200.000.000	130	2.089.230.769	5.969
Sacro Monte Varallo-Sesia (Vercelli), 1995	Bravi M., Curto R. (1996)	"special nature"	Alpine	CV	2.130.789.870	123.381	17.270	106.539.493.500	213.078.987.000	30	3.551.316.450	4.113
Simeto Oasi (Catania), 1988	Signorello G. (1990)	natural reservoir	coastal	CV	nd	nd	35.133	nd	nd	nd	nd	nd
Pollino Park (Calabria e Basilicata), 1995	Bernetti I., Romano S.(1996)	national park	Appennine	CV	5.752.630.000	274.000	20.995	287.631.500.000	575.263.000.000	294.600	976.346	1
Portoselvaggio (Lecce), 1989	De Fano G., Grittani G. (1992)	"natural park tackled"	Mediterranean	ZTCM	425.027.625	39.875	10.659	21.251.381.250	42.502.762.500	425	50.003.250	94
Orecchiella (Lucca), 1988	Marinelli A., Casini L., Romano D. (1990)	natural park	Appennine	ITCM	675.384.240	167.340	4.036	33.769.212.000	67.538.424.000	5.218	6.471.677	32
				CV(WTP)	4.329.085.800	167.340	25.870	216.454.290.000	432.908.580.000	5.218	41.482.233	32
				CV(WTA)	10.833.926.280	167.340	64.742	541.696.314.000	1.083.392.628.000	5.218	103.813.015	32
Umbra Forest (Foggia), 1984	Marinelli A., Romano S. (1987)	national park	Mediterranean	ZTCM	682.771.716	360.873	1.892	34.138.585.800	68.277.171.600	4.989	6.842.771	72
Dolomiti Bellunesi,1987	Gatto P. (1988)	regional park	Alpine	ZTCM	18.718.800	7.600	2.463	935.940.000	1.871.880.000	8.000	116.993	1
				CV(WTP)	29.556.400	7.600	3.889	1.477.820.000	2.955.640.000	8.000	184.728	1
Abetina Reale (Reggio Emilia), 1988	Gatto P. (1988)	State forest	Appennine	ZTCM	33.680.000	10.000	3.368	1.684.000.000	3.368.000.000	852	1.976.526	12
				CV(WTP)	38.160.000	10.000	3.816	1.908.000.000	3.816.000.000	852	2.239.437	12
Cansiglio (Belluno), 1987	Tosi V. (1989)	forest	Pre-alpine	ZTCM	4.272.341.024	468.664	9.116	213.617.051.200	427.234.102.400	4.230	50.500.485	111
				CV(WTP)	1.424.269.896	468.664	3.039	71.213.494.800	142.426.989.600	4.230	16.835.342	111
				CV(WTA)	4.628.057.000	468.664	9.875	231.402.850.000	462.805.700.000	4.230	54.705.165	111
Calamento Valey (Trento), 1987	Tosi V. (1989)		Alpine	ZTCM	2.120.379.750	214.722	9.875	106.018.987.500	212.037.975.000	4.000	26.504.747	54
				CV(WTP)	587.264.670	214.722	2.735	29.363.233.500	58.726.467.000	4.000	7.340.808	54
				CV (WTA)	2.120.379.750	214.722	9.875	106.018.987.500	212.037.975.000	4.000	26.504.747	54
Cimoliana Valey (Pordenone), 1987	Tosi V. (1989)		Pre-alpine	ZTCM	352.369.360	38.020	9.268	17.618.468.000	35.236.936.000	6.826	2.581.082	6
				CV(WTP)	103.984.700	38.020	2.735	5.199.235.000	10.398.470.000	6.826	761.681	6
				CV(WTA)	363.889.420	38.020	9.571	18.194.471.000	36.388.942.000	6.826	2.665.466	

Tab. 1 (cont) - Economic Evaluation of Italian Parks and Natural Areas

The good	Author(s)	type of good	type of environment	valuation method	use value	n° trip for year	per capita value	capital value at 2% rate	capital value at 1% rate	surface (ha)	capital value per ha (2%)	n° trip for year ha
Vico Lake basin (Viterbo), 1991	Romano D., Carbone F. (1993)	partial natural reservoir	Appennine	ITCM	293.304.000	33.000	8.888	14.665.200.000	29.330.400.000	4.500	3.258.933	7
				CV(ticket)	207.108.000	33.000	6.276	10.355.400.000	20.710.800.000	4.500	2.301.200	7
				CV(club)	307.989.000	33.000	9.333	15.399.450.000	30.798.900.000	4.500	3.422.100	7
				CV (season ticket)	174.240.000	33.000	5.280	8.712.000.000	17.424.000.000	4.500	1.936.000	7
Mis Valey (Belluno), 1993	Tempesta T. (1995)	national park	Alpine	ITCM	422.040.000	60.000	7.034	21.102.000.000	42.204.000.000	nd	nd	nd
				ZTCM	426.240.000	60.000	7.104	21.312.000.000	42.624.000.000	nd	nd	nd
				CV(WTP)	203.640.000	60.000	3.394	10.182.000.000	20.364.000.000	nd	nd	nd
Lessinia Natural Park (Vicenza), 1994-95	Bettinazzi R. (1994-95)	regional park	Pre-alpine	ITCM	1.064.969.700	203.900	5.223	53.248.485.000	106.496.970.000	10.500	5.071.284	19
				ZTCM	981.166.800	203.900	4.812	49.058.340.000	98.116.680.000	10.500	4.672.223	19
Appennino Tosco-Romagnolo, 1991	Romano D., Rossi M. (1994)	national park	Appennine	CV(WTP)	121.132.000	11.000	11.012	6.056.600.000	12.113.200.000	nd	nd	nd
Ninfa Garden (Latina), 1995	Venzi L., Rivetti M. (1990)	botanic garden		ZTCM	2.500.000.000	40.000	62.500	125.000.000.000	250.000.000.000	nd	nd	nd
Anapo Valey (Siracusa), 1990	Signorello G. (1994)	regional reservoir		CV(logit)	285.660.000	30.000	9.522	14.283.000.000	28.566.000.000	nd	nd	nd
				CV(PAVA)	230.370.000	30.000	7.679	11.518.500.000	23.037.000.000	nd	nd	nd
Rosandra Valey (Trieste), 1981	Merlo M. (1982)	State forest	Alpine	ZTCM	61.666.473	28.589	2.157	3.083.323.650	6.166.647.300	210	14.682.494	136

Tab. 1 (cont) - Economic Evaluation of Italian Parks and Natural Areas

The good	Author(s)	type of good	type of environment	valuation method	use value	n° trip for year	per capita value	capital value at 2% rate	capital value at 1% rate	surface (ha)	capital value per ha (2%)	n° trip for year ha
Tarvisio (Udine), 1981	Boatto V., Defrancesco E., Merlo M. (1984)	State forest	Alpine	ZTCM (summer turists)	5.001.502.275	308.259	16.225	250.075.113.750	500.150.227.500	23.000	10.872.831	13
				ZTCM (winter turists)	4.175.984.673	308.259	13.547	208.799.233.650	417.598.467.300	23.000	9.078.228	13
				ZTCM (turists in hotel)	14.198.409.540	308.259	46.060	709.920.477.000	1.419.840.954.000	23.000	30.866.108	13
				ZTCM (turists in camping)	12.527.954.019	308.259	40.641	626.397.700.950	1.252.795.401.900	23.000	27.234.683	13
Akamas (Cipro), 1989	De Fano G., Grittani G. (1992)	peninsula	Mediterranean	CV		271.400				17.000		
Campogrosso (Trento), 1998, pessimistic estimate	Notaro S., Signorello G. (1999)	alpine pasture	Alpine	CV (open)	310.389.190	49.370	6.287	15.519.459.500	31.038.919.000	500	31.038.919	98,74
				CV (BS)	433.369.860	49.370	8.778	21.668.493.000	43.336.986.000	500	43.336.986	98,74
				CV (DB)	342.331.580	49.370	6.934	17.116.579.000	34.233.158.000	500	34.233.158	98,74
				CV (MB)	259.291.240	49.370	5.252	12.964.562.000	25.929.124.000	500	25.929.124	98,74
Campogrosso (Trento), 1998, optimistic estimate				CV (open)	351.883.390	55.970	6.287	17.594.169.500	35.188.339.000	500	35.188.339	111,94
				CV (BS)	491.304.660	55.970	8.778	24.565.233.000	49.130.466.000	500	49.130.466	111,94
				CV (DB)	388.095.980	55.970	6.934	19.404.799.000	38.809.598.000	500	38.809.598	111,94
				CV (MB)	293.954.440	55.970	5.252	14.697.722.000	29.395.444.000	500	29.395.444	111,94

Note: Actual Values for 1995

Every alternative must be described from at least two attributes, among which an evident tradeoff has to subsist; one of the attributes is the price, varying in relationship to the level of offer.

3. ECONOMIC EVALUATION OF ITALIAN ENVIRONMENTAL RESOURCES

Empirical studies in environmental economic valuation started in eighties and grew only during last years. The whole production consists of about 40 studies. Only 22 deal with outdoor recreation in Italian national/natural parks and areas. Table 1 shows 45 estimates of net economic value (26 estimated using the Contingent Valuation Method and 19 using the Travel Cost Method) reported in these outdoor recreation demand studies, updated in actual Liras.

4. RESULTS

Summary estimates has been calculated employing one of the L estimator (linear order statistics estimators), the α trimmed mean, and four different M estimators (maximum likelihood type estimators), Huber, Hampel, Andrews and Tukey's biweight estimators. Both families are part of the bigger "Robust Estimators" family.

Robust Estimators allow having a protection against uncertainty or acknowledge about the mechanism of data generation (5). They are not sensitive to violations of assumptions around the way the data have been produced (6). The underlying logic of this type of techniques is to minimize the role developed by potential great errors relative to the statistic of interest, involving only small losses in efficiency. The common procedure of these estimators consists in removing observations that don't seem produced by the model that produced all others.

Univariate robust procedures estimating the central tendency of a distribution follow this general form:

$$MIN \sum_{i=1}^{n} w_i (x_i - \overline{x})^p \qquad (2)$$

where w_i is the weight, x_i the ith observation, $\overline{x}$ the estimate of central tendency and p defines the distance function to minimize.

L estimators result from a linear combination of ordered sample statistics:

$$\tilde{\mu} = \sum c_i x_{(i)} \tag{3}$$

where c_i, weights, are smaller in extremes.

α trimmed mean is one of L estimators. After having removed the α percentage of bigger and smaller offers, giving them zero weigh, one to other observations and value two to the distance function, the average of the observations is calculated. That is, α trimmed mean is defined as:

$$\overline{X} = \frac{X_{(n\alpha)+1} + \ldots + X_{n-(n\alpha)}}{n - 2[n\alpha]} \quad \text{for } 0 \leq \alpha \leq ½ \tag{4}$$

where:

X_1 = smaller offer;

X_n = bigger offer.

α trimmed means estimated values for Italian Parks and Natural Areas are presented in table 2.

Tab. 2 - Per-capita and per hectare estimate of Italian Parks and Natural Areas calculated employing L estimators.

Per capita values (Italian £)	Per hectare values (Italian £)	L estimator employed
12.259	2.684.797	Mean
11.343	1.257.476	α trimmed mean 5%
10.389	364.984	α trimmed mean 10%
9.729	332.929	α trimmed mean 15%
9.131	322.223	α trimmed mean 20%
8.290	302.190	α trimmed mean 25%

Source: our elaboration.

If the function is not normal, as in this case, M estimators can be used. M estimators follow the logic of underweight observations on the base of the distance from the center of the distribution.

Some of these always give positive weights, smaller if the distance from the core of the data increases, some others assign zero weight.

Assuming that the sample originates from a continuous distribution, the principle is to estimate μ through $\tilde{\mu}$ to minimize:

$$\sum_{i=1}^{n} \rho(x_i - \tilde{\mu}) \tag{5}$$

where $\rho(\;)$ is a non-constant function of real values.

Four main different M estimators exist:

the Huber estimator $\psi(t) = \begin{cases} \frac{1}{2}t^2 & |t| \le \kappa \\ \kappa|t| - \frac{1}{2}\kappa^2 & |t| > \kappa \end{cases}$ (6)

where t is the absolute value of the error and κ is chosen by the researcher;
the Hampel estimator

$$\psi(t) = \begin{cases} t & |t| \le a \\ a \operatorname{sgn} t & a < |t| \le b \\ a(c \operatorname{sgn} t - t)/(c - b) & b < |t| \le c \\ 0 & |t| > c \end{cases} \tag{7}$$

the Andrews estimator

$$\psi(t) = \begin{cases} sin(t/d) & |t| \le d\pi \\ 0 & |t| > d\pi \end{cases} \tag{8}$$

the Tukey's biweight estimator

$$\psi(t) = \begin{cases} t(1 - t^2)^2 & |t| \le 1 \\ 0 & |t| > 1 \end{cases} \tag{9}$$

The main advantage of M estimators is that they underweight observations depending of the distance from the center of the distribution, and not symmetrically as α trimmed means do.

M estimated values for Italian Parks and Natural Areas are presented in table 3.

Tab. 3 - Per-capita and per hectare estimate of Italian Parks and Natural Areas calculated employing M estimators.

Per capita values (Italian £)	Per hectare values (Italian £)	M estimator employed
7.715	369.684	Huber
6.656	337.079	Tukey biweight
6.899	361.218	Hampel
6.639	337.291	Andrew

Source: our elaboration.

Range for per-capita and per hectare values for Italian Parks and Natural Areas are presented in table 4.

Tab. 4 - Range of the per-capita and per hectare values for Italian Parks and Natural Areas
Note: This table doesn't take into consideration 5% α trimmed mean that doesn't exclude from the analysis one very big extreme observation.

Per capita values (Italian £)	Per hectare values (Italian £)
6.639 - 10.389	337.079 - 369.684

Source: our elaboration.

It is evident that the choice of the estimator to use could be of substantial importance in relation to the value that will be obtained. Nevertheless, the amounts indicated have not to be read as the exact value of the environmental good, but as the indication of it's order of magnitude.

In fact all methods used to estimate the economic value have drawbacks, because they are not founded on market values but on social utility. For example, in TCM it is in general impossible to introduce all variables that determine the value of the good. The CVM overcomes big part of TCM limits, but it also has drawbacks. Biases can derive, as example, from sample arguments, the system of aggregation adopted, the free-rider behavior, hypothetical arguments, etc.

5. CONCLUSIONS

In the next future, the valuation activity will be more and more turned to the evaluation of environmental resources, as natural parks and area. More, it will be used not only for exclusive finalities of research, as has happened until today. It will be employed to answer to specific solicitations that will emerge from the real system of the economy and from Courts. This will happen if decision makers try to pursue the objective of sustainable development through a spread use of benefits-cost analysis, a progressive adjustment of the traditional tools of environmental politics to the rules of market, a development of a national system of economic and environmental integrated accounting, a generalized application of the objective responsibility for environmental damages.

Evaluation techniques will be a useful tool to give concreteness to the sustainable development concept. In this study a summary measure of Italian national and natural parks and areas have been estimate. Estimates obtained using a linear order statistics estimator and four maximum likelihood type estimators achieve similar values, for both per capita and per hectare values. Therefore amounts presented can represent the order of magnitude of these environmental values and could be transferred to estimate the benefits deriving from the creation of a new park (or protected area) in Italy, to compare with costs this creation implicates.

6. REFERENCES

Barnett, T. Lewis Outliers in statistical data, 3a ed., Chichester, England, John Wiley, 1994.

Bernetti, S. Romano, La valutazione dei progetti di sviluppo turistico nei parchi naturali, Genio Rurale 4 (1996) pp. 31-43.

Bettinazzi R., Valutazione e remunerazione dei benefici ambientali del parco naturale regionale della Lessinia, Bachelor Dissertation, Facoltà di Agraria, Padova, 1994-95.

Bishop R.C., T.A. Heberlein, Measuring Values of Extra Market Goods: Are Indirect Measures Biased?, American Journal of Agricultural Economics, 61 (5) (1979) pp. 926-930.

Bishop R.C., T.A. Heberlein, Simulated markets, hypothetical markets, and Travel Cost Analysis: alternative methods of estimating outdoor recreation demand, Staff Paper Series 187 (1980), Department of Agricultural Economics, University of Wisconsin.

Boatto V., E. Defrancesco, M. Merlo, La funzione turistico ricreativa della Foresta di Tarvisio. Valutazione economico-sociale, Università degli Studi di Padova, Istituto di Economia e Politica Agraria, 1984.

Bravi M., R. Curto, Stima di beni pubblici con il metodo della valutazione contingente: finalità d'uso e valori, Genio Rurale 2 (1996) pp. 56-62.

Cooper J.C., M.W. Hanemann, Referendum Contingent Valuation: How many Bounds are Enough? USDA Working Paper, (1995).

De Fano G., G. Grittani, La valutazione monetaria di due beni ambientali: il parco naturale di Portoselvaggio e la penisola di Akamas, Genio Rurale 6 (1992) pp. 9-20.

Gatto P., La valutazione economica del paesaggio forestale e del verde urbano, Monti e Boschi (1988) pp. 28-34.

Kennedy P., (3rd Ed.), A Guide to Econometrics, The MIT Press, Cambridge, Massachusetts, 1992.

Marinelli, L. Casini, D. Romano, Valutazione economica dell'impatto aggregato e dei benefici diretti della ricreazione all'aperto di un parco naturale della Toscana, Genio Rurale 9 (1990) pp. 51-

Marinelli, S. Romano, Analisi della domanda di ricreazione all'aperto in foresta. Aspetti metodologici ed applicativi, Studi di Economia e Diritto 2 (1987) pp. 123-153.

Merlo M., Una valutazione della funzione ricreazionale dei boschi, Rivista di Economia Agraria 2, (1982) pp. 385-397.

Notaro S., G. Signorello, Elicitation Effects in Contingent Valuation: Comparison among Multiple Bounded, Double Bounded, Single bounded and Open Ended Approaches, Ninth Annual Conference of the European Association of Environmental and Resource Economists, Oslo 25-27 June 1999

Romano D., F. Carbone, La valutazione economica dei benefici ambientali: un confronto fra approcci non di mercato, Rivista di Economia Agraria 1 (1993) pp. 19-58.

Romano, M. Rossi, La valutazione economica del trekking sull'Appennino tosco-romagnolo: un confronto fra approcci non di mercato, Aestium June-December (1994) pp. 171-191.

Signorello G., La stima dei benefici di tutela di un'area naturale: un'applicazione della "contingent valuation", Genio Rurale 9, (1990) pp. 59-66.

Signorello G., Valutazione contingente della "disponibilità a pagare" per la fruizione di un bene ambientale: approcci parametrici e non parametrici, Rivista di Economia Agraria 2, (1994) pp. 219-238.

Tempesta T., La stima del valore ricreativo del territorio: un'analisi comparata delle principali metodologie, Genio Rurale 12, (1995) pp. 15-34.

Tosi V., I servizi turistico-ricreativi dei boschi: esperienze nel Triveneto, Annali ISAFA, Trento, (1989) pp.103-265.

Venzi L., M. Rivetti, La valutazione di un giardino con peculiari caratteristiche architettoniche e paesaggistiche, Genio Rurale 9, (1990) pp. 76-85.

Welsh M.P., R.C. Bishop, Multiple bounded discrete choice models, in Benefits & Costs Transfer in Natural Resource Planning (John Bergstrom, Ed.), Western Regional Research Publication, W-133, Sixth Interim Report, Department of Agricultural and Applied Economics, University of Georgia (1993) pp. 331-352.

Environmental and Human Impacts on Coastal and Marine Protected Areas in India

R. KRISHNAMOORTHY, J. DEVASENAPATHY, M. THANIKACHALAM & S. RAMACHANDRAN
Institute for Ocean Management, College of Engineering, Anna University, Chennai 600025, India

Key words: Coastal and Marine Protected Areas, Remote Sensing and GIS Applications, Mangroves and Coral Reefs Management, Linking to Social Science and Community Participation

Abstract: The Indian coastline is about 7500 km in length and many ecologically sensitive coastal areas were declared as protected areas (marine biosphere reserves, national parks, wild life sanctuaries, reserve forests, Ramsar sites, etc). Based on the analysis of multidate remote sensing satellite data, the degradation sites within the protected areas have been identified. The study revealed that the remote sensing and GIS technology tools are more suitable to map and monitor the natural resources especially the mangroves and coral reefs in coastal and marine protected areas with reasonable accuracy in a most cost-effective way. This paper highlights the suitable sensor and techniques to study the protected areas. The major causes (i.e. biophysical and human induced activities) and driving forces for ecosystem degradation were analysed using remote sensing data.

1. INTRODUCTION

1.1 People of the Coasts

Coastal environments are the interface between land and sea and, at their broadest level of definition, cover approximately 8 % of the earth. They are exceptionally diverse and productive, particularly in shallow water tropical

G. Visconti et al. (eds.), Global Change and Protected Areas, 373–392.

regions. The majority of the world's known marine species reside within nearshore zones or depend on coastal habitats for part of their life cycles. The coastal zone of India is rich in estuaries, beaches, nearshore environment that includes mangroves, coral reefs, seagrass beds and algal communities and many small island ecosystems. The West Coast starts from Lakphatin Kutch on the Pakistan border to Cape Camerin and the East Coast extends from this point to the active delta of Ganges in Bangladesh. According to World Resources Institute ranking based on the criteria of existing population, road and pipeline densities 71 coastal districts in India fall under high threat potential [1]. There are 11 cities with a population more than 1 million along the Indian coastline. There are 53 coastal districts and six Union Territories, including islands of Andaman & Nicobar and Lakshadweep. Nearly 50 % of the country's total population reside in these districts. Population density in these areas ranges from less than 25 to more than 700 persons per sq.km. According to the Anthropological Survey of India, there are about 337 communities in India whose primary occupation is marine or inland fishing. For the last two decades attempts have been made to identify sensitive areas and unique habitats to preserve, protect and propagate the living and non-living resources, through *in-situ* and/or *ex-situ* measures. However, due to increasing pressure on land, increasing population growth, and urbanizing activities, the efforts made so far have not been able to make any significant impact.

1.2 Protected Areas

Presently India has 54 National Parks (of this 4 of them are Marine Biosphere Reserves) and 372 Sanctuaries. This network of National Parks and Sanctuaries cover more than 87,835 sq.km and accounts for 2.7 % of the total area of the country. The major Wildlife Sanctuaries, Reserves and Parks in India are given in Table 1.

The Marine Parks in India are Sunderban, Gulf of Kutch, Wandoor and Gulf of Mannar (Figure 1). These areas possess the biologically rich ecosystems like mangrove forests, coral reefs, seagrass meadows, and estuaries. In India 45 mangrove species are recorded so far from its coast, of this all those mangrove species are coming under endangered (IUCN criteria). Similarly the sensitive coral reefs are also under pressure due to the human impacts over the ecosystem. The highly productive (6,000 gC/M2/ yr.) seagrasses and their diversity are also decreasing due to coastal processes and alteration of the habitat by human activities. The seagrass depended organisms like *Dugong dugong* (Endangered), Seahorse (*Hippocampus kuda*) are rare in population. How are we going to protect the coastal resources without understanding the natural process working on the

Table 1: Major Wildlife Sanctuaries, Reserves and Parks in India

Major Wildlife Sanctuaries, Reserves & Parks	State
Wandoor Marine National Park*	Andaman
Coringa Wildlife Sanctuary*	Andhra Pradesh
Krishna Wildlife Sanctuary*	Andhra Pradesh
Neelapattu Sanctuary*	Andhra Pradesh
Manas Wildlife Sanctuary	Assam
Kaziranga National Park	Assam
Hazaribagh Wildlife Sanctuary	Bihar
Palamau Game Reserve	Bihar
Gulf of Kutch Marine National Park*	Gujarat
Gir Sanctuary / National Park	Gujarat
Velavadar National Park	Gujarat
Bandipur National Park	Karnataka
Dachigam Wildlife Sanctuary	Kashmir
Periyar Wildlife Sanctuary	Kerala
Madhav National Park	Madhya Pradesh
Kanha National Park	Madhya Pradesh
Bandhavgarh National Park	Madhya Pradesh
Sanjay Gandhi National Park	Maharashtra
Taroba National Park	Maharashtra
Malvan Wildlife Sanctuary*	Maharashtra
Similipal National Park	Orissa
Bhitarkanika Wildlife Sanctuary	Orissa
Bharatpur (Keoladeo Ghana Bird Sanctuary)	Rajasthan
Sariska & Ranthambhore National Parks	Rajasthan
Vedantangal Bird Sanctuary	Tamil Nadu
Point Calimere Wildlife Sanctuary*	Tamil Nadu
Mundathurai Tiger Sanctuary	Tamil Nadu
Pulicat Sanctuary*	Tamil Nadu
Gulf of Mannar Biosphere Reserve*	Tamil Nadu
Valley of Flowers National Park	Uttar Pradesh
Corbett National Park	Uttar Pradesh
Sunderbans Wildlife Sanctuary*	West Bengal
Jaldhapara Wildlife Sanctuary	West Bengal
Halliday Island Sanctuary*	West Bengal
Lothian Island Sanctuary*	West Bengal
Sajnakahli Sanctuary*	West Bengal

* Protected areas that posses a coastal/marine component

coastal ecosystems and ignoring our (human) impact on the sensitive habitats? We should think of using the modern technology like satellites to monitor the resources and identify the problem sites in protected areas during different seasons/periods.

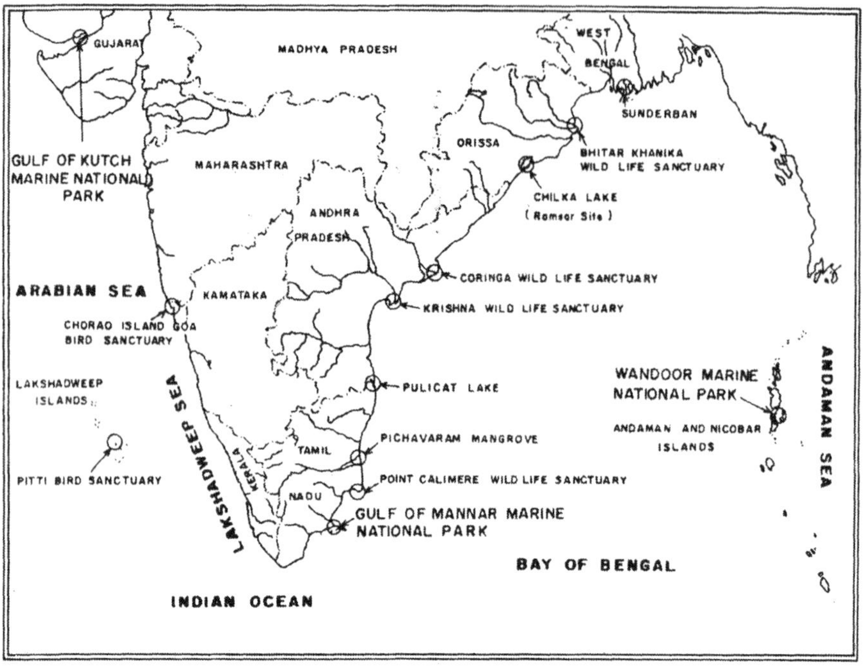

Figure 1: Location of Marine Parks, Wildlife sanctuaries and Reserve Forests in India

Marine Protected Areas (MPAs) are a well-established part of most marine conservation programmes and are likely to remain so. The approach to management of MPAs varies depending on the type of site, the aims and the circumstances under which protection is being achieved. There is also recognition for the need to improve the effectiveness of existing MPAs. The major issues are (i) site selection, (ii) legislation (iii) involving local community, and (iv) management plans [2].

Data collection and generation from all available sources and using remote sensing and GIS tools are very vital in the site selection process. Data should be collected and evaluated for consistency and adequacy and stored in easily retrievable form for analysis, such as maps and in organised computerised GIS databases. Remotely sensed data have been used to study land cover, vegetation types, physiography/landforms and human interventions at fine to coarse spatial scales.

Such information provides better knowledge on potential, extension, composition, evolution and rate of transformation of habitats and ecosystem. Attempts were made earlier to study National Parks in India using remote sensing data to analyse the spatial pattern of landscape and vegetation types [3].

2. OBJECTIVES AND BACKGROUND

Considerable amount of work has been carried out in India to map and monitor the coastal resources during the last two decades. Apart from continuos updating, more larger scale mapping are being carried out to generate theme maps from high resolution IRS data especially using LISS-II, LISS-III and PAN sensor data. National programmes on mangrove wetlands, Marine Biosphere Reserves and coastal critical habitats are being carried out with funding support from International agencies like GEF, World Bank, CIDA, etc. The present study is aimed to highlight the advantages of remote sensing and GIS technology tools to analyse the biophysical and anthropogenic impacts on mangroves and coral reef resources in coastal and marine protected areas.

3. MANGROVE RESOURCES

3.1 Status of Indian Mangroves

The mangroves constitute an important coastal resource in India and mainly function as the most ideal spawning, breeding and nursery grounds for economically important fish and crustaceans. Fairly large percentage of the coastal population is depending on mangroves for their livelihood. The State Forest Departments and the Ministry of Environment and Forests (Federal Government) manage mangroves in India, either jointly or independently. Mangrove areas come under the Reserve Forest category, which are all completely protected areas. The Ministry of Environment and Forests (MoEF) has also set up a National Mangrove Committee, which provides policy and guidance. Indian mangroves are reported as being most diverse with 45 recorded mangrove and associated species. Because of the economic value of mangroves and the land they occupy, many mangrove areas have been destroyed over the period 1963-1977. The estimate by MoEF [4] in 1994 is at about 4250 sq.km area of mangroves in India. The largest stretch of mangroves in the country lies in the West Bengal where the Sunderbans cover an area of about 2119 sq.km (Indian part). The Andaman and Nicobar Islands account for an additional 966 sq.km (18 % of the total mangrove area) and about 380 sq.km area of mangroves are in Orissa and Andhra Pradesh under the three Wildlife Sanctuaries. The species diversity is more in Orissa (Bhitarkanika and Mahanadhi), whereas, a minimum number of 7 species occur in the Gulf of Kutch. There are two mangrove sites in Tamil Nadu - Pichavaram and Muthupet, having a total area of about

Table 2: IRS and TM data for mangroves mapping

Sensor	Applications	Remarks
LISS-I (72.5 m)	Demarcation of mangrove areas and its associated categories together	Both digital analysis and visual interpretation are found to be not highly suitable for accurate demarcation of mangrove areas
LISS-II (36.25 m)	Possible to demarcate mangrove areas from other coastal categories	The spatial resolution is not sufficient to do mangrove zonation studies. Digital analysis results show more mixed pixels
LISS-III (23.5 m)	More accurate demarcation of mangrove areas and zonation of mangroves	This sensor data is found to be more useful to demarcate the regeneration (natural/artificial) areas. This sensor is having additional band in the short wave infrared region which is more suitable to study the mangrove vegetation cover
PAN (5.8 m)	Not useful for mangrove studies	It provides more information on infrastructure, urban settlements, etc. The spectral merge of LISS-II / LISS-III and PAN is found to be more useful for mangroves mapping
TM (30 m)	Similar to LISS-III sensor data	Digital analysis especially band ratioing found to be more useful to demarcate mangrove areas. Especially the ratio of bands 4/7 shows the zonation of mangroves

20 sq.km. Pichavaram mangrove is of estuarine type and it was declared as a Reserve Forest in 1897. A total number of 15 hamlets (7 farming and 8 fishing hamlets) with a population about 10,000 exist around Pichavaram. The Muthupet mangrove is a lagoon type and it was declared as Reserve Forest in 1911. Eight revenue villages surround this area with a number of fishing and farming hamlet [5]. The Krishna and Godavari mangrove wetlands are located on the Andhra Pradesh coast and each area has a Wildlife Sanctuary. Both the mangrove areas receive surplus amount of freshwater for the entire period of the Southwest monsoon as well as during the later period of Northeast monsoon. Bhitarkanika and Mahanadhi mangrove areas located in the northern and southern end of the Mahanadhi delta in Orissa coast. During 1975, the Bhitarkanika mangrove forest and the associated wetlands including Gahirmata beach occupying an area of about 70,000 ha, were declared as Wildlife Sanctuary. The Sunderban mangroves are having a status of Biosphere Reserve. The Wandoor National Marine Park area in Andaman is having both mangroves and coral reefs. In the West Coast of India about 400 sq.km area of mangroves occur in Gulf of Kutch

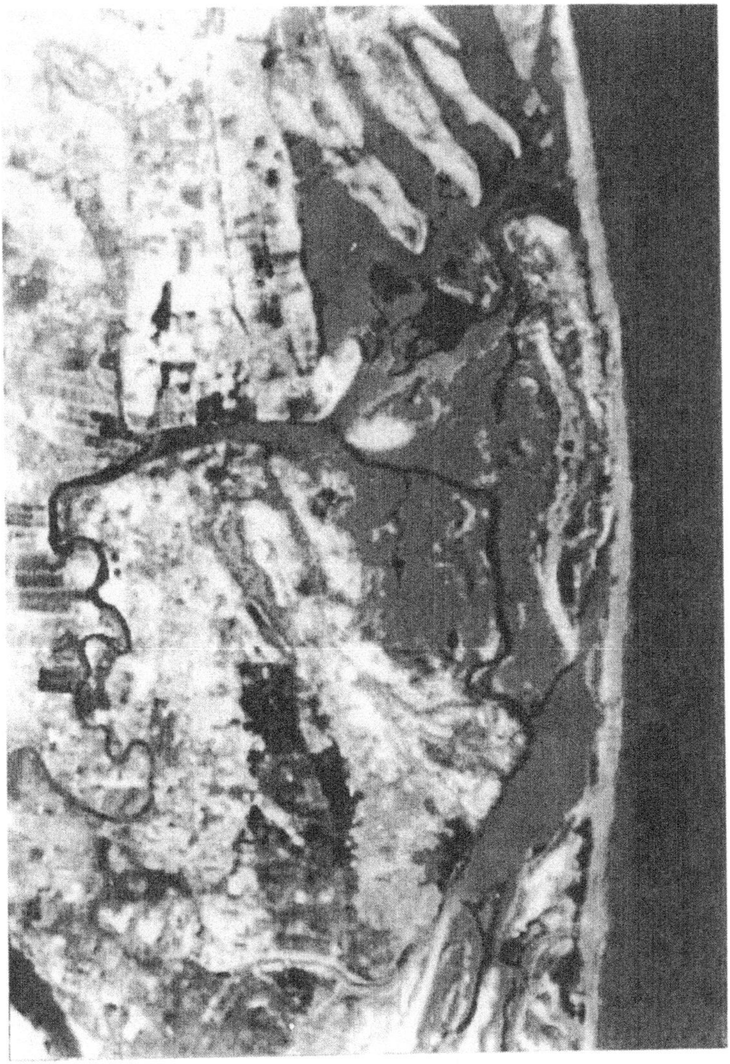

Figure 2: Mapping of mangrove areas and other associated coastel categories using IRS LISS-III imagery of Pichavaram area.

and lesser areas in the coastal area of Goa, Karnataka and Kerala, dominated with the family Avicenniaceae.

3.2 Remote Sensing Data Analysis

IRS-1A is the first of the series of operational earth resources satellite of India. It was launched on 17 March 1988. IRS-1A had a payload of two sensors, namely, LISS-I and LISS-II with a ground resolution of 72.5 m and

Table 3: Image interpretation keys for demarcation of mangrove categories

Tone	Texture	Pattern	Remarks
Brick red	Medium to course	Irregular	Dense canopy of mangroves with species like *R. mucronata, R. x. lamarckii, R. apiculata, E. agallocha, A. officinalis, Sonneratia,* etc.
Dark red	Fine to medium	Irregular	Mangrove canopy with moderate density. Species like *A. marina, A. alba, C. decandra,* etc. will form this category
Less red	Fine	Irregular	Especially *S. maritima, S. monoica,* etc. will for this category
Dark grey	Medium to course	Irregular	Degradation due to natural felling (changes in topography, soil and water quality, etc.). These areas will be generally inside the core mangroves having saline water during high tide and also some sprouting due to regeneration.
Dark grey mixed with light red	Fine	Irregular/R egular	Natural regeneration areas will have irregular pattern. Mangrove afforestation areas will have regular pattern.
Light to medium grey	Fine	Regular	The converted mangrove areas for agriculture will be having some kind of regular pattern.
Dark brown	Course	Irregular	Forest plantations especially Casuarina, Eucalyptus, etc. which will occur close to mangrove areas

36.25 m respectively.

The second generation of satellite, IRS-1C was launched in December 1995 and the indigenously developed PSLV-C1 rocket launched IRS-1D in September 1997 (http://www.nrsa.gov.in). IRS-1D is a follow-on satellite to the IRS-1C to ensure continuous availability of data with high spatial resolution. Both IRS-1C and 1D are having the sensor LISS-III multispectral (23.5 m resolution) and PAN single band (5.8 m resolution). Initial studies were carried out for the mangrove wetlands of Tamil Nadu and Andaman & Nicobar Islands using remote sensing and GIS techniques. Multidate satellite data have been used to analyse the changes in area of mangrove vegetation cover and *in-situ* spectral studies to assess their health, biomass, etc.[6]. The following table shows the capability of IRS and TM satellite sensors for mangrove areas mapping and monitoring.

Figure 2 shows the mangrove area of Pichavaram and other coastal categories on IRS LISS-III FCC imagery.

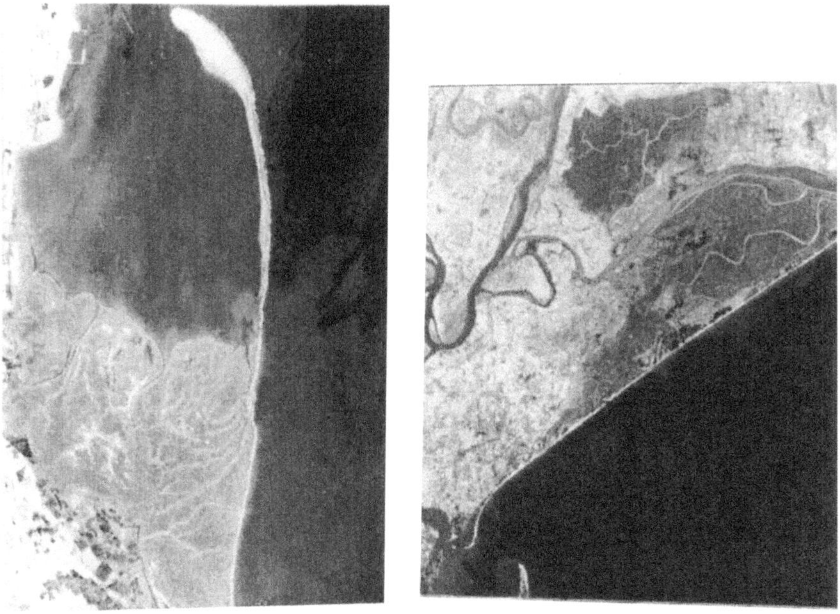

(a) Coringa Wildlife Sanctuary (b) Bhitarkanika Wildlife Sanctuary

Figure 3: IRS LISS-III data applications in mangrove areas mapping and monitoring

The mangrove area appears in thick red colour, brackishwater aquaculture areas in blue colour with a regular pattern, emerging mudflats in dark grey and the forest plantations in dark brown. High spatial resolution and inclusion of more infrared region, the LISS-III sensor data of IRS-1C and ID are found to be useful in demarcation of mangrove areas more clearly as shown in Figures 3 a & b. Very narrow channels (less than 1 m width) within the mangrove areas can also be traced on LISS-III imagery (Figure 3 a) by visual interpretation. All the other categories like mangrove degradation areas (light to dark grey colour), aquaculture areas adjacent to mangrove areas (blue colour with regular pattern) and the agriculture areas (light red colour) outside the mangrove area could be demarcated on the LISS-III imagery. The variations in red colour intensity within the mangrove area are mainly due to species zonation and variations in canopy density. The degradation of mangroves has been more towards the land ward side i.e. adjacent to the areas of aquaculture.

The discharge from aquaculture sites leading to changes in water and soil quality in the mangrove swamps may be the reason for degradation. Similarly the LISS-III imagery of Orissa (Figure 3 b) shows the land use categories upto level-III.

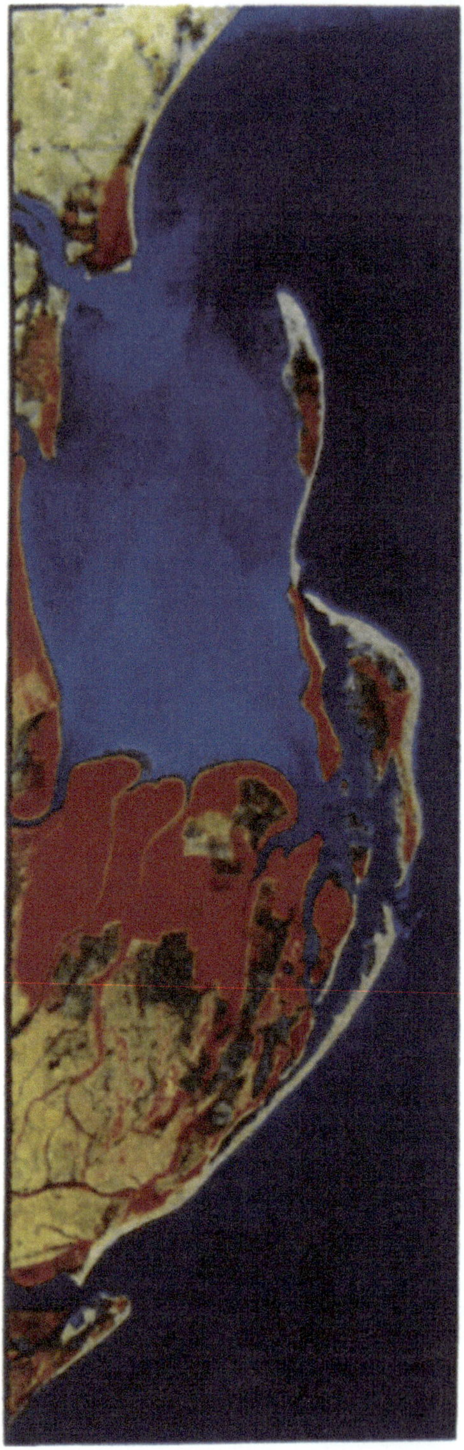

Figure 4:
IRS LISS-11 imagery of Mahanadhi mangrove areas (healthy mangroves in red colour; agriculture/fallow lands in light yellowish colour; converted mangrove areas for agriculture/aquacuiture in light bluish colour, casuafina piantations are in dark brown colour and beach sand in white colour) Data Source: NRSA

The advancement of agriculture and aquaculture activities around the mangrove areas could be assessed using this sensor data. Many villages surrounded this area have full time agricultural activities. Many repatriates after the Bangladesh war settled around the mangrove areas and converted the mangrove areas for agriculture. The fishing activity is completely banned after this area has been declared as Wildlife Sanctuary. Mangroves in the southern part of Mahanadhi delta are facing very severe threat due to conversion for agriculture. Figure 4 shows the LISS-II imagery with healthy mangroves in dark red, casuarina plantations in brown and the agriculture areas in light to dark grey colours. The advancement of agriculture activities and the converted mangrove areas are clearly appeared on the satellite image. Based on visual analysis of satellite data for mangroves mapping, it is possible to demarcate the healthy mangroves, mangrove zonations, degradation due to biophysical and anthropogenic causes based on the image tone, texture and pattern. Few examples are given in Table 3.

4. CORAL REEFS

4.1 Indian Coral reefs

Coral reefs are biologically diverse in comparison with other ecosystems. India has a total of 1270 sq.km of coral reefs. The major area of reefs in Andaman and Nicobar Islands (1200 sq.km) followed by Gujarat (130 sq.km) and Lakshadweep (71 sq.km). Tamil Nadu has 10 sq.km area of reef [7]. Gulf of Mannar, Wandoor in Andaman and Gulf of Kutch are declared as Marine National Parks. It was proposed by the Government of India in 1976 that the area initially be considered for the establishment of India's first National Park off the coast of Tamil Nadu with 21 islands which are mostly of coral origin.

The Government of Tamil Nadu notified the intention of setting up of a Marine National Park in Gulf of Mannar for the protection of Wildlife and its environment and finally this area received the status of a Biosphere Reserve in 1989. Recently the Gulf of Mannar marine Park recorded three Dugong species from the wild (accidentally captured in the net) but the authorities cannot assess the population and its habitat and movement. The Australian Great Barrier Reef Authority used aerial photography to monitor the surface dwelling Dugong species and also estimated its population. There was also a limitation in using aerial photographs to estimate the population of Dugong mainly due to the occurrence of Dolphins. The major advantage of this study is to survey the vast area within short time.

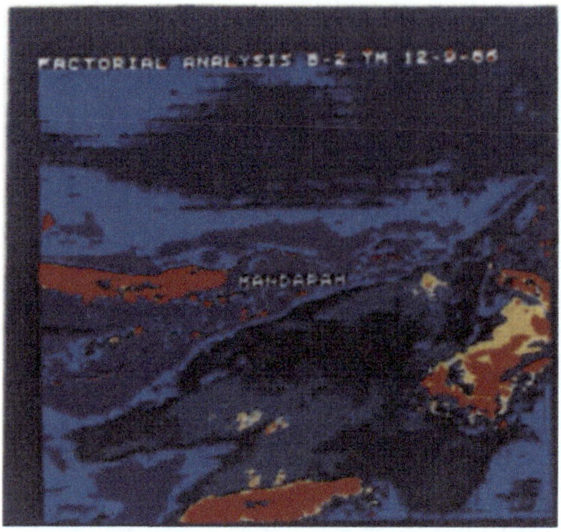

Figure 5: Coral reef zonations based on TM data analysis (reef flat in red colour and deep reef in yellow colour).

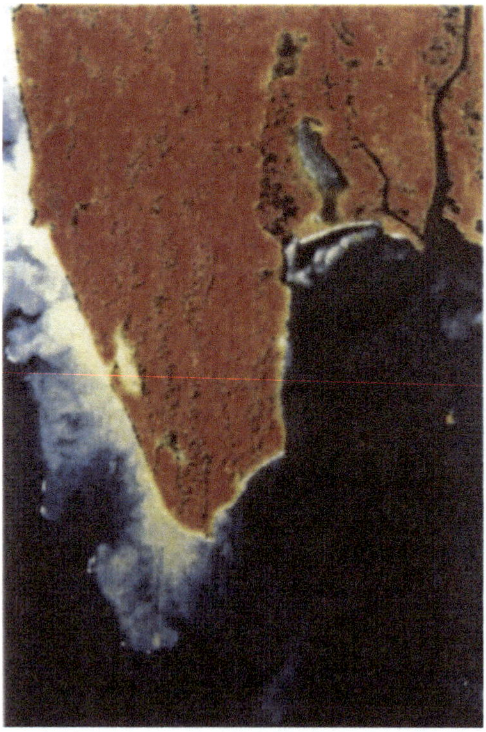

Figure 6: SPOT FCC of part of South Adaman shows coral reefs at shallow depth in light blue and deep reefs in turquoise blue colours.

The applications of remote sensing data for coral reef resources mapping and monitoring are explained below.

4.2 Remote Sensing Applications in Coral reefs mapping and monitoring

Remote sensing data was used to map Great Barrier Reef Marine Park by devising a standard classification system for classifying and labeling geomorphological information on reef covers and zonation. The interpretation of remote sensing imagery for coral reef mapping involves four steps (i) detection of features, (ii) recognition and identification of features, (iii) analysis and delineation of patterns, and (iv) classification [8]. Coral reefs mapping of the Gulf of Kutch, Gulf of Mannar and Palk bay, Lakshadweep and Andaman and Nicobar Islands has been carried out using remote sensing data.

High resolution optical remote sensing data were used for coral reef zonation studies primarily of a geomorphological rather than an ecological nature. Making comparisons between the ability of different sensors for coral reef studies is difficult because of differences in reef terminology and study sites [9]. Coral reefs areal extent and changes in the total area are vital input for the conservation and management of the reefs. Using remote sensing data the coral reef area at Gujarat has been estimated as 94 sq.km in 1999, this denotes a reduction of 36 sq.km over a period of 15 years. However after 1988 after declaring this area as Marine National Park, the coral reef are has increased [10]. The coral reef maps prepared using satellite imagery serve as a baseline data for monitoring the reef area and defining the bioreserves boundaries. Monitoring of reef areas would be possible using multidate satellite data with reasonable accuracy. Optical remote sensing data are found to be more suitable when compared to microwave data. The high resolution optical sensors especially TM, SPOT and IRS LISS-II were compared with airborne SAR data for coral reef mapping in Andaman and Nicobar Islands. Optical sensors are found to be suitable to demarcate living and non-living corals and its depth of occurrence (shallow or deep) based on their tonal characteristics [11]. The GIS analysis based on integration of relevant spatial and non-spatial data are useful to study the impact of biophysical factors like shoreline erosion/accretion [12].

Validation of coral reef maps by ground truthing has proved that reefs occur up to 13 m depth in less turbid coastal waters and this could be demarcated from satellite imagery by visual interpretation. Digital analysis is required for more accurately demarcating reef areas from suspended sediments. Especially the Factorial Analysis of TM band 2 data were found to be more suitable to demarcate reef flat and deep reefs as shown in Figure

5. Visual interpretation is sufficient to demarcate reef areas/categories where the coastal water is less turbid. Figure 6 shows the SPOT FCC of South Andaman where the fringing reef (shallow and deep reefs) could be demarcated by visual interpretation. The Wandoor Maine Park in South Andaman has extensive reef areas. Using satellite data the reef areas can be further classified into fringing reef, platform reef, patch reef and coral pinnacles. The extensive occurrence of coral reefs towards Bay of Bengal is mainly due to tidal wave circulations were aid the growth of corals. Field visits to the Andamans and Gulf of Mannar were made to verify the reef areas mapped using satellite imagery. Discussions were held with local officers, community and scholars to understand the current status of reefs. It was observed that the reefs in Wandoor as well as in Gulf of Mannar are facing the problem of shoreline erosion and sand deposition over reef. Especially the changes in coastal land cover in Wandoor leads to sand/silt deposition over reef. Figure 7 shows the conversion of rain forest area into agriculture adjacent to Wandoor Marine Park. Nearly 2500 ha of rain forest is being logged every year for wood industries. Due to topographical set up and the absence of perennial rivers in Andamans sand deposition occurs over reef area.

The human impact is more in Gulf of Mannar. Coral reefs are being quarried for chemical and lime industries (Figure 8). Excessive exploitation of corals around the islands has very much affected and altered the ecology of the islands, resulting in the loss of certain land portions of islands and their associated flora and fauna from the ecosystem.

The LISS-III sensor data of 23.5 m resolution is found to be more suitable to demarcate different reef categories, reef with seagrass and seaweeds, etc. Figure 9 shows the coral reef map derived from LISS-III imagery of Kadamat Island in Lakshadweep.

5. DISCUSSION

The major advantages of remote sensing data in the study of mangrove and coral reef resources in protected areas have been discussed in the previous sections. Aerial photographs and satellite data have widespread use for more than 20 years in India. From the perspective of social science, one important reason of using remotely sensed data is to gather information on the context that shapes social phenomena and related issues of human-environment interaction at spatial and temporal scales. Some have expressed the hope that remotely sensed data could be used to study human population to update the census reports [13]. The distribution of human population across the Earth's surface has been identified as one of the key data sets

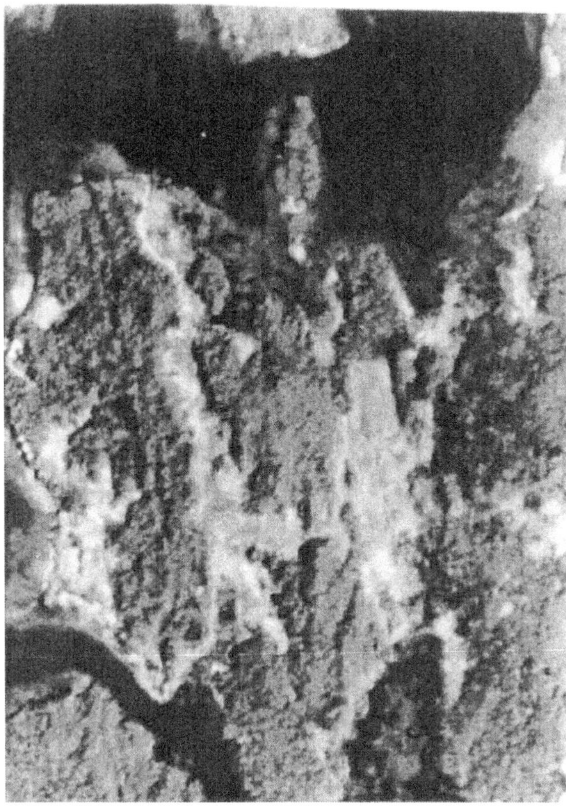

Figure 7: SPOT data shows the coastal land use/cover changes adjacent to Wandoor NationalMarine Park in Andaman.

Figure 8: Collection of coral reefs from Gulf of Mannar by the local people.

required for global change research. The locations and areal extents of cities and towns can be extracted from high spatial resolution remotely sensed data, such as Landsat, SPOT or IRS. Time series data from the Defense Meteorological Satellite Programme (DMSP) Operational Linescan System (OLS) have been used to derive georeferenced inventories on human settlements for Europe, North and South America and Asia [14]. Remote sensing data can be used to analyse human activity, urbanisation, road development, etc. Changes in vegetation density may be related to the effects of different agricultural practices which may be linked to the effects of local polices on land use. Especially in coastal and marine protected areas, remote sensing data plays a vital role in the preparation of management plans and monitoring the activities. The preventive, protective and conservation measures taken in the National Marine Park of Gulf of Kutch have resulted in the restoration of the area under reef significantly. The living coral area has increased to 20 – 30 % after establishing the Marine Park in this area [15]. It was felt by many organisations and experts in the country that there is an urgent need for remote sensing education to social scientists who can utilise the outputs in community based conservation and management activities. The impact of implementation of coastal regulations in protected areas could be assessed using remote sensing data since the IRS LISS-III sensor proved to be useful to identify and demarcate the areas of natural/artificial regeneration in degraded mangrove areas. Some of the NGOs (http://www.mssrf.org) in India are using remote sensing and GIS tools for the preparation of resource atlas and database for Joint Mangrove forest Management (JMM) plans specifically to identify suitable sites for mangrove of forestation and also selecting sites to plan for alternate fuel wood. Under the JMM programme, the social scientists, field staff and local Forest Department officials were given training on the use of remote sensing data and GIS outputs for mangrove forest mapping, identification of degradation sites and new regeneration sites [16]. User awareness programmes on remote sensing applications to coastal and marine studies are being conducted frequently under the Department of Space programme on National Natural Resources Management Systems (NNRMS) in India. The Institute for Ocean Management, Anna University (http://www.annauniv.edu) is one of the regional centres to conduct the user training and awareness programmes on coastal and marine applications. This Institute has recently initiated a training programme on Integrated Coastal Zone Management funded by the Department for International Development (DFID), UK jointly with the University of Newcastle. Geographic Information System (GIS) and remote sensing are often designed to assist decision-makers in spatial planning. A multi-objective decision support system (MODSS) was developed for coastal resource management in Rayong province in Thailand

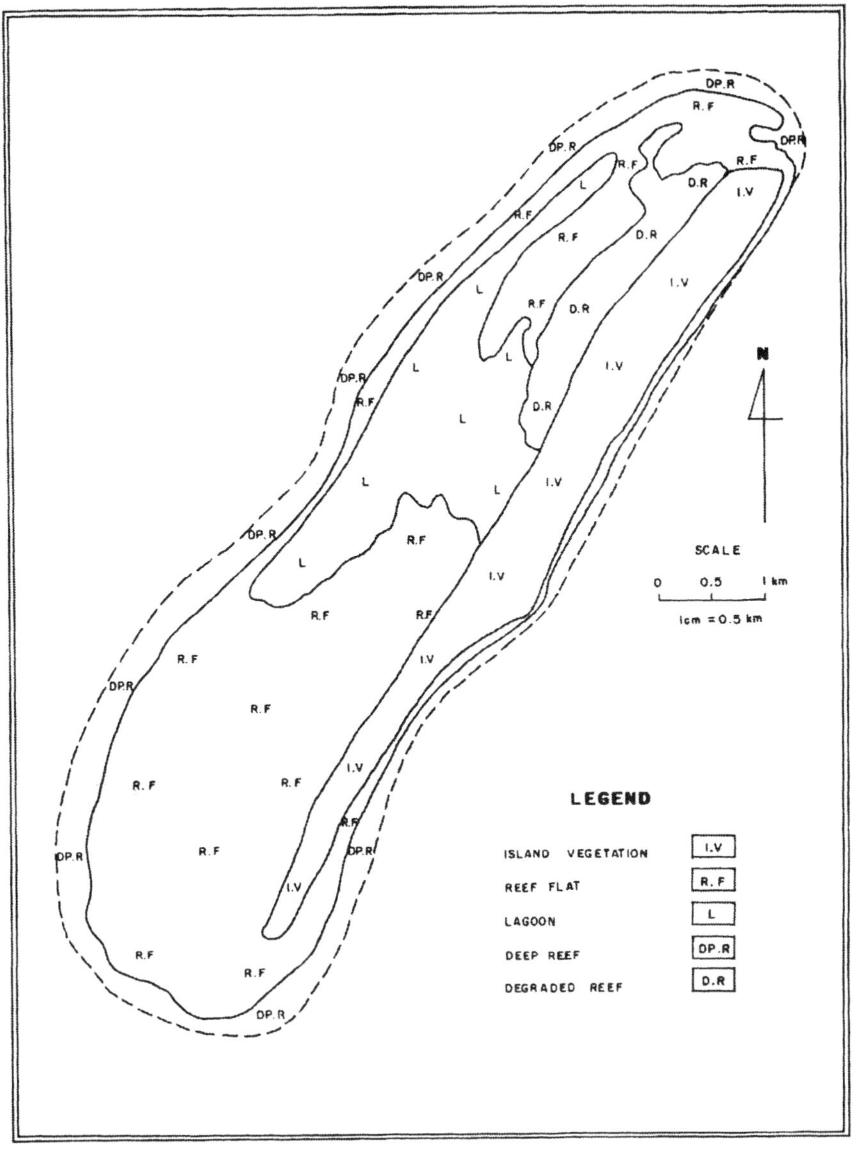

Figure 9: Coral reef categories derived from visual interpretation of IRS LISS – III Imagery (Kadhamat Island in Lakshadweep).

as a tool to help decision-makers in effectively coordinating coastal zone resource development [17].

The GIS is found to be useful in storing and integrating coastal zone data. A GIS based Coastal Zone Information System (CZIS) has been developed for Rameswaram Island in Gulf of Mannar Biosphere by integrating conventional, remote sensing and socio-economic data which served as a DSS for coral reef management [18].

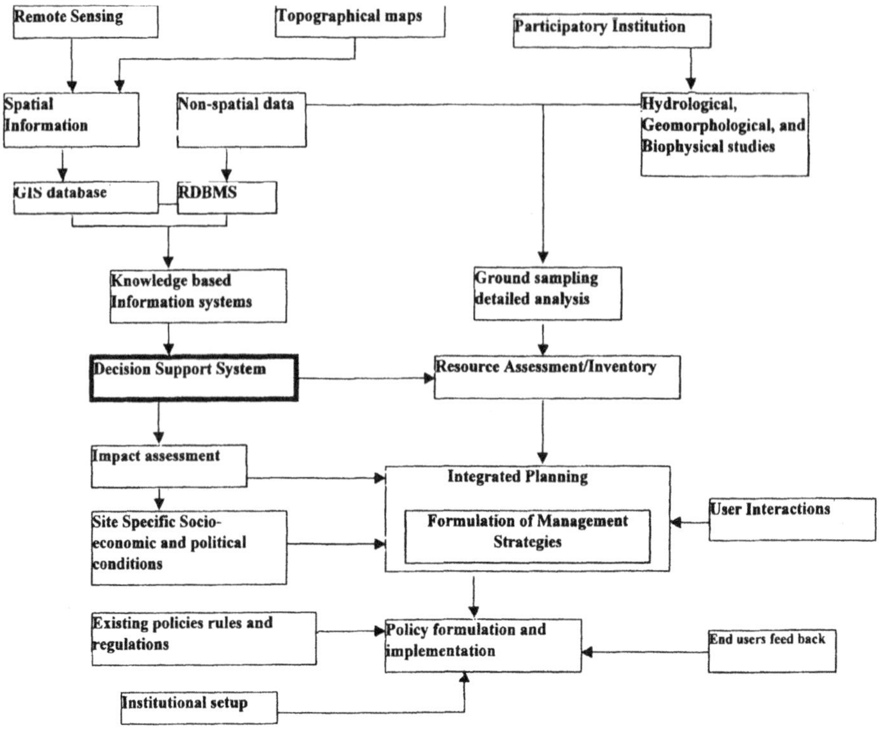

Figure 10: Systems view of Critical Habitat Information System.

Recently the Department of Ocean Development initiated a national programme on GIS based information system for critical coastal habitats which includes many protected areas along the coastal zone of India. Multidate remote sensing data are being used to generate temporal spatial data and to study the changes in coastal resources. The field data, mainly on biophysical and socio-economic data are also being stored in the RDBMS to develop DSS as shown in Figure 10. Under this programme the GIS database will be developed for most of the coastal and marine protected areas in India before the end of 2002.

6. ACKNOWLEDGMENTS

The authors are grateful to the Department of Ocean Development, Space Applications Centre for the funding support and guidance to carry out mangroves and coral reefs mapping under the MARSIS national programme. The authors are thankful to the National Remote Sensing Agency (NRSA) for providing the satellite data products. One of the authors (R. K) is grateful to Prof. M. S. Swaminathan for the assignment as Remote Sensing and GIS expert in the UNDP/GEF and ICEF programmes. Thanks are also due to Dr. Shailesh Nayak of SAC for his very valuable input.

7. REFERENCES

Bryant D., E. Rodenburg, T. Cox and D. Nielsen, Coastlines at Risk: An Index of Potential Development – Related threats to Coastal Ecosystems, WRI Indicator Brief, 1995, 8 pp.

Elvidge C. D., K. E. Baugh, V. R. Hobson, E. A. Kihn, H. W. Kroehl, E. R. Davis and D. Cocero, Satellite inventory of human settlements using nocturnal radiation emissions: a contribution for the global toolchest, Global Change Biology, 1997, Vol.3, pp. 387-395.

Green E. P., P. J. Mumby, A. J. Edwards and C. D. Clark, A review of remote sensing for the assessment and management of tropical coastal coastal resources, Coastal Management, 1996, Vol. 24, pp. 1-40.

Gubbay S., Marine Protected Areas – Past, Present and Future, In: S. Gubbay, (Ed.), Marine Protected Areas: Principles and Techniques for Management, Chapman & Hall publication, 1995, pp. 1-14.

Hoon V., Coral reefs of India: Review of their extent, condition, research and management status, In: Proces. Regional Workshop on the Conservation and Sustainable Management of Coral Reef, MSSRF, 1997, pp. B-1 to B-25.

Krishnamoorthy R., A. Bhattacharya and T. Natarajan, Mangroves and Coral reef mapping of South Andaman Islands through Remote Sensing, In: M. S. Swaminathan and R. Ramesh (Eds.), Sustainable management of coastal ecosystems, MSSRF & Anna University, 1993, pp. 143-151.

Kuchler D. A., Reef cover and Zonation Classification System for use with Remotely Sensed Great Barrier Data – User Guide and Handbook, Unpublished, GBRMPA, Townsville, 1983.

Looijen J., N. Pelesikoti and M. Staljanssens, ICOMIS: a spatial multi-objective input decision support system for coastal resource management, ITC journal, 1995, Vol.3, pp.202-216.

Mo E.F, Wetlands of India – a directory, Ministry of Environment & Forests, Govt. of India, New Delhi, 1994, 150 pp.

MSSRF (M. S. Swaminathan Research Foundation), Annual Report, Chennai, 1995, 90 pp.

MSSRF, Annual Report, 1998.

Ramachandran S. and R. Krishnamoorthy, Remote Sensing and the Application of GIS for Coral Reef Management, In: Biodiversity of Gulf of Mannar Marine Biosphere Reserve, MSSRF, Chennai, 1998, pp.49-55.

Ramachandran S., R. Krishnamoorthy, J. Devasenapathy and S. Sundaramoorthy, Remote Sensing Integrated GIS based Spatial Decision Support System for Coastal Zone Management, Paper in preparation.

Ramachandran S., Report on case studies on coastal fragile areas, Vol.1. West Coast, Submitted to Central Pollution Control Board, Govt. of India, New Delhi. 1997.

Ramachandran S., S. Sundaramoorthy, R. Krishnamoorthy, J. Devasenapahty and M. Thanikachalam, Application of remote sensing and GIS to coastal wetland ecology of Tamil Nadu and Andaman and Nicobar group of islands with special reference to mangroves, Current Science, Vol.75, No.3, pp.236-244.

Rindfuss R. R. and P. C. Stern, Linking remote sensing and social science: The need and the challenges, In: D. Liverman, E. F. Moran, R. R. Rindfuss and P. C. Stern (Eds.), People and Pixels: Linking to Remote Sensing and Social Science, National Academy Press, Washington D.C., 1998, pp.1-27.

Roy P. S., S. Tomar and C. Jeganathan, Biodiversity characterisation at landscape level using satellite remote sensing, NNRMS Bulletin, Dept. of Space, Bangalore, pp.12-18.

SAC, A Summary of Report on Coastal Studies Report, Scientific Note, SAC/RSA/RSAG/DOD-COS/SN/06/93, Ahmedabad, India, 1994.

Past Climate Change and the Generation and Persistence of Species Richness in a Biodiversity Hotspot, the Cape Flora of South Africa

GUY FRANKLIN MIDGLEY AND REUBEN ROBERTS.
Climate Change Research Group, Ecology and Conservation, National Botanical Institute, Cape Town, South Africa. Fax + 21 797 6903 Tel + 21 762 1166 Email: Midgley@nbict.nbi.ac.za

Key words: Species pump, speciation, paleoclimate, modelling, Fynbos, Succulent Karoo

Abstract: The physiognomically distinct Fynbos and Succulent Karoo Biomes are associated with the winter or all season rainfall region of southwestern South Africa. Fynbos (the more mesic) is co-dominated by small trees, shrubs and reed-like plants. Woody stemmed succulent shrubs, mainly small leaved, dominate the Succulent Karoo. Both biomes contain spectacular floristic richness. Have past climates determined present species richness in this region? We employed GIS-based bioclimatic modelling techniques to address this. After deriving bioclimatic envelopes for both Fynbos Biome (Fynbos Envelope; FYE) and Succulent Karoo (SKE), we used modelled paleoclimate reconstructions to map the distribution of the envelopes at 6, 12 and 18 kBP in the southern and western Cape. At the LGM, FYE possibly had a considerably greater extent along the west coast and the western interior, replacing much of the SKE. At 18 kBP SKE was limited to two separate areas, an extensive plain, the Knersvlakte (South) and montane Richtersveld (North). SKE expanded between 12 and 6 kBP to its present range, and FYE contracted into its current montane distribution. These reconstructions concur with the palynological record of indicator groups of both biomes, with current patterns of relictual Fynbos vegetation, and with Succulent Karoo centres of endemism. Climatic oscillations during the pleistocene probably shifted these biomes iteratively along a North/South axis on the west coast of southern Africa. Mutual vegetation replacements are likely to have accelerated sympatric and allopatric speciation, giving rise to high β and γ species turnover.

G. Visconti et al. (eds.), Global Change and Protected Areas, 393–402.

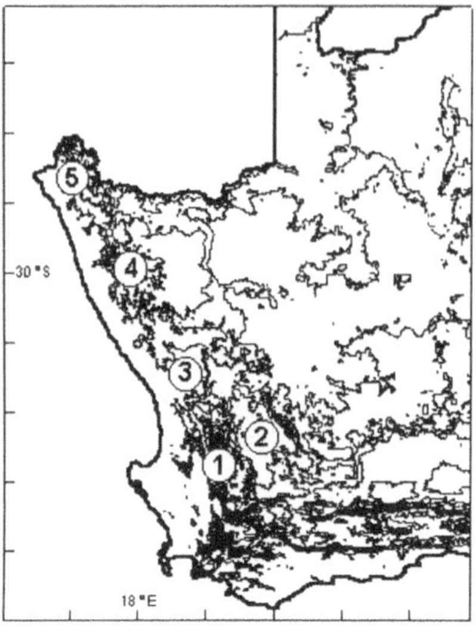

Figure 1: Topography of the study region, the southern and western Cape, South Africa, represented by 300m interval altitude isolines. Key mountain ranges and plains are as follows: 1 – Cederberg mountains, 2 – Tanqua karoo plain, 3 – Knersvlakte plain, 4 – Kamiesberg mountains, 5 – Richtersveld mountains.

1. INTRODUCTION

A pair of floristic hotspots unique to southern Africa and the world is centred on the mountains of the Cape Fold Belt and adjacent arid semi arid plains (See Figure 1 for topographic details). This area is covered by two physiognomically distinct vegetation biomes [1], the Fynbos (montane and mesic coastal lowland) and the Succulent Karoo biomes (arid montane and arid and semi-arid interior lowland). Both biomes are extraordinarily rich in species by world standards [2], with species: genus ratios more akin to island than mainland floras, and very high β (habitat) and γ (landscape level) species richness. This intriguing pattern has been the subject of many attempts at explanation within the Fynbos [3,4], and more recently, the Succulent Karoo Biome [5].

This paper is the first to question whether a common cause, namely past climate change, underlies the exceptional levels of plant richness in both of these juxtaposed biomes. Previous explanations of the unusual species richness in this region have been rooted in a view of the world in climatic equilibrium. Indeed, in a recent volume entitled "Species diversity in space and time" [6], we were unable to find mention of the phrase "climate change". We wish to challenge this world view.

Exploring this question is also useful in a conservation context, because, in order to understand how future climate change may drive changes in plant diversity it is very enlightening to look at the paleohistorical record of biodiversity change, especially during the late pleistocene and holocene. This reveals (albeit incompletely) the results of a "natural experiment" in global warming which has taken place on a grand scale. In different parts of the world, reconstructions of past vegetation have mainly been done by directly investigating the paleobotanical record at key sites [7,8].

However powerful though, paleobotanical studies suffer from being point-based subsamples of local floras, with all the attendant biases this implies. Also, using paleobotanical information to reconstruct past climates leads finally to a circular argument if the objective is to assess how much climate change has impacted vegetation distribution and diversity. In this paper we use an alternative method, bioclimatic modelling, in conjunction with physical rather than proxy-based paleoclimatic reconstructions, to gain insight into the role of climate change in driving vegetation shifts in the adjacent biodiversity hotspots of the southwestern Cape region. We develop from these insights a potential explanation for the exceptional levels of plant species richness in this area.

2. METHODS

Paleobioclimatic modeling is particularly challenging because general circulation models (GCM's) provide paleoclimatic data that is limited in scope and spatial resolution. This demands that the development of climatic envelopes for the modelled vegetation entities be based on quite rudimentary parameters, such as mean annual rainfall, mean annual temperature and potential evapotranspiration.

3. DERIVATION AND APPLICATION OF VEGETATION ENVELOPES

We used coverages of vegetation types classed as Fynbos and Succulent Karoo Biomes, digitised from [9]. GIS techniques were employed to define climatic envelopes for the vegetation types, using the approach broadly described in [10] and [11]. We used an agroclimatological database for South Africa [12], which provides high resolution climatic data (1' x 1' pixel size) for the entire region. Our analysis was constrained to only temperature- and rainfall-based climate criteria.

These were minimum temperature of the coldest month (T_{min}) mean annual rainfall (MAR) and a water balance estimate (WB) given by the difference between MAR and an estimate of potential evaporation (PE) [13], such that WB = (MAR – PE). We determined iteratively the values of the most significant envelope limitant (which was WB), so that the error between mapped vegetation distribution and modelled distribution was minimised, and the current distribution of the two biomes did not overlap.

Given the strict limitations on the number and nature of climatic factors permitted for envelope derivation, we were able to model the distribution of both vegetation types with a relatively high degree of accuracy (>75% of 1'x1' [1.5kmx1.5km] cells correctly identified, Figure 2).

After past climate scenarios (rainfall and temperature) had been reconstructed (see following sections), we were able to map the geographical shift in the derived envelopes with the changing climate.

4. TEMPERATURE RECONSTRUCTION

Temperature scenarios were developed using output from the GCM CCM0 [14,15,16].

GCM-derived temperature data (for months of January and July) for 0, 6, 12 and 18 kBP were downscaled to a 1' x 1' scale by a spline technique. Paleotemperature data (6, 12 and 18 kBP) were subtracted from current CCM0-predicted data (0 kBP) to derive a temperature difference (ΔT) for each 1' x 1' pixel each month. ΔT was then subtracted from "present" data given by the ACRU climate database to derive reconstructed paleotemperatures for January and July. ΔT was also subtracted from present T_{min} data to derive an estimate of past T_{min}.

5. RAINFALL RECONSTRUCTION

CCM0-derived rainfall data bear extremely poor comparison with the current rainfall patterns in southern Africa. We used instead a model [17], which suggests a latitudinal shift of the cyclonic frontal systems which sweep south of the country, producing mainly winter rainfall. We found a significant linear relationship between MAR and latitude (MAR[mm] = -2004.17 – 71.95*LAT[°], $r^2 = 0.6$) within the study region, and used this to approximate a latitudinal shift of cyclonic influence. As a first approximation, we applied a modest shift (~2.4° N at the LGM), half the upper bound predicted [17] (~5° N at the LGM).

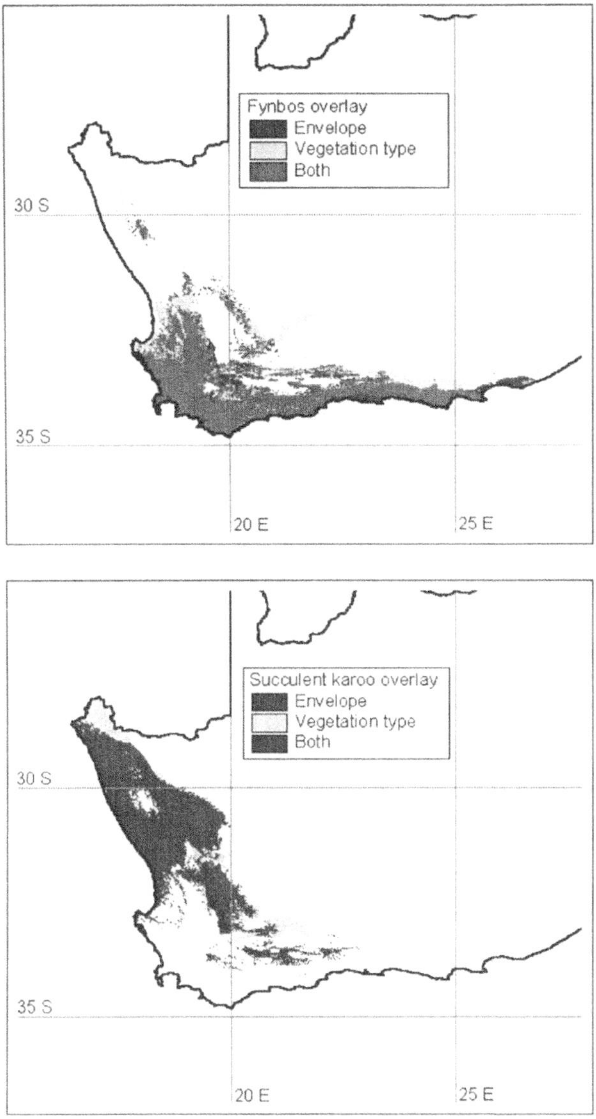

Figure 2: Comparison between model predicted envelope and actual vegetation type distributions for the Fynbos and Succulent Karoo biomes. The legends annotated mean the following: "Both" refers to agreement between model and reality, "Envelope" refers to areas the model falsely identifies as included in the vegetation type, and "Vegetation type" refers to areas which model omits, but are included in the vegetation type.

A summary of the rainfall and temperature scenarios applied is given in Table 1.

Table 1 Summary of climate scenarios (temperature and rainfall) used in this study.

Time	Temperature Change	Rainfall change
6 k BP	+0.6°C	-50mm (~0.8°S)
12 k BP	-0.4°C	+50mm (~0.8°N)
18 k BP	-1.8°C	+150mm (~2.4°N)

Temperature changes given are averages for the entire region of study. Rainfall changes given are associated with the approximate latitudinal shifts of the cyclonic belt, given in brackets.

6. RESULTS

Our reconstructed paleodistributions of the vegetation envelopes suggest the following patterns of envelope, and by implication, vegetation shifts (Figure 3): The FYE was of far greater extent, and the SKE of far lesser extent in the study region by the LGM. At the LGM, the SKE appears to have occupied a relictual retreat in the northern Cape, associated with the relief of the Richtersveld mountains, and interestingly, remained situated in the arid plain of the Knersvlakte and dotted in the northern Tanqua Karoo, completely surrounded by the FYE. With the southward movement of the cyclonically-influenced climatic belt, and some warming by 12 kBP, the FYE retreated southward and contracted onto a relictual outpost associated with the montane relief of the Kamiesberg range. The SKE expanded comprehensively from its glacial relictual outposts in the northern Tanqua Karoo, Knersvlakte and the Richtersveld, and increased its dominance somewhat at 6 kBP with a further southward movement of the cyclonic belt and increased warming. This situation has remained relatively unaltered to the present, with a possible shrinking of the SKE within some valleys between the mountains of the Cape Fold Belt, occupied by the FYE, all associated with some cooling from the climatic optimum of 6 kBP.

7. DISCUSSION

Our results reveal that climate shifts may have been fundamental in controlling vegetation distributions in the past, but that these interact with major features of relief, such as mountains and interstitial plains, to create a mosaic landscape which can harbour and preserve biodiversity generated by climate change itself. Although alluded to [3,4,18], the role of climate has not yet been fully appreciated as a driver of speciation in the western Cape region. Climate oscillations, within relatively well-defined upper and lower bounds (glacial and interglacial conditions), have taken place in this part of the world for at least the past 500 000, and probably longer than 1 000 000 years [17]. The key to biodiversity retention has been that the climatic oscillations have been bounded, and conditions did not rapidly exceed lethal limits for much of the flora. Each of the two biomes in question has therefore migrated repeatedly into space once occupied by the other, probably replacing one disturbance regime (e.g. fire) with another (e.g. drought). It is small wonder that these biomes display such a floristic relatedness [19], even though their dominants are physiognomically so distinct.

Biome replacements with climate change are likely facilitated by disturbance events of intermediate frequency (15 to 20 years), namely recurring fire (Fynbos Biome) and recurring drought (Succulent Karoo Biome). However, in both biomes it appears that refugia (often, but not only montane) allow core areas to remain within each envelope throughout both glacial and interglacial phases. In the context of the sequence of pleistocene glacial-interglacial oscillations, we suggest that the Fynbos Biome is currently in retreat, occupying a refugium afforded by altitudinal buffering by the mountains of the Cape Fold Belt.

In terms of speciation patterns, one would expect a rich and complex interplay of factors to influence the course of speciation events in each biome. With the knowledge of the role of oscillatory climate change in mind, many intriguing and confusing patterns begin to make sense. High levels of β and γ diversity in Fynbos may well be due to climate-driven altitudinal shifts in ecological zones, driving sympatric speciation within landscapes (boosting β diversity), and allopatric speciation when change is sufficient to isolate landscapes (boosting γ diversity). The role of secondary contact [20] in both biomes may also obscure speciation sequences, as some climate changes (e.g. short duration of interglacials) may encourage this form of speciation. For example, in Scandinavia, allopolyploidization is thought to have given rise to Saxifraga osloensis within the last 14 000 years in an area of contact between S. tridactylites and S. adscendens which may have come about as a result of holocene warming [21].

The apparently recent astonishing explosion of speciation [5] in the Mesembryanthemaceae may be explained by the retreat of Fynbos type vegetation across vast expanses of semi arid landscape, thus re-opening a novel environment for this group, and other succulent leaved forms well adapted to this bioclimate. The Richtersveld and Knersvlakte centers of endemism for a major group of the Mesembryanthemaceae, the Ruschioideae, can be associated with its isolated persistence in these areas through the last glacial period. Both apparent sympatric and allopatric speciation events have occurred in Mesembrynthemaceae situated in the Knersvlakte [22].

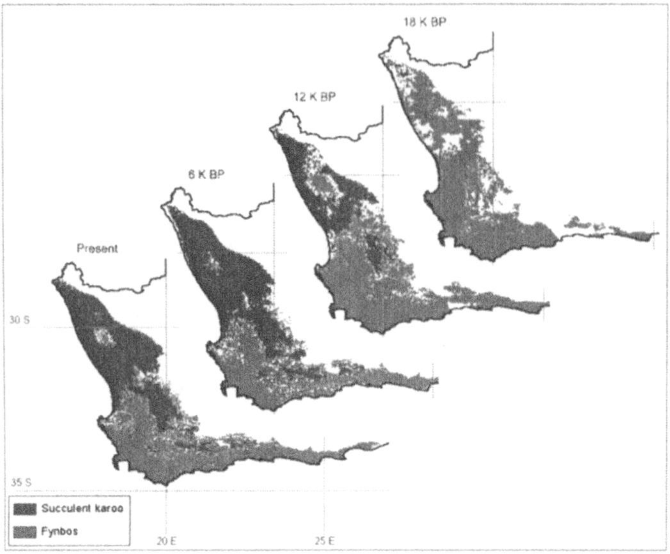

Figure 3: Current and reconstructed past distributions of the Fynbos and Succulent Karoo envelopes.

It is possible that sympatric speciation occurred during long periods of glacial isolation, marked by subtle climate shifts, and allopatric speciation has, and continues to accompany the dispersal of remnants from centers such as the Knersvlakte as the world warmed.

How does our model fit with the pollen record, the ultimate independent test of our proposition? Unfortunately, there is a paucity of comprehensive studies we can call upon. We cite pollen histories of key lineages (Restionaceae, Ericaceae for Fynbos, Mesembryanthemaceae [also termed Aizoaceae], for Succulent Karoo) from three studies to provide a highly parsimonious independent test. Shi et al [23] identified pollen of the Ericaceae and Restionaceae which originated from the north of 30º 30'S, and have a continuous record dating from around 20 kBP. This trace shows significant representation of both FYE diagnostic elements prior to 18 kBP, but these tail off to vestigial traces by about 15 kBP, a pattern in good agreement with our reconstruction. Scott et al [24] present an incomplete record spanning roughly between 12 kBP and 8 kBP in the Richtersveld mountains. They identify Aizoaceae (Mesembryanthemaceae) pollen to be continuously present in this region, but gaining significantly in importance from around 11 kBP, again consistent with our model. Finally, Scott et al [25] show for a continuous trace from before 18 kBP, that Aizoaceae (Mesembryanthemaceae) are absent in the Cederberg area (32º 30'S) until about 9 kBP, in broad agreement with our prediction of a significant expansion of the SKE into this area between 12 and 6 kBP. In conclusion, we suggest that in the western and southern Cape, climatic oscillations of intermediate amplitude over at least the past million years have caused biome and component plant species distributions to shift, and probably fragment during contraction phases, favouring both sympatric and allopatric speciation. Subsequent expansion may also have increased the likelihood of secondary contact, favouring hybridisation and associated speciation. The existence of landscape scale refugia has allowed species richness to be retained during phases of retreat. This is a classic example of a "species pump" previously invoked to account for tropical system biodiversity levels [26]. In this case, the species pump operates in both biomes, and may explain the enormous floristic richness of both. This region provides an outstanding living laboratory for the study of the effects of climate change on species diversity and speciation processes.

8. ACKNOWLEDGEMENTS

This work was partially supported by a grant from the Center for Applied Biological Sciences, a branch of Conservation International, and by the National Botanical Institute. We gratefully acknowledge the support of the Abruzzi Parks Network of Italy in their support of GFM's attendance of this conference.

9. REFERENCES

Comes H.P. and J.W. Kadereit, Trends in Plant Science 3 (1998) 432-438.

Cowling R.M. and C. Hilton Taylor, Strelitzia 1 (1994) 31-52.

Cowling R.M., P.W. Rundel, P.G. Desmet and K.J. Esler, Diversity and Distributions 4 (1998) 27-36.

Cowling R.M., S. Afr. J. Sci. 83 (1987) 106-112.

Davis M.B., In: H.E. Wright Jr. (Ed.) Late Quarternary Environments of the United States (Vol 2) The Holocene, Longman, London. 1984.

Department of Environmental affairs and Tourism, Pretoria, 1998.

Flenley J., TREE 8 (1993) 119-120.

Goldblatt P., Annals of the Missouri Botanic Gardens 65 (1978) 369-436.

Huntley B. and H.J.B. Birks, An Atlas of Past and Present Pollen Maps for Europe: 0 – 130 000 Years Ago, Cambridge University Press, Cambridge, 1983.

Ihlenfeldt H.D., Annu. Rev. Ecol. Syst. 25 (1994) 521-546.

Kutzbach J., CCMO General Circulation Model output Data Set, NOAA/NGDC Paleoclimatology Program, Boulder, Co. 1994.

Kutzbach J.E. and P.J. Guetter, Journal of the Atmospheric Sciences 43 (1986) 1726-1759.

Linder H.P., In: E.S. Vrba, (ed.) Species and Speciation, Transvaal Museum Monograph no. 4, Transvaal Museum, Pretoria 1985, pp. 53-57.

Low A.B. and A.G. Rebelo, Vegetation of South Africa, Lesotho and Swaziland,

Rosenzweig M.L., Species Diversity in Space and Time, Cambridge University Press, Cambridge, 1995 .

Russel Gibb G.E. s, Bothalia 15 (1987) 213-227.

Rutherford M.C. and R.H. Westfall, Mem. Bot. Surv. S. Afr. 54 (1986) 1-98.

Rutherford M.C., M. O'Calaghan, L.W. Powrie, J.L. Hurford and R.E. Schulze, Environmental Software 11 (1996) 113-121.

Rutherford M.C., M. O'Callaghan, J.L. Hurford, L.W. Powrie, R.E. Shulze, R.P. Kunz, G.W. Davis, M.T. Hoffman and F. Mack, J. Biogeog. 22 (1995) 523-531.

Schulze R.E., South African Atlas of Agrohydrology and Climatology, University of Natal, Pietermaritzburg. 1997.

Scott L., Historical Biology 9 (1994) 71-81.

Scott L., M. Steenkamp and P.B. Beaumont, Quarternary Science Review 14 (1995) 937-947.

Shi N., L.M. Dupont, H.J. Beug and R. Schneider, Veget. Hist. Archaeobot. 7 (1998) 127-140.

Stebbins G.L., Variation and Evolution in Plants, Columbia University Press, New York. 1950.

Thornthwaite C.W., Geographical Review 38 (1948) 55-94.

Tyson P.D., S. Afr. J. Sci. 95 (1999) 194-201.

Wright H.E., J.E. Kutzbach, T.Webb III, W.F. Ruddiman, F.A. Street Perrot and P.J. Bartlein (Eds.) Global Climates since the Last Glacial Maximum. University of Minnesota Press, Minneapolis, Mn. 1993.

The World Network of Biosphere Reserves: a Flexible Structure for Understanding and Responding to Global Change

MARTIN F. PRICE
Mountain Regions & Conservation Programme, Environmental Change Unit, University of Oxford, 11 Bevington Road, Oxford OX2 6NB, UK

Key words: global change, biosphere reserve, research, monitoring, exchange

Abstract: The processes of global change affect both the biogeochemical systems of the Earth and its human population. Such changes may be either cumulative or systemic, and there are complex interactions between the diverse processes of change. These interactions are increasingly a focus of international research programmes. One structure with considerable potential for understanding and responding to global change is UNESCO's World Network of Biosphere Reserves, with 356 sites in 90 countries. These have a three-fold zonation and three functions: conservation, development, and logistic support. While not all biosphere reserves fulfil the functions adequately, the periodic review process which began in 1998 means that more are likely to do so in future. The elements of partnership, regional approaches, and exchange which characterise biosphere reserves could be of particular value in an era of global change, both at the regional scale – through partnerships, research, development, and information sharing - and also at continental and global scales, particularly through the Biosphere Reserve Integrated Monitoring Programme (BRIM) for research and monitoring, and through exchanges for education and training.

1. AN ERA OF GLOBAL CHANGE

As we approach the beginning of the third millennium, both the biogeochemical systems of our planet and its human population are increasingly subject to diverse processes of change. In one way or another, many of these processes affect the entire planet and are referred to as "global change". Since the 1970s, social scientists have used this term to refer to

G. Visconti et al. (eds.), Global Change and Protected Areas, 403–411.

changes in international social, economic, and political systems. This anthropocentric use continues, but another, geocentric, definition has emerged since the early 1980s: "global change" as changes in the Earth's atmospheric, biological, geological, and hydrological systems (i.e., the biosphere [1]. Natural scientists recognise that such changes have taken place throughout the evolution of the biosphere. Now, however, the combined effects of the activities of billions of human beings on many of these systems have become equal to or greater than the effects of natural processes (i.e., natural variability). A particular concern is that many rates of change appear to be more rapid than at any other previous time in human history – and in some cases, the history of the Earth – and that such rapid rates of change may lead to highly unpredictable future states with no past or current analogues.

Two types of geocentric, or global <u>environmental</u> change (GEC), can be recognised [2]. The first are cumulative changes that take place at discrete locations around the globe but, when combined, have global importance. These include acid deposition, deforestation, desertification, and other processes leading to changes in the quantity and quality of land for agriculture and forestry and to decreased biodiversity at all scales. The second type are systemic global changes, i.e., those which occur throughout global systems. These include modifications of the stratospheric ozone layer and of the climate system, deriving from increased concentrations of certain atmospheric gases and likely to result in a further systemic change: sea-level rise.

The complex processes of change in the Earth's biogeochemical systems are intricately linked to those taking place in social, economic, and political systems, which may also be cumulative or systemic. While international research programmes on global change initially considered either biogeochemical systems (the International Geosphere-Biosphere Programme: IGBP) or human systems (the International Human Dimensions of Global Environmental Change Programme: IHDP), there is increasing overlap between these two global programmes, initially through the Land Use and Land Cover Change (LUCC) project and increasingly in other projects and initiatives. As the Executive Director of the IGBP noted in 1998, "We in the biophysical sciences need new paradigms and fresh approaches to addressing global change issues, especially as we are moving into the impacts/sustainability domain of GEC research... Perhaps we need to abandon our view that humans are passive recipients of changes in the biophysical world and recognise that they have, to a greater or lesser extent, developed complex, adaptive systems that strongly modify their interactions with the environment... the most spectacular progress in tackling GEC issues will come when natural and social scientists realise that the most creative approach is to adopt and integrate the most appropriate elements of each "world view"" [3].

2. THE WORLD NETWORK OF BIOSPHERE RESERVES

To understand global change requires appropriate structures for research, monitoring, and the exchange of information and expertise. To respond to global change necessitates flexible and sustainable approaches to the use and management of resources at regional scales, combined with opportunities for demonstration, education and training. Many of these needs could be answered by the World Network of Biosphere Reserves, which currently includes 356 sites in 90 countries, nominated by national governments and designated since 1976 by the United Nations Educational, Scientific, and Cultural Organisation (UNESCO). These sites have become the primary focus of UNESCO's Man and the Biosphere (MAB) programme, which began in 1971 and had as its objective, from the start, the interdisciplinary approach recognised as essential for tackling GEC issues [4].

Biosphere reserves have three characteristics that, at least in principle, differentiate them from other 'protected areas':

- they are part of an international network of sites designated by UNESCO, rather than by national governments;
- their outer boundary is flexible, rather than being legally defined;
- the land and water they contain is administered and managed by more than one agency or owner.

In fact, only the first of these characteristics defines all biosphere reserves; many have fixed outer boundaries and/or are administered and managed by one agency or owner. The conceptual model of a biosphere reserve (Fig. 1) includes three zones, whose characteristics are defined in the Statutory Framework of the World Network of Biosphere Reserves, adopted by the General Assembly of UNESCO in 1995 [5]:

a) a legally-constituted core area or areas devoted to long-term protection, according to the conservation objectives of the biosphere reserves, and of sufficient size to meet these objectives;
b) a buffer zone or zones clearly identified and surrounding or contiguous to the core area or areas, where only activities compatible with the conservation objectives can take place;
c) an outer transition area where sustainable resource management practices are promoted and developed.

Within this model, while the core and buffer zones often fit within the IUCN categories [6], the reserve as a whole – and usually the outer transition area, if it exists – cannot be so neatly categorised; it is now well-recognised that biosphere reserves are not 'protected areas' which fit into the six-fold system of categories developed by IUCN-The World Conservation Union [7]. Similarly, the Statutory Framework does not use the terminology of 'protected areas', recognising that the transition area is not necessarily legally protected, and that the flexible outer boundary means (at least in theory) that a biosphere reserve's specific geographic extent should not be precisely defined.

Over time, the biosphere reserve concept has evolved [8]. This is a reflection of global trends in conservation, with the emergence of approaches such as ecosystem management [9, 10] and bioregionalism [11], together with increasing emphasis on the need for involving local people and other stakeholders in conservation [12, 13]. Parallel to these trends has been the development and widespread acceptance of the concept of sustainable development, particularly through the 1987 report of the World Commission on Environment and Development [14] and the United Nations Conference on Environment and Development in 1992, which included the endorsement of 'Agenda 21', a "plan for action into the 21st century" by the heads of state or government of most of the world's nations.

The latest version of the biosphere reserve concept is found in the Statutory Framework. While two of the original functions – 1) 'conservation' of biological and landscape diversity; and 2) 'logistic support' for research, monitoring, demonstration, education, and training – have remained, there has been an increasing emphasis on the third 'development' function, which links to the expectation that all stakeholders should be involved in planning and managing activities in and around biosphere reserves. These three functions are brought together in the integrated goal for each biosphere reserve stated in article 4 of the Statutory Framework: that it "should provide an opportunity to explore and demonstrate approaches to conservation and sustainable development at the regional scale".

For a variety of reasons, including boundaries which are inappropriate for the current suite of functions, and the lack of willingness and/or resources necessary to foster stakeholder involvement in developing and implementing regional-scale management plans, many of the biosphere reserves designated in the 1970s and 1980s do not fulfil this goal. To encourage national governments to ensure that the biosphere reserves on their territories do contribute to this goal, as well as the broader objectives of the World Network (e.g., exchange of information, cooperation in management practices, comparative research and monitoring), the Statutory Framework requires a periodic review of all biosphere reserves. The aim is to ensure that all member sites of the World Network do fulfil the expected functions, both regional and global; and it is anticipated that the review process, which began in 1998, will gradually lead to the removal of 'non-functional' biosphere reserves from the network. This is preferably done by states themselves – a process which has already begun – in the last resort, the International Coordinating Council of the MAB programme can ask UNESCO's Director-General to inform a state that a biosphere reserve has been 'de-listed'. Thus, over time, it appears likely that the Network as a whole will move closer to achieving its objectives, as 'non-functional' sites leave, and new sites join.

Analysis of prerequisites for the successful implementation of the biosphere reserve concept has been undertaken in a number of countries. Drawing on North American experiences [15, 16], basic principles are that the impetus for development should preferably not come from the principal regional land management agency, but rather that many stakeholders should be involved from the outset – including not only local agencies and organisations, but also external groups with a strong interest in the region – and that they work cooperatively to develop and implement management strategies. The implementation of these principles faces many challenges, not least because of the different sets of concepts, decision-making structures, and technologies among the diverse stakeholders. In addition, there is often a need for the reform of existing institutions or the establishment of new ones, and adequate funding is essential. Similar conclusions have been drawn from studies of the biosphere reserves of France [17] and the UK [18].

At the broader regional and global scales, the research and monitoring functions are being facilitated by the compilation of information in standardised formats through the Biosphere Reserve Integrated Monitoring Program (BRIM), which began with a methodology for recording records of flora and fauna (MABFlora and MAB Fauna), but now also includes a database from permanent plots (BioMon) [19]. Increasing consideration is being given to including socio-economic data in BRIM. In addition, the regional networks of biosphere reserves for various parts of the world provide means for both coordinators of biosphere reserves, and scientists those working in them, to exchange experiences and information through meetings and field visits.

3. THE POTENTIALS OF BIOSPHERE RESERVES IN AN ERA OF GLOBAL CHANGE

The elements of partnership, regional approaches, and exchange which characterise biosphere reserves mean that they could be of particular value in an era of global change. At the regional scale, the criteria in the Statutory Framework that, for an area to be qualified for designation as a biosphere reserve, it "should encompass a mosaic of ecological systems..., including a gradation of human interventions.... (and) have an appropriate size to serve the three functions" are particularly relevant. From the Convention on Biological Diversity at the global scale, to the Pan-European Biological and Landscape Diversity Strategy and the European Union's Natura 2000 programme, down to national and smaller-scale biodiversity action plans, policies for the conservation of biological and landscape diversity now recognise that regional approaches are necessary; and these implicitly concern significant areas of land (and sometimes water). In particular, agreements on mechanisms for maintaining biological diversity, for instance through the establishment of conservation corridors and networks, may be more easily reached when structures for regional cooperation already exist.

Changes in climatic regimes, as well as land use changes driven by economic and political forces, are likely to be evident at the regional scale, and ecosystems with different levels of previous human impact will often be affected in different ways. Thus, fully in keeping with the stated functions of biosphere reserves, the different zones could be used to examine and experiment with changes in both natural and anthropogenic systems. In the core area, the emphasis would be on monitoring change. In the buffer zone(s), activities would focus on conservation responses to a changing environment. In the transition area – recognising that its outer boundary is not fixed, and can be defined in relation to specific needs – a variety of sustainable resource management practices could be tested and developed, and the most appropriate being promoted both locally and further a field.

Such regional-scale approaches to conservation, resource management, and demonstration require effective partnerships involving multiple stakeholders; and these are also required for a biosphere reserve to fulfil its functional criteria. One basis for such partnerships can be the sharing of information and data, for instance through a geographic information system (GIS) at the regional scale of the biosphere reserve. Such partnerships will be of increasing importance as the impacts of global environmental change emerge, for instance as habitats are fragmented, potential limits of cultivation move (typically polewards or up-slope). Such environmental impacts may have considerable potential to lead to conflicts between groups of stakeholders, and may be exacerbated by both long-established and new economic, social, and political factors. Thus, long-term trends in both physical and policy climates need to be taken into consideration when developing approaches to sustainable development. In this context, an appropriately-constructed GIS may be invaluable in developing and evaluating alternative scenarios for long-term resource management.

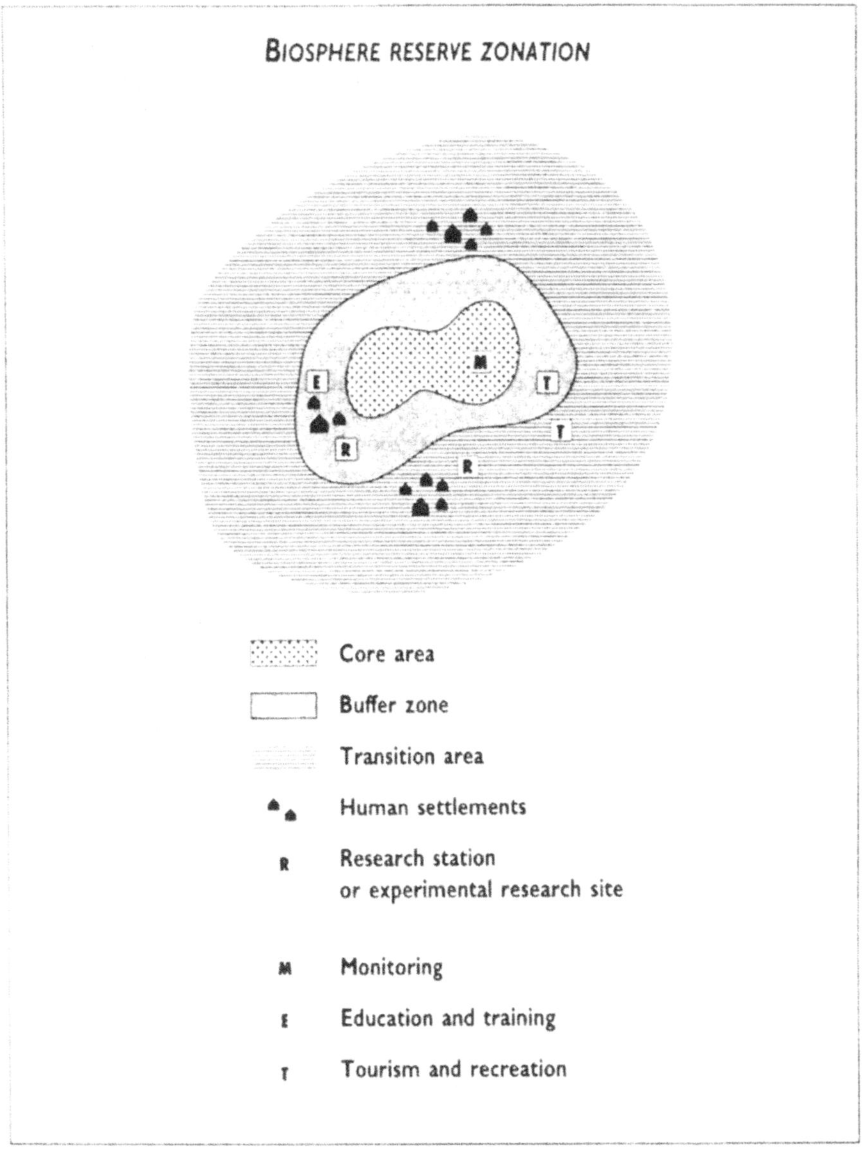

Figure 1: Biosphere reserve zonation. From reference [17] with permission.

In conjunction with suitable models, GIS can be used to consider many environmental components (location of treeline, forest fires, wildlife habitat, availability of grazing and water resources and snowpack, etc.) that may change in response to climate change. These issues could potentially alter the relationships between resource management practices in a number of economic sectors.

At the scale of continents or the globe, the World Network of Biosphere Reserves offers particular benefits with regard to the possibilities for comparative research, experimental work, and the compilation and use of the various elements of BRIM. While such possibilities have been recognised since the establishment of the first biosphere reserves, global change could give an added impetus to their 'logistic' function, built around not only research and monitoring, but also education and training. In the latter regard, one of the key values of the network is the opportunities for exchange of reserve coordinators, scientists, and other stakeholders in order to understand processes of global change – and sustainable responses to them – in comparable locations around the world.

In summary, the World Network of Biosphere Reserves should play a global role in increasing our knowledge of the processes of global change and sustainable responses to these processes. The members of the Network are spread widely around the world, across a great diversity of ecosystems and landscapes; partnerships between stakeholders are growing; and governments increasingly recognise the potential of these sites as models for conservation and sustainable development at a regional scale. Growing out of a recognised need for interdisciplinary and international collaboration in the early 1970s, biosphere reserves have significant roles to play at the beginning of a new millennium.

4. REFERENCES

Aley, J., (Ed.), Ecosystem Management: Adaptive Strategies for Natural Resources Organizations in the Twenty-first Century, Taylor and Francis, Philadelphia and London, 1999.

Bioret F. et al., A Guide to Biosphere Reserve Management: A Methodology Applied to French Biosphere Reserves. MAB Digest 19, UNESCO, Paris, 1998.

Borrini-Feyerabend, G., (Ed.) Beyond Fences: Seeking Social Sustainability in Conservation. IUCN, Gland, 1997.

Bridgewater, P, A. Phillips, M. Green and B. Amos, Biosphere reserves and the IUCN system of protected area categories. Australian Nature Conservation Agency/World Conservation Union/UNESCO MAB Programme, Canberra, 1996.

http://www.mabnet.org/brim/

IUCN, Guidelines for protected area management categories. IUCN, Gland, 1994.

McGinnis, M.V., (Ed.), Bioregionalism, Routledge, London, 1999.

McNeely, J.A., (Ed.) Expanding partnerships in conservation. Island Press, Washington DC, 1995.

Peine, J.D., (Ed.), Ecosystem Management for Sustainability: Principles and Practices Illustrated by a Regional Biosphere Reserve Cooperative. CRC Press, Boca Raton, 1999.

Price, M.F. et al., Review of UK Biosphere Reserves. University of Oxford, Oxford, 1999.

Price, M.F., Biosphere reserves: a flexible framework for regional cooperation in an era of change. In: D.L. Peterson and D.R. Johnson, (Eds.) Human Ecology and Climate Change, Taylor and Francis, Philadelphia and London, 1995, pp. 261-277.

Price, M.F., Global change: defining the ill-defined. Environment 31 (1989) 18-20, 42-44.

Price, M.F., Humankind in the biosphere: the evolution of international interdisciplinary research. Global Environmental Change: Human and Policy Dimensions 1 (1990) 3-13.

Price, M.F., People in biosphere reserves: an evolving concept. Society and Natural Resources 9 (1996) 645-654.

Steffen, W., Putting the human dimensions into IGBP. IHDP Update 2 (1998) 2.

Turner, B.L. et al., Two types of global environmental change: definitional and spatial-scale issues in their human dimensions. Global Environmental Change: Human and Policy Dimensions 1 (1990) 14-22.

UNESCO, The Statutory Framework of the World Network of Biosphere Reserves, UNESCO, Paris, 1996.

World Commission on Environment and Development, Our Common Future. Oxford University Press, Oxford and New York, 1987.

Yaffee, S.L. et al., Ecosystem Management in the United States, Island Press, Washington DC 1996.

The Role of a Global Protected Areas System in Conserving Biodiversity in the Face of Climate Change

LEE HANNAH
Conservation International
This work was funded in part by a grant from the Center for Applied Biodiversity Science (CABS) of Conservation International.

Key words: Biodiversity, Global Modeling.

Abstract Impending global change requires levels of coordination, long-range planning and funding never before required of protected areas. To meet this need a truly global system of protected areas is required. This system will require major funding and must ensure the persistence of biodiversity over decades and centuries. It will need to include corridors of semi-natural areas, as well as core parks and reserves. It must be able to ensure the persistence of biodiversity across international boundaries and in spite of international differences in scientific expertise and financing ability. Conservation of species and habitat types has been traditionally framed in a static, two dimensional view, in which reserves are intended to protect these in their present locations. Yet overwhelming evidence exists for major shifts in vegetation in the recent, mid-term and log-term past. The interplay of climate changes and habitat destruction threatens the existing system of isolated reserves. Climate change will create new ranges for many species. Large areas converted to human use, such as agriculture and housing will cut off existing populations from these new ranges. A truly global protected area system does not yet exist. Currently there is a global group of national systems, some of which are connected through information network. Cross border protected areas have existed to protect shared wildlife resources, particularly in Africa, for decades. IUCN regional reviews have provided information to improve protected area network coverage based on regional needs. None of these efforts fully incorporates climate change considerations nor promotes connectivity in the log-term on a global scale. Only a global protected area system which creates corridors between reserves and protects historic locales

G. Visconti et al. (eds.), Global Change and Protected Areas, 413–422.

of habitat suitability can minimize biodiversity loss in the face of climate change and habitat destruction. Currently more than half the habitable surface of the planet is dedicated to human uses. Climate change will be driving vegetation shifts across this landscape. Species with strong dispersal ability will fill new ranges, while slower dispersing species will be unable to STS jump the gap CCH to a new habitat Species with less dispersal ability will become locally extinct, and eventually large-scale loss of biodiversity will occur. A global system will require a global modelling effort, prioritisation, and novel finance mechanisms. A coordinates global modelling system will be required to determine likely biological changes in response to climate change. Base don model results, priority areas for international investment can be selected. Since biodiversity hotspots are predominantly in the temperate countries, a new international protocol and new mechanisms of finance will be required. Finally, because available land for protected areas is nearly exhausted, global financing will have to buy back land and convert it from human use to natural or semi-natural cover in key corridor areas. This paper present a preliminary proposal for such a system.

1. INTRODUCTION

Protected areas need to be designed to accommodate climate change. It is now well accepted that human-induced climate changes are highly likely on a time scale of 10-100 years (IPCC, 1996). It is also recognized that natural climate changes would be expected in the absence of human influence on a time scale of 50-500 years (Crowley and North, 1991). Extensive palaeoecological evidence indicates that vegetation ranges will shift in response to these changes (Huntley, 1998). Reserve design has historically not focused on the medium- and longer-term implications of climate change. The picture of pristine ecosystems representing a long-term ideal steady state of nature is a misconception based on the short time span of human life. If we are conserving biodiversity forever, we must design protected areas capable of withstanding climate change and the biological changes that go with it.

Rob Peters first explored the implications of climate change for protected areas (Peters and Darling, 1985). At the time of Peters' analysis, it was believed that human-induced climate change would occur on a time scale much faster than historical changes, and that plants would therefore not be adapted to shift range in time to keep pace. We now believe that this may not be correct (Lovejoy, 1999). Climate shifts of the general magnitude and speed expected in the near future due to human influence may have occurred naturally at least twice in the past 20,000 years (Crowley and North, 1991). Twelve or more relatively rapid shifts have occurred in the past 500,000 years (Crowley and North, 1991).

So natural vegetation, if not adapted to relatively rapid climate changes, has at least survived many such shifts in the past. Peters' conclusion that large-scale mechanical shifts of vegetation might be the only solution for conservation in the face of human-induced climate changes is therefore probably not correct.

Protected areas systems intended to conserve biodiversity in perpetuity should be designed to accommodate natural climate changes as well as human-induced climate change. Because human-induced climate change is likely to include warming and natural climate change is expected to include cooling, the range of design parameters is challenging. Climate change encompasses rainfall and seasonality changes in addition to warming and cooling, further complicating the design picture.

Connections, corridors, protected agricultural or forestry landscapes and adaptive management are all tools which can be useful in designing protected areas systems which respond to climate change. In the past, there is strong evidence for species range shifts in response to climate change, and little evidence for *in situ* adaptation (Huntley, 1998).

Connections or corridors between protected core areas can help accommodate vegetation dispersal and range shifts. Non-core conservation areas can provide forestry or agricultural landscapes which are managed to be "semi-permeable" to vegetation dispersal in response to climate change.

Combined, the concept of connectivity and conservation management of human landscapes are key ingredients in developing a system which allows natural vegetation shifts to occur as climate changes. Since the exact magnitude, nature and timing of climate change is not known, adaptive management will be key to implementing a system which uses both natural and semi-natural landscapes in a unified system to respond to change.

To make effective use of these tools, we must change the way we think about protected areas systems. We must no longer think of protection as providing representation of a static, two-dimensional pattern of vegetation types. We must instead envision a mosaic of vegetation associations shifting in time and design systems which conserve the processes of this change as well as the present patterns of biodiversity.

2. PAST CLIMATE SHIFTS AND BIODIVERSITY RESPONSES

The view that nature is dynamic, not static, has grown in stature within the past decade. Works like Botkin's "Discordant Harmonies" (1990) have developed the notion of a dynamic nature which requires dynamic management.

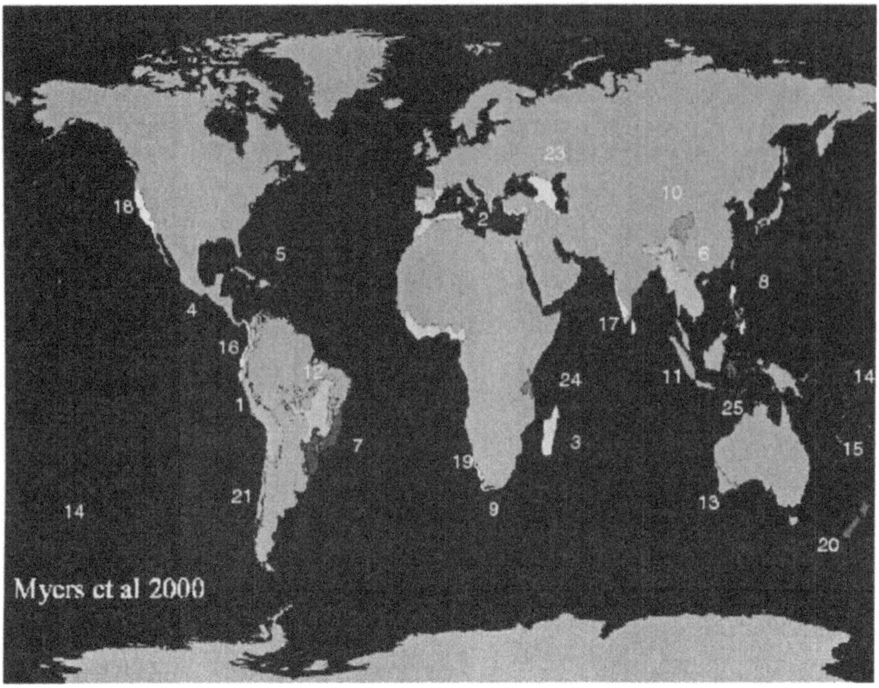

Fig. 1. The Global Biodiversity Hotspots. Each Hotspot has a minimum of 1,500 endemic plants, and has lost at least 75% of its original habitat. 1. Tropical Andes 2. Mediterranean Basin 3. Madagascar/Indian Ocean Islands 4. Mesoamerica 5. Caribbean Islands 6. Indo-Burma 7. Atlantic Forest of Brazil 8. Philippines 9. Cape Floristic Region of South Africa 10. Mountains of South Central China 11. Sundaland 12. Brazilian Cerrado 13. Southwest Australia 14. Polynesia and Micronesia 15. New Caledonia 16. Choco/Darien/Western Ecuador 17. Western Ghats & Sri Lanka 18. California Floristic Province 19. Succulent Karoo 20. New Zealand 21. Central Chile 22. Guinean Forests of West Africa 23. Caucasus 24. Eastern Arc Mountains, Coastal Forests of Kenya and Tanzania 25. Wallacea

The misconception of vegetation types as fixed entities has been replaced, at least in some quarters, by the conviction that vegetation associations are fluid and changing. What has not been widely recognized is the primary importance that these climate change-induced changes have had in producing current patterns of diversity and endemism. The consequence of dozens of global temperature swings in the past (and the associated changes in rainfall, storm patterns and seasonality) is vegetation which has been driven into and out of association and across landscapes. These shifts have resulted in patterns of diversity and endemism which reflect past climate changes.

The concept of climate change refuges (more usually referred to as "Pleistocene refugia") has been explored by many authors (Prance, 1987). Refugia theory as originally proposed, held that areas of exceptional endemism may be explained as areas in which stable climate during the Quaternary permitted species persistence and diversification. More recent theories hold that ecotones and areas of climate dynamism may be sources of diversity and endemism. In both theories, change within limits produces centers of endemism which are different for different taxonomic groups, based on the complex set of climatic variables and habitat requirements acting on each individual species.

When these centers of endemism are added up on a global scale, "hotspots" emerge. "Biodiversity Hotspot" is a term developed by Norman Myers to refer to large areas of exceptionally high plant endemism. Myers originally identified over 10 such areas worldwide in the late 1980's and early 1990's (Myers, 1988; 1990). Myers' analysis has been expanded by Myers, Mittermeier and a large group of taxonomic specialists to include 25 areas worldwide (Mittermeier et al, 1998). These 25 areas comprise 1.4% of the land surface of the earth, yet they account for over 45% of all plant species *as endemics*. Counting other, non-endemic species, it is believed that the hotspots contain as much as two-thirds of all plants on earth (Mittermeier et al, 1998). While the hotspots have been identified based on plant endemism, there is good global agreement with endemism in vertebrates and other groups. The 25 global biodiversity hotspots are illustrated in Figure 1.

At a global scale, the hotspots correlate very well with areas which provide 'buffering' from climate change. This would be expected based on refuge theory. If areas which contain high levels of endemism on a regional scale are associated with relatively stable climate refuges, the hotspots of global endemism should be associated with features that limit impacts of climate instability. On a global scale, these features are primarily mountains. Through adiabatic cooling and warming, mountains provide refuge from temperature changes. Orographic rainfall moderates drying associated with climate change. Slope allows rapid migration and fragmented structure provides abundant microsites. Figure 2 illustrates areas of extended slope over a 25km radius analogous to mountains, and their associated high overlap with hotspots.

The climate change buffering of mountains is a dynamic process. Climate change refuges of sufficient scale to be called hotspots are broad areas in which temperature, precipitation and other climate changes can be accommodated with minimal shifts in vegetation ranges. Adiabatic temperature change in mountains does not ensure a stable temperature at a given point, it ensures that a similar temperature can be found within a short distance. Similarly for rainfall and microhabitat conditions, mountains ensure that similar conditions are nearby and within the dispersal range of plants when climate changes. In flat land, the next area of similar rainfall, temperature and microhabitat may be hundreds of kilometers away during climate change. In mountains, the next similar area is likely to be tens or fractions of a kilometer away, depending on the severity of the climate change. This allows species lineages to persist and diversify in mountains. It promotes diversification by presenting suites of new microhabitats and new climatic conditions which often equate to vacant niches. Modeling of vegetation shifts in response to climate change have born out this relationship. In South Africa, a unique juxtaposition of two hotspots exists. One is an arid hotspot, the Succulent Karoo, the other a mesic hotspot, the Cape Floristic Region. The two lie next to each other along adjacent north-south and east-west trending mountain systems at the southwestern tip of Africa. Modeling has shown that under past climate changes, these two hotspots have migrated, one giving way to the other, and always uniquely staying within the mountain zone (Midgley and Roberts, this volume). Figure 3 in Midgely and Roberts (this volume) shows this relationship. This historical movement within the mountainous area is believed to have created the tremendous diversity and endemism of the region by providing a setting in which similar temperature, soil, rainfall and microhabitat were always within the dispersal range of the plants of the two hotspots, yet in which these same species were constantly presented with opportunities to radiate into slightly new temperature, rainfall and microsite conditions (Midgely and Roberts, this volume).

3. IMPLICATIONS FOR PROTECTED AREAS DESIGN

Present protected areas, isolated and in static locations, are poorly equipped to respond to climate change. Core areas are only part of the conservation solution when vegetation is a shifting mosaic and not a static assemblage of stable communities. An interlocking system of protected core areas and managed connections is required to fully respond to climate change. Core areas in this system can include existing protected areas. Modeling such as the South Africa example presented above can suggest areas in which additional core areas should be located. Areas which are believed to have contained relict vegetation types during climate extremes are critical core areas which should be maintained in as natural a state as possible. To respond to change, these core areas need to be connected. Connections are essential to let plants move in response to changing climatic ranges. In some cases, the connections need to be corridors of natural habitat. In other cases, landscapes under human uses such as agriculture or forestry can be managed to make effective connections which are at least "semi-permeable" to plant dispersal. These areas of managed connections may need to be relatively large in comparison to core areas. They will need to be designed based on the specific biology and dispersal mechanisms of the vegetation.

The vegetation shifts that must be accommodated have several distinctive characteristics.

Climate change-driven vegetation shifts:

- cross political boundaries,
- generate biodiversity 'hotspots',
- involve land outside parks,
- cross land not currently in natural vegetation, and
- may be rapid (50-100 years).

The South Africa example illustrates most of these characteristics. The model results show that vegetation shifts may cross political boundaries. The Succulent Karoo in cooler times from 15,000-10,000 years before present was centered in what is now Namibia. It is now centered in northern South Africa, but still has limited distribution in Namibia. As stated above, the unique properties of mountains have generated the hotspots, and past shifts have crossed large landscapes, including land now dominated by human uses.

A protected areas system designed to accommodate climate change needs to respond to each of these characteristics. Regional international coordination is required to address change across political boundaries. The system needs to have special focus on the mountainous areas which have generated the hotspots, both because of their high levels of endemism and because conservation in the face of climate change will be most effective in these areas. Land outside parks must be included in the strategy in the form of connections or corridors between core areas. Non-natural vegetation will have to be specifically managed to accommodate both human use and conservation compatibility if vegetation shifts associated with climate change are to successfully cross broad landscapes. Since the nature and magnitude of climate change cannot be exactly predicted, this landscape management will have to be active and adaptive.

The addition of core areas, creation of connections and active management all require major new expenditures not in current park management agency budgets. Reserving land for uses compatible with conservation may require compensation for land owners. Restoration of land to natural or semi-natural habitat will require compensation and physical measures which may be costly. Focusing on hotspots can help make this additional expenditure highly cost-effective, but major additional funding is required to provide these tools. As a first estimation, a doubling of protected areas budgets may be required in non-hotspot areas which are generally temperate and have higher national protected areas system budgets. In the hotspots, most of which are tropical and fall in countries with relatively low protected areas systems budgets, an order of magnitude increase in protected areas system spending or more may be required.

In summary, a protected areas system responsive to climate change requires:

- regional coordination,
- hotspot focus,
- integration of all land uses,
- proactive land use change, and
- major increases in funding.

4. ROLE OF A GLOBAL SYSTEM

A problem termed '*global change*' might be expected to require a solution which is both *global* and *dynamic*. Present protected areas strategies are neither. A truly global protected area system does not yet exist. Currently there is a global group of national systems, some of which are connected through information network. Cross border protected areas have existed to protect shared wildlife resources, particularly in Africa, for decades. IUCN regional reviews have provided information to improve protected area coverage based on regional needs. None of these efforts fully incorporates climate change considerations nor promotes connectivity in the long-term on a global scale.

Protected areas responsive to climate change require regional modeling efforts, prioritization, and novel finance mechanisms. A coordinated modeling system will be required to determine likely biological changes in response to climate change. Based on model results, priority areas for international investment can be selected. Since biodiversity hotspots are predominantly in the tropics and financial resources predominantly in the temperate countries, a new international protocol and new mechanisms of finance will be required. Finally, because available land for protected areas is nearly exhausted, financing will be needed to buy back land and convert it from human use to natural or semi-natural cover in key corridor areas.

A global system may be required to meet these goals. Financing and regional coordination are aspects which may best be addressed with worldwide cooperation. Regional coordination may require overall coordination to ensure that there is continuity between regions. Financing of the magnitude required will require north-south transfers that would require

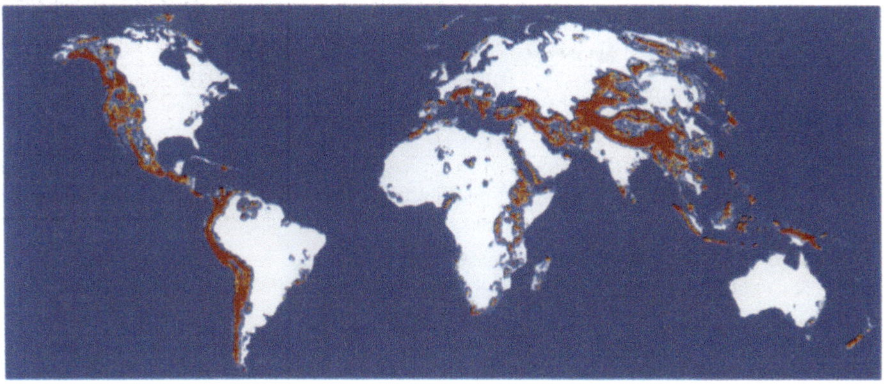

Fig. 2. Extended Slope Areas of the world (see text for definition). Warmer colours are higher slope. These mountainous areas offer climate diversity and proximity of micro-habitats which help retain biodiversity during climate change.

new international agreements. While all possible functions should occur at the local, national or regional level, only the best global scientific advice and a unified global funding effort can fully address a problem of such complexity.

A Global Protected Areas System suitable for responding to climate change can then be envisioned. The global aspect of the system would emphasize planning, coordination and funding support.

Cross-border cooperation and system design would take place at the regional level.

At the national and local levels, core area management and conservation management of connecting natural and semi-natural landscapes would be designed and implemented. This system would be much different from the largely national systems of today. It would emphasize existing core areas, core areas added for climate change response, and broad areas of forestry and agricultural landscape presently little involved in conservation management. It would be a dynamic system, capable for the first time of responding to the global and dynamic processes which have shaped the distribution of the biodiversity we seek to conserve.

5. REFERENCES

Botkin, D. 1990. *Discordant Harmonies* (pp3-14) Oxford University Press. New York

Crowley and North, 1991. *Paleoclimatology* (pp62-66) Oxford University Press. New York.

Huntley, 1998. The Dynamic Response of Plants to Environmental Change and the Resulting Risk of Extinction. In: *Conservation in a Changing World* pp69-87. Eds G. Mace, A. Balmford and J. Ginsberg. Cambridge University Press.

IPCC, 1996. *Climate Change 1995:Impacts, Adaptations and Mitigation of Climate Change.* Intergovernmental Panel on Climate Change. Cambridge University Press. New York.

Lovejoy, 1999. Linkages Between Climate Change and Biodiversity (presentation) in *Conserving Biodiversity in the Face of Uncertainty* (conference and web-cast). American Museum of Natural History. New York.

Midgely and Roberts, this volume.

Mittermeier, R., N. Myers, J. Thomsen, G. da Fonseca and S. Olivieri, 1998. Biodiversity Hotspots and Major Tropical Wilderness Areas: Approaches to Setting Conservation Priorities. *Conservation Biology* 12(3)516-520.

Myers, N. 1988. Threatened Biotas: Hotspots in Tropical Forests. *The Environmentalist* 8:178-208.

Myers, N. 1990. The Biodiversity Challenge: Threatened Hotspots Analysis *The Environmentalist* 10:243-246

Peters, R. and J. Darling. 1985. The Greenhouse Effect and Nature Reserves. *Bioscience* 35(11) 707-717.

Prance, G. 1987. Biogeography of Neotropical Plants. In: *Biogeography and Quarternary History in Tropical America* Eds T. Whitmore and G. Prance. Oxford University Press.

Section 4

The Abruzzi Parks: A Case Study

The Strong Reduction Phase of the Calderone Glacier During the Last Two Centuries: Reconstruction of the Variation and of the Possible Scenarios With GIS Technologies

LEANDRO D'ALESSANDRO (*), MAURIZIO D'OREFICE(**), MASSIMO PECCI (***) CLAUDIO SMIRAGLIA (****) & RENATO VENTURA (**)

(*) Dept. of Earth Science, University of Chieti "G. D'Annunzio", Madonna delle Piane – Via dei Vestini – 66013 Chieti Scalo, ITALY - Tel + 39-871-564056; Fax + 39-871-564050

(**) Italian Geological Survey, Via Curtatone, 3 – 00185 Rome ITALY - Tel + 39-6-44442444; Fax + 39-6-4465622

(***) ISPESL - DIPIA (Higher Institute for Occupational Safety and Health - Department of Production Plants and Interaction with the Environment), Via Urbana, 167 - 00184 Rome ITALY- Tel + 39-6-4714261; Fax + 39-6-4744017, Italian Glaciological Committee

(****) Dept. of Earth Science, University of Milan, , ITALY - Tel + 39-2-23698230; Fax + 39-2-70638261; Italian Glaciological Committee

Key words: Glaciology, Variations of the Calderone glacier, Geographic Information Systems in glacial geomorphology, Little Ice Age, Glacier Mass Balance

Abstract: The paper presents the state of the art of the works in progress; in particular Prof. L. D'Alessandro co-ordinates geomorphologic topics, Dr. Massimo Pecci field activities and data elaboration in GIS environment and Prof. C. Smiraglia glaciological themes. The Calderone glacier is now confined into a deep central mountain valley of the Gran Sasso d'Italia, with steep walls, and does not show movements along the borders and along the front. The little apparatus is characterised by a reduction phase since the end of the "Little Ice Age" Auct., particularly strong since the end of the last decade. During the nineties a set of multidisciplinary researches started to evaluate the role of the Glacier like an indicator of the effects of human activities and finally of regional and global climatic change. In this paper the authors present the analysis performed to evaluate the variations, particularly in surface area and volume, since the end of the Little Ice Age up to now. The use of a Geographic Information System (GIS), modelling available data, regarding both the morphology of the ice surface (georeferenced from recent topographic

G. Visconti et al. (eds.), Global Change and Protected Areas, 425–433.

maps and reconstructed from historic documents and old topographic maps) and the morphology of the bed rock (from Ground Penetrating Radar surveys), allows the 3D-reconstruction of the volumetric variation during time. The "time sample points" in the modelling processes include the end of the LIA, the end of the past century and the present century for a total of six 3D-reconstruction. The resulting values, reconstructed and/or calculated since the end of the LIA, of about 50.000 squared metres of ice surface area and of about 4.000.000 cubic metres in volume represent the loss up to now (1990). The obtained results are finally discussed also in order to apply and to generalise the presented methodology in calculating the mass balance of particular type of glaciers and in order to evaluate the probable living (surviving) times of the glacier.

1. INTRODUCTION

The monitoring of the Calderone Glacier, the southernmost in Italy and in Europe (Fig. 1), has been started again in 1994 [1, 2] and new interesting fields of study have been carried out.

The studies involved up to now:

- field surveys of glaciological, geophysical, geomorphologic, radio-chemical and meteo-climatic features;
- data, images and available cartography collection;
- data elaboration and restitution on GIS.

A full geo-data base has been implemented in GIS environment for a first reconstruction of variation in thickness and surface area during the last two centuries [3]. The reconstructions have been performed starting from the six better described most significant and truly data source.

In this paper the authors present the further elaboration regarding variations in volume during the last two centuries, hypothesising the glacier confined by the frontal/lateral moraine inside the cirque. In fact the historical and iconography research [4] gave interesting suggestion and confirmation in order to the presence of little, active, tongues since the half of the last century.

2. SOURCES OF DATA

The data used for elaboration in GIS environment have been divided into two main groups:

- descriptions and pictures for extrapolated reconstruction;
- cartographic data (maps restituted in different scale) directly georeferenced in GIS environment.

3. DATA INCLUDED IN SCIENTIFIC DESCRIPTION

The only kind of data source taken into account for the reconstruction of the variations of the Glacier is the description of the first ascent to the Corno Grande Peak through the Calderone glacier in July 1794 by Orazio Delfico from the northern slope [5].

In the manuscripts the following detailed description (in old Italian language with translation in English) is given:

- "... Così a stento, ed adagio andando avanti giunsi in un esteso ripiano, quasi intieramente circondato dal alte rocche, che ne formano come una maestosa conca. Per indicarne in qualche modo l'elevazione basterà il dire, ch'essa è continuamente coverta di neve non eguale in durezza al gelo, ma ben solida, e ferma per non ricevere alcuna impressione dalle più forti pedate dei contadini che mi accompagnavano...."
- ("....In this way going up slowly and laboriously I reached a wide terrace-plain almost completely surrounded by high peaks, forming a majestic circular depression. To define the elevation it is sufficient to say that it is always covered by snow, very hard and not equal but similar to ice and so solid to not allow the foot-print of accompanying mountaineers").

Starting from this description and the pictures of painters and drawers of the last century a reconstruction of the glacier has been performed taking into account the morphologic limits and edges.

4. CARTOGRAPHIC DATA

The sources of cartographic data used in the elaboration are reported with the year of surveying and general information:

- (**1884-1885**) First 1:50.000 topographic map of the Corno Grande area (contour lines 20 m), surveyed on the field by IGM (Military Geographic Institute) in 1884-1885 and mechanically enlarged at a scale 1:20.000 [6] with a contour lines distance of 10 metres (Projection Flaamsted modified, with the origin of the co-ordinates at the intersection between the Meridian passing through Napoli and the 40° Parallel).
- (**1916**) Detailed Topographic sketch of the Calderone Glacier (scale 1:5.000) surveyed in September 1916 the 4th by Marinelli & Ricci [7], in local co-ordinates system.
- (**1934**) Detailed topographic map, surveyed originally at a scale 1:1.000 in September 1934 the 8th and the 9th by Sforzini, Tonini & Tonini [8], referenced to the IGM main topographic point of Corno Grande Peak

and, indirectly, to the International Ellipsoid oriented to Rome (M.Mario). Co-ordinate U.T.M. are referenced to the Fuse 33.

- **(1960)** Detailed photogrammetric map, surveyed originally at a scale 1:1.000 in August 1960 the 30th by Balducci, Pesavento, Di Fazio & Tiberio [8] and restituted by I.R.T.A. – Milan (1961), referenced to the IGM main topographic point of Corno Grande Peak and, indirectly, to the International Ellipsoid oriented to Rome (M.Mario). Co-ordinate U.T.M. are referenced to the Fuse 33.
- **(1990)** Detailed (1:750 scale) topographic map of the Calderone Glacier, surveyed by Gellatly *et alii* [9] in local co-ordinates system.: each data elaboration of the presented study is referenced to this more recent map, integrated by the 1:10.000element of the Corno Grande area of 1982 [10] - Orthophotomap of Abruzzo Region, drawn starting from the photo survey of 1981 by Compagnia Generale Riprese Aeree – Parma; contour lines distance is 10 metres; Reference International Ellipsoid ED 1950). Co-ordinate Gauss-Boaga are referenced to the Fuse East. Co-ordinate U.T.M. are referenced to the Fuse 33.

5. DATA ELABORATION

The volume of the glacier in each considered period has been calculated through the following steps:

1. definition of the shape and related georeferencing and cartographic restitution of the DEM (Digital Elevation Model) of the calcareous bedrock;
2. calculation of each volume for the six considered periods, obtained by a difference in volumes between the DEM of the glacier for each considered period and the DEM of the bedrock.

Regarding point 1, the shape and the morphology of the bedrock has been reconstructed starting from the DEM of the surface already available and integrated by in situ surveys and the topographic detailed map (scale 1:750 [9]). Particularly Digital Elevation Model (DEM) has been generated from 1:10.000 scale topographic map, both for the reconstructed and the georeferenced ones, used GRID and TIN module of ARC/INFO.

The reconstruction of the ice thickness distribution for the whole glacier has been possible by the availability of data of thickness obtained from GPR surveys [11]. The GPR profile surveyed during summer 1998 in the lower and in the middle sector of the glacier restituted a thickness in ice ranging from about 4 to about 20 metres; in the upper sector of the glacier the thickness has been reconstructed according to the morphologic limits and the hypothesising an homogeneous behaviour of the glacier.

Figure 1 – Location of the Calderone Glacier (Abruzzi Region – Central Apennines)

In the 3-D view from NE of the studied area (Fig. 2), including the Calderone Glacier (in black thick line) and the distribution of the sample points (thickness of ice in bracket) used in the reconstruction of bedrock is shown, distinguishing between measure points and reconstructed ones. The elaboration has been performed starting from ground data of Abruzzo Region [10].

Regarding point 2, the reconstruction (for 1794 and 1884) and the georeferencing (for 1916, 1934, 1960 and 1990) has been performed in the following phases [3]:

- *reconstruction* of the probable maximum growth of the glacier in surface area, thickness and volume during the "Little Ice Age", based on the description of DE MARCHI in 1573 [12] and mainly on the description of DELFICO in 1794 [5]. Particularly following the description of DELFICO [5], the lower part of the glacier has been considered like a plane, directly linked and levelled in thickness to the frontal moraine at its maximum level (about 2.700 m asl), *hypothesising the glacier confined by the frontal/lateral moraine inside the cirque.* The reconstruction of the upper part of the glacier has been possible using the topographic restitution of IGM 1884-85 [13]. It has been possible any georeferencing of the obtained results;

- *reconstruction* of the distribution of surface area and related thickness and of volume, based on the first true topographic map at a scale 1:50.000 (contour lines 20 m) surveyed on the ground by IGM in 1884-85 [6]. The reconstruction has been also possible and true, considering the distribution of "indicative" contour lines along the glacier and checking the extrapolation by historic photographic images, mainly belonging to Rome Italian Alpine Club and private inventories. Unfortunately a real and useful georeferencing has not be possible, due to the intrinsic error in surveying method and related displacement of the check points of the area ("TICS" in GIS environment), with respect to the original position;
- *georeferencing* of the perimeter, the area, the volume and the related DEM of the morphologic surface of the glacier based on the available topographic and numeric map of 1916, 1934, 1960 and 1990. The georeferencing of data has been also implemented and controlled by the availability of several photographs, gently given by private collections, mainly [13], [8], [14], [15].

The results of spatial analysis in GIS environment are presented and summarised by the 3D graphic of the variations of ice volumes of the considered time intervals of Fig. 3a and 3b and in table 1.

6. DISCUSSION, PERSPECTIVES AND SCENARIOS

The variations drawn in Fig. 3a and 3b and tabulated in table 1 have been obtained in GIS environment. In data elaboration the starting hypothesis considered the glacier confined by the frontal/lateral moraine inside the cirque, but historical research and iconography [4] is given further, interesting elements of discussion. Particularly a new view of the whole area of the Gran Sasso Mountain Range during the maximum of LIA could be proposed, including a climate cooler than today with severe conditions, coupled with the presence and the activity of at least one ice tongue.

Furthermore the data elaboration on GIS has been affected by a minimum error of about the 10 %, due to georeferencig procedures and by an higher error, due to the extrapolation performed. In this way the variation in surface area are not similar to the ones in volume, huge in the studied period, in confirmation of a strong reduction of the apparatus in the last century.

Such a condition could be strictly linked to the loss of the hypothesised tongues, completely and quickly melted in few decades, during the second half of the last century.

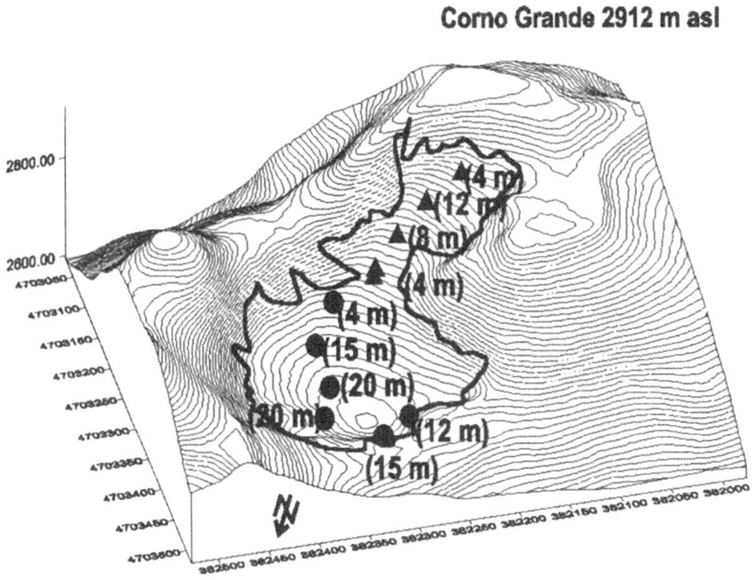

Fig. 2 – Recontruction of the variation in volume: DTM including the distribution of sample point from GPR (full circle) and reconstructed ones (full triangle)

This revolutionary perspective of study, if confirmed by further and quantitative results, besides the descriptions and the iconography already collected, induces to think about a warming trend more severe than in other geographic zones. In this perspective the presence of the Calderone glacier is now due to particularly favourable (lucky) conditions, mainly morphologic and topographic, and a "complete" apparatus with two well developed tongues, active since the second half of the last century, has been melted by severe climatic conditions.

The scenario for the future, widely confirmed by the trend pointed out in the past, includes realistically the melting of the whole glacier in few decades (2-3), supposing the continuity of the current climatic regime, the maximum surveyed thickness in about 20 m and the annual average ablation rate of about 1-2 m per year (for severe summer). The truth and the accuracy of such a scenario is partially confirmed by the evidence of a developing subdivision of the glacier since the 1997, in correspondence of the morphologic narrow centre.

7. ACKNOWLEDGMENTS

We want particularly to thank A. F. Gellatly, J. M. Grove and R. Latham for the data utilisation of the Calderone glacier topographic map, surveyed at

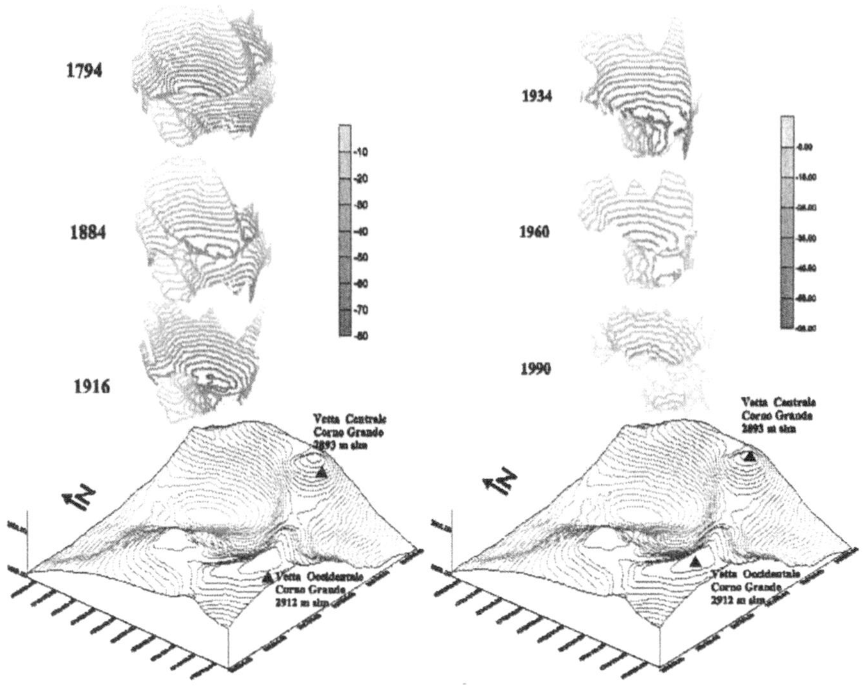

(a) (b)

Fig. 3 – Reconstruction of the variations in volume: period 1794-1916 (a) and period 1934-1990 (b), hypothesising the glacier inside the cirque.

Table 1 – Variations of ice surface area and volume and related reduction starting from LIA

YEAR	AREA (m^2)	REDUCTION IN AREA, STARTING FROM LIA (m^2)	VOLUME (m^3)	REDUCTION IN VOLUME, STARTING FROM LIA(m^3)
1794	104257	0	4332207	0
1884	90886	7444	3382166	1369003
1916	63335	7127	3368485	2100598
1934	59713	10749	2461529	3007554
1960	60030	38300	1729934	3021235
1990	52586	51671	360931	3971276

a scale 1:750 in 1990, Luca Mazzoleni for the presence and support at the Franchetti Hut, Adolfo Esposito, Anna Claudia Cartoni, Michela Mazzali, Enrico Bernieri and Antonella Balerna for helping us in digging trenches of snow.

8. REFERENCES

Capaldi F., Postcard collection, (1900- 1930)L'Aquila

D'Orefice M., L. Ledonne, M. Pecci, C. Smiraglia, C. and R. Ventura, Geogr. Fis. e Din. Quat., 18, (1), (1996), 253-256.

D'Orefice M., M. Pecci, C. Smiraglia, C. and R. Ventura, Artic Alpine Res. (send for the press)

De Marchi F., Cronaca della prima ascensione sulla vetta del Gran Sasso d'Italia. Carte 7-14 Cod. Man. Magliabechi, cl. XVII, a. 3, Atti Bibl. Naz., II, I, 277-280, 1573 Firenze.

De Sisti G., A. Marino, M. Pecci M. (in press) – Indagini Georadar sul Ghiacciaio del Calderone del Gran Sasso d'Italia: primi dati relativi alla ricostruzione dello spessore di ghiaccio residuo. 17° Convegno Nazionale Gruppo Nazionale di Geofisica della terra solida, 10-12 Novembre 1998, CNR, Roma

Delfico O. 1794. 'Osservazioni di Orazio Delfico su di una piccola parte degli Appennini dirette a sua eccellenza il Signor Marchese D. Filippo Mazzocchi, Presidente del Sacro Regio Consiglio', Bibl. Prov. L'Aquila.

Di Filippo G., M. D'Orefice, R. Graziani, M. Pecci, F. Silvestri and C. Smiraglia, C, Proceedings VII Congresso Nazionale Società Italiana di Ecologia, Napoli 11-14 Settembre (1996), 75-78.

Gellatly A. F., C. Smiraglia, J. Grove, M, R. Latham (1994) – Journ. Glac. 40 (136 – 1994) 486-490.

Graziosi S., Photograph collection, (1965 – 1984) L'Aquila.

Istituto Topografico Militare, Carta del Gran Sasso d'Italia Bibl. Ist. Geogr. Mil., 1885, Firenze

Marinelli O. & L. Ricci,, Rivista Geografica Italiana, 23, 399-405. (1916), Roma.

Marsili B., Photograph collection. (1930 – 1968) Pietracamela (Teramo).

Ortophoto map of Abruzzo region, Section: "Corno Grande" - scale 1:10.000 – (1982)

Pecci M.,.. Proceed. Congr. "Global Change and Protected Areas. L'Aquila 8-13 Sept. 1999,

Tonini D., Bollettino Comitato Glaciologico Italiano, II Serie, 10, (1963) 71-135

Digital Geomorphologic Cartography of the Top Area of the Gran Sasso D'Italia Mountain Group (Central Apennine, Italy)

LEANDRO D'ALESSANDRO (*), MAURIZIO D'OREFICE(**), MASSIMO PECCI (***) & RENATO VENTURA (**)

(*) Dept. of Earth Science, University of Chieti "G. D'Annunzio", Madonna delle Piane – Via dei Vestini – 66013 Chieti Scalo, ITALY - Tel + 39-871-564056; Fax + 39-871-564050

(**) Italian Geological Survey, Via Curtatone, 3 – 00185 Rome ITALY - Tel + 39-6-44442444; Fax + 39-6-4465622

(***) ISPESL - DIPIA (Higher Institute for Occupational Safety and Health - Department of Production Plants and Interaction with the Environment), Via Urbana, 167 - 00184 Rome ITALY- Tel + 39-6-4714261; Fax + 39-6-4744017, Italian Glaciological Committee

Key words: Geomorphologic Mapping, Geographic Information Systems in geomorphology, Gran Sasso d'Italia, central Apennine

Abstract: The top area of the Gran Sasso d'Italia and of the whole Apennines Chain, including the Corno Grande peak, the Corno Piccolo peak and the Calderone glacier, is particularly important from the geomorphologic point of view, due to the activity of many processes and to the related presence of interesting landforms. Furthermore the whole top area is included in the integrally protected zone of the National Park of the Gran Sasso and Laga Mountains. In the last five years many studies have been carried out on the geomorphologic features of the top area, particularly regarding the Calderone glacier (monitoring and evolution). The preliminary synthesis of the study and of the results obtained after the management of territorial data on a dedicated GIS (Geographic Information System) is represented by the elaboration and the availability of the following thematic maps: Shaded Relief of the surface from the DEM (Digital Elevation Model), based on topographic data ranging from 1:750 to 1:10.000 scale; Slope Aspect Map (automatic output in GIS environment), Slope Angle Map (automatic output in GIS environment), Geomorphologic Map (including hydrography and basic lithology) surveyed at a scale 1:5.000. The landforms surveyed and mapped are related to the main following processes: morphotectonics, gravitational, hydrology - fluvial-glacial and watershed, karst, glacial, periglacial and anthropic. Mainly the problems related to the applied methodology in GIS environment are presented, furthermore the results of such a kind of activity have been drawn

G. Visconti et al. (eds.), Global Change and Protected Areas, 435–444.

in preliminary scheme, waiting for the publication of the final cartography on the Italian Geological Survey Bullettin, as an example of geomorphologic cartography in high mountain environment.

1. INTRODUCTION

The digital cartography of the top area of the Gran Sasso d'Italia (Fig. 1) is presented. The area is included between the Corno Grande peak and the Corno Piccolo peak, also containing a geomorphologic rarity - the Calderone Glacier - the southernmost in Europe and the only one in the centro-mediterranean area. The survey data have been collected, georeferenced and analysed in GIS (Geographic Information System) environment. The preliminary drafts of base, derived and thematic maps are presented. The final maps are going to be published on the Italian Geological Survey Bullettin as an example of geomorphologic cartography and, at the same time, a contribution to the production of the official Italian geomorphologic map.

In this perspective the problems related to the management of a georeferenced data base, aimed at the production of a geomorphologic map, are discussed and analysed.

2. DIGITAL CARTOGRAPHY

The use of computer aided tools in the field of traditional cartography has radically modified the techniques of map projecting and producing, even maintaining the following base concept:

- **congruence** (each information included in the map must not be in contradiction with any other);
- **readability** (univocal interpretation);
- **veracity** , (correspondence between qualitative information and the reality).

The first step in the process is represented by the numerical transformation of any physical object to be drawn in the map, through the available digital devices (digitizer, raster and vector scanner). Everything representing a particular information on the earth surface is converted into a simple geometric shape (point, line and polygon): established the projection and the reference system, the Cartesian co-ordinates and the attributes are defined in order to distinguish any object from similar ones. An informative system memorising these objects is also able to store and to draw for any object numerical information (coordinates describing the geometry of the

Figure 1 – Location of the studied area

representation), qualitative information (codes describing the typology) and topological information ("belonging to", "joining to" and "including").

Furthermore such a system allows to make univocal the metric content because of:

- the elimination of subjectivity elements introduced in measuring operation in traditional cartography passing from the drawing to the co-ordinates;
- the elimination of the consequences, due to the deformation of the support;
- the univocal definition of the qualitative content due, to the substitution of the draw-interpretation with a code-reading.

3. QUALITY PARAMETERS

In order to characterise the global quality of the whole product it is important to define:

- **the completeness** (measure of exceeding or lacking information);

- **the updating capability** (percent of changing between the survey and the elaboration);
- **The genealogy** (information about sources and elaboration).
- In order to characterise the local quality of a single cartographic object is important to define:
- **the metric accuracy**, (difference between the position of a point on the map and the real position into the used cartographic reference system);
- **the resolution** (dimension of the minimum represented object);
- **the semantic accuracy** (correspondence between the real world and the qualitative attribute associated to the object);
- **the logic congruence** (absence of incongruence observable independently of a check on the real world like intersecting contour);
- **the geometric congruence** (absence of errors in shape and position not directly observable).

The enounced parameters are at present moderately used in data describing, because generally the description includes only the geometric content and the associated attributes at the nominal scale, at the source and at a metric accuracy.

4. NUMERICAL OPERATIONS

The production of a map generally combines data coming both from field surveys and from existing cartography (topographic map and base cartography).

This kind of cartography is naturally rich of information and the addition of further information would be only redundant, in fact the principal activity of the cartographer is the cartographic reduction maintaining the informative content. The most important operation in this phase is the attribution of each element to the appropriate informative layer, containing similar objects from the geometric and typological point of view.

Each theme is decomposed into the topological "primitive" and successively coded with the attribution, defined inside the geographic data base. In this way for the topographic map, basic for further elaboration, the following informative layers are created:

- Contour lines (for instance with height and line-type attributes)
- spot heights,
- road net
- hydrological net
- inhabited centres;
- toponomy.

Figure 2 – Shaded Relief of the studied area

In order to make possible the overlay, each informative layer must be georeferenced, using control points common to all layers. Furthermore several "libraries" of symbols must be filled up for several geometric primitives for the plot and the final print of the map.

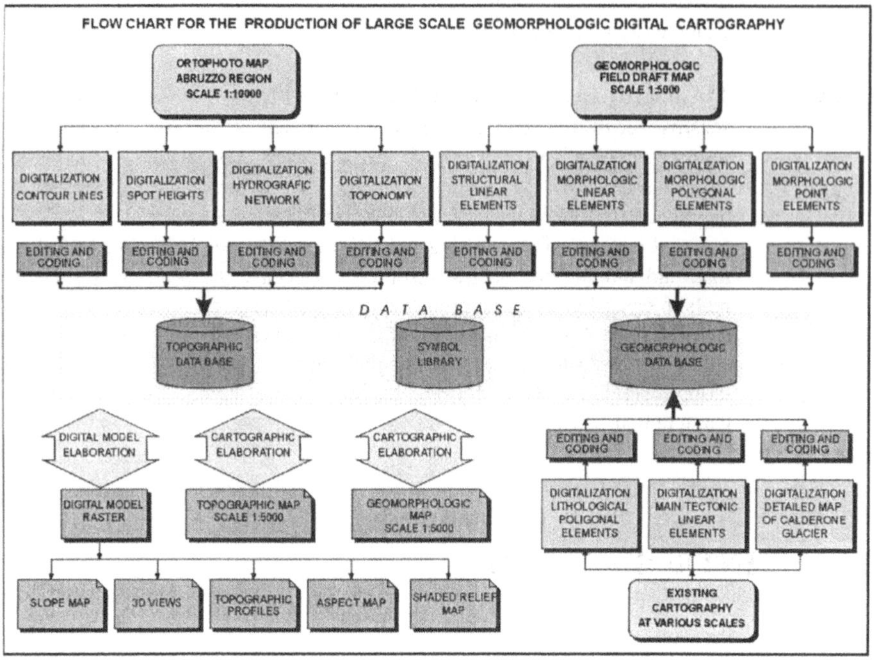

Figure 3 – Methodology used in producing the digital morphologic cartography

5. GENERATING A DIGITAL TERRAIN MODEL

Starting from the digitised contour line many algorithm are able to generate a Digital Terrain Model (DTM) or a Digital Elevation Model (DEM), considering the elevation as a casually distributed set of points, constructing a triangular model (TIN) or exploring the topologic and morphologic properties of the contour lines. The best method, in author's view, allow to add constraints to be respected by the same model. This technique allows to draw morphologic situation impossible to represent with the classic interpolation, but at the same time it needs a best definition and accuracy in the data entry (spot eighths, edges, drainage lines, depressions and peaks). Furthermore the model gives a good accuracy of the ground both from the morpho-hydrologic and the numerical point of view.

The algorithm has been developed by Hutchinson [1, 2, 3, 4] at the "Centre for resource and environmental studies" of the Australian National University –Camberra" and recently integrated inside the used GIS software [5]. The algorithm basically must set up the following constraints:

1. The guaranty of a connected drainage pattern, imposing a drainage condition on the single cell and successively removing depressions and peaks;

2. The guaranty of a correct representation of the edges and drainage lines as deduced from the contour lines, pointing to the angular points of the same contour lines.

Generated the model the quality control has been performed applying the following tests:

1. Contour lines: overlaying between contour lines generated by the model and contour lines of topographic maps
2. Ipsometric analysis: identifying incongruences inside the digital model by plotting or analysing the frequencies histogram of the spot heights
3. Control point: the estimation of the RMS of the model has been performed re-calculating the elevation of a set of control points with well-known heights
4. Drainage net: the calculation of potential drainage path is performed starting from the DEM. The correct position and composition of the drainage pattern is basic for the use of the model in the hydrological field.

Starting from the Digital Model, a typical raster product, it is possible to interpolate contour lines, to generate cross sections and 3D views, to calculate slope map and aspect map. Mainly it is possible to calculate real lengths with a development in altitude.

6. SPATIAL ANALYSIS

The spatial analysis is a particular kind of data analysis with a set of mathematical structures describing complex objects in a synthetic and significant way. The main problem concerns the translation of scientific language in mathematical formalisation (descriptions logically coherent, comparable and objective). The main limitations are the following:

- the correspondence between the conceptual lack of scientific language and the intrinsic significance of the models;
- The organisation of spatial data in the operative steps includes the aggregation in significant configuration, but, at the same time, the erasing of information;
- The degree of error, always active, in data collecting, like area data of surface area without punctual co-ordinates of location or in simplifying real objects in categories (from ordinal data in nominal data);

The modelling in GIS environment creates the geographic data base, performing the phenomenon in a set of information layers, inter-related, and homogeneous for typology and topology.

In this way the cartographic data base includes two different data sets: vector and raster: vector describes the spatial object in a continuous and accurate manner; raster describes the spatial object in a less accurate and discrete manner (the precision is function of the cell dimension). The overlaying operation, fundamental in geo-data set management is, in this way, slow but accurate in vector modality and fast in raster modality. Typical examples of raster elaboration are the calculation of slope angle and slope aspect, starting automatically from the digital model of the terrain and using different algorithm in calculating:

1. the slope angle, like angular coefficient of the plane passing through a points set homogeneously distributed around the examined node;
2. the slope aspect, like the direction (respect to the geographic north) of the perpendicular projection to the same plane in horizontal position.

7. DIGITAL CARTOGRAPHY OF THE TOP AREA OF THE GRAN SASSO D'ITALIA

In the specific case of the geomorphologic map of the top area of the Gran Sasso d'Italia the first step was the numerical transformation of the following already available map:

- ortophoto map of Abruzzo region [6], "Corno Grande" (scale 1:10.000);
- detailed map of the Calderone glacier (scale 1:750), obtained from the restitution of the topographic survey carried out in July 1990 [7];
- preliminary and field draft (scale 1:5.000), obtained from the enlargement, of the base at a scale 1:10.000, after ground topographic control in correspondence of the vertexes of IGMI's net and containing field data.

The digital report (maps) obtained performing the spatial analysis on geo data set of the Gran Sasso d'Italia Mountain Group are:

1. Digital Elevation Model (DEM), calculated with a resolution of the cell of 5 metres and the calculated Shaded Relief of Fig. 2 representing the morphology of surface in the studied area;
2. Topographic profiles, calculated for the Calderone Glacier at the end of the topological overlaying between the DEM and the profile trace;
3. Slope map, obtained from the classification of the dipping angle in degree of the following class: (0°-20°), (20°-25°), (25°-30°), (30°-35°), (35°-40°), (>40°);
4. Aspect map, obtained from the classification of the slope aspect value in 8 classes, each wide 45°;
5. Geomorphologic map (scale 1:5.000), including in the legend of Gruppo di Lavoro per la Cartografia Geomorfologica [8] the main following processes:

- Morpho-tectonics,

- gravitational,
- hydrology, fluvio-glacial and watershed,
- karst,
- glacial,
- periglacial
- anthropic.

In the flow chart of Fig. 3 is summarised the methodology used for the production of the digital geomorphologic cartography

8. CONCLUSIONS

The geomorphologic map of the top area of the Gran Sasso d'Italia and the related outputs have been produced with GIS technologies starting from the geo data set of field work and elaboration.

The map is also an example of utilisation of the "standard" legend proposed by GRUPPO DI LAVORO PER LA CARTOGRAFIA GEOMORFOLOGICA [8], concerning the recommended symbols and the informative (not only graphic) methodologies using GIS technologies. Furthermore the work has pointed out both the limits of the methodology, concerning mainly graphics solution for overlaying processes and themes, and the advantages, deriving from o the use of GIS technologies, concerning :

- the georeferencing and successive interpolation of territorial data or extrapolation of quantitative data;
- the up-dating of new data (almost in real time);
- the quick print of the updated map
- the "easy" correction and integration of the data, directly from terminal;
- the possibility in dividing data into several and different informative layers;
- the availability of maps in automatic.

9. REFERENCES

ARC/INFO Version 7.1.2

Gellatly A. F., C. Smiraglia, J. Grove, M, R. Latham (1994) – Journ. Glac. 40 (136 – 1994) 486-490.

Gruppo di Lavoro per la Cartografia Geomorfologica, Carta geomorfologica d'Italia - 1:50.000, guida al rilevamento. Serv. Geol. Naz. Quaderni serie III, 4, (1994), pp 1-42.

Hutchinson M. F., *Journ. Hydrol.*106 (1989) 211- 232.

Hutchinson M. F., *Proc. Third Inter. Symp. Spatial Data Handling*, August 17-19, Sydney. Inter. Geog. Union, Columbus, Ohio (1988).

Hutchinson M. F.,. Third International Conference/Workshop on Intergrating GIS and Environmental Modeling, NCGIA, University of California, Santa Barbara(1996).
Hutchinson M. F.., T. I. Dowling, *Hydrological Processes*, 5(1991) 45-58.
Ortophoto map of Abruzzo region, Section: "Corno Grande" - scale 1:10.000 – (1982)

The Late Pleistocene and Holocene Temporary Lakes in the Abruzzo Parks and the Central Apennines

CARLO GIRAUDI
ENEA , CR. Casaccia, P.O. Box 2400, 00100 Roma A.D.

Key words: Paleolake, Hydrology, Mountain Environment.

Abstract: In a number of closed basins in the Central Apennines (mainly in the Abruzzo and Gran Sasso-Laga National Parks and in the Velino-Sirente Regional Park) lacustrine sediments, produced by temporary lakes that have now disappeared, have been found. These sediments have been dated to the Last Glacial Maximum and the Holocene by the ^{14}C method. A number of lakes have formed and then disappeared several times in the last 20,000 years, while others survived until historical times. The number of lakes decreased in periods characterized by a negative hydrologic balance and increased when the balance was positive. The variations in the number of temporary lakes are mostly due to climatic causes and can be correlated with the oscillations of Lake Fucino level. Only in the last 4000 years divergent trends occurred. The cause of this could be the uman impact on the mountain environment. Of the twenty-four non-perennial lakes studied, currently only three reappear in spring and disappear with the onset of summer.

1. INTRODUCTION: TEMPORARY LAKES, A TESTIMONY OF PAST ENVIRONMENTAL VARIATIONS

The non-perennial lakes studied in the present work are located in mountain areas of the Central Apennine and are often situated within protected areas (Gran Sasso - Laga and Abruzzo National Parks and the Velino-Sirente Regional Park - Fig. 1). The occurrence of various lakes during the Upper Pleistocene and the Holocene has been treated in a series of works. Some of these lakes, varying in size between a few km^2 to a few

G. Visconti et al. (eds.), Global Change and Protected Areas, 445–458.

hundred m^2, disappeared at the end of the Upper Pleistocene, and others during the course of the Holocene.

A few depressions, occupied in the past by these lakes, are still flooded in the spring, during the melting of the snows, or after particularly wet periods.

Many of the ancient lakes identified lay in areas covered by glaciers during the Last Glacial Maximum (LGM), or in periglacial zones, at altitudes of more than 1400 m. Today, in these areas the anthropic impact is generally small, and limited moreover to stockraising.

Today, the presence of closed depressions is not sufficient for the formation of lakes: lakes form only if the climatic and environmental conditions are favourable. The balance between the amounts of water entering the depressions (rainfall, surface runoff, groundwater, etc.) and those leaving them (through permeability, outflow, evaporation, etc.) has to be a positive one. Hence the balance is determined by local geological conditions and by the climate.

If lakes have disappeared in places where the morphological and geological situation has not changed, the cause has to be ascribed to the climate.

In a number of places the lakes disappeared because of the changed morphological conditions: the sills of some closed depressions which contained lakes were eroded, or the depression became filled by sediments.

The non-perennial lakes identified may be classified in three groups:

- lakes in depressions of glacial and periglacial origin;
- lakes in depressions of tectonic origin;
- lakes dammed by alluvial fans.

2. LAKES IN DEPRESSIONS OF GLACIAL AND PERIGLACIAL ORIGIN

The highest parts of the Apennine chain were occupied by many glaciers during the LGM: the equilibrium line altitude in the Central Apennine went down to as low as 1600-1900 m according to the location and exposure of the glacial valleys [1;2]. The glacial tongues in some cases reached elevations of around 1000 m, while in several places they went below 1500 m. Lakes formed both due to the damming of the valleys by the glacial tongues or moraines. During the glacial recession the lakes formed between the glacial tongue and the frontal moraine just abandoned, or between moraine ridges, while others formed in dead ice depressions. In two cases lakes formed in depressions produced by the melting of the permafrost present in the rock glaciers.

The chronological pattern of the lacustrine deposits identified has been

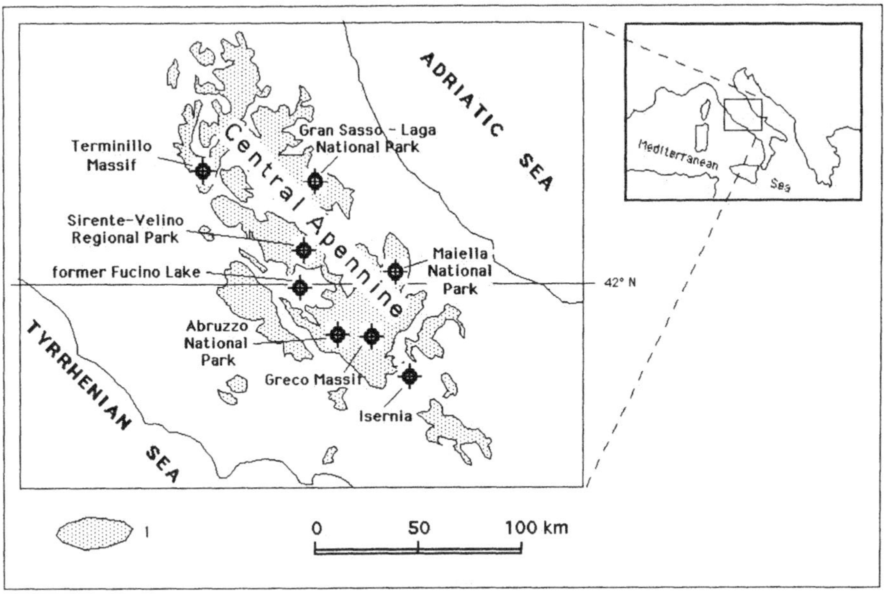

Fig. 1 - Map of the studied area

based on two fundamental elements:

- dating by the ^{14}C method (not calibrated) of the organic matter in the sediments;
- the age of the glacial deposits covered by the sediments; the dating of the glacial phases in the Central Apennine is fairly well defined, especially for the Abruzzi area [2].

Some depressions of glacial origin have been occupied by lakes also during the course of the Holocene.

Fig. 2 and Table 1 show the origin, the ages and the ^{14}C dating of the sediments relating to each studied former lake.

2.1 Campo Imperatore : Gran Sasso-Laga National Park

– The lacustrine deposits at the mouth of Val Coppone

Val Coppone ends in a closed depression dammed by a moraine, located at 1657 m, having an area of about 5000 m^2. A number of boreholes made at the bottom of the depression [2] have evidenced the lacustrine series: a sample of the deepest sediments gave an age of 22,680±530 years with the ^{14}C method, while a stratigraphically higher sample gave an age of 22,350±300 years ^{14}C B.P.

In the higher part of the lacustrine sediments there is a tephra layer.

Table 1: ^{14}C datings in Fig. 2 and in the text

	14C datings	Material	Laboratory Number	Author(s)
1	2750±110	peat	B0-175	Giraudi (1997)
2	4400±60	peat	BO-175	Giraudi (1997)
3	12,300±300	Neapolitan Yellow Tuff	======	Alessio et al., (1973)
4	1880±60	soil	Beta-129541	present paper
5	6090±80	soil covered by lacustrine sed.	---------	Cinti et al. (1992)
6	5790±70	soil covered by lacustrine sed.	BO-287	Giraudi (1997)
7	3310±240	soil covered by lacustrine sed.	---------	Cinti et al. (1992)
8	3175±65	soil covered by lacustrine sed.	BO-289	Giraudi (1997)
9	3260±50	soil covered by lacustrine sed.	BO-288	Giraudi (1997)
10	1575±60	soil covered by lacustrine sed.	BO-290	Giraudi (1997)
11	920±40	colluvial sediments	---------	Cinti et al. (1992)
12	4520±65	lacustrine sediments	Ua-10984	Giraudi (1998)
13	4020±80	soil covered by alluvial sed.	Beta-95779	Giraudi et al. (in press)
14	3440±60	soli covered by lacustrine sed.	Ua-10985	Giraudi (1998)
15	22,350±300	lacustrine sediments	BO-250	Giraudi & Frezzotti (1997)
16	22,680±530	lacustrine sediments	BO-251	Giraudi & Frezzotti (1997)
17	8035±140	lacustrine sediments	UD-399	Giraudi & Frezzotti (1997)
18	3740±90	lacustrine sediments	BO-177	Giraudi & Frezzotti (1997)
19	12,850±200	peat	UD-318	Frezzotti & Giraudi, (1989)
20	6110±180	alluvial silt	UD-384	Giraudi & Frezzotti (1995)
21	5640±60	lacustrine sediments	BO-233	Giraudi (1998)
22	3375±75	soil covered by lacustrine sed.	BO-291	Giraudi (1997)
23	6530±70	lacustrine sediments	Ua-11331	Giraudi (1998)
24	1560±60	soil covered by lacustrine sed.	Beta-95780	Giraudi et al.(in press)
25	11,760±160	lacustrine sediments	UD-383	Giraudi & Frezzotti (1995)
	4690±220	soil covered by alluvial sed.	UD-326	Frezzotti & Giraudi (1989)
	6110±180	alluvial silt	UD-384	Giraudi & Frezzotti (1995)
	6920±150	soil	UD391	present paper
	17,840±200	soil covered by aeolian sed.	UD-385	Giraudi & Frezzotti (1997)
	21,450±250	soil covered by debris	BO-254	Giraudi & Frezzotti (1997)

This layer could correspond to that of the Neapolitan Yellow Tuff reported in various areas of Central Italy [3] dated at 12,300±300 years B.P. [4].

- The synglacial lacustrine sediments between the moraines of Coppe di Santo Stefano and the Lake Pietranzoni.

In the Coppe di Santo Stefano area, at an elevation of about 1600 m, lacustrine sediments occur [5] between the moraines deposited by the glacier during both the LGM (Campo Imperatore Stade [2]) and the initial recessional stages.

Lacustrine deposits occur a few hundred metres upstream of the frontal moraine. The lacustrine sediments are overlain by fairly fine sandy gravel and then by coarse gravelly sand of fluvioglacial origin.

An analogous statigraphy may be observed about 1.5 km further upstream, behind a recessional frontal moraine (Piano Pietranzoni Stade; age more than 17,840±200 years B.P. and less than 21,450±250 years B.P. [2]) at an altitude of ca. 1620 m.

Another lacustrine deposit overlies glacial debris. The lacustrine deposit is overlain in its turn by fluvioglacial sediments.

At the Lake Pietranzoni, lacustrine sediments lie on glacial debris of another recessional moraine (Lago Pietranzoni Stade, age more than 17,840±200 years B.P. and less than 21,450±250 years B.P. [2]). These sediments are covered, towards the West, by other, coarser lacustrine deposits, and then by fluvioglacial sediments. These lakes could not form again at the present because they were typically connected with the presence of the glacier.

Between the moraine ridges that form this area there are a lot of closed depressions, produced by the melting of the dead ice, at elevations of 1600÷1640 m. Inside these depressions lacustrine deposits have been observed in a number of cases. The lakes could therefore have been formed either by the melting of the dead ice, or during the course of favourable climatic periods: nevertheless there are no ^{14}C datings available.

- The lacustrine sediments on the inactive rock glacier of Sorgente Fontari.

In a closed depression having a surface area of a few hundred square metres, on the body of the rock glacier of Sorgente Fontari, at an elevation of approximately 1910 m, have been evidenced sediments connected with three different lacustrine phases: the first starting at 8035±140 years B.P., and the second at 3740±90 years B.P. (^{14}C) [2].

The formation of the depression and the development of the first lake must date from the moment of melting of the interstitial ice among the debris, and thus from the moment of the disappearance of the permafrost.

- The lacustrine sediments on the inactive rock glacier of Fontari-southern slope of Mt. Aquila

Lacustrine deposits overlying a thin layer of soil [5] have been found in a depression at an elevation of about 1930 m, on another inactive rock glacier at Fontari, at the base of the southern slope of Mt. Aquila. The organic matter contained in a sample taken from the bottom soil, dated with the ^{14}C-AMS method, has indicated an age of 3440±65 years B.P. At present the lake could not form as the sill has been eroded.

2.2 Monti della Meta - Abruzzo National Park

- Le Forme area

Sediments of a lake, which occupied a depression dammed by moraines dated to the initial recessional phases of the LGM [6], have been found in the area of Le Forme, at an elevation of about 1400 m.

In the bottom part of this deposit two tephra layers have been found: the more recent one has been attributed to the Neapolitan Yellow Tuff; the age of this eruption is 12,300±300 years B.P. [4] In spring the depression is occupied by an ephemeral in which sedimentation does not take place.

- Mt La Meta, Biscuri area

Lacustrine sediments have been found in a small basin inside a depression between moraine ridges, at Biscuri, located at an elevation of about 1875 m, close to the base of the north-eastern slope of Mt La Meta [6]. The moraines around the basin are the most recent ones on the massif and can be dated as Late Glacial [7; 1].

The lacustrine sediments are more recent than the Neapolitan Yellow Tuff and, based on the correlation with other lacustrine deposits present in the area, should be dated ca. 6000-7000 years B.P. [6].

- Mt Cavallo, Upper Venafrana Valley

In the upper part of the Venafrana valley, behind a stadial frontal moraine, there is a small closed basin formed by lacustrine and alluvial sediments [6]. In this plain a very small ephemeral lake forms in the spring.

A fragment of wood found at approximately 3.6 m below ground level has been dated 4400±60 years B.P. by ^{14}C method, while a sample taken from a peat level at about 3.3 m below ground level has been dated 2750±110 years B.P.

2.3 Mount Terminillo Massif

- Lower Organo Valley

Lacustrine sediments have been found [8] behind the frontal moraines of the Organo valley, at an elevation of about 1250 m. These have been cryoturbated and covered by fluvioglacial deposits, and have therefore the same age of the first phases of glacial recession. The lake could not form again at present because the erosion of the sill.

- Upper Organo Valley

Lacustrine sediments have been found [8] in a small depression on the most recent moraine of the Organo valley, at an elevation of about 1850 m. These sediments contain the ^{14}C-AMS date of 6530±70 years B.P.

- Prato Comune

Lacustrine sediments more than 4 m thick have been found [8] at Prato Comune, at an elevation of ca.1680 m, behind a stadial frontal moraine. Dating with the ^{14}C-AMS method of a sample of peat obtained at a depth of 4 m has indicated an age of 4520±65 years B.P. The lake could not form again at present because it has been filled in and because of the erosion of the sill.

2.4 Mount Greco Massif

- Val Macchione

Lacustrine sediments, lying on an area of approximately 10,000 m^2 behind a recessional moraine, have been found in Val Macchione, at an elevation of about 1850 m.

The lacustrine sediments are covered by a soil dated 6920±150 years B.P. by the ^{14}C method and contain the tephra layer of the Neapolitan Yellow Tuff, dated 12,300±300 years B.P. [4]. The lake could not form again at present because of being filled in and due to the erosion of the sill.

3. LAKES IN DEPRESSIONS OF TECTONIC ORIGIN

In many places of the Central Apennines closed depressions of tectonic origin exist. The quaternary activity of the faults has locally produced the blocking of a number of catchment basins, stopping the runoff and forming tectonic depressions with endorheic drainage.

The presence of these depressions led to the development of various lakes during the course of the Quaternary.

The sediments have been dated both by ^{14}C method and by correlation with sediments of known age.

3.1 Gran Sasso-Laga National Park

- The lacustrine sediments of the upper Venaquaro Valley

A lacustrine deposit has been found [9] in the Upper Venacquaro valley, on the northern slope of the Gran Sasso, at an elevation of about 1930 m. The sediments lie in a depression of about 20.000 m^2 dammed by a fault scarp.

The age of the bottom of the thin lacustrine deposit is 11,760±160 years B.P.(^{14}C). In spite of the fact that said depression still exists, the younger sediments are alluvial and colluvial: their bottom is dated 6110±180 years ^{14}C.

3.2 Velino-Sirente Regional Park

– The lacustrine sediments of the eastern portion of the Piano di Pezza

Study of the stratigraphy of the late-Pleistocene and Holocene sediments exposed on the Piano di Pezza has revealed the presence of five lacustrine formations [10]. The lake that has deposited them, presently extinct, once had a maximum surface area of about 1 km^2 and reached a depth of 8-10 m.

Overall the lake of Piano di Pezza must have been present and had considerable depths during five periods:

- at a not well-defined time prior to the LGM;
- during the course of the LGM, when it attained its maximum depth, reaching the level of the sill and producing a surficial outflow which drained towards the Altipiano delle Rocche.
- in a period between ca. 5790±70 and 3260± 60 years B.P.;
- in a period appreciably more recent than 3175 ± 65 and older than 1575 ± 65 years B.P.;
- in a period subsequent to 1575 ± 65 years B.P.

The lake was shallow or desappeared in the intervals between the periods listed.

It is probable that the lake disappeared at least between 3260±60 and 3175±65 years B.P., when aeolian sediments were deposited.

– The lacustrine sediments of the eastern portion of Campo Felice

In the eastern portion of Campo Felice, at the elevation of 1525-1540 m, the lacustrine sediments occupy an area of approximately 1 km^2 [11].

Boreholes made in the area have shown that the lacustrine sediments, containing a few alluvial and pedogenetic horizons, have a thickness of several tens of metres. The surficial sediments are dated to the Upper Pleistocene and Holocene, as they contain a tephra layer that should correspond to the late-glacial or to the Holocene tephra layer found in the same area. Furthermore in a limited part of the plain the lacustrine sedimentation ended after the development of a soil dated at 1880±60 years B.P.

– The lacustrine sediments of the western portion of Campo Felice

In the western portion of Campo Felice there is a depression with a surface area of about 1.5 km^2, at the same elevation as the one in the eastern portion. The boreholes made [11] have shown, as in the preceding case, the presence of tens of metres of lacustrine sediments. The age of the most surficial sediments should be analogous to that of the sediments in the eastern portion, since the same tephra layer has been found. In the spring a shallow ephemeral lake forms, which may reach a size of more than 0.5 km^2, in which no sedimentation takes place.

3.3 Mount Greco Massif

– The lacustrine sediments of the Piano di Aremogna

Lacustrine sediments have been found in a part of the southern depression of the Piano di Aremogna, having an area of about 0.5 km^2, located at an elevation of about 1400 m [12]. These are heteropic with the fluvioglacial sediments dated to the LGM. At the top of these sediments there is a peat dated 12,850±200 years B.P. with the ^{14}C method, and a tephra layer, attributed to the eruption of the Neapolitan Yellow Tuff, dated 12,300±300 years B.P.

– Piano delle Cinquemiglia

In the southern depression of the Piano delle Cinquemiglia [13] there are lacustrine sediments lying on an area of about 2 km^2.

There are no radiometric datings of these sediments, but they are partly heteropic and partly younger than the fluvioglacial fans dated to the LGM.

4. ALLUVIAL FAN-DAMMED LAKES

Alluvial dammed lakes have developed in various parts of the studied area. In fact, alluvial fans have dammed up some valleys, producing lakes of various size. In one case, the development of two fans has dammed part of a flat area, forming a depression.

4.1 Campo Imperatore - Gran Sasso-Laga National Park

– The Holocene lacustrine sediments of the northern portion of the little valley that cuts across the Lake Pietranzoni-Coppe di Santo Stefano area

Lacustrine sediments are present in the little valley that cuts across the area of Lake Pietranzoni-Coppe di Santo Stefano, at an elevation of 1620÷1630 m [5]. These sediments, towards the SE, overlie the alluvial sediments of a fan. The fan dammed the valley and a lake thus formed, stretching for at least 0.5 km to the NW.

A paleosoil underlying the lacustrine sediments has been dated 12,100±100 years B.P. with the ^{14}C method.

The bottom of the lacustrine deposits has an age of 6090±70 years B.P.

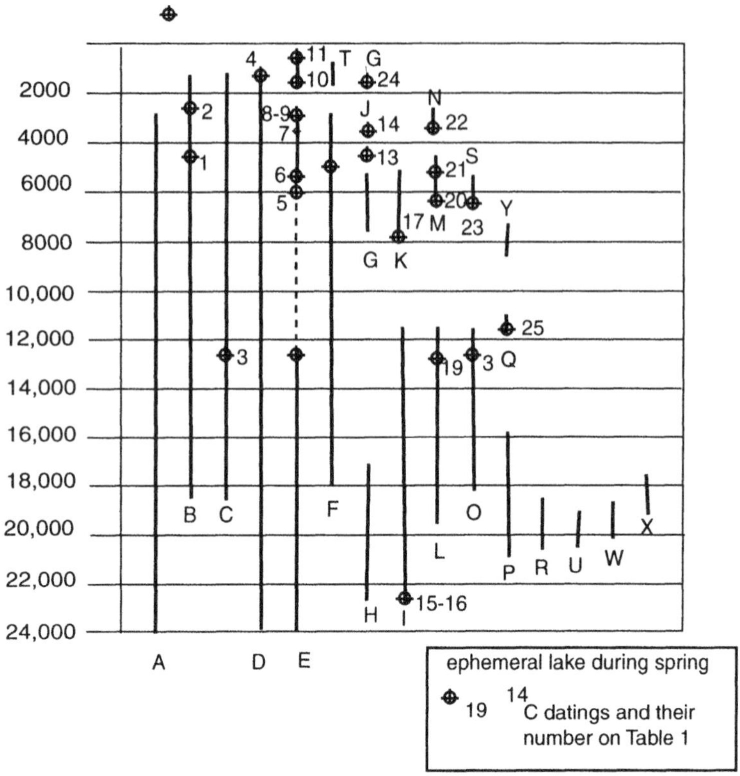

Fig. 2 - Name, age and origin of the temporary lakes

A= Campo Felice West, lake in tectonic basin;
B= Mt. Cavallo, moraine-dammed lake;
C= Mt. Meta. Le Forme, moraine-dammed lake;
D= Campo Felice East, lake in tectonic basin;
E= Piano di Pezza, lake in tectonic basin;
F= Terminillo Il Prato, moraine-dammed lake;
G= La Piana di Isernia, lake dammed by alluvional fans;
H= Piano Cinquemiglia, lake in tecnonic basin;
I= Campo Imperatore, Valle Coppone, ice/moraine-dammed lake;
J= Campo Imperatore, southern slope of M. Aquila lake in a depression on a rock glacier;
K= Campo Imperatore, Fontari, lake in a depression on a rock glacier;
L= Piano di Aremogna, lake in tectonic basin;
M= Campo Imperatore, valley cut through the moraines, lake dammed by alluvial fans;
N= Piano di Pezza West, lake dammed by alluvial fan;
O= Piano di Aremogna, Val Macchione, moraine-dammed lake;
P= Campo imperatore, Pietranzoni, moraine-dammed lake;
Q= Valle Venacquaro, lake in tectonic basin;
R= Mt. Terminillo, Valle Organo, moraine-dammed lake;
S= Mt. Terminillo, Valle Organo, lake in a depression on moraine
T= Campo Imperatore, Valle Cortina, lake dammed by alluvial fan;
U= Campo Imperatore, moraine-dammed lake;
W= Campo Imperatore, moraine-dammed lake;
X= Campo Imperatore, moraine-dammed lake;
Y= Mt Meta, Bisturi, moraine-dammed lake.

and the intermediate part an age of 5640±60 years B.P. At present the lake could not be formed again because of the erosion of the sill.

– The lacustrine sediments of the Cortina Valley

In the western portion of the Cortina Valley there are lacustrine sandy-silty sediments [1;5]. The lake, about 2 km long and quite narrow, in which the sediments were deposited, was produced by the damming of the Cortina Valley by the fan formed by the Fornaca Valley stream.

The sediments are heteropic with the alluvium present upstream. These alluvium is more recent than 5640±60 years ^{14}C B.P.

In view of the relations with the barely pedogenized deposits - or not pedogenized at all - of the fan damming the Cortina Valley, it is probable that the top sediments were very young. In this valley was present, probably, the "huge lake" which existed at Campo Imperatore according to popular tradition [14].

4.2 Piane di Isernia

The area consists of a depression with an area of slightly less than 1 km^2 having a flat bottom, at an elevation of about 460 m.

Two lacustrine deposits have been found in this area [15]. The first one is overlain by a soil dated 4020±80 years B.P. by the ^{14}C method.

The second one has been dated by the ^{14}C method at 1560±70 years B.P. Due to the erosion of the sill, the lake could not at present be formed again.

4.3 Velino-Sirente Regional Park

– The lacustrine sediments of the western portion of the Piani di Pezza

In the western portion of the Piani di Pezza, at an elevation of about 1470 m, lacustrine sediments lies on an area whose size it is hard to define [10;16].

These sediments overlie a soil dated 3375±65 years B.P. The lacustrine sediments are located upstream of a detritic-alluvial fan which has dammed a small valley. In this area, as the depression has been filled in, the lake could not form again.

5. DISCUSSION

An examination of the Fig. 2 shows that the number of temporary lakes has varied considerably in the course of time and that none of these lakes are now present, except as ephemeral spring lakes.

Of the twenty-four non-perennial lakes examined, only thirteen could be reformed. In the others the depressions have become filled in and/or the sill has been eroded. Therefore the environmental variations that have occurred in some basins have gradually reduced the number of depressions potentially able to become lakes again.

About the variation in the number of lakes during the period examined one can observe:

- in the final phases of the upper Pleistocene between about 20,000 e 12,000 years B.P., there was a considerable number of lakes, many of them formed in that period and fed by the meltwater from the glaciers;
- in the period between 11-12,000 and 7-8000 years B.P. the number of lakes dropped;
- in the period between ca. 7000 and 5000 years B.P. the number of lakes raised; some lakes reformed in basins from which they disappeared;
- from 5000 to shortly before 4000 years B.P. a decrease occurred in the number of lakes;
- between shortly before 4000 and 3000/2500 years B.P. the number of lakes raised again;
- the number of lakes decreased dramatically between ca. 2500 and 1700 years B.P.;
- between ca. 1700 and 1000 years B.P. the number of lakes raised;
- between 1000 B.P. and the present, the lakes completely disappeared.

The disappearance of the lakes in the last 1000 years and their dramatic reduction between ca. 2500 and 1700 years B.P. cannot be explained by the filling-in of the basins and the incision of the sills. In the last 3-4000 years such events occurred only in three places.

If the formation of the temporary lakes corresponds to positive phases of the hydrologic balance, and their disappearance to negative ones, then the variations in humidity should be linked to the climate and have been recorded also in other places.

The studied area includes Lake Fucino, one of the largest lakes in Central Italy, reclaimed in the 19th century. Studies on the Holocene variations in the level of the lake [17], based on many ^{14}C datings, archaeological and historical datings, have led to the fairly detailed reconstruction of the oscillations in the level of this lake (Fig. 3). These oscillations were due essentially to the climate.

Therefore there should be elements of correlation between the number of temporary lakes and the oscillations of Lake Fucino. Comparing the graph shown in Fig. 3 with that in Fig. 2, various periods are observed during which increases in lake Fucino level were matched by an increased number of temporary lakes, and, vice versa, when decreases in lake Fucino level were matched by a decrease in the number of temporary lakes.

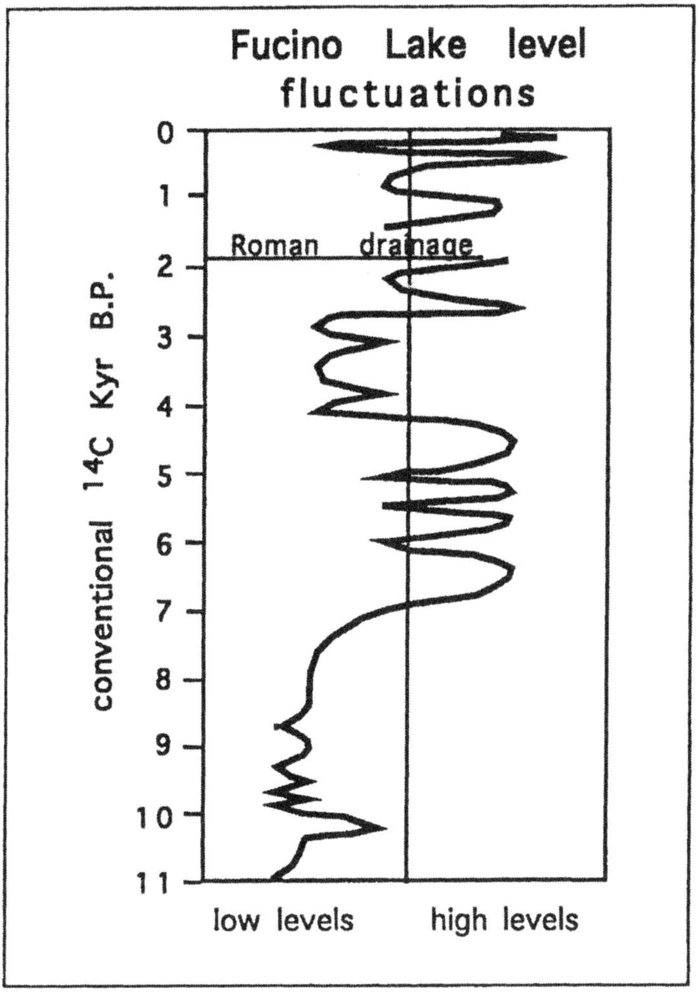

Fig. 3 - Holocene fluctuations of the Fucino lake level

For these periods the variation in the number of temporary lakes is therefore due to climatic causes.

In other periods discordant trends occurred:

- ca. 3-4000 years B.P. the number of temporary lakes increased, but the level of Lake Fucino remained low;
- between 3000 and 2000 years B.P. there was a strong decrease in the number of temporary lakes, whereas the level of Lake Fucino experienced at least one strong increase;
- during the course of the Little Ice Age, the level of Lake Fucino increased considerably, reaching perhaps its highest levels throughout the Holocene, but the temporary lakes did not form again.

During these phases, therefore, the variations in the number of temporary lakes did not coincide with the climatic variations.

It may be just a coincidence, but the trends differ as from ca. 4000 years B.P., that is, more or less from the start of the intensive exploitation of the mountain by communities of shepherds [18].

The likelihood is that, due to the modifications of the plant cover and of the soils produced by this anthropic impact, the small basins of the temporary lakes reacted differently to the climatic variations.

It is possible that the disappearance of the temporary lakes in the studied areas was, at least partly, caused by the intensity of the pastoral use of the mountains.

Even if many of the areas studied are presently protected and form part of natural parks, a rebalancing of the ecological conditions existing prior to the anthropic impact seems unlikely. This might possibly come about in the long term, but in limited areas and only if climatic conditions do not evolve towards a higher degree of aridity.

6. REFERENCES

Alessio M., Bella F., Improta S., Cortesi C. and Turi B., Radiocarbon 15 (1973), 165-178.

Cinque A., Liccardo C., Palma B., Pappalardo L., Rosskopf C. & Sepe C., Geogr. Fis. Dinam. Quat., 13 (1990), 121-133.

Cinti F.R., Pantosti D., D'Addezio G. & De Martini P.M.Atti 11° Convegno del Gruppo Nazionale di Geofisica della Terra Solida, Roma, (1992), 273-285.

Demangeot, J., Centre Recherche et Documentation Cartographiques Mémories et Documents, Numéro hors series (1965), Paris, 403 pp.

Frezzotti M. and Giraudi C.,Memorie Società Geologica Italiana 42 (1989), 5-19.

Frezzotti M. and Narcisi B.,Quaternary International 34-36 (1996), 147-153.

Giraudi C. & Frezzotti M., Quaternary International 25 (1995), 81-93.

Giraudi C. & Frezzotti M., Quaternary Research, 48 (3), (1997), 280-290.

Giraudi C. , Il Quaternario, 11(2),(1998), 217-225.

Giraudi C., Palaoklimaforschung-Palaeoclimate Research, 25 (1998), 1-18, Gustav Fisher Verlag Ed.

Giraudi C., Atti del convegno "Geofisica della terra solida". I (1987), 111-116, Roma.

Giraudi C., Galadini F.& Galli P., Il Quaternario (in press).

Giraudi C., Giornale di Geologia, sez. 3, 60 (1998), 67-82.

Giraudi C., Il Quaternario, 11(1), (1998), 1-5.

Giraudi C.,Il Quaternario, 10(1), (1997), 93-100.

Giraudi C., Il Quaternario, 10(2), (1997), 191-200.

Jaurand E., Thèse pour le Doctorat dès Lettres de l'Université de Paris I Panthéon-Sorbonne, 1994.

Radmilli A.M. (1981) - Storia dell'Abruzzo dalle origini all'Età del Bronzo. 451 pp., Ed. Giardini, Pisa.

The Travertine Deposits of the Upper Pescara Valley (Central Abruzzi, Italy): A Clue for the Reconstruction of the Late Quaternary Palaeoenvironmental Evolution of the Area

[1]MIRIAM LOMBARDO, [2]GILBERTO CALDERONI, [3]LEANDRO D'ALESSANDRO AND [3]ENRICO MICCADEI

[1]*Via della Brianza 11, I-00161 Roma, Italy. Tel.: 39.6.44.23.09.28; fax: 39.6.86.32.58.82*

[2]*Dipartimento di Scienze della Terra - Università di Roma "La Sapienza", P.le A. Moro 5, I-00185 Roma Italy. Tel.: 39.6.49.91.45.80; fax: 39.6.49.91.45.78*

[3]*Dipartimento di Scienze della Terra - Università di Chieti "Gabriele D'Annunzio", Madonna della Piane, I-66013 Chieti, Italy. Tel. 39.871.35.56.153; fax: 39.871.33.56.453*

Key words: Meteogene travertine, late Quaternary, climate changes, upper Pescara valley, palaeoenvironmental evolution

Abstract: Three main phases of meteogene travertine growth have strongly conditioned the late Quaternary landscape evolution along the upper valley of Pescara river. A comparison between the ages of travertine deposits, carried out with ^{14}C and U/Th methods, and the oxygen-isotope record deep-sea core, has been performed. The travertine formation in the study area is controlled mainly by the palaeoclimatic fluctuations, and the periods of growth are connected with the improvement climatic phases

1. INTRODUCTION

A research dealing with the Quaternary environmental evolution of the areas encompassed by the Gran Sasso and Maiella National Parks has been recently undertaken by the Departments of Earth Science of the Universities "G. D'Annunzio" (Chieti) and "La Sapienza" (Roma). Travertines are of great concern in that their deposition strongly depended upon the landscape evolution of the upper Pescara river valley.

G. Visconti et al. (eds.), Global Change and Protected Areas, 459–464.

Table 1 – Travertine deposits ages

Travertine deposit	Age (yrs. BP)
Sulmona plain	
Fonte S. Callisto	37080*-35900
Popoli	45000-36000***
Popoli gorges	
Decontra	45100-37500
Bussi Station	
"left" 1st phase	121100-114000
"left" 2nd phase	76350*-34000**
"right"	76350-34000***
Barrage Travertine System	23000-6000****

*age estimates on the basis of the local travertine deposition rate (1,7 mm/yr)
**[6]
***interval estimates by correlation
****interval estimates by correlation [4]

The planned major objective of the study consisted in the characterization on chronological, sedimentological, geomorphological and geochemical basis of the most significant travertine deposits widespread throughout the area. Travertine deposits outcrop both in the Northern side of the Sulmona plain, the most external between the intermontane graben and half-graben basins of Central Apennines Belt [1], and along the Popoli Gorges (Fig. 1).

2. THE TEMPERATURE OF TRAVERTINE-DEPOSITING WATERS

A distinction can be made between meteogene travertines, whose carrier carbon dioxide originated in the epigene and soil atmospheres, and thermogene travertines, mostly associated with hot springs, whose carrier gas comes predominantly from thermal activity involving oxidation, decarbonation and other deep outgassing processes in tectonically active regions [2]. Meteogene travertines appear to show a stronger dependence upon climatic factors, since the carrier carbon dioxide originates mainly from soil. Elevated concentrations and input rates of biogenic carbon dioxide in soil covering the surface of limestone aquifers, has been recognized as a main factor for calcium carbonate concentration in karstic ground water and ted travertine deposition.

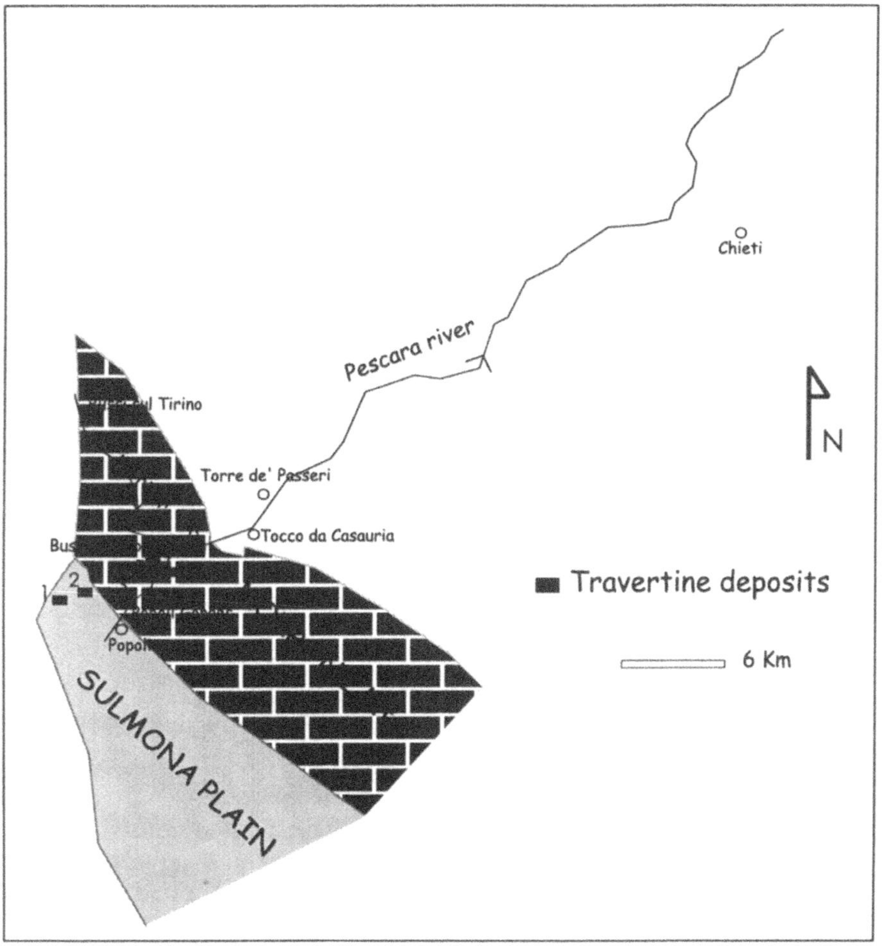

Fig. 1: Distribution of travertine deposits along the upper valley of Pescara river. 1)F.te S. Callisto, 2) Popoli, 3) Decontra, 4) Bussi Station "right", 5) Bussi Station "left", 6) Thalweg Pescara.

Therefore, meteogene travertines timing is strictly linked up with periods of soil production and meteorological water supply, providing a clue for a climatic conditions dependent processes [3].

A distinction can be made between the two types of travertine (meteogne and thermogene) in terms of lithologic features of the primary deposits, as a result of temperature and degree of calcium carbonate parent waters supersaturation. Since primary travertine textures mainly result from carbonate encrustation on templates of organism, their biogenic imprints and degree of diversity significantly decrease as parent waters approach high temperatures, inducing a simplification of life composition [3].

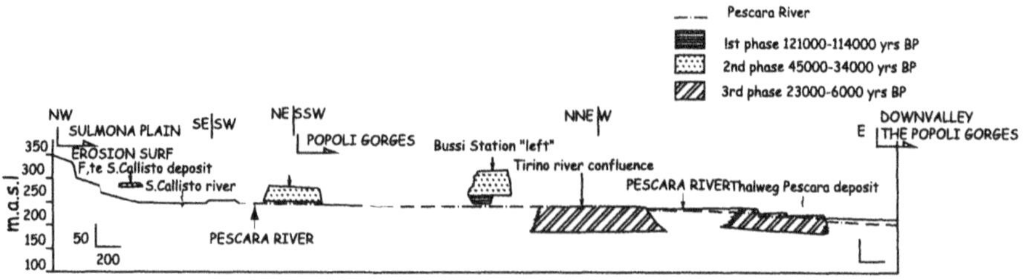

Fig. 2: Schematic distribution of travertine deposits along the upper Pescara river valley (longitudinal section of the valley).

The degree of the lithofacies diversity and of biogenic imprint by primary travertine texture in the study area, point out that the upper Pescara valley deposits are formed in ambient temperature waters (meteogene travertines). This interpretation agrees with $\delta^{18}O$ and $\delta^{13}C$ readings concerning the travertine deposits in central Abruzzi Apennines Belt [4].

3. THE MAIN PHASES OF TRAVERTINE FORMATION

Regarding the outcroppings so far investigated it resulted that travertine deposition, first triggered during the upper Pleistocene, lasted, with a discontinuous pattern, up to the beginning of the Holocene.

In particular the results of radiometric dating, carried out with ^{14}C and U/Th methods, point to three main phases of a huge blooming of travertine formation, separated by significantly long lasting pauses, locally marked by erosion (Table 1).

The travertines referred to the oldest depositional phase, spanning in age from 121,000 to 114,000 yrs. ago, were mostly laid down over the central reach of Popoli gorges (1st phase Bussi Station "left"), that is according to the longitudinal profile of Pescara river. Field evidences suggest that as a result of such travertine accumulation the valley experienced some partial damming.

Further, it is reasonable to infer that such natural damming by constraining the linear erosion caused the development of a wide erosion surface, at some 300 m a.s.l., on the N reach of the plain of Sulmona (Fig. 2).

Later on, from 45,000 and 34,000 yrs. ago, further travertine features were deposited on the N and NE carbonate edges of the Sulmona basin (e.g.,

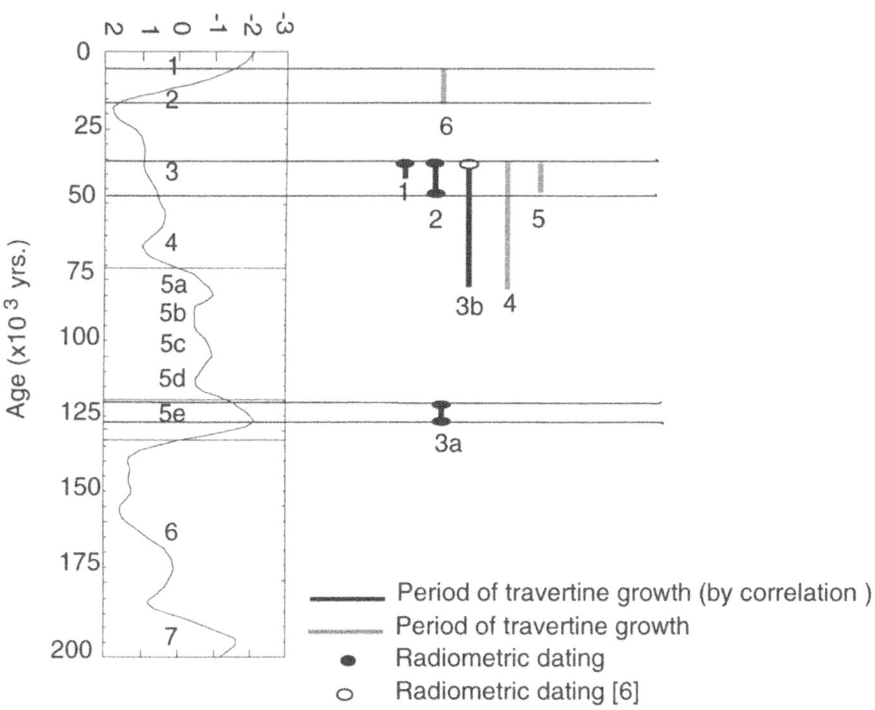

Fig. 3: Comparison between the ages of travertine deposits, carried out with ^{14}C and U/Th methods, and the oxygen-isotope record deep sea core.
1) F.te S. Callisto, 2) Decontra, 3a) Bussi Station "left"(1st phase), 3b) Bussi Station "left" (2nd phase), 4) Bussi Station "right", 5) Popoli 1, 6) Barrage Travertine System of Popoli Gorges (including Thalweg Pescara deposit).

at Popoli) and again along the gorges of Popoli (e.g., at Decontra and nearby the railway station of Bussi) (Fig. 1).

Regarding the last deposited travertines it is noticed that they occur on the valley floor of Popoli gorges, just below a thin alluvial layer (Fig. 2). Although such features are lacking of any radiometric readings, they could have been reliably correlated with their analogs cropping out downvalley the Popoli gorges, which were assigned an age bracketed between 23,000 and 6000 yrs. ago [4]. It is noteworthy that the travertines belonging to such group were found all over the reach of Pescara river valley, confined within the Popoli gorges. Because of their spread occurrence was responsible of frequent dammings of this reach of the valley, a series of shallow lakes, marshy spots and waterfalls resulted (Barrage Travertine System [3]).

4. DISCUSSION AND CONCLUSIONS

The reconnaissance survey over the area failed to recognize any travertine younger than 6000 yrs. In this respect it has to be stressed that the timing of the end of travertine deposition in the study area could reflect a global, rather than a local event, in that it matches previous reports for central Italy as well as Europe [5].

There is general agreement that such large scale and almost synchronous ending of travertine formation does reflect a significant climate change occurred in middle Holocene. This view receives further support by the comparison between the profile of the oxygen isotope in deep sea core sediments and that for the studied travertines, pointing out that the phases of enhanced formation of travertine fit with those of climate ameliorations (Fig. 3).

5. REFERENCES

G. Calderoni, G. Cilla, F. Dramis, D. Esu, M. Magnatti & M. Materazzi, Il Quaternario 9 (C) (1997), 481-492

C. Carrara, Il Quaternario (in press)

G. Cavinato & E. Miccadei, Il Quaternario 8 (1) (1995), 12-140

J. Imbrie, J.D.Hays, A. Mcintyre, A. C. Mix, J.J. Morley, N.G. Pisias, W. Prell & N.G. Shackleton, In: A. Berger, J. imbrie, J. Hays, G. Kukla, and B Saltzman (Eds.), Milankovitch and Climate, Part 1, Reidel, Boston, 1984, pp. 269-305

M. Lombardo, unpublished Ph.D. Thesis, "La Sapienza" University of Rome, 1999

A. Pentecost, Quaternary Science Reviews 14 (1996), 1005-1028

C. Violante, V. Ferreri & B. D'argenio, I.A.S. 15th Reg. Meet., 13-15 April, Ischia, Guide Book to the Field Trips, 1994

The Protected Areas System for the Conservation and for an Eco-Compatible Development of the Territory: The Maiella National Park

GIACOMO CAVUTA
Università "G. D'Annunzio" di Chieti, Dipar.to di Economia e Storia del Territorio (Pescara).

Key words: Conservation, Natural Parks.

Abstract: The National Park of Maiella streches in the South East of Abruzzo and includes the Maiella and Morrone relief: the Pescara river valley borders to the north, the Aventino valley and the Sub-Apennine hills to the east and Fossa di Caramanico to the West. Up to now the development of these areas has been promoted only by non-local investors financial resources; as a matter of fact this type of investment has brought significant and adverse effects on the social fabric and on the environment since the local populations reached a sudden wealth and as a consequence they definitively abandoned the traditional activities in favour of those to mass tourism related. More generally speaking it can be said that the economic benefits from protected areas can be derived from the non-remarkable character of the natural resources. The creation and management of national parks may bring a social and an economic welfare to the local population and implement the development of those mountainous areas which are not easily reachable, not suitable for industrial set dements and therefore subject to degradation and migration phenomena.

1. A NEW CONCEPT OF PARK

The concept of "national park" was born in America at the end of last century, as governmental laws were necessary too garantee the conservation of natural resources in extended up right lands. On 1st March 1872 President Grant set up Yellowstone Park, the first national park in the world. A lot of countries also in Europe such as Italy followed that example. The basic principle of protective areas is strictly naturalist: as the park land saves the

G. Visconti et al. (eds.), Global Change and Protected Areas, 465–473.

original integrity of its natural habitat it becomes also a sort of museum where nature itself reigns with the beauty of its landscape and with its biological and aesthetic values. For that reason it must be protected. Therefore the park concept favours the principle of the complete conservation of ecological-naturalistic origin. When the number of parks all over the world increased and the environmental culture spread out, the idea of park and the consequent concept of safeguard became more complex as it also included the idea of man with his identity as land inhabitant. In fact the same concept of nature grows larger by considering man as integral part of it, as part of its ecological system which consists of culture and history. Thus the park becomes an instrument which protects and exploits both natural resources and the human ones inside it. Consequently conservation and development must be harmonized within the park. Nowadays according to the above mentioned conceptual evolution the parks themselves are more and more oriented to become instruments of territorial management and of service. Parks allow to control the anthropic interference and land not only to preserve but also to promote those activities capable to induce development process in marginal areas, such as the updating of those functions which has been in crisis for a long time but not yet completly disappeared, and the creation of actions compatible with the environment that, in the meantime, can offer economical and employmental outlets. In fact the economic efficacy of protection arises from the evolution of the resources. Consequently the creation and the management of natural parks may provoke a social and economic welfare to the local communities. In this way the park becomes a new instrument of development above all in those regions characterized by the deterioration and migration phenomena. At the beginning the parks were considered only as places where one could relax during free time or as places to be studied for their natural aspects. In other words they were seen as something a part from the rest of the land they belonged to and from the economy of that land. Nowadays by the above mentioned evolution the parks have lost their characteristics of "the sanctuary of the nature" to become "parks with many aims", those are parks based on the integration of environmental, economic and urban functions during free time.

2. PRESERVATION AND DEVELOPMENT

The natural resources theory, concerning singleness and irreducibility aspects, states that it would be better not to go on with development even if it seemed facilitated by a positive costs-benefits analysis. As the benefits from resources are uncertain it would be easy even to lose the resource itself:

as in case of drugs that come out from an unknown plant. All of that is true also for resources of particular landscape beauty which could be destroyed forever or, at least, for a long time (Di Plinio, 1994). A decision is irreversible also when its effects may be changed to sustain very high costs. Every effort would be useless also for scientific research about a natural habitat of extraordinary landscape value, that is even where it seems possible to use knowledge in order to shorten the time necessary to revive the original situation. If they wish a recreational service market, there will always be a certain number of citizens who consider that reinstatement as an unsatisfactory remedy to the alteration of the natural environment. The importance of that market segments grows up as much as income, education and urbanization rise. A lot of scholars have proved that the best growth rate is lower when, in a perspective future, there will be possibility to obtain some information that the development would remove in a not reversable way. The information consists of benefits which derive from conservation therefore their value, called option value, is only hypothetically accessible. In fact that value consists of earnings by the upkeep of conservation options or by the development in the future. That idea is very considerable for such areas as the Italian national parks, because they contain extraordinary natural habitat and unique animals and plant species. Economists indicate two basic reasons to set up national parks: 1-the presence of exterior diseconomy in the natural resource conservation sector; by setting parks up the authorities can protect resources, the cost of which is not completly considered by citizens. As the parks are public goods they can be used by a person without reducing the possibility to do the same for other persons. Therefore government intervention is necessary to garantee the efficiency of the resources and to operate price discrimination for goods in order to marginally evaluate each consumer. 2- even if exterior diseconomy and public goods are absent, the market cannot ensure the best inter-temporal location of the resources for non reproducible goods in a context characterized by a technical progress. In this way, further generations will receive a lower stock of natural goods then the wished ones (Barbieri and others, 1991). In fact if the technical progress gradually reduces the cost of production and, consequently, the value of the reproducible goods to the «unique» ones. Whatever evaluation of nowadays value will reveal the relative values which will be shown in the future. Anyway, future benefits of conservation, of major empiric importance and less unreliability may be growing in order to justify a conservation policy. They come from a more and more growing demand for protected nature. All of these theorical concepts have important consequences for the advices that an economist gives about "park policy". In fact the demand for environmental protection may be underestimated both when the demand itself is promoted by losing park service underpaid and when it is justified as a governmental finance to the park. Anyway it does not mean not to imitate an enterpreneurial model of national park to such an extended to gain a considerable amount of money by self-financing. But all of that must be carried out by methods which favours a sufficient location of resources (Barbieri and others, 1991). At this point it is necessary to mention how the setting up of a national park may cause distribution problems. In fact as soon as beneficts are spread out, the costs must be spread out too; therefore costs must be sustained by the whole national community. For benefits we mean : better life quality, also in marginal areas, and the possibility to enjoy local animals and plants typical of specific places, which the garantee that those benefits will be present also in the future. For those reasons local population must be taught the right eco-compatible forms of development and whenever profits are not gained they must be given allowances by the government to offset the missed earnings. A good number of traditional economists prefer to follow the cost-benefit analysis, because that represents an effective instruments to promote protection-development balance and to accept the parks also from a social point of view. If we consider the application of that concept to the Regional Carso Park, we can realize the convenience of setting up a park by comparing the operating costs with the monetary flows and with the induced tourism. We will not analyse in detail the significative social benefits of the park setting for the local population both as caution and because of difficulties to classify them. We just point out that an improvement in life quality and new kinds of employment represent a good chance to migration and a way to balance generational relationship in those depressed areas like the nowadays national park ones. In fact to avoid migration of young people from original land it is sufficient to improve the economic outlooks of the protected areas, and at the same time to strenghten the cultural and relation possibilities; it would be an undertaking with positive effects on tourism itself. It is time now to evaluate the application of the costs-benefits methodology to the national parks. First of all it is important to remember the high costs to carry out a protected area and the missed earnings for the local population because of the environmental bonds which must be respected. There are, of course, perspective benefits by exploiting the compatible resources and the park aims. Even if those benefits are less then those coming out the traditional development methods. Anyway, we cannot deny the real nature of the problem, that is the perspective necessity to compensate those who pay the costs of a missed development which is denied because not compatible with the park itself. The amount of compensation is the other problem: they may be more expensive where there are populations both with per capita income higher than the average national one and with per capita income lower then the average national one. By the way using cost-benefit analysis may not be profitable to promote the acceptance of the park by local populations, probably because they will be more easly convinced by concrete results then by a theorical research. It is necessary to reconcile in a concrete way the use of a specific area for leisure time and relax with the use of it for scientific and protective aims. Division into zones is the method accepted all over the world to solve the "quarrel". The above consideration have tried to explain the quarrel between protection and development, which will be always present whenever there are areas of great naturalistic interest (Vallega, 1994). As regards the necessity or not to set up national parks it is more conflictual the discussion about bonds and compatible activities. That conflict must be solved for the parks themselves above all in those areas where opposite groups of interest are present. The general lines to solve the conflicts are: 1 – by distinguiscing areas absolutily protected and areas where is possible to consider a more acceptable balance; 2 – traditional developing system may be replaced by others both profitable and compatible with the particular park environment. As there are bonds for financial, organizing and enterpreneurial capacity still operating, it is essential for those who manages the park organization to promote, indirectly, their own institutional purpose. Consequently it is necessary to set up enterpreneurial model of National Park; 3 – the two above mentioned items may be insufficient, that is they are economic activities which may be denied because no compatible. Therefore if there are bonds they must be offset.

3. ECONOMICAL ASPECTS

Primary sector was, and is, of main importance for the economy of those lands with limited external relationship. The consequence of that isolation, plus other serious problem of rural economic depauperation are responsible of: the ageing of the work forces, the parcellization of lands and of productive structures and the lackness of new techniques for farming and rearing. For what concerns the productive asset the Maiella's territory is different from its piedmont land. The mountain part exploits high pastureland and productive woods. The piedmont part is concentrated on cow rearing and zootechnical breeding together with the different way of agricultural land exploitation, above all extensive farming. There was a considerable decrease of pasture activities whereas once they were important such as the famous transhumance migratory phenomena, long the long-out of roads called "tratturi" which must be followed by sheep during their moving. The ancient plenty of herds, anyway, reduced as time went by: nowadays the regional sheep rearing is of 460.000 animals, on the contrary in the last century there were 1,5- 2 million animals. This sector should be relaunched for the development of the mountain land of the park area as it is considered the cultural estate of the people of that land. In the last few years, the sylviculture, the third income, a once rewarding sector, was suffered from a decrease due to excessive use of the wood estate which is responsible for unproductivity, deterioration and for environmental damages. By the short description of the above reported classical economies we realize the situation of impoverishment that both all the Maiella's centres and the Abruzzo mountains in general have suffered from in the last fourty years. A possible solution to limit the economic degeneration of the lands taken into consideration in this research is about a planning of the traditional activities on a large broad, by respecting the bonds imposed by the setting up of the national park. For the Maiella mountain area, industry has never been a leading sector, even if there are good leveled industries both from an employment point of view and from an income one. Some of them have local characteristic and are more useful for regional realities, some others show a national and international character such as the pasta factories in Fara San Martino, where very esteemed pasta qualities are produced and exported all over the world. That kind of industry made that Maiella's centre one of the most thriving in that area. As a matter of fact the industrial area of Fara San Martino with its four pasta factories, and other industries, represents the standard rate of the economy of the centres located to the south-east slope of the Maiella mountain. As you can see on table 1, other significative presences in the industrial area in Chieti province are: Guardiagrele, with a very ancient and traditional copper working and wrough iron working; Pennapiedimonte and Lettopalena with the Maiella stone working (very important is the caving work); Rapino with its ancient and traditional pottery working and Pretoro with its wood working. Among L'Aquila province centres, within the park, Sulmona and Pratola Peligna are two important industrial centres for the whole area where there are located Fiat manufacture, many other metallurgical industries and sugared almond industries, all of them well known all over the world for the good quality of this products. In Pescara province nearly all the industrial centres are present in the urban centres along the valley: Manopello where there is copper working and wrough iron working; Popoli, Scafa and Tocco da Casauria with their alcohol industries. In the Park district, a part for those few industrial realities, in economy is essentially of craftsman kind which, if it is appropriately exploited for example tourism, it will be an attraction both for migrants who might be employed in tourism as craftsman and for local mountain populations. The structural painting above described shows the actual social economic collapse of the mountain area which have been characterized by deterioration and migration for many years. That is why a " re-reading " of territory is needed in order to create both a concrete development in those marginal areas, and to give a new role to the populations who live there, since that area have got very high environmental capacity which have not been adequately exploited since now. The developing model proposed for that area is an eco-compatible one, which offers the possibility to join economical growing needs to the conservation of natural habitat. Of course, all of that can be carried out not only through the natural resources of the same territory. Up to now, the development model followed in these areas is based almost exclusively on the use of foreign capital for tourism, as a matter of fact that kind of investment has caused a lot of changes in local populations, as they give immediate earnings. But if we analyse them in detail they show mass touristic activities and consequently the complete forsaking of the traditional occupations, environmental deterioration and a social balance degradation. The exogenous capital had a great power by abusing the urban local laws. It is sufficient to mention the estate which were sold for the second property houses, or the investment to set hotels, the profits of which, go back to where the capital come from. Thus entire winter resorts are set without taking the

N.	L'AQUILA	Industry	Commerce	Other activities	Istitut.	Total experts
1	ATELETA	53	37	19	38	147
2	CAMPO DI GIOVE	39	107	45	28	219
3	CANSANO	0	11	2	16	29
4	CORFINIO	70	32	35	57	194
5	PACENTRO	50	49	32	37	168
6	PESCOCOSTANZO	68	104	20	195	387
7	PRATOLA PELIGNA	364	473	310	288	1.435
8	RIVISONDOLI	54	90	48	30	222
9	ROCCACASALE	41	71	10	21	143
10	ROCCAPIA	7	17	6	10	40
11	ROCCARASO	88	261	105	224	678
12	SULMONA	2.883	1.771	2.594	1.460	8.708
	Tot. L'Aquila province	**3.717**	**3.023**	**3.226**	**2.404**	**12.370**
N.	**PESCARA**	**Industry**	**Commerce**	**Other activities**	**Istitut.**	**Total experts**
1	ABBATEGGIO	80	14	4	5	103
2	BOLOGNANO	269	67	43	45	424
3	CARAMANICO TERME	169	155	83	146	553
4	LETTOMANOPELLO	107	136	68	79	390
5	MANOPELLO	1.038	326	117	260	1.741
6	POPOLI	212	416	193	704	1.525
7	ROCCAMORICE	2	33	45	53	133
8	SALLE	76	9	7	12	104
9	SANT'EUFEMIA A MAIELLA	24	36	13	12	85
10	SAN VALENTINO	61	61	40	81	243
11	SCAFA	690	279	191	166	1.326
12	SERRAMONACESCA	18	29	7	30	84
13	TOCCA DA CASAURIA	686	257	107	79	1.129

N.	CHIETI	Industry	Commerce	Other activities	Istitut.	Total experts
	Tot. Pescara province	**3.432**	**1.818**	**918**	**1.672**	**7.840**
1	CIVITELLA MESSER RAIM.	43	29	24	22	118
2	FARA SAN MARTINO	612	69	88	63	832
3	GAMBERALE	17	13	9	7	46
4	GUARDIAGRELE	1.379	666	405	592	3.042
5	LAMA DEI PELIGNI	72	85	40	45	242
6	LETTOPALENA	10	14	15	9	48
7	MONTENERODOMO	35	30	50	33	148
8	PALENA	85	82	52	71	290
9	PALOMBARO	179	53	27	29	288
10	PENNAPIEDIMONTE	73	20	29	29	151
11	PIZZOFERRATO	33	69	32	33	167
12	PRETORO	197	138	9	139	483
13	RAPINO	144	73	37	52	306
14	TARANTA PELIGNA	31	22	23	16	92
	Tot. Chieti province	**2.910**	**1.363**	**840**	**1.140**	**6.253**
	TOTAL PARK AREA	**10.059**	**6.204**	**4.984**	**5.216**	**26.463**

Source: processing on 1991 ISTAT data

urban planning into account, by exploiting landscape resource in a not controlled way. Infact after a few years, as they are effected by a savage urbanization they lose their attractive power; at that point, a land is destroyed in a irreversible manner (Giacomini and Romani, 1992). On the contrary development aim is to satisfy human needs, therefore it aims to reach anthropo-economic purpose. Whereas the conservation aim is to be sure that the environment shows such development and at the same time it has the capacity to maintain it for a long time.

4. BIBLIOGRAPHY

AA.VV., Con il parco nel duemila. Tutela e sviluppo compatibile nel comprensorio del Gran Sasso d'Italia, Centro Servizi Culturali, L'Aquila, Reg. Abruzzo, 1994.

Barbieri G., Canigiani F. E Cassi L., Geografia e ambiente, Torino, UTET, 1991.

Biondi M., Bologna M. A. E Osella B. G., I nuovi Parchi nazionali in Abruzzo, in «Boll. Club Alpino Ital .- Sez. de l'Aquila», L'Aquila, 1992, n. 25, pp. 51-54.

Cardinale B. E Cavuta G., Economia e territorio: il Parco nazionale del Gran Sasso e dei Monti della Laga, in «Notizie dell'economia», Teramo, 1995, n. 5-6, pp. 63-78.

Cavuta G., Alcune riflessioni socio-economiche sull'eco-sviluppo. Il caso dell'Abruzzo, in «Quaderno Dottorato di ricerca in Geografia Politica», n. 4, Trieste, Dip. di Scienze Pol., 1994, pp. 54-69.

Cavuta G., Parks Project and Compatible Development for the Abruzzo Mountains, in Scaramellini G. (a cura di), Sustainable Development of Mountain Communities, Milano, Guerini e Ass., 1995, pp. 195-204.

Di Federico G., Parco nazionale della Maiella, Chieti, BAG Ed., 1993.

Di Plinio G., Diritto pubblico dell'ambiente e aree naturali protette, Torino, UTET, 1994.

Giacomini V. e Romani V., Uomini e parchi, Milano, F. Angeli, 1992.

Landini P. e Leone V., Ipotesi di un parco naturale nella duna di Lesina. Un approccio interdisciplinare, in «Memorie della Soc. Geogr. Ital.», XXXIII, Pisa, Pacini Editore, 1984.

Landini P., Sistema ambientale e sistema regionale: un approccio geografico al caso barese, in Conferenza "Bari: città-ambiente. Inquinamento aria, suolo, acqua.", Bari, 1982.

Massimi G. e Cardinale B., Pathways of Development and Environmental Compatibility in the Abruzzo Mountains: the Marsica Fucense as a Case Study, in SCARAMELLINI G. (a cura di), Sustainable Development of Mountain Communities, Milano, Guerini e Ass., 1995, pp. 99-115.

Pellegrini M. e Crisante C., Parco nazionale della Maiella, Roma, WWF, 1992.

Vallega A., Geopolitica e sviluppo sostenibile. Il sistema mondo del secolo XXI, Milano, Mursia, 1994.

Environmental Protection and Social Protection: The Sirente-Velino Regional Park.

MARINA FUSCHI
Dip. di Economia e Storia del Territorio- Fac. di Economia-V.le Pindaro 42 Pescara

Key words: environment, sustainable development, park, region, population, economy.

Abstract The commonly accepted principle of a sustainable development reinforces the global-local dyad at the various stages of the developing process of the territory organization in its different forms. The establishment of protected natural areas which ca be considered as a sort of environmental welfare in Italy and mainly Abruzzo, is the best example of a global fruition of environmental resources. The case of Sirente-Velino Natural Park has been analysed its physical and human peculiarities and it can be considered as an example of global planning founded on consensus and on the widest popular willingness / receptiveness / involvement. The inherent challenge is to accept/promote the environmental enhancement as a unique in the territory making, bound to return the cultural identity and to express the peculiarity and quality of the place.

1. ENVIRONMENTAL PROTECTION AND SOCIAL PROTECTION: THE SIRENTE-VELINO REGIONAL PARK.

In the last decades an improved and mature interest in the social-environment relationship has conveyed to economic policy actions less related to classical economic laws not related to land. The learn and manufacturing methodologies together with an increasing importance to services have moved the attention to a less dispersable kind of development by determining a dematerialization of the society which takes much more care to safeguard life quality testable through environmental quality. It is meaningful the attempt to recover the unitary concept of environmental

G. Visconti et al. (eds.), Global Change and Protected Areas, 475–487.

value within the daily uncertainty and confusion about environmental subject by starting from the plurality of public bodies on different grounds. If all of that, on one side, harmonizes the cultural-landscape valence the wealth and urban one as well (Giannini, 1973), on the other side there is the request for an institutional order which is target-oriented to the development of a unitary function. Of course, this function cannot leave out the citizens participation in a territorial planning, according to the British examination in public and on condition to bound suitably risks of "overload signs of interest". The same planning, by leaving the aim to rationalize and optimize the use of the resources, typical of the 30s, is directed to the research of a new balance between economic policy laws and natural laws: it consists of the supporting development concept ratified during the United Nations Conference in Rio, in 1992. When the idea of productivity was exceeded and that of efficiency was accepted, the achievement of completeness of the ecosystem, the intra-generational and the inter-generational social equality is proclaimed (Vallega, 1994). Therefore, if on one hand the supporting development needs long-term perspectives, on the other hand it raiset the global-local dyad during the different stages of changes in the developing process and during its forms of territorial organization. The new developing-environment culture runs through the supporting-developing way which is considered like a model of balanced order for the territorial system. The institution of natural protected areas, among which the most representative one is the Park, is one of the most demanding aspects in the ecological subject considered like a process of global planning. It passes over the easy protectionist or preventive aim to lead to territorial management kinds which are considered like answers for concrete needs of economical, social and cultural order. There is no territorial policy without solution to social problems, therefore the Park must be considered like an "open system" linked to other systems and interchangeable with them, never disjunctive by the surrounding land of which it represents the climax of the global fruition of environmental resources. Consequently it is time to go beyond of those concepts and laws responsible of the functional efficiency of many national and regional parks, above all in those highly populated countries. Infact, the high anthropic land level must consider the "park" as a moment of synthesis between natural landscape conservation and human development. Just in Italy a great number of fervent enterprices (see the National Bill n. 394 dated 1991) both theorical and practical, both legal and urbanistic has brought to a particular kind of environmental welfare state which created two national parks (Gran Sasso-Laga and Maiella), a regional park (Sirente-Velino) and a lot of sheltered areas with different defence levels in Abruzzo. At the regional level it is interesting to stress the fact that there is, in particular, the will to evolve in order to gain further goals of credibility and utility, just like the natural regional Sirente-Velino Park which builts its global planning both on the popular approval and on the popular complete availability. We can say, from a political point of view, that the Park was projected by town-administrators who, in 1979, realized that within the idea of environmental defence there was also the possibility for a new social, economic rebalance policy for their areas. It is interesting to emphasize the very short times necessary for the Park Law to became operative (Bill n. 54-July 1989) both on the institutional side and on the administrative side: with the settlement of the Park Body by the Regional Council in July 1992. The Park (60.000 ectares, 5.5 % of the regional territory) is located in the centre of the Appennino mountain range among the west slope of the Gran Sasso to the north/north-east, the Fucense basin to the south and the Lazio region to the west. Its territory consists of morphological different areas such as the mid-Aterno Valley, the Rocche tableland and the Subequana Valley.

The density of population in the Sirente-Velino Park which lies within L'Aquila province, concerns both the areas of twenty-two communes[1] and four mountain communities (Sirentina, Giovenco Valley, Marsica I and Amiternina). According to a demographic study based on census data concerning the period 1921-1991 (see table 1), we can see a great decrease of the residential population, that is to say, a continous depopulation which has become a constant element, as time went by, for the territorial physionomy of the inland areas. Particular conditions such as sparely fertile lands, the extreme parcelling out of property, the continental characteristic of the territory, the harsh weather, and deficiency road conditions are responsable of the rural exodus for most of the population. At the beginning of the twentieth century the appeal of many European and overseas centuries, the attraction of the close urban centres of Rome and Pescara, the nearness to the Funcense Basin with the Avezzano industrial centre (which was not influenced by depopulation) all these factors reinforced a negative migratory settlement which was superior then the natural one up to the 7Os. Later on the latter became so negative as to be superior to the back-migration result (Massimi, 1990). Anyway some communes such as Celano and Magliano de' Marsi were influenced by the above mentioned back-migration. Between 1921 and 1951 the only centres which presented a positive constant trend were Aielli (+16.7%), Celano (+18.81%) and Pescina

1 The institutive Regional Park Act n. 54, 13 July 1989 located 18 communes (Acciano, Aielli, Castel di Ieri, Castelvecchio Subequo, Celano, Cerchio, Fagnano Alto, Fontecchio, Gagliano Aterno, Goriano Sicoli, Magliano de' Marsi, Massa d'Albe, Molina Aterno, Ovindoli, Rocca di Cambio, Rocca di Mezzo, Secinaro, Tione degli Abruzzi); further Regional Laws joined them other 4 communes (L'Aquila, Ocre, Pescina e Collarmele). During the present research we will not consider L'Aquila commune because it is both capital of a province and administrative centre, therefore it is not comparable with others because of its social-economical reality.

(+22.36%). On the contrary during the following decade, 1951-1971, there was a decrease of the population extent in all the Park communes and in particular in those ones located in the more mountainous areas where there was a reduction of more than half of the whole population such as in Fagnano Alto (-51.25%), Fontecchio (-55.68%), Gagliano Aterno (-50.22%) and Tione degli Abruzzi (-49.25%). Between 1971 and 1991 the position of those communes near the Fucense Basin such as Celano (+6.05%), Magliano de' Marsi (+13.50%), Massa d'Albe (+6.7%) and Pescina (+6.07%) became stronger, on the contrary Aielli and Collarmele recovered the ratio (exactly: +4.17% and +3.44%) in the sole decade 1981-1991. Of course, the Fucino area continued to be a sort of pole even if the positive accessibility to the centres located near both the A25 Rome-Pescara motorway and the railway has determinated commuting towards Rome. During the same decade the migration settlement came positive again in certain centres such as Aielli (+120), Celano (+48), Cerchio (+13), Collarmele (+29), Fontecchio (+123), Magliano de' Marsi (+300), Massa d'Albe (+148), Ocre (+126), Ovindoli (+25) and Pescina (+229). On the contrary the natural migration presents a negative settlement as regards births and deaths which testifies the gradual ageing process of the population (except for the centres of Celano +416, Collarmele +6, Magliano de' Marsi +9 and Pescina +10). The old age index (see table 3) regarding the period between 1971 and 1991 (see table 2) shows a structural ageing of the population due not only to the average life extention, but also to the weak social economic structure of the area which goes on to turn out young people. The economic dependence index (see table 4) streighten farther on the image of a double-speed territory highly generalized by the Marsica Fucense communes, "it corresponds to the depression of the Fucino area and to its mountainous areas which are set all around it and which are more than 2000 metres high" (Fondi, 1970, p. 465), which show the most favourable index (compare table 3 and table 4). Obviously the territorial reality of the Park is lacking of its own economic structure which will be able to keep alive endogenous path of development which can be in this way in a great demand in the neighbouring areas. At this point it would be better "to create" a regional identity of the communes based on the dual concept of agriculture and tourism which should be carried out by following rules and choices with the exclusive aim to respect environment. Of course it does not mean to condamn the territory to keep still, on the contrary it is meant to project it towards new developing scenarios based on the acceptance of nature as the main productive factor. Always the main economic activity of that area was of agro-wood-pastoral kind and in this way it reflected the reality of a Region that in 1936 had the highest percentages of employed people in agriculture with its 75% of the whole active population. After the second world war agriculture became a marginal aspect by determining a great rural exodus in the Abruzzo, as a consequence there were both a migration from inland areas and a desertion of the territories. All of this justifies the permanence within the Park area both of a poor economy and of subsistence based above all on the cereals and fodder farming, while it does not seem considerable the contribution of the wooden ones. AUS (Agricultural Used Surface) analysis reports the existence of the communes with no agricultural possibility for more than 80% of the territory such as Fagnano Alto and Tione degli Abruzzi. On the contrary Ovindoli, Rocca di Cambio and Rocca di Mezzo show good relative values in terms of agriculture surface used as permanent laws and pasture land. (see table 5). The tableland area is rich in extensive lawns cultivated with forage which, by cattle rearing, can give life to a good industry which transforms zootechnical products produced by family business and located in Rocca di Mezzo. Once more according to both AUS considerations and income value, the communes of the Marsica Fucense merge because of good farming of potatoes, fodder, cereals and sugar beets (see table 6). Both industrial and service activities cover a marginal aspect because it regards craftsman characteristics which showed their streighten in its model flexibility that is in its efficiency in the decade 1970-1980. Nowadays they suffer from structural limits present in the inability to adeguate themselves to a correctional and modified model. The new competitive reality based on innovation-information acquisition and on effective service, points out the environment strategic role in which small industries work. All of this proves of the weakness of the territorial system which is, in this way, characterized by a poor inter-connection and reduced weight of the same employmental role, with its total 7.058 personnel per 2.410 local units[2]. Except for some communes such as Celano, Pescina and Magliano de' Marsi which provide more than 60% of the personnel and about 50% of the local units. It is important to stress the high industrial level of Aielli (119.8) and of the services of Ovindoli (131.1)[3], versus a widespread pulverization in the less developed areas or in territories with regression in development. Among those areas there are the Aterno Valley communes (Acciano, Castel di Ieri, Fagnano Alto, Gagliano Aterno, Goriano Sicoli, Tione degli Abruzzi, Secinaro, Castelvecchio Subequo, Molina Aterno and Fontecchio) which are excluded from both the infrastructural process and from the generic equipment. But at the same time they are characterized by a decline in population, by activity index and service structures. On the contrary the Marsica Fucense communes (Celano, Pescina, Magliano de' Marsi, Aielli and Cerchio) are the most equipped from the socio-economic aspect and for this reason they are more effected by the growth and development

[2] ISTAT data, 1991 census.
[3] CRESA data, 1993 report.

phenomena. In the same way those communes, the limits of which are marked by the Rocche tableland, have started a revitalizing process of the environment thanks both to the rebirth of commerce as a complementary economic activity and to tourism which utilizes landscape beauty, above all snow resources. That district has not been able to organize the tourism supply which represents a sort of "attraction" based on infrastructures, even if there are certain important elements which can satisfy the urban demand of Rome, Naples, Pescara and L'Aquila by keeping alive in this way also a weekend tourism. In fact it is a sort of tourism based on the use of the second property house as the high percentage of the non residential houses proves in some centres such as Ovindoli (85.15%), Rocca di Cambio (81.62%) and Rocca di Mezzo (79.81%) (see table 7). Therefore it is a tourism of migration origin which, if well organized, might bring about synergic effects because they are based, by considering the increased demand for the environment, by those tourist who are more and more directed to those realities which have unchanged their natural characteristics at all. The analysis about the average per capita income in 1997 (see table 8) shows a sort of uniformisation among the Park communes, the data of which goes from 15 million of Secinaro to more than 18 million of Celano, versus 20 million of the Regional average data. All of those figures show the further confirmation of a territorial reality kept alive only marginally by a large scale income. At the same time, in fact, they points out a territorial aspect which finds in the work basins of Rome, Avezzano, L'Aquila and Sulmona, and above all in the old-age pension incomes, the possibility to give historical continuity to the Park area. Therefore the Park is considered a breaking moment, a qualifying factor for the whole area: the dual concept agricolture-tourism might represent the territorializing process capable to give life to a new culture through the environmental exploitation. The whole Abruzzo region has to keep up the following challenge: the A.R.V.E. (Abruzzo as the Green Region of Europe) which implies the image of the Sirente-Velino Park as a sort of linking corridor between the National Abruzzo Park and the Gran Sasso Park. The research for the territorial unity through the environmental resource starts, first of all, from the beginning of the social processes and than from laws and plannings. Therefore it is added to a settlement planning produced by history on a homogeneous physical support. Thus it would be necessary to analyse the cultural identity of the local communities involved and call for a territorial policy which can express the quality and the originality of that area. Consequently we are forced to wonder if there is a real popular will which can see the Park as an effective planning instrument and which would be able to combine landscape conservations to human development. Education and information represents the two basic conditions to build an homogeneous origin. On a sparely populated but very much aged area (in 1991 the residential was 44 inhabitants per Km²) the awakening process takes young people into consideration for the state of cultural development which takes care of the environment. But at the same time it relies on the whole population to move from a merely welfare economy to the characterization of the territory. If, on one hand, the social economic deterioration of that area has lessned the use of natural resources, on the other hand it has ensured the landscape preservation in order to set up a "product"' which considers the necessity of market more and more attracted by environmental values. For that reason it would be necessary to reconvert agriculture by taking into much consideration the quality of proper trade mark which expresses the authenticity and the ecological compatibility of the products which will not be considered as means of sustenances any more but as means of "enjoyment". By farm holiday it will be possible to exploit "niche" products (truffles, mushrooms) and start ancient farming again (grass, medical plants, wild fruit, beekeeping) even by direct marketing.

Tab. 1 Resident Population in communes of the Sirente-Velino Park from 1921 to 1991.

Commune	1921	1931	1936	1951	1961	1971	1981	1991
Acciano	2.179	2.006	1.906	1.703	1.311	991	742	538
Aielli	2.383	2.562	2.658	2.778	2.225	1.562	1.414	1.473
Castel di Ieri	1.459	1.360	1.273	1.100	852	631	518	437
Castelvecchio Subequo	2.404	2.440	2.417	2.409	2.118	1.811	1.547	1.448
Celano	9.302	9.531	10.146	11.052	10.389	10.271	10.429	10.893
Cerchio	2.259	2.709	2.657	2.602	2.152	1.777	1.770	1.735
Collarmele	1.705	1.702	1.688	1.662	1.313	1.069	1.016	1.051
Fagnano Alto	2.369	2.031	2.008	1.641	1.195	800	649	499
Fontecchio	1.368	1.272	1.257	1.126	739	499	438	469
Gagliano Aterno	2.081	1.926	1.646	1.157	849	576	478	396
Goriano Sicoli	1.558	1.556	1.556	1.367	1.133	754	751	685
Magliano de' Marsi	3.927	3.848	3.842	4.058	3.812	3.081	3.188	3.497
Massa d'Albe	2.442	1.787	1.832	1.795	1.466	1.209	1.151	1.291
Molina Aterno	962	972	940	909	783	722	653	554
Ocre	1.509	1.421	1.329	1.252	1.053	957	887	984
Ovindoli	2.954	2.324	2.420	2.421	1.900	1.386	1.244	1.204
Pescina	4.914	5.671	5.943	6.013	5.438	4.430	4.460	4.699
Rocca di Cambio	1.225	1.059	847	950	796	618	506	447
Rocca di Mezzo	4.168	3.434	3.166	2.976	2.632	2.032	1.722	1.531
Secinaro	1.956	1.934	1.757	1.585	1.335	924	765	558
Tione degli Abruzzi	1.918	1.836	1.713	1.413	1.082	717	596	485
Total	55.042	53.381	53.001	51.969	44.573	36.817	34.924	34.874

Source: ISTAT

But farm holiday will be determinant above all to give agriculture its social role back through the safeguard of the rural cultural heritage which will become in this way the conjunction between urban reality and rural habitat. It will be necessary to stimulate and to boost enterpreneurship on the local population by facilities and subsidies and by promoting cooperation among young people, in order to relaunch and to qualify tourism and to give to the whole territory the real look of a protected and preserved nature. The naturalistic, cultural and artistic capacity of that area qualify the territory according to the general and coordinated political viewpoint: the safeguard of a minor architecture (Castello di Celano, Castello di Ocre, Chiesa di Santa Maria in Porclaneta Valley) just to say some of the ancient buildings; the discovering again of a certain number of archeological area proper of Italic, Roman and High-Medieval eras (Alba Fucens, the hypogeum cemetry in Castelvecchio Subequo area, a fort in Mount Secine near Aielli) all of them are well innate with the mountainous Abruzzo environment because they testify the ancient way of living; the exploitation of the natural richness with its numerous 2000 m high mountains and because of a severe weather those mountains can garantee abundant snowfalls till late spring season; the exploitation of the territorial *environmental suitability*: those territories present themselves as "faunistics corridors" with a qualified vegetable cover according to a botanic point of view (birch, ash tree, hornbeam, beechwood). It would be necessary to carry out a touristic policy capable to join green tourism to white tourism, social-cultural tourism to farm holiday in order to safeguard the whole territory and recover man-nature relationship for a better life quality and a defence. In other words it would be necessary to carry out a territorial policy based on zonation by distinguishing *core zones* and *buffer zones* according to the Park structure. The Park will be depicted as a local product daily generated by the help of the whole community (Giacomini and Romani, 1991).

Tab. 2 Population divided into years of age in the communes of the Sirente-Velino Park.

		1971			1981			1991	
Commune	<14 years of age	>65 years of age	total	<14 years of age	>65 years of age	total	<14 years of age	>65 years of age	total
Acciano	155	146	991	96	207	742	39	199	538
Aielli	261	271	1.562	244	303	1.414	253	321	1.473
Castel di Ieri	97	120	631	56	147	518	53	144	437
Castelvecchio Subequo	334	263	1.811	218	322	1.547	191	338	1.448
Celano	2.531	1.191	10.271	2.348	1.229	10.429	2.009	1.656	10.893
Cerchio	381	217	1.777	366	279	1.770	309	349	1.735

Collarmele	231	169	1.069	193	171	1.016	179	219	1.051
Fagnano Alto	116	184	800	75	233	649	42	199	499
Fontecchio	76	111	499	48	150	438	38	201	469
Gagliano Aterno	98	146	576	56	156	478	31	138	396
Goriano Sicoli	117	147	754	114	186	751	98	185	685
Magliano de' Marsi	720	372	3.081	664	496	3.188	602	597	3.497
Massa d'Albe	250	197	1.209	184	209	1.151	206	271	1.291
Molina Aterno	153	116	722	114	130	653	73	119	554
Ocre	183	155	957	168	152	887	153	194	984
Ovindoli	261	206	1.386	242	250	1.244	178	263	1.204
Pescina	936	529	4.430	799	635	4.460	765	929	4.699
Rocca di Cambio	96	106	618	70	121	506	45	112	447
Rocca di Mezzo	382	416	2.032	274	420	1.722	160	391	1.531
Secinaro	187	155	924	115	176	765	36	187	558
Tione degli Abruzzi	111	174	717	69	189	596	37	180	485
Total	7.676	5.391	36.817	6.513	6.161	34.924	5.497	7.192	34.874

Source: ISTAT

Tab. 3 Old Ages Index in the communes of the Sirente-VelinoPark.

Commune	1971	1981	1991
Acciano	94,19	215,62	510,25
Aielli	103,83	124,18	126,87
Castel di Ieri	123,71	262,50	271,69
Castelvecchio Subequo	78,74	147,70	176,96
Celano	47,05	52,34	82,42
Cerchio	56,95	76,22	112,94
Collarmele	73,16	88,60	122,35
Fagnano Alto	158,62	310,66	473,80
Fontecchio	146,05	312,50	528,94
Gagliano Aterno	148,97	278,57	445,16
Goriano Sicoli	125,64	163,15	188,77
Magliano de' Marsi	51,66	74,69	99,16
Massa d'Albe	78,80	113,58	131,55
Molina Aterno	75,80	114,03	173,01
Ocre	84,70	90,48	126,80
Ovindoli	78,92	103,30	147,75
Pescina	56,52	79,47	121,44
Rocca di Cambio	110,41	172,85	248,88
Rocca di Mezzo	108,90	153,28	244,37
Secinaro	82,88	153,04	519,44
Tione degli Abruzzi	156,75	273,91	486,48
Average Value	97,25	160,03	254,24

Source: Author's estimates using data from ISTAT

Tab. 4 Economic Dependence Index in communes of the Sirente-Velino Park.

Commune	1971	1981	1991
Acciano	43,62	69,02	79,33
Aielli	51,65	63,09	63,85
Castel di Ieri	52,41	64,44	82,08
Castelvecchio Subequo	49,18	53,62	57,56
Celano	56,83	52,20	50,70
Cerchio	50,72	57,33	61,09
Collarmele	59,79	55,83	60,95
Fagnano Alto	60,00	90,32	93,41
Fontecchio	59,93	82,50	103,91
Gagliano Aterno	73,49	79,70	74,45
Goriano Sicoli	53,88	66,52	70,40
Magliano de' Marsi	54,90	57,20	52,18
Massa d'Albe	58,66	51,85	58,60
Molina Aterno	59,38	59,66	53,04
Ocre	54,60	56,44	54,47
Ovindoli	50,82	65,42	57,80
Pescina	49,41	47,39	56,37
Rocca di Cambio	48,56	60,63	54,14
Rocca di Mezzo	64,67	67,51	56,22
Secinaro	58,76	61,39	66,57
Tione degli Abruzzi	65,97	76,33	80,97
Average Value	56,06	63,73	66,10

Source: Author's estimates using data from ISTAT

Tab. 5 Agricultural Used Surface in the communes of the Sirente-Velino Park.

Commune	Territorial Surface (hectare)	Half-active	Permanent farming	Permanent lawns and pasture lands	Total AUS	AUS/Tot. Surface
Acciano	3.236	145,80	44,98	272,80	463,58	0,14
Aielli	3.470	620,66	35,51	980,38	1.636,55	0,47
Castel di Ieri	1.879	534,76	20,52	264,20	819,48	0,44
Castelvecchio Subequo	1.923	378,86	43,14	523,44	945,44	0,49
Celano	9.177	4.205,80	135,06	4.519,00	8.859,86	0,96
Cerchio	2.011	1.110,90	35,95	21,00	1.167,85	0,58
Collarmele	2.370	908,78	14,50	912,53	1.835,81	0,77
Fagnano Alto	2.448	154,29	13,91	140,05	308,25	0,12

Fontecchio	1.689	147,15	8,05	472,33	627,53	0,37
Gagliano Aterno	3.336	468,27	4,27	1.296,81	1.769,35	0,53
Goriano Sicoli	2.177	552,65	38,35	863,84	1.454,84	0,67
Magliano de' Marsi	6.796	1.496,05	22,15	1.614,88	3.133,08	0,46
Massa d'Albe	6.847	1.160,52	1,20	940,20	2.101,92	0,31
Molina Aterno	1.184	356,65	14,40	20,80	391,85	0,33
Ocre	2.354	284,96	11,76	304,76	601,48	0,25
Ovindoli	5.884	224,87	10,65	2.106,89	2.342,41	0,40
Pescina	3.751	1.532,15	72,20	1.287,58	2.891,93	0,77
Rocca di Cambio	2.762	81,89	-	1.116,66	1.198,55	0,43
Rocca di Mezzo	8.714	453,10	1,05	2.502,70	2.956,85	0,34
Secinaro	3.205	241,15	37,00	795,70	1.073,85	0,33
Tione degli Abruzzi	4.024	90,70	9,09	300,62	400,41	0,09
Total	79..237	15.149,96	573,74	21.257,17	36.980,87	0,47

Source: ISTAT, 1990, with our partial processing data

Tab. 6 Agricultural Regions of the Sirente-Velino Park.

Agricultural Regions	More Profittable farming	Value (mill. for hect.)
Rocca di Mezzo Tableland	Lawn	4,110
Subequana Basin Mountain and of the Mid-Aterno	Half-active	3,720
Fucino Tableland	Half-active	15,900

Source: Ufficial Bullettin of the Abruzzo Region, n. 5 special issue, 15 March 1991

Tab. 7 Houses in the communes of the Sirente-Velino Park.

Commune	Residential Houses	Non-Residential Houses	% Non-Residential Houses
Acciano	237	297	55,62
Aielli	531	360	40,40
Castel di Ieri	184	227	55,23
Castelvecchio Subequo	585	209	26,32
Celano	3.897	1.160	22,94
Cerchio	617	133	17,73
Collarmele	377	221	36,96
Fagnano Alto	244	229	48,41
Fontecchio	161	180	52,78
Gagliano Aterno	184	242	56,81
Goriano Sicoli	288	203	41,34
Magliano de' Marsi	1.213	601	33,13
Massa d'Albe	515	516	50,05
Molina Aterno	250	17	6,37
Ocre	339	207	37,91

Ovindoli	461	2.644	85,15
Pescina	1.850	403	17,89
Rocca di Cambio	215	955	81,62
Rocca di Mezzo	650	2.570	79,81
Secinaro	280	299	51,64
Tione degli Abruzzi	221	243	52,37
Total	13299	11916	47,25

Source: ISTAT, 1991

Tab. 8 Per Capita Income in the communes of the Sirente-Velino Park.

Commune	Per Capita Income
Acciano	15.429
Aielli	16.629
Castel di Ieri	15.643
Castelvecchio Subequo	16.561
Celano	18.313
Cerchio	16.338
Collarmele	16.408
Fagnano Alto	15.386
Fontecchio	15.170
Gagliano Aterno	15.456
Goriano Sicoli	16.144
Magliano de' Marsi	17.571
Massa d'Albe	16.967
Molina Aterno	16.317
Ocre	16.971
Ovindoli	16.491
Pescina	17.528
Rocca di Cambio	16.247
Rocca di Mezzo	17.097
Secinaro	14.992
Tione degli Abruzzi	15.370
Average Value	16.335

Source: CRESA

2. BIBLIOGRAPHY

CARDINALE B. e FUSCHI M., *Environmental Protection in Abruzzo*, in BESANA A. (ed.), *Urban and Regional development in Italy and in Poland*, Atti VIII Seminario italo-polacco di geografia, Trento 22-26 settembre 1997, Trento, Ed. Colibrì, 1998, pp. 207-221.

CRESA, *Rapporto sulla economia abruzzese*, various years.

DI DONATO F., *La montagna abruzzese: dalla tradizione alla innovazione*, in BERNARDI R., SALGARO S. e SMIRAGLIA C. (eds.), *L'evoluzione della montagna italiana fra tradizione e modernità*, Bologna, Pàtron, 1994, pp. 321 -329.

DI DONATO F., *Milieu naturel et evolution de la presence humaine dans les montagnes des Abruzzes*, in "Atti VII Seminario Italo-Polacco, Wierzba 26 sett.-3 ott. 1993",Varsavia, PAN IGiPZ, 1995, pp. 109-119 .

FONDI M., *Abruzzo e Molise*, Torino, UTET, 1970.

FUSCHI M., *Il carsismo in Abruzzo: fra determinismo e sviluppo*, in BERNARDI R., SALGARO S. e SMIRAGLIA C. (eds.), *L'evoluzione della montagna italiana fra tradizione e modernità*, Bologna, Pàtron, 1994, pp. 331-341.

FUSCHI M., *Proposta di classificazione delle aree intermedie* (della montagna abruzzese), in BERNARDI R. (ed.), *La montagna appenninica italiana: conoscere per gestire*, Bologna, Pàtron, 1999, in printing.

GIACOMINI V. e ROMANI V., *Uomini e parchi*, Milano, Angeli, 1992.

GIANNINI M.S., *<<Ambiente>>: saggio sui diversi suoi aspetti giuridici*, in "Riv. Trim. Dir. Pubbl.", Milano, 1973, pp. 15-53.

LANDINI P., *Abruzzo. Un modello di sviluppo regionale*, in "Boll. Soc.Geog.Ital.", Roma, 1997, pp.3-15.

LANDINI P., *Il ruolo della montagna nel modello di sviluppo regionale*, in BERNARDI R. (ed.),
La montagna appenninica italiana: conoscere per gestire, Bologna, Pàtron, 1999, in printing.

MASSIMI G. (ed.), *Temi e problemi del territorio abruzzese*, Sambuceto, Arti Grafiche Galvan, 1990.

MASSIMI G., *Caratteri dominanti nelle aziende agricole dei comuni abruzzesi: orientamenti per un riordino del settore*, in "Boll. Soc. Geogr. Ital.", Roma, 1994, pp. 165-176.

PAOLINI E., *Il progetto Abruzzo Regione Verde d'Europa*, Penne, Ed. Cogecstre, 1993.

REGIONE ABRUZZO e COMUNITA' MONTANA SIRENTINA, *Piano del Parco naturale del Sirente-Velino*, L'Aquila, Gruppo di pianificazione territoriale, 1978.

SALVATORI F. e LANDINI P. (eds.), *Abruzzo. Economia e territorio nel Nord del Mezzogiorno,* Pescara, Libreria dell'Università Editrice, 1993.

VALLEGA A., *Geopolitica e sviluppo sostenibile. Il sistema mondo del secolo XXI*, Milano, Mursia, 1994.

Protected Areas Management: an Example of Application in the Gran Sasso Park

LORETTA GRATANI[1], MARIA FIORE CRESCENTE[1], ALESSANDRA ROSSI[2], ANNA RITA FRATTAROLI[2]
[1] *Dipartimento di Biologia Vegetale, Università degli Studi di Roma "La Sapienza", P.le A. Moro 5, 00185 Rome, Italy*
[2] *Dipartimento di Scienze Ambientali, Università degli Studi dell'Aquila, Via Vetoio, Coppito, 67100 L'Aquila*

Key words: Campo Imperatore, LAI, Plant structure, Plant biomass

Abstract: This study refers to structural characteristics and plant biomass of high mountain grassland at Campo Imperatore (Gran Sasso d'Italia). Plant biomass was measured by total harvesting in June (presampling in 1990 and sampling in 1991), corresponding to the period of peak standing crop. Spatial distribution from 1440 to 2000 m a.s.l. was studied

1. INTRODUCTION

Stand productivity is controlled by the interactions of involved species, in which both type and size play an important role, since they both change with climate and environmental conditions.

Changes in plant and stand processes are mediated by the local state of disturbance and one would expect that variation in structure could have the effect of altering processes in terrestrial ecosystems [1] [2] and consequently in plant productivity. Climate induced increases in disturbance could, in turn, significantly alter total biomass and compositional response of forests.

New sensitivity tests carried out on vegetation confirm that climate induced increases in disturbance could significantly alter total biomass [3]. An increase in disturbance frequency is likely to increase the rate at which natural vegetation reacts to future warming [3].

G. Visconti et al. (eds.), Global Change and Protected Areas, 489–493.

Tab. 1. Distribution of vegetational types in 11 biomass classes (g m^{-2})

< 70	*HELIANTHEMO CA.NI-PLANTAGINETUM HOLOSTEI*
141 - 210	*GALIO MAGELLENSIS-FESTUCETUM DIMORPHAE*
281 – 350	*SESLERIETUM APENNINAE POLIGALO MAJORIS-SESLERIETUM NITIDAE*
351 – 420	*POO ALPINAE-FESTUCETUM CIRCUMMEDITERRANEAE*
491 - 560	*CARICI HUMILIS-SESLERIETUM APENNINAE TARAXACO APENNNINI TRIFOLIETUM THALII KOELERIO SPLENDENTIS-BROMETUM ERECTI*
561 – 630	*LUZULO ITALICAE-FESTUCETUM MICROPHYLLAE CARICETOSUM KITAIBELIANAE*
771 – 840	*CIRSIO ACAULIS-SESLERIETUM NIYTDAE*
841 - 910	*POO ALPINAE-FESIUCETUM CIRCUMMEDITERRANEAE POETOSUM VIOLACEAE*
1541-1610	*LUZULO ITALICAE-FESTUCETUM MICROPHYLLAE*
3291- 3360	*DAPHNO OLEOIDIS-JUNIPERETUM ALPINAE ARCTOSTAPHYLETOSUM UVAE-URSI*
141 - 350	DOLINA PLANT COMMUNITIES *CARICI HUMILIS-SESLERIETUM APENNINAE KOELERIO SPLENDENTIS-BROMETUM ERECTI POO VIOLACEAE-NARDETUM STRICTAE FESTUCETOSUM CIRCUMEDITERRANEAE*

Monitoring communities is also important in order to establish the ecosystem tolerance threshold to anthropic perturbation, natural events and climatic change.

2. STUDY AREA

The study area is the mountain grassland at Campo Imperatore (Gran Sasso d'Italia) [4] [5]. The plateau is constituted by contiguous plains from 1430 to 2000 m a.s.l. [6] [7].

In the subalpine bioclimatic belt extending from 1800 to 2000 m a.s.l. the following plant typologies were considered:

- shrub structures (Daphno oleoidis-Juniperetum alpinae arctostaphyletosum uvae-ursi Blasi et allii 1990)
- xerophyte grassland with Sesleria apennina Ujhelyi (Seslerietum apenninae Migliaccio 1970, Em. Bonin 1978)

- mesophyte grassland with Festuca nigrescens Lam. ssp. microphylla (St.-Yves) Mgf.-Dbg. and Carex kitaibeliana Degen (Luzulo italicae-Festucetum microphyllae caricetosum kitaibelianae Biondi et allii 1995)
- mesoacidophilous grassland with Festuca nigrescens Lam. ssp. microphylla (St.-Yves) Mgf.-Dbg. e Luzula italica Parl. (Luzulo italicae-Festucetum microphyllae Biondi et allii 1995)
- mesoacidophilous grassland with Taraxacum apenninum (Ten.) Ten. and Trifolium thalii Vill. (Taraxaco apennini-Trifolietum thalii Biondi et allii 1992)

In the mountain bioclimatic belt, developing from 1440 to 1800 m a.s.l., the following typologies were considered:

- xerophyte grassland with Koeleria splendens Presl. and Bromus erectus Hudson (Koelerio splendentis-Brometum erecti Biondi et allii 1992)
- mesophyte grassland with Poa alpina L. and Festuca circummediterranea Patzke (Poo alpinae-Festucetum circummediterraneae Biondi et allii 1992)
- xerophyte grassland with Carex humilis Leyser and Sesleria apennina Ujhelyi (Carici humilis-Seslerietum apenninae Biondi et allii 1988)
- mesophyte grassland with Cirsium acaule (L.) Scop. and Sesleria nitida Ten. (Cirsio acaulis-Seslerietum nitidae Biondi et allii 1992)
- mesophyte grassland with Festuca circummediterranea Patzke and Poa violacea Bellardi (Poo alpinae-Festucetum circummediterraneae poetosum violaceae Biondi et allii 1992)
- xerophyte grassland with Polygala major Yacq. and Sesleria nitida Ten. (Polygalo majoris-Seslerietum nitidae Biondi et allii 1995).
- In the alluvial fans Helianthemo cani-Plantaginetum holostei Biondi et allii 1992 (sin. Plantago holostei-Helianthemetum cani) was analysed.

In the unconsolided gravel *Galio magellensis-Festucetum dimorphae* Feoli e Chiapella 1983 was analysed.

Doline plant communities were analysed from the base to the top.

3. MATERIAL AND METHODS

Plant biomass and plant structure measurements included: plant height, plant diameter, plant fresh weight, plant dry weight, stand density and leaf area index (LAI).

LAI was measured by the "LAI 2000 Plant Canopy Analyser" LICOR Inc., Nebrasca, USA, according to [8]. LAI 2000 Plant Canopy Analyser provides an accuracy within 15% of directly determined LAI [9].

Plant biomass, defined as dry weight per unit of soil, was measured by the direct method [10].

4. RESULTS

Green shoot biomass ranged from 44.0 to 3314.3 g m^{-2} and *Graminaceae* were the most important component of plant biomass. The root/shoot ratio, always above 1, might result from a higher proportion of the photosynthate to roots.

The highest plant biomass values of the association of the mountain belt were attributable to the dead stand components, which made up 28% of total biomass. The highest plant biomass values of the subalpine belt were due to the woody structure of *Juniperus communis* ssp. *nana* and *Arctostaphylos uva-ursi*. Mosses and lichens were directly related to side exposure and soil evolution.

The biomass data were used to draw the "Plant Biomass Map of Campo Imperatore (Gran Sasso d'Italia)" (Fig. 1). 11 biomass classes were defined; they delimit: 1) high biomass (from 1541 to 3360 g m^{-2}) corresponding to steeply soils and favourable exposition; 2) mean biomass (from 491 to 910 g m^{-2}); 3) low biomass (from <70 to 420 g m^{-2}) corresponding to very inclined slopes or Nord exposition sites.

5. DISCUSSION

Plant biomass and structure result from interactions of several variables acting on vegetation at different levels. Geomorphology and soil depth also have an import role. By the elevation, microhabitats differences between northern and southern slopes are of obvious importance for plant growth.

The reduction of plant height decreases LAI.

Plant aboveground biomass, height and LAI are significantly correlated and the relationship between vegetation structure and microtopography provides a synthetic image of the mosaic phenomena occurring within these grassland systems.

The regression analysis of LAI versus biomass and of LAI versus height are highly positive ($y = 0.001x + 0.887$, $r = 0.84$ and $y = 0.117x + 0.223$, $r = 0.86$ respectively). They emphasises the interaction among the variables.

The "Plant Biomass Map of Campo Imperatore (Gran Sasso d'Italia)" shows the distribution of quantitative data on the territory. Different associations are included in the same biomass class, by the importance of topography, soil depth and soil evolution. The long term monitoring is be easily achieved by LAI measurements which may be converted into biomass values by their relationship. This map allow us to follow changes in biomass from year to year and to define livestock cattle capacity for vegetation, which are prerequisite data for any range management project.

6. REFERENCES

Biondi E., F. Taffetani, S. Ballelli, M. Allegrezza, A.R. Frattaroli, R. Calandra, Carta Fitoecologica del Paesaggio di Campo Imperatore (Gran Sasso d'Italia) (1995) S.E.L.C.A., Firenze.

Biondi E., M. Allegrezza, S. Ballelli, R. Calandra, M.F. Crescente, A.R. Frattaroli, L. Gratani, A. Rossi, Bollettino dell'A.I.C. 86 (1992) 85-98.

Gratani L., M.F. Crescente, A.Rossi, A.R. Frattaroli, Coll. Phytosoc. 21 (1993) 579-587.

Gratani L., M.F. Crescente, A.Rossi, A.R. Frattaroli, La Carta della Biomassa Vegetale dei Pascoli di Campo Imperatore (Gran Sasso d'Italia) (1994) Stampa Borgia, Roma.

Milner C. and R.E. Hughes, (Eds.), Methods for the Measurements of Primary Productivity of Grassland, IBP Handbook No 6, Blackwell Sci., 1968.

Overpeck J.T., D. Rind, R. Goldberg, Nature 343 (1990) 51-53.

Shao G., H.H. Shugart, T.M. Smith, Vegetatio 121 (1995) 135-146.

Shugart H.H., G. Shao, W.R. Emanuel, T.M. Smith, In: B. Huntley, W. Cramer, A.V. Morgan, H.C. Prentice, J.R.M. Allen (Eds.), Past and Future Rapid Environmental Changes: the Spatial and Evolutionary Responses of Terrestrial Biota, Springer-Verlag, Berlin Heidelberg, pp. 265-272.

Welles J.M., J.M. Norman, Agron. J. 83 (1991) 818-825.

Welles J.M., S. Cohen, J. Exp. Bot. 47 (1996) 1335-1342.

The Main Invasive Alien Plants in the Protected Areas in Central Italy (Abruzzo)

Invasive alien organisms

LORETTA PACE *& FERNANDO TAMMARO**
** Dipartimento di Scienze Ambientali -Università degli Studi dell'Aquila- Via Vetoio Loc.Coppito 67010 L'AQUILA, ITALY (tel.0862/433247, fax 0862/433205)*
*** Sistematic Botany, Educational Sciences Faculty –Via Verdi, University L'AQUILA, ITALY.*

Key words: Invasive alien plants of Abruzzo, Italy, *Senecio inaequidens*.

Abstract: The main invasive alien plants in the protected areas of the Abruzzo region have been recently pointed out. The areas taken into account have been the ones of the National Park of Abruzzo, National Park of Gran Sasso and Monti della Laga, National Park of Maiella, and the Regional Park of Sirente-Velino. The main plants have been found in the basal areas (400-1000 mt.), in uncultivated fields and urban surroundings (road and path margins, dumps, etc.). The most wide spread are the following: *Asteraceae*: *Aster squamatus* (Sprengel) Hieron.(Neotropical), *Conyza* Less. (*C. bonariensis* (L.) Cronq., *C. canadensis* (L.) Cronq.), *Helianthus tuberosus* L. (North America), *Xanthium spinosum* L. (South America), *Artemisia verlotiorum* Lamotte (East Asia), *Senecio inaequidens* DC. (South Africa); *Amarantaceae* (*Amaranthus chlorostachys* Willd., *A. retroflexus* L., *A. deflexus* L.). The most invasive plant is *Senecio inaequidens* DC. (*Asteraceae*) which can be found in different areas, from the uncultivated calcium-rich fields to dumps, in the dry pastures and degraded oak-woods. A very common plant is also *Isatis tinctoria* L. (*Cruciferae*), an invasive plant coming from the South-East of Asia. It is very widespread in the ex-cultivated fields as it was introduced as a dyeing plant in the XVI century. *Robinia pseudoacacia* L. (*Leguminosae*) (native from Carolina and Virginia) and *Ailanthus altissima* (Miller) Swingle, *Simaroubaceae* (originated of China and Molucche Isles) are the most common trees. Phytosociological studies have been carried out on *Senecio inaequidens* , *Isatis tinctoria, Artemisia verlotiorum* and *Amaranthus* species with the purpose of defining their quantitative and qualitative presence both in urban surroundings and in natural environments. In the protected areas taken into examination the presence of invasive alien plants is of low entity. In the

G. Visconti et al. (eds.), Global Change and Protected Areas, 495–504.

last decades *Senecio inaequidens* has represented a potential danger since it spreads very quickly and replaces the native flora. Phenological studies of *Senecio inaequidens* have shown a long period of flowering and seeds production.

1. INTRODUCTION

The main alien plants of tropical-subtropical origin or coming from hot Mediterranean countries have been individualized and reported. The place of origin, ecological characteristics and the territories where the plants have been detected (L'Aquila=AQ, Teramo=TE, Chieti=CH, Pescara=PE) are listed:

Amaranthus caudatus L., native of paleotropical countries, cultivated and rarely sub-spontaneus (AQ);

Amaranthus chlorostachys Willd., native of tropical and sub-tropical America, ruderal plant (AQ,PE,TE);

Amaranthus cruentus L., native of tropical America, ruderal plant (TE);

Amaranthus deflexus L., native of South and Eastern America, ruderal plants (AQ,CH,PE,TE);

Ambrosia coronopifolia Torrey et Gray, native of Nord-America, ruderal plant (PE,TE);

Artemisia verlotiorum Lamotte, native of West-Asia, found in cool and shaded soil (AQ,PE,TE);

Aster squamatus (Sprengel) Hieron, native of Central and Southern America found in uncultivated and ruderal fields (CH,PE,TE);

Mesembryanthemum edule L., native of Northern Africa, found on sea shores (PE);

Cenchrus incertus Curtis, native of tropical America, found on shores (PE,TE);

Chenopodium ambrosioides L., native of Tropical America, found on ruderal fields (PE,TE);

Conyza albida Willd., native of Southern America, found in pebbly river-bed and uncultivated lands (CH,PE,TE);

Conyza bonariensis L. Cronq., native of Central America, found in uncultivated lands(AQ,CH,TE,PE);

Crypsis aculeata (L.) Aiton, native of Paleo sub-tropical countries, found on wet shores (PE,TE);

Datura innoxia Miller, native of Central America, found in uncultivated fields (AQ,PE,TE);

Datura stramonium L., native of America, found in uncultivated fields (AQ,PE,TE);

Table 1: List of species with most frequencies in the detected areas.

Biolog. Form	Chorol. Type	**Vegetation of *Senecio inaequidens* communities**	1	2	3	4	5	6	7
		Relev. Area(mq)	40	40	40	50	40	50	60
		Altitude(m)	610	640	650	700	1000	700	750
		Slope(°)	5	4	.-	.-	.-	.-	5
		Esposure	SE	SE	.-	.-	.-	.-	S
		Covering(%)	80	100	100	50	100	100	50
		Relev. Date (1999)	May 13	May 13	May 20	May 20	Jun 10	July 20	July 24
T scap	Sud-Afric	Senecio inaequidens DC.	2	2	2	2	2	1	1
T scap	Euri-MeditTuran	Avena barbata Potter	2	2	1	.	.+	.	.+
H scap	Centro Europ-Pont	Anthemis tinctoria L.	1	1	1	1	2	.	.
H scap	Circumpol	Artemisia vulgaris L.	2	2	.	.+		1	.
H scap	Paleotemp	Dactylis glomerata L.	.	3	1	.	1	.	.+
H scap	Subcosmopol	Artemisia absinthium L.	1	1	.	.+	3	.	.
H scap	Subcosmopol	Rumex crispus L.	.+	3	.	.+		.	.
H scap	Euri-MeditTuran	Dasypyrum villosum	.	1	.	1	1	1	.
T scap	Cosmopol	Convolvulus arvensis L.	1	.	.		.+	.+	.+
H bien	SE-As(steppica)	Isatis tinctoria L.	3	.+	.	2	.+	.	.
H bien	Subcosmopol	Malva sylvestris L.	.+	.	.	.+		.	.+
H rhiz	Circumbor	Agropyrum repens (L.) Beauv.	.	.	.	.	2	.+	1
T scap	Subcosmopol	Chenopodium album L.	.	.	.	.	.	1	2
H scap	S-Medit	Foeniculum vulgare Miller	.+	.	.	.	.	.	.+
T er	Subcosmopol	Fumaria officinalis L.	.	.	.	.	.	.+	.+
T scap	Eurasiat	Galium aparine L.	2	.	.	.	.	.	.+
T scap	Paleotemp	Mercurialis annua L.	.	.	.	.	.	.+	.+
T er	Subcosmopol	Echinochloa crus-galli(L.)Beauv.	.	.	.	.	.	1	.+
T scap	E-Medit(Archeofita)	Papaver rhoeas L.	1	.	.	.	.	.	.+
H scap	Circumbor	Rumex acetosa L. subsp. acetosa	.	.	.	.	.	1	.+
T scap	Cosmopol	Senecio vulgaris L.	.	.	.	.	.	.+	.+
T scap	Subcosmopol	Setaria viridis (L.)Beauvais	.	.	.	.	.	3	1
T scap	Subcosmopol	Sonchus oleraceus L.	.	.	.	.	.	.+	1
T scap	Nordameric	Amaranthus retroflexus L.	.	.	.	.	.	4	1
T scap	Euri Medit-Turan	Bromus sterilis L.	.	.	.	1	.	.	.+
H scap	Europ	Echium vulgare L.	.	.	.	.+	.1	.	.+

T scap	Euri-Medit-S-Siber	Lactuca serriola L.	.	.	.	.+	.	1	.
T scap	Paleotemp	Medicago lupulina L.	.	1	.	.+	1	.	.
T scap	Eurimedit	Rapistrum rugosum(L.)All.	1	.	.	.+	.	.	.
		subsp. linneanum Rouy et Fouc.					.		
T scap	Steno-Medit	Reichardia picroides(L.)Roth	.	.	.	.+	.	.+	.
T scap	Europ	Reseda lutea L.	.	.	.	.+	.	.	.+
T scap	Cosmopol	Stellaria media(L.)Vill.subsp. media	1	.	.	.+	.	.	.
H scap	Paleotemp	Tussilago farfara L.	.	1	.	1	.	.	.
T scap	Medit-Turan	Alyssum campestre(L)L. subp. campestre	1	.	.	.	.	.	.
T scap	Sudameric	Amaranthus deflexus L.	.	.	.	.	.	1	.
T scap	Eurimedit	Anthemis cotula L.	.	.	.	.	.	.+	.
H bien	Eurasiat temp	Arctium lappa L.	.	2	.	.	.	.	.
H scap	Paleotemp	Arrhenatherum elatius(L.) Beauvais	.	4	.	.	.	.	.
		subsp. elatius					.		
T scap	Subcosmopol	Bromus hordeaceus L.	2	.	.	.	.	.	.
H scap	Medit-Turan	Cardaria draba(L.)Desv.	.+	.	.	.	.	.	.
H bien	Euri-Medit-Turan	Carduus pycnocephalus L.	1	.	.	.	.	.	.
H scap	Eurasiat	Chelidonium maius L.	.	.	.	.	.	.	.+
T scap	Euri-Medit	Chenopodium vulvaria L.	.	.	.	.	.	.	.+
T scap	Subcosmopol	Cirsium arvense (L.) Scop.	.	.	.	.	.	.+	.
H bien	Subcosmopol	Conium maculatum L.	.	.	.	.	.	.+	.
T scap	Americ. Sett.	Conyza canadensis(L.)Cronquist	.	.	.	.+	.	.+	.
T scap	Euri-Medit	Crepis sancta(L.)Babc.	.+	.	.	.	.	.	.
H rept	Termo-Subcosmop	Cynodon dactylon(L.) Pers.	.	.	.	.	.	.	.+
T scap	Subcosmopol	Erodium cicutarium(L.) L'Her.	1	.	.	.	.	.	.
T scap	Euri-Medit-Turan	Heliotropium europaeum L.	.	.	.	.	.	.	1
T scap	Circumbor	Hordeum murinum L.	3	.	.	.	.	.	.
T scap	Euriasiat	Hyoscyamus niger L.	.	.	.	.	.	.	.+
T scap	Euri-Medit	Legousia speculum-veneris(L.)Chaix	.+	.	.	.	.	.	.
H rept	Cosmopol	Polygonum aviculare L. subsp. aviculare	.	.	.	.	.	.	1
H scap	Circumbor	Fallopia convolvulus(L.)A:Love	.	.	.	.	.	.	.+
T er	Eurimedit	Rubus ulmifolius Schott	.	.	1	.	.	.	.
H scap	Paleotemp	Silene latifolia Poiret	.	1	.	.	.	.	.
		subsp.alba(Miller)Greuter et Burdet	.	.	.	1	.	.	.
T scap	Euri-Medit	Sinapis alba subsp.alba	.+	.	.	.	.	.	.
H bien	Euri-Medit	Verbascum sinuatum L.	.	.	.	.	.	.	.+
H bien	Cosmopol	Verbena officinalis L.	.	.	.	.	.	.	.+

T scap	Subcosmopol	Veronica arvensis L.	.+	.	.	.	.	.	.
T scap	Sud-Americ	Xanthium spinosum L.	.	.	.	.	.	.	.+
T er	Cosmop sinan	Solanum nigrum L. subsp. nigrum	.	.	.		.	.	.+
T scap	Subcosmopol	Daucus carota L.		.	.	.+	.1	.	.
T scap	Subcosmopol	Cichorium inthybus L.	.	.	.	1	.	.	.
T scap	Submed-Subatl	Diplotaxis tenuifolia(L.)DC.				.	.1		
T scap	Eurasiat	Linaria vulgaris L.	.	.	.	.	1	.	.
T er	Europ-Caucas	Lepidium campestre(L.)R.Br.	.	.	.	.	1	.	.
T er	Eurosib	Picris hieracioides L.					.+		
H scap	W-Medit	Plantago cynops L.				.+	.		
H scap	E-Med S-Siber	Chondrilla juncea L.					.+		

Rilev. 1,2,3,	Uncultivated field near L'Aquila (Coppito)	
Rilev. 4	Dumps near S.Pio delle	Camere (Aq)
Rilev. 5	Ruderal near Calascio (Aq)	
Rilev. 6:	Uncultivated field near L'Aquila (S.Elia)	
Rilev.7:	Left over material near S. Pio delle Camere (Aq)	

Eleusine indica (L.) Gaertner, native of tropical and subtropical countries, found on roads and trampled places (CH);

Euphorbia prostrata Aiton, native of America and Africa, found on ruderal fields (TE);

Hibiscus trionum L., aboriginal of Paleo tropical and Paleo sub tropical countries, found on moist grounds (AQ);

Isatis tinctoria L., native of South-East Asia, found on cultivated and uncultivated fields (AQ,PE,TE);

Lepidium sativum L., native of East Africa, found in uncultivated fields (PE);

Myagrum perfoliatum L., native of South-East Asia, found in uncultivated moist places (AQ,PE,TE);

Oxalis articulata Savigny, native of Central America, found in uncultivated places, in and roads (TE);

*Paspalum paspalodes (*Michx.) Scribner, neo-tropical, which become sub-cosmopolite, found on riversides and in moist places (CH,PE,TE);

Phalaris canariensis L., native of North West Africa and Canary Islands, found in marginal roads and fields (AQ,CH,PE,TE);

Senecio inaequidens DC., native of South Africa, found in ruderal fields (AQ,PE);

Setaria italica (L.) Beauv., possibly native of tropical Asia, found in ruderal fields (AQ,CH,TE);

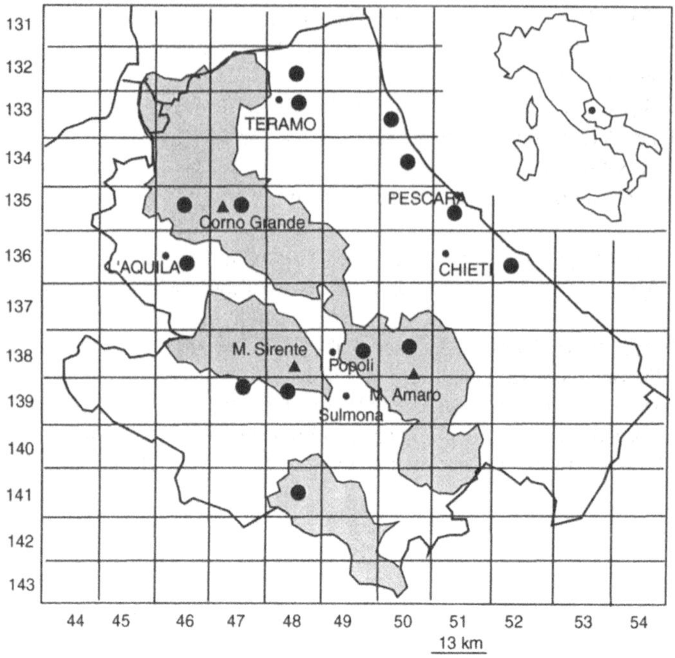

Fig.1- Distribution map (Gauss-Boaga system of Middle-European cartography) of *Artemisia verlotiorum* Lamotte, native of West Asia, found in cool and shaded soil. Map symbol **O** represents the geographic areas where the plants are found. This data is calculated from bibliographic research and personal observations.(shaded = Protected areas of Abruzzi).

Sorghum halepense (L.) Pers., native of North Africa and South West Asia, found in cultivated and grass grown fields (AQ,CH,PE,TE);

Xanthium italicum Moretti, native of South and North America, found in uncultivated fields and riversides (AQ,PE);

Xanthium spinosum L., native of South America, found on ruderal fields (AQ,CH,PE,TE);

2. OBSERVATIONS ON THE ECOLOGY OF THE SENECIO INAEQUIDENS

In order to know the qualitative and quantitative composition of the community where the alien plant *Senecio inaequidens* predominates, phytosociological relevées have been carried out.

Other relevées are effectuated also where the following species are noted: *Artemisia verlotiorum, Aster squamatus,Datura stramonium, Conyza bonariensis, Cenchrus incertus, Xanthium italicum.*

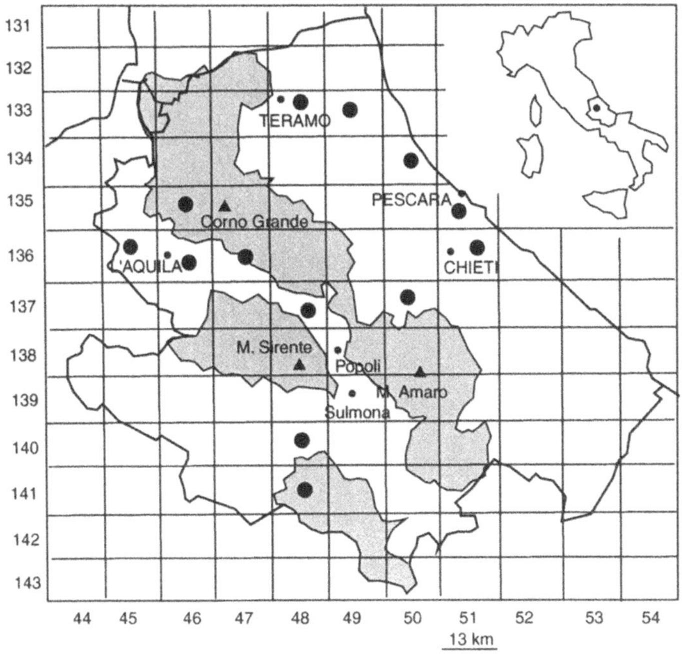

Fig.2 - Distribution map (Gauss-Boaga system of Middle-European cartografy) of *Senecio inaequidens* DC, native of South Africa, found in ruderal fields. Map symbol **O** represents the geographic areas where the plants are found. This data is calculated from bibliographic research and personal observations.(shaded = Protected areas of Abruzzi).

The main alien plants of Central Italy are found in areas naturalistically degraded; they penetrate particularly in the nitrophilo ruderal vegetations of the shores, of the road edges, of the uncultivated fields and also in the places where left over materials are collected.

Examining the phytosociological relevées in *Senecio inaequidens* phytocoenosis (Tab.1) , we can observe an high degree of annual plants (Therophytic species). They reach up to 50% of the total presence. We can also observe that the Mediterranean species predominates.

3. SENECIO INAEQUIDENS PHENOLOGY

Senecio inaequidens, shows, in the surroundings of L'Aquila (721 m. a.s.l.), a long period of flowering, from about April to November. In the same period it produces a great number of capitula and achenes .

Fig.3 - *Senecio inaequidens*, particular of the capitula. This plant develops itself due to the special characteristics of flowering and for the great number of capitula and achenes that it produces during the 8 months of the year (foto C.Visca).

In fact, it has been verified that each plant produces generally nine capitola (Fig.2) for each stem of flower, and each capitulum produces about 70-80 achenes.

The high production of seeds by a single plant is of potential danger to colonize many naturalistic degraded areas, especially where the presence of left over material, uncultivated fields, marginal roads etc., is wide spread. Moreover, this plant produced hairy achenes that are dissipated easily by the wind. After germination this plant produces a very strong rhizome which gives a long life to the plants. Furthermore, it has the possibility of a vegetative reproduction. Similar characteristics are found in the *Artemisia verlotiorum,* a native plant from West Asia. It produces dense communities of herbaceous plants with a wood basal stem, they have strong rhizomes. It is found near the river-beds. This flowering period is up to seven months (to May from October).

4. CONCLUSIONS

This study of alien plants in the protected areas of the Abruzzo region, native from the territories with tropical or subtropical climate, has showed

Fig.4 - *Senecio inaequidens* in urban surroundings, Coppito L'Aquila. The large quantity of seeds, that this plant produced, is able to colonize degraded areas where the presence of left over material, uncultivated fields, marginal roads, etc., is abundant (foto C.Visca).

that the presence and the diffusion of these plants depend mainly by environmental degeneration and not due to climatic change.

However, the diffusion of alien plants could be explained by their biological characteristics (very long phenological period, long rhizomes, great number of seeds), which are not present in the largest part of local flora. Their largest diffusion could be due to the latest climatic characteristics of the Abruzzo region.

5. REFERENCES

Articles in journals:

Anzalone B. - Il "Senecio inaequidens" DC. in Italia.Giorn.Bot.Ital., 110: 437-438 (1975), 1976.

Chapters in books:

Pirone G.- La vegetazione alofila della costa abruzzese (Adriatico Centrale). Fitosociologia, 30: 233-256, 1995.

Pirone G., Ferretti C. - Flora e vegetazione spontanee della città di Pescara. Fitosociologia, 36, 1999.

Viegi L., Cela Renzoni G., D'Eugenio M.L., Rizzo A.M.- Flora esotica d'Italia: le specie presenti in Abruzzo e Molise (revisione bibliografica e d'erbario). Archivio Botanico Italiano 66 1/2, 1990.

Books:
Corbetta e coll. - Flora e vegetazione, Studi preliminari al piano del Parco Regionale Sirente-Velino, 1999
Giglio E. - Ecologia del paesaggio nella Conca Aquilana - Tesi di dottorato di ricerca in Scienze Ambientali "ambiente e uomo in Appennino" VIII ciclo, 1995.
Pirola A.- Elementi di Fitosociologia - Coop.libr.Univ., Bologna, 1970.
Pignatti S. - Flora d'Italia, vol 1/2/3, Edagricole, 1986.
Pirone G., Frattaroli A.R., Corbetta F.- Vegetazione, cartografia vegetazionale e lineamenti floristici della Riserva Naturale "Sorgenti del Pescara" (Abruzzo- Italia), Comune di Popoli, Centro Stampa, Roma, 1997.
Tammaro F.- Il paesaggio vegetale dell'Abruzzo. Cogecstre Edizioni, 1998.

The Historical and Iconographic Research in the Reconstruction of the Variation of the Calderone Glacier: State of the Art and Perspective

MASSIMO PECCI
ISPESL - DIPIA (Higher Institute for Occupational Safety and Health - Department of Production Plants and Interaction with the Environment), Via Urbana, 167 - 00184 Rome ITALY- Tel + 39-6-4714261; Fax + 39-6-4744017, Italian Glaciological Committee

Key words: Glaciology, Variations of the Calderone glacier, Historical and iconographic research in glacial geomorphology, Little Ice Age.

Abstract: The Calderone glacier is characterised by a reduction phase since the end of the "Little Ice Age" (LIA) Auct., particularly strong during the last decade. In fact during the nineties a set of multidisciplinary researches started to evaluate the role of the Glacier like an indicator of the effects of human activities and finally of regional and global climatic change. The apparatus is now confined into a deep mountain valley of the Gran Sasso d'Italia Range, with steep walls, and does not show movements along the borders and along the front. The aim of the historical research is not only the collection of the available images and literature, but mainly the evaluation and the evidence of the past dimension of the studied glacier in length, surface area, thickness and volume. In this paper the state of the art is presented, giving information about sources, consistence and importance of collected data. The documentation collected is proposed in the perspective of giving a contribution to the confirmation or to the discussion of the performed reconstructions.

1. INTRODUCTION

Since the ancient Roman time the so called *Fiscellus mons* (the Gran Sasso at present) has been well recognisable, at least for the majesty, and perhaps Pontano in 1513 [1] in the *Meteorum liber* describes in the chapter *De fontibus et fluminibus* an "*horrida cautes indigena vocitant Cornu*" (the

G. Visconti et al. (eds.), Global Change and Protected Areas, 505–512.

Corno Grande peak at present) that few years later Mercatore in 1598 [2] still calls *Fiscellus mons* in the map "*Abruzzo et Terra di Lavoro*". But only De Marchi, an infantry captain, in 1573 [3] explored "scientifically" the range, reached the top and accurately described the ascent, included the glacier, seen from the top but not walked. During the Enlightenment both the figures of naturalist and scientist coincided with that one of the mountaineer; in this period a systematic exploration of the Gran Sasso coupled with an increasing knowledge started by Orazio Delfico: he climbed the top in July 1794, going up from the northern slope for the first time, passing through and fully describing the Calderone glacier [4].The last century transmitted a reach and interesting production of scientific and technical description (many unpublished or unknown) and valuable pictures and paintings and, finally, the first detailed cartography (scale 1:20.000) and photographs of the top, including the glacier. Since the beginning, the present century shows a less interest for archives research, increasing, on the opposite, for georeferencing in GIS environment, due to the availability of many topographic maps and teledetected photographs and images.In this perspective a synthetic state of the art of the archives research (more than five years long and still in progress) is presented, passing through the most significant steps, mostly oriented for the scientific discussion or the formulation of new hypotheses on the evolution of the Calderone glacier and related changes, since the end of the LIA (Little Ice Age). The main data sources have been: the Provincial Libraries of Teramo city "M. Delfico" and of L'Aquila city; the State Archives of L'Aquila city, the libraries of the Italian Alpine Club (Sections of Rome and L'Aquila) and of local environmental Associations.

2. HISTORICAL DESCRIPTIONS

The following passages, translated from the original language, represent the four fundamental step, in the author's opinion, for a new knowledge of the Calderone glacier'e evolution, regarding:

- the discovery in 1573;
- the first description in 1794;
- the first full description of the whole glacial environment of the north slope in 1845;
- the description of a "surviving" glacial environment at the lower heights still in 1928.

1. The description of the discovery (in old Italian language and given from the top of the mountain) is by Francesco De Marchi [3], during the first ascent of the Gran Sasso Peak in August 1573, from the southern slope:

"*...Dico che non vi è Fontana nessuna, ma che vi è bene un gran vallone tra il Monte di Santo Nicola et il Corno Monte, dove sempre vi è la nieve alta quindeci o venti piedi e più in alcun luocho dove la nieve e ghiaccio sta perpetuamente. Et quest'è una quantità d'un grosso miglio di lunghezza e di larghezza più di mezo miglio della qual sempre puoco o assai se ne disfà...*" ("... I say that no Fountain is there, but a great valley between Santo Nicola Peak and Corno Peak, where there is always snow of about 15 or 20 foots of thickness, and anywhere snow and ice fill it perpetually. The valley dimensions are of about one Italian mile (about 1500 metres) of length and of about half an Italian mile of width").

2. The description of the first ascent to the Corno Grande Peak through the Calderone glacier in July1794 by Orazio Delfico [4] from the northern slope in which the following detailed description (in old Italian language) is given: "*... Così a stento, ed adagio andando avanti giunsi in un esteso ripiano, quasi intieramente circondato dal alte rocche, che ne formano come una maestosa conca. Per indicarne in qualche modo l'elevazione basterà il dire, ch'essa è continuamente coverta di neve non eguale in durezza al gelo, ma ben solida, e ferma per non ricevere alcuna impressione dalle più forti pedate dei contadini che mi accompagnavano....*" (....In this way going up slowly and laboriously I reached a wide terrace-plain almost completely surrounded by high peaks, forming a majestic circular depression. To define the elevation it is sufficient to say that it is always covered by snow, very hard and not equal but similar to ice and so solid to not allow the foot-print of the accompanying mountaineers").
3. The full description of the northern slope's environment is given by Raffaele Quartapelle in summer 1845 [5]; the following extracted two passages are particularly interesting in order to evaluate the more wide presence of the ice (first passage) and, above all, the presence (?) of an ice fall at the terminus of the glacier (second passage). In the lower part of the Valle delle Cornacchie (about 2200 m asl): "*....Sormontati sassi, e calpestati inclinati banchi di neve ridotti quasi alla consistenza di durissimo gelo, dell'altezza di più decine di piedi, si veggono scorrere al basso di essi ruscelli di acqua prodotti dalla loro liquefazione ...*" (....Climbing boulders and walking up inclined layers of snow, hard in consistence like strong ice and thick many tens of foot, many streams of melting water flow down.....). In the cirque of the Calderone glacier: "*... Giunto nella conca, si trova questa formata di un grande spazio quasi circolare di più moggia di estensione, coverta di duro banco di neve, sotto cui si vede correre un gran ruscello. Questo cerchio è adorno di tante cime; ma verso N. E. mancano, per essersi nei più remoti tempi giù dirupate, formando colle loro rovine il sopraddetto brecciaio (Valle delle*

Cornacchie, n. d. r.). Verso E.N.E. il ruscello si precipita, e prima di questo suo abbandonamento si vede scorrere in mezzo a massi di geli di cui uno gli forma una volta, e l'altro gli serve di sostegno. Nella distanza di sei palmi, prima di giù andare, la volta manca e si vede fluire sul saldissimo sottoposto gelo; quindi forma la sua cascata, si perde e non più si vede, se non all'estremità del bosco, ove riuscendo forma il principio del fiume Mavone..." (....I reached the basin, almost circular and wide many hundred of metres, covered by hard snow, with a big stream flowing below. The circular basin is surrounded by many peaks, excluding the N. E. side where the rocks, falling down in the past, generated the big debris cone ("brecciaio" – Valle delle Cornacchie,). Before falling down towards E.N.E., the stream flows between boulders of ice, forming a sort of tunnel. In six palms length , before falling down and lacking the vault, the stream flows on the strong ice below and then falls down, invisible as far as the wood below, generating the Mavone river....).

4. The following passage, extracted from a page of the newspaper "L'Italia Centrale", edition of the 3rd of November 1928, is the published report by Pasquale Fabbri of an autumn excursion to the Ice Cave at 1600 m asl in the N. E. face of the Corno Grande: "*...Della Grotta, o meglio, Ponte di ghiaccio formatosi sotto questa parete, l'arco è ora in gran parte spezzato, ma non tarderà a ricongiungersi ai primi freddi invernali...*". ("....The arch of the Cave, or better, the ice Bridge formed below this rock wall, is now partially broken, but will regenerate during the winter cool....").

3. PAINTINGS, IMAGES AND PHOTOGRAPHS

Starting from the first sketch of De Marchi [3] a rich production of images regarding the Gran Sasso has been collected year by year from several archives. With specific reference to the Calderone Glacier and its evolution in the last two centuries, the following images are particularly interesting:

- The Corno Grande from Isola del Gran Sasso, originally painted in colour by Edgar Lear the 30th September 1843 and reproduced in B/W in Fig. 1. In the painting the Gran Sasso shows at the end of the summer a look completely different from the present, with a development of processes related to snow and ice to the lower heights, with persistence of snow (and ice?) not only as far as the Arapietra edge (about 2000 m. asl), but also interesting the East face of the Corno Grande, as far as about the height of 1500 m asl. The period almost corresponds with

Fig. 1 – Isola, Monte Corno (Edward Lear, 1843), particular of a colour reproduction (Edward Lear "Il viaggio come avventura estetica", Finarte, Milano 1994) of the Corno grande East face in September 1843 with evident snow and ice (?) as far as the toe of the slope

the description of Quartapelle [5].

- Still interesting, even if contemporary of B/W photographs, is "Il Gran Sasso d'Italia", originally painted in colour by Vincenzo Alicandri in summer 1916, as recognisable by the green colour of grass. The painting well shows the wide snow covered landscape, still in summertime, even if reduced respect to Lear's time.
- The picture of a postcard [6], taken during the summer at the beginning of the century, well shows the look of the Glacier with a calculated reduction in thickness of about 40 m in the upper sector of the glacier [7].

4. CARTOGRAPHY

The most significant documents is the first 1:50.000 topographic map of the Corno Grande area (Projection Flaamsted modified, with the origin of the co-ordinates at the intersection between the Meridian passing through Napoli and the 40° Parallel contour lines 20 m), surveyed on the field by ITM (Military Topographic Institute) in 1884-1885 and mechanically enlarged at a scale 1:20.000 with a contour lines distance of 10 metres [8]. Using this map as the topographic base, in 1887 G. E. Frietzsche published the topographic map of the Gran Sasso d'Italia at a scale 1:80.000 [9], with the enlarged window of the top area at a scale 1:25.000 (reproduced in Fig. 2a). In the area reproduced is well recognisable the "Brecciaio" of Quartapelle [5], corresponding to the upper part of the Valle delle Cornacchie and the Calderone glacier clearly deviating eastward, at least interesting the stream of melted water. The detailed Topographic sketch of the Calderone Glacier (scale 1:5.000) surveyed the 4th of September 1916 by Marinelli e Ricci [10] is particularly interesting due to the indication of the exact height of the frontal moraine of 2712 m asl in the central sector and 2744 m asl in the eastern sector [10]. The values clearly indicate an higher and a wider moraine than the present one, but mainly the eastern values is higher about 40 m than the present and close to the present edge (less than 2750 m asl). The first 1:25.000 topographic map of the Corno Grande area [10] was surveyed from aereophotographs and controlled on the field by IGM in 1954-55, (geographic Co-ordinate referenced to the International Ellipsoid oriented to Rome-M.Mario U.T.M-Fuse 33, contour lines distance of 25 metres - Fig 2b).

5. DISCUSSION AND PERSPECTIVES

The activity of the Archives research collected many tens of documents and only the most significant has been presented in order to stimulate the scientific discussion and to focus the future researches.

The main focal points can be synthesised as following:

- the snow and ice covered environment was strongly wide diffused, reaching about the wood limit, since the beginning of the present century;
- the preferential direction of development (and movement?) of the glacier during the LIA was not Northward, but E. N. E. directed (towards the "Big East Face " of the Corno Grande), as described by Quartapelle [5] and cartographed in the official map of 1884 [8] and later [9], since the first dozen of years of the present century;
- at the beginning of the century [8] the width and mainly the height of the frontal moraine were sensibly superior than today, reaching almost the edge of the "Anticima della Vetta Orientale, 2750 m asl".

The elements collected induce to verify in the future the position and the maximum advance of the probable two tongues of the glacier: the main one

(a) (b)

Fig. 2 – Comparison at the same scale (1:25.000) between the morphology of the Calderone glacier area surveyed by Frietzsche [8] in 1887, after ITM, 1855 [8], (a) and (b) by IGM in 1955 [11]. In heavy lines the main treks of the older map are drawn and with thick arrows the discussed area is shown.

but less wide in the Vallone delle Cornacchie and the suspended one on the edge of the Anticima della Vetta Orientale, still active during the LIA.

In this perspective the field of research, performing the more appropriate methodologies, concerns:

- the detection of the trim lines and possibly of their age;
- the individuation of position and , possibly, age of suspended glacial deposits referable to LIA;
- the reconstruction of effective thickness, surface area and volume of the whole glacier during the maximum of LIA;
- the reconstruction of the climate of the time;
- the evolution, if possible and if existing, of the anthropogenic contribution to the climate warming, especially sensible in the central Mediterranean area and synchronous to the industrial development of the last century.

The Archives research seems to confirm a reduction phase of the Calderone Glacier, stronger than in Alpine environment, at least active starting from the second half of the last century.

Furthermore the glacial apparatus seems to have been complete of tongues until the beginning of the present century and an huge volume of ice has been melted, probably modifying the hydrogeological circulation and causing gravitational phenomena, both on rock and earth slopes with long distance and long term evolution.

The better will be the reconstruction of the geomorphologic evolution of the area, the greater will be the possibility in modelling the scenarios for the Calderone and for similar glaciers.

6. REFERENCES

Capaldi F. private postcard collection, L'Aquila.

D'Alessandro L., M. D'Orefice, M. Pecci, C. Smiraglia & R. Ventura, Proc. Climate Change and Protected Areas, (1999).

De Marchi F., Carte 7-14 Cod. Man. Magliabechi, cl. XVII, a. 3, Atti Bibl. Naz., II, I, 277-280, Firenze,1573.

Delfico O., Osservazioni di Orazio Delfico su di una piccola parte degli Appennini dirette a sua eccellenza il Signor Marchese D. F. Mazzocchi, Pres. del S. R. Cons., Bibl. Prov., L'Aquila, 1794.

Frietzsche G. E., Carta topografica del Gran Sasso d'Italia, Ist. Cart. Ital., Rolla, Roma, 1887.

Istituto Geografico Militare Italiano (IGMI), Tavoletta "Gran Sasso d'Italia", Firenze, 1955.

Istituto Topografico Militare, Carta del Gran Sasso d'Italia, Bibl. Ist. Geogr. Mil., Firenze,1885

Marinelli O. & L. Ricci, Rivista Geografica Italiana, **23**, 399-405, (1916), Roma.

Mercatore, Abruzzo et Terra di Lavoro, scala 1:750.000, 1598.

Pontano, Meteorum Liber - De fontibus et fluminibus, Aldo Manuzio (Ed.), Venezia, 1513.

Quartapelle R., Manuale pel viaggiatore naturalista al Gran Sasso d'Italia, Giuseppe Marsili Ed., Teramo, Bibl. Prov. Teramo (unpublished report),1849.

Numerical Experiments to Study the Possible Meteorological Changes Induced by the Presence of a Lake.

BARBARA TOMASSETTI, GUIDO VISCONTI, TIZIANA PAOLUCCI, ROSSELLA FERRETTI AND MARCO VERDECCHIA.
Department of Physics, University of L'Aquila Scientific and Technology Park of Abruzzo, Italy

Key words: Land use, regional climate

Abstract: The Lake Fucino was the largest reservoir of fresh water in the Abruzzo Region until it was drained at the end of last century. The surface of the lake was about 150 square km. Temperature and precipitation historical records show appreciable changes in these variable that could be related to the draining of the lake. Changes in the vegetation around the lake are also recorded especially concerning the existence of plantations of olive trees. The setting of the lake is peculiar being at the center of a closed valley. To assess the possible effect of the lake on the local climate and the meteorology regime we have carried out a simulation using the climate version of the MM5 limited area model. This model includes a code for the soil atmosphere interactions. Also the model has been updated for the land use/land cover at very high resolution (1km). The simulations are carried out at different resolution and use the nesting technique. Preliminary results seem to indicate an overestimation of the changes with respect those present in the historical records.

1. INTRODUCTION

The lake of Fucino was the largest reservoir of fresh water in the Abruzzo region until it was drained at the end of the last century (1873).The goal of this work is to study the possible local changes on both the climate and meteorological regime. A recent study by Schar et al.(1999), asses the sensitivity of the regional climate to the soil-moisture availability through a precipitation processes feedback; effects of the land-use on the weather were

G. Visconti et al. (eds.), Global Change and Protected Areas, 513–521.

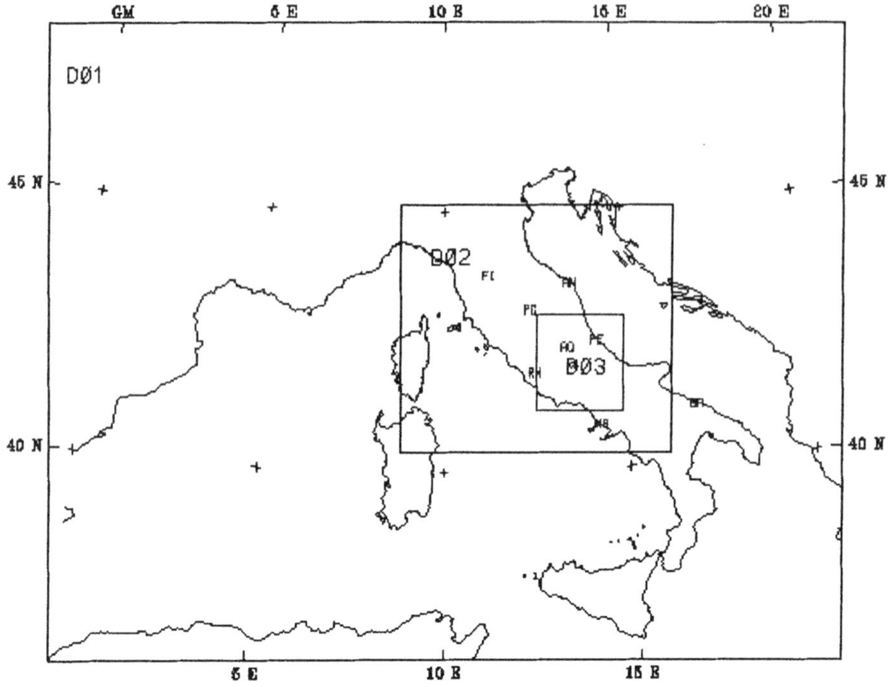

Figure 1
The three different model domains. The largest has a 27 km resolution, the intermediate 9 km resolution and the smallest 3km resolution.

studied by Pielke and Dalu (1991) while effects on the climate were evaluated by Seath and Giorgi (1996).

The effects of the lake on the local circulation (Avissar and Pan, 1999, Changnon et al.,1972) and on the regional climate (Bates et al., 1993, Scott and Huff, 1996, etc.) it is well known and this study is developed within the same frame work. The MM5 is used to simulate few typical local meteorological situation with and without the lake.

2. MODEL CHARACTERISTICS

For this study the MM5 from NCAR/PSU (Dudhia, 1993 and Grell et al., 1994) is used, the model configuration is the one used operationally at PSTd'A/University of L'Aquila (Paolucci et al.,1999): 24 sigma levels not equally spaced; MRF (Troen and Mahrt, 1986) planetary boundary layer parameterization; Kain-Fritsch (Kain and Fritsch, 1993) cumulus convection

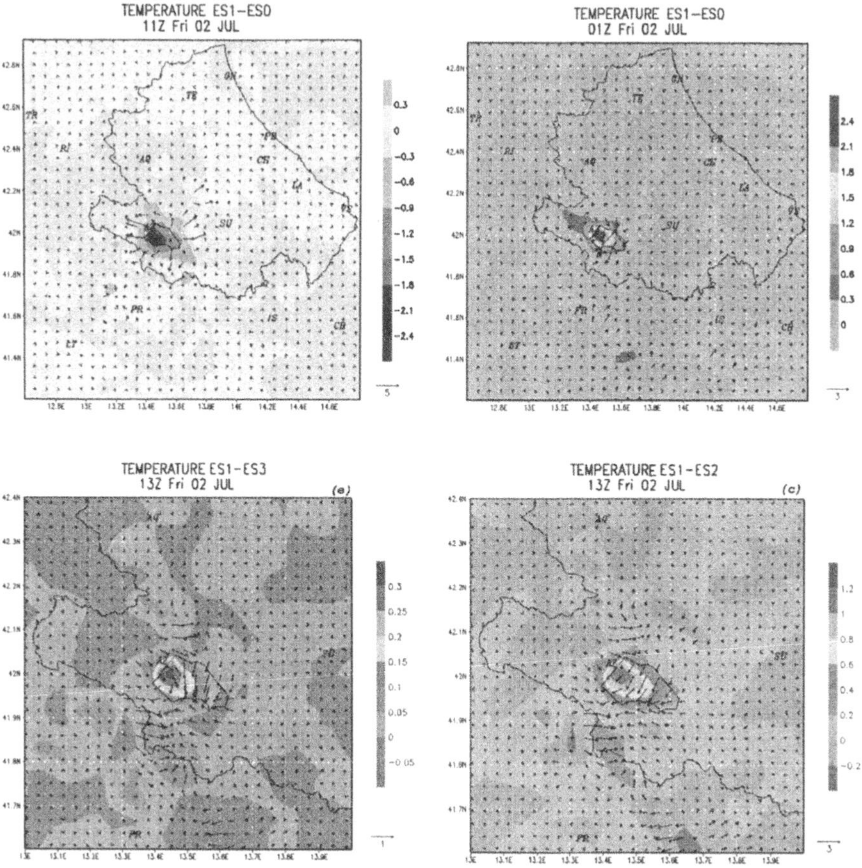

Figure 2

The temperature differences of a number of simulations. At the top left it is shown the daytime difference with and without the lake(ES1 – ES0). At the top right the same difference is shown for the night. The bottom left figure shows the temperature difference at night with the lake at two different temperatures 24 and 23 C (ES1 – ES3), and at the bottom right the lake is at 24 and 15 C (ES1 – ES2).

Name Run	Vegetation	Lake	LST
ES0	NO	NO	NO
ES1	NO	YES	24°C
ES2	NO	YES	15°C
ES3	NO	YES	23°C
ES4	NO	YES	15°C
ES6	YES	YES	NO
EP0	NO	NO	NO
EP3	NO	YES	20°C

parameterization associated to an explicit configuration of cloud water and rain. Three domains two way nested staring with a grid size of 27Km (Fig.

1), and ECMWF data analyses are used to initialize the model and the Boundary Conditions are upgraded every 6 hr. Several model simulation are performed to test the sensitivity to presence of the lake and to its surface temperature. A control simulation without the lake (ES0) is performed as reference. A simulation inserting the lake with the associated land-use characteristic (ES1) is performed to study the change on the local climate induced by the basin.

Several simulations to asses both the sensitivity to the lake surface temperature (LST) and to its diurnal variation are performed:

ES1 setting LST=24°C; ES2 setting LST = 15°C; ES3 setting LST = 23°C and finally ES4 setting the initial LST = 15°C and forcing it to vary by 1°C during the day. Furthermore a simulation adding the vegetation around to lake (ES6), to roughly simulate a feedback processes produced by a possible increase in the precipitation occurrences with an associated vegetation, is performed. To better analyze the model response the previous experiments are performed for a case with weak forcing. The initial condition for ES0 - ES6 are all the same starting at 1200 UTC July 1 and ending at 1200UTC July 3, 1999. Finally two more experiments, with a strong forcing (July 23, 1999) producing precipitation in the area nearby the lake, are performed: a control run without the lake (EP0) and another with the lake (EP3) using LST=20°C. Both simulations starting at 1200 UTC on 23 July 1999 and ending 48 hr later.

3. METEOROLOGICAL SITUATION

A case of high and low pressure system during the winter and the summer season are analyzed: on July 1, 1999 a typical anticyclonic circulation characterized the mediterranean area associated with a weak south-eastward wind, no precipitation was detected at that time; on July 23, 1999 a cyclonic circulation entered the mediterranean area developing in a cyclogenesis over the south Tyrrenian sea. Strong easterly wind advecting warm and humid air toward the eastern Italian coast produced heavy precipitation.

4. SENSITIVITY TESTS

4.1 Sensitivity to lake surface temperature

The great thermal inertia of the water usually acts to reduce the diurnal

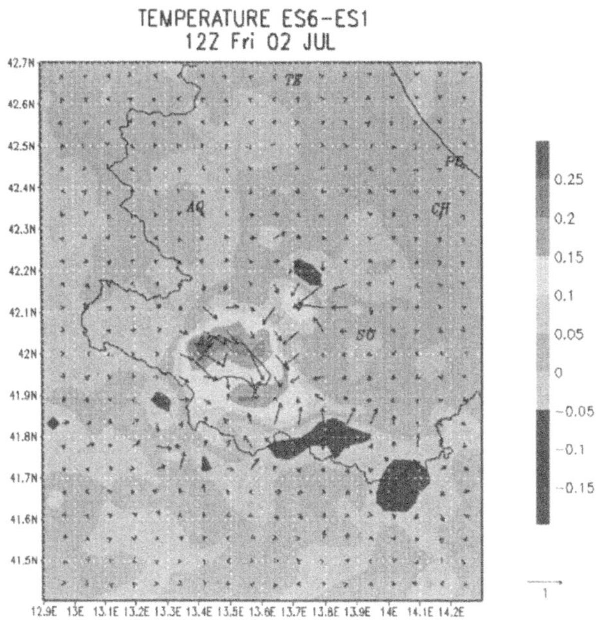

Figure 3
As figure 2 for ES6-ES1 case

and increase the nocturnal temperature, over and around the lake. In addition, evaporation from the lake is a large source for atmospheric moisture. These effects are well reproduced by the model where the surface temperature differences for ES1-ES0 are reported (Fig. 2). These feature enhances during the night. A strong sensitivity to the different initial LST is found: using ES1 as reference, response to the LST variation is inferred by analysing the ES1-ES2 and ES1-ES3. The results show an increase in the breeze regime strength proportional to the LST differences. The same structure is found for the night with an even stronger breeze regime (not shown).The simulation performed forcing the LST to have a diurnal variation does not produce large differences with respect to the control run (ES4-ES2); the differences in the temperature are of order of 1/10°C.

The effect of small temperature variations during the day time is probably more evident on the long term. Therefore this last experiment allow to keep LST constant during the next experiments.

4.2 Feedback on the vegetation.

The last experiment is performed to study feedback processes by the vegetation that was surrounding the lake.

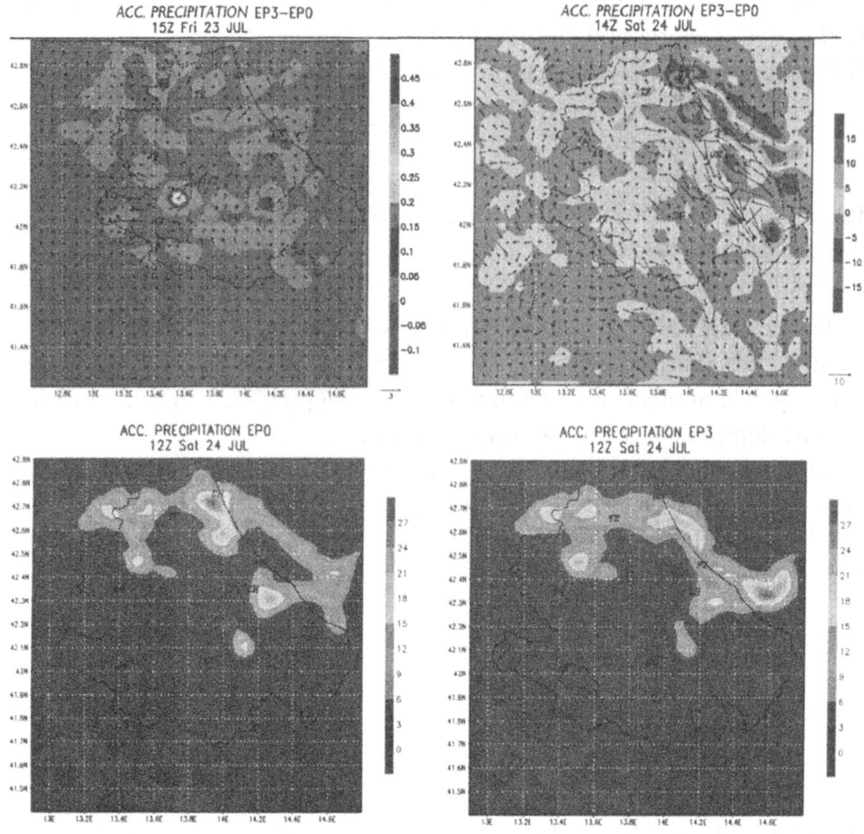

Figure 4

The accumulated precipitation differences are shown at the top. For July 23 with the lake and without the lake (at left) and the same for July 24. The bottom figures show the accumulated precipitation without the lake (left) and with the lake (right)

The increase in the surface covered vegetation would produce a temperature increase in the area around the lake (Avissar and Pan, 2000); the model correctly reproduces this effect, as shown in Fig. 3 which refer to the ES6-ES1 case.

5. EFFECTS ON THE LOCAL METEOROLOGICAL REGIME

The increase of the soil-moisture availability produces both an increase in the precipitation locally and on a larger area (Schar et al., 1999) generating the so called soil-precipitation feedback.

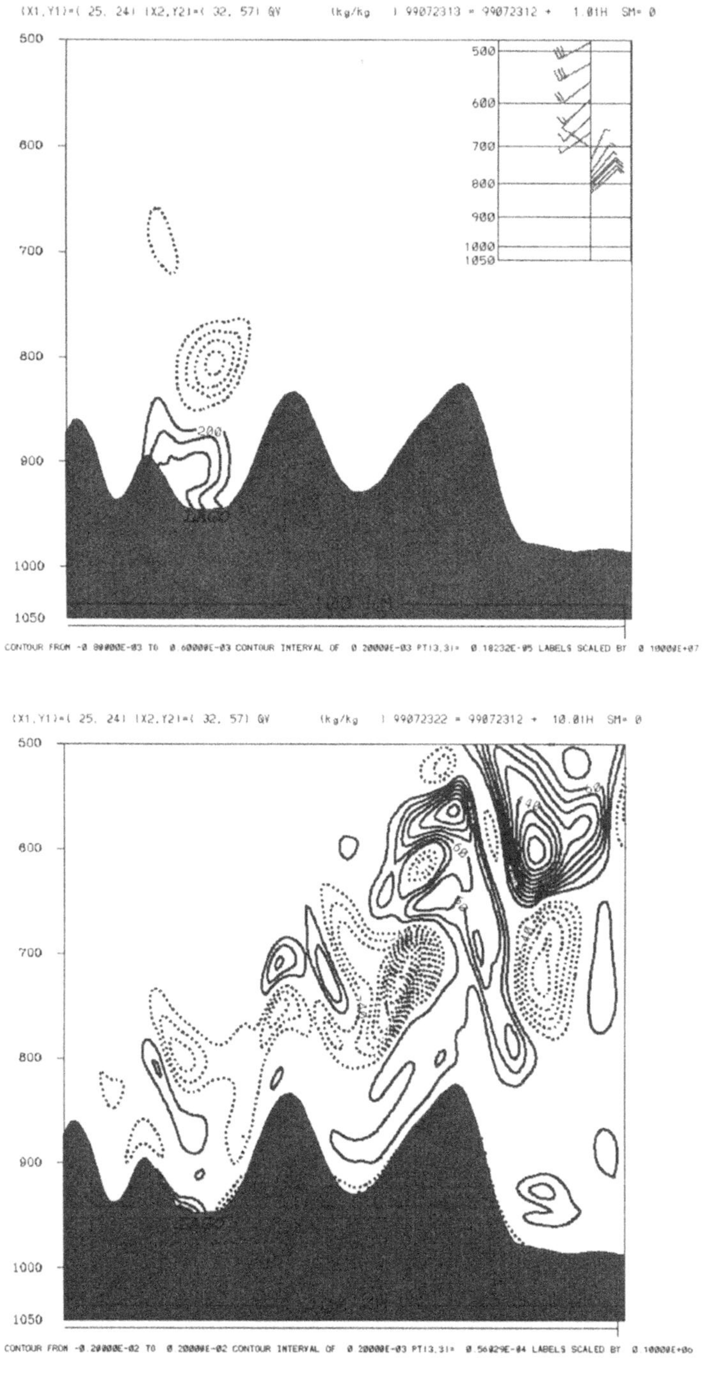

Figure 5
Cross section of the water vapor advection

Similarly the presence of the lake may induce an increase in occurrence of the precipitation in the area surrounding the lake, but also an increase of the water vapour content that may be advected away from the lake itself.

The lake-precipitation feedback is shown by EP3-EP0. There is a first stage of enhancement of the orographic precipitation close to the lake (Fig. 4), whereas in the following hours the north-eastward advection of moisture produces an increase of the precipitation over the sea side. The 24 hr accumulated precipitation for EP0 and EP3 clearly show the strong effects produced by the lake on both are al extentand the amount of the precipitation.

Finally the cross-section of the EP3-EP0 along the precipitation pattern shows the advection of the water vapor. At the initial stage the EP3-EP0 shows an increase of the water vapor content only on the lake area (Fig. 5), forced to move upward by the mountain and moving toward the sea.

6. CONCLUSIONS

A set of sensitivity test are carried out using the MM5. The results show a strong response of the model to the presence of the lake as it was expected. Furthermore the initial LST seems to be important to correctly reproduce the lake breeze regime, whereas the diurnal variation of the LST does not influence the local circulation. The feedback vegetation produced by the presence of vegetation has been also verified.

An experiment is has performed to evaluate the impact of the lake on the meteorological regime.

The results confirm the role of the soil-precipitation feedback mechanism. Indeed the effect involve not only the area close to the lake, but the whole region around the lake.

7. REFERENCES

Avissar R. e H. Pan, 2000 Simulation of the summer Hydrometeorological Processes of Lake Kinneret. Journal of Hydrometeorology, 1 95-109

Bates G.T., F. Giorgi e S.W. Hostetler, 1993Toward the Simulation of the Effects of the Great Lakes on Regional ClimateMonthly Weather Review, 1993 1373-1387

Changnon, S.A. e D.M.A. Jones, 1972 Review of the Influences of the Great Lakes on Weather Water Resources Research 8 360-371

Dudhia J., 1993: A non hydrostatic version of the Penn State-NCAR mesoscale model: Validation tests and simulation of an Atlantic cyclone and cold front. Mon. Wea. Rev., {121, 1493-1513.

Grell G.A, J. Dudhia and D.R. Stauffer, 1994: A Description of the Fifth-Generation Penn State/NCAR Mesoscale Model (MM5). NCAR Tech. Note NCAR/Tn-398+STR Natl. Cent. for Atmos. Res., Boulder Colo.

Kain J.S., and J.M. Fritsch, 1990: A one-dimensional entraining/detraining plume model and its application in convective parameterization. J. Atmos. Sci., 47, 2784-2802.

Paolucci T., L. Bernardini, R. Ferretti and G. Visconti, 1999: MM5Real-Time Forecast of a Catastrophic Event on May, 5 1998. Il Nuovo Cimento, 727-736, 1999.

Pielke R.A, Dalu G.A 1991: Non linear influence of land-use on weather and climate Journal of Climate, 4, 1053-1069

Scha, C., D. Luthi, U. Beyerle and E. Heise, 1999 Soil-Precipitation Feedback: A Process Study with a Regional Climate Model Journal of Climate 12 722-741

Seath A. e F. Giorgi 1996, Three-dimensional model study of organized mesoscale circulations induced by vegetation.Journal of Geophys. Research, 101 7371-7391

Scott R.W. e F.A. Huff, 1996Impacts of the Great Lakes on Regional Climate Condictions. Journal Great Lakes Res., 22 845-863

Troen I. e L. Mahrt, 1986: A simple model of the atmospheric boundary layer: Sensivity to surface evaporation. Boundary Layer Meteorology, 37, 129-148, 1986.

Advances in Global Change Research

1. P. Martens and J. Rotmans (eds.): *Climate Change: An Integrated Perspective.* 1999
ISBN 0-7923-5996-8
2. A. Gillespie and W.C.G. Burns (eds.): *Climate Change in the South Pacific: Impacts and Responses in Australia, New Zealand, and Small Island States.* 2000
ISBN 0-7923-6077-X
3. J.L. Innes, M. Beniston and M.M. Verstraete (eds.): *Biomass Burning and Its Inter-Relationships with the Climate Systems.* 2000 ISBN 0-7923-6107-5
4. M.M. Verstraete, M. Menenti and J. Peltoniemi (eds.): *Observing Land from Space: Science, Customers and Technology.* 2000 ISBN 0-7923-6503-8
5. T. Skodvin: *Structure and Agent in the Scientific Diplomacy of Climate Change.* An Empirical Case Study of Science-Policy Interaction in the Intergovernmental Panel on Climate Change. 2000 ISBN 0-7923-6637-9
6. S. McLaren and D. Kniveton: *Linking Climate Change to Land Surface Change.* 2000
ISBN 0-7923-6638-7
7. M. Beniston and M.M. Verstraete (eds.): *Remote Sensing and Climate Modeling: Synergies and Limitations.* 2001 ISBN 0-7923-6801-0
8. E. Jochem, J. Sathaye and D. Bouille (eds.): *Society, Behaviour, and Climate Change Mitigation.* 2000 ISBN 0-7923-6802-9
9. G. Visconti, M. Beniston, E.D. Iannorelli and D. Barba (eds.): *Global Change and Protected Areas.* 2001 ISBN 0-7923-6818-1

KLUWER ACADEMIC PUBLISHERS – DORDRECHT / BOSTON / LONDON

GPSR Compliance

The European Union's (EU) General Product Safety Regulation (GPSR) is a set of rules that requires consumer products to be safe and our obligations to ensure this.

If you have any concerns about our products, you can contact us on ProductSafety@springernature.com

In case Publisher is established outside the EU, the EU authorized representative is:

Springer Nature Customer Service Center GmbH
Europaplatz 3
69115 Heidelberg, Germany

Batch number: 10799853

Printed by Printforce, the Netherlands